W0071753

THE CHRISTIAN INVENTION OF TIME

Time is integral to human culture. Over the last two centuries people's relationship with time has been transformed through industrialization, trade and technology. But the first such life-changing transformation – under Christianity's influence – happened in late antiquity. It was then that time began to be conceptualized in new ways, with discussion of eternity, life after death and the end of days. Individuals also began to experience time differently: from the seven-day week to the order of daily prayer and the festal calendar of Christmas and Easter. With trademark flair and versatility, world-renowned classicist Simon Goldhill uncovers this change in thinking. He explores how it took shape in the literary writing of late antiquity and how it resonates even today. His bold new cultural history will appeal to scholars and students of classics, cultural history, literary studies and early Christianity alike.

SIMON GOLDHILL is Professor of Greek at the University of Cambridge and a Fellow of King's College, as well as Foreign Secretary of the British Academy. He is one of the best-known classicists of his generation who has lectured all over the world, and he has appeared on TV and radio from Canada to Australia. His books have been translated into ten languages and have won three international prizes.

GREEK CULTURE IN THE ROMAN WORLD

SERIES EDITORS

JAŚ ELSNER, University of Oxford
SIMON GOLDHILL, University of Cambridge
CONSTANZE GÜTHENKE, University of Oxford
MICHAEL SQUIRE, King's College London

Founding Editors
SUSAN E. ALCOCK
JAŚ ELSNER
SIMON GOLDHILL

The Greek culture of the Roman empire offers a rich field of study. Extraordinary insights can be gained into processes of multicultural contact and exchange, political and ideological conflict, and the creativity of a changing, polyglot empire. During this period, many fundamental elements of Western society were being set in place: from the rise of Christianity, to an influential system of education, to long-lived artistic canons. This series is the first to focus on the response of Greek culture to its Roman imperial setting as a significant phenomenon in its own right. To this end, it will publish original and innovative research in the art, archaeology, epigraphy, history, philosophy, religion and literature of the empire, with an emphasis on Greek material.

Recent titles in the series:

The Christian Invention of Time: Temporality and the Literature of Late Antiquity
Simon Goldhill

The Moon in the Greek and Roman Imagination: Myth, Literature, Science and Philosophy
Karen ní Mheallaigh

The Resurrection of Homer in Imperial Greek Epic: Quintus Smyrnaeus' Posthomerica and the Poetics of Impersonation
Emma Greensmith

Oppian's Halieutica: Charting a Didactic Epic
Emily Kneebone

Preposterous Poetics: The Politics and Aesthetics of Form in Late Antiquity
Simon Goldhill

The Aesthetics of Hope in Late Greek Imperial Literature: Methodius of Olympus' Symposium and the Crisis of the Third Century
Dawn LaValle Norman

THE CHRISTIAN INVENTION OF TIME

Temporality and the Literature of Late Antiquity

SIMON GOLDHILL

University of Cambridge

CAMBRIDGE
UNIVERSITY PRESS

CAMBRIDGE
UNIVERSITY PRESS

University Printing House, Cambridge CB2 8BS, United Kingdom

One Liberty Plaza, 20th Floor, New York, NY 10006, USA

477 Williamstown Road, Port Melbourne, VIC 3207, Australia

314–321, 3rd Floor, Plot 3, Splendor Forum, Jasola District Centre,
New Delhi – 110025, India

103 Penang Road, #05–06/07, Visioncrest Commercial, Singapore 238467

Cambridge University Press is part of the University of Cambridge.

It furthers the University's mission by disseminating knowledge in the pursuit of
education, learning, and research at the highest international levels of excellence.

www.cambridge.org
Information on this title: www.cambridge.org/9781316512906
DOI: 10.1017/9781009071260

© Simon Goldhill 2022

This publication is in copyright. Subject to statutory exception
and to the provisions of relevant collective licensing agreements,
no reproduction of any part may take place without the written
permission of Cambridge University Press.

First published 2022

A catalogue record for this publication is available from the British Library.

Library of Congress Cataloging-in-Publication Data
NAMES: Goldhill, Simon, author.
TITLE: The Christian invention of time : temporality and the literature of late antiquity /
Simon Goldhill.
DESCRIPTION: Cambridge ; New York : Cambridge University Press, 2022. | Series: Greek
culture in the Roman world | Includes bibliographical references and index.
IDENTIFIERS: LCCN 2021030353 (print) | LCCN 2021030354 (ebook) | ISBN 9781316512906
(hardback) | ISBN 9781009071260 (ebook)
SUBJECTS: LCSH: Time perception – History – To 1500. | Time – Religious aspects –
Christianity. | Time – Social aspects – History – To 1500. | Time – History – To 1500. | Time –
Philosophy – History. | Classical literature – Themes, motives. | Time in literature.
CLASSIFICATION: LCC CE25 .G65 2022 (print) | LCC CE25 (ebook) | DDC 115–dc23
LC record available at https://lccn.loc.gov/2021030353
LC ebook record available at https://lccn.loc.gov/2021030354

ISBN 978-1-316-51290-6 Hardback

Cambridge University Press has no responsibility for the persistence or accuracy of
URLs for external or third-party internet websites referred to in this publication
and does not guarantee that any content on such websites is, or will remain,
accurate or appropriate.

Augustine of Hippo (Aug.)

    Civ. Dei — *De civitate Dei* (On the City of God)

    Con. — *Confessions*

    De doctr. christ. — *De doctrina Christiana* (On the Christian Doctrine)

    De Gen. contra Manichaeos — *De Genesi contra Manichaeos* (Against the Manichaeans on Genesis)

    De ver. relig. — *De vera religione* (On the True Religion)

    Enarr. in Ps. — *Enarrationes in Psalmos* (Explanations of the Psalms)

    In Evan. Joh. Tract. — *In Evangelium Joannis Tractatus* (Commentary on the Gospel of John)

    Serm. — *Sermones* (Sermons)

Basil of Caesarea

    Ep. — *Epistles*

Cassian

    Inst. — *Institutes*

Cassiodorus

    De an. — *De anima* (On the Soul)

Clement of Alexandria (Clem.)

    Misc. — *Miscellanea*

    Paed. — *Paedagogus* (The Tutor)

    Strom. — *Stromateis*

Cor. — Corinthians

Cyprian (Cypr.)

    Ep. — *Epistles*

    Laps. — *De lapsis* (On the Fallen)

Cyril of Alexandria

    Comm. ad Ioh. — *Commentary on John's Gospel*

    Contra Jul. — *Contra Julianum* (Against Julian)

    De ador. — *De adoratione* (On the Adoration)

    Expos. in Psal. — *Expositio in Psalmos* (Commentary on the Psalms)

    Glaphyr. in Pent. — *Glaphyra in Pentateuchum* (Elegant Comments on the Pentateuch)

    In Joh. — *In Johannum* (Commentary on John)

    In xii proph. — *In xii prophetos* (On the Twelve Prophets)

Dan. — Daniel

Deut. — Deuteronomy

Dio (of Prusa/Chrysostom)
   Or.   *Orationes* (Speeches)
Dio. Sic.   Diodorus Siculus
DL   Diogenes Laertius
Epict.   Epictetus
Epiphanius
   De mens. et pond.   *De mensuris et ponderibus* (On Weights and Measures)
Eun.   Eunapius
Euripides (Eur.)
   Heracl.   *Heracleidae* (Daughters of Heracles)
   Or.   *Orestes*
Eusebius (Eus.)
   Hist. Eccl.   *Historia Ecclesiastica* (History of the Church)
   Praep. Evang.   *Praeparatio Evangelica* (Preparation for the Gospel)
Ex.   Exodus
Galen
   Libr. Ord.   *De ordine librorum suorum* (On the Order of His Books)
Gen.   Genesis
Gregory of Nazianzus
   Apor.   *Aporrhēta* (Ineffable Matters)
   Carm.   *Carmina* (Poems)
   Ep.   *Epistles*
   Or.   *Orationes* (Speeches)
   Poem. Arcan.   *Poemata Arcana* (Poems on the Mysteries)
Gregory of Nyssa (Greg. Nyss.)
   Cant.   *Cantus* (Hymns)
   Contra Eun.   *Contra Eunomium* (Against Eunomios)
   Trid. Spat.   *De tridui inter mortem et resurrectionem domini nostri Iesu spatio* (On the Three-Day Space between the Death and Resurrection of our Lord Jesus)
   Vita Mos.   *Vita Mosis* (Life of Moses)
   HA   *Historia Augusta*
Her.   Herodotus

Hesiod
  Theog.   *Theogony*
  W&D   *Works and Days*
Homer (Hom.)
  Il.   *Iliad*
  Od.   *Odyssey*
Irenaeus
  Adv. her.   *Adversus hereses* (Heresies)
Isocrates
  Phil.   *Philip*
Jerome (Jer.)
  Adv. Ruf.   *Adversus Rufinum* (Against Rufinus)
  Ep.   *Epistles*
  Praef. In Pent.   *Praefatio In Pentateuchum* (Preface to
              the Translation of the Pentateuch)

Josephus
  AJ   *Antiquitates Judaicae* (Jewish Antiquities)
  Vit.   *Vita* (Life)
Lactantius (Lact.)
  Inst.   *Institutes*
Longinus
  De sub.   *De sublimitate* (On the Sublime)
Lucian
  Ver. Hist.   *Vera Historia* (True History)
Macrobius
  Sat.   *Saturnalia*
Man.   Manilius
Marinus
  Vit. Procl.   *Vita Procli* (Life of Proclus)
Martin
  Dial.   *Dialogus*
  Mat.   Matthew
Nonnus
  Dion.   *Dionysiaca*
  Par.   *Paraphrase*
Num.   Numbers

Origen (Orig.)

   Cels.   *Contra Celsum* (Against Celsus)

   Jo.   *Homily on John*

Orosius

   Hist.   *Historiae* (Histories)

Ovid

   Met.   *Metamorphoses*

Paulinus

   Ep.   *Epistles*

   Vit. Ambr.   *Vita Ambrosii* (Life of Ambrose)

Paus.   Pausanias

PG   Patrologia Graeca

Phil.   Philippians

Philo

   De somn.   *De somniis* (On Dreams)

   De vit. contemp.   *De vita contemplativa* (On the Contemplative Life)

   Opif.   *De opificio mundi* (On the Creation of the World)

   Vit. Mos.   *Vita Mosis* (On the Life of Moses)

Plato

   Prot.   *Protagoras*

   Rep.   *Republic*

   Theaet.   *Theaetetus*

   Tim.   *Timaeus*

Pliny the Elder

   NH   *Naturalis Historia* (Natural History)

Pliny the Younger

   Ep.   *Epistles*

Plutarch

   De E   On the E at Delphi

   De facie in orbe lunae   On the Face of the Moon

   Praef. Chron. Eus.   *Praefatio, Chronicon Eusebii* (Preface to the Chronicon of Eusebius)

Prudentius (Prud.)

   Apoth.   *Apotheosis*

   Cath.   *Cathemerinon* (Hymns for the Day)

Ditto.	*Dittochaeon* (Double Shepherd's Staff)
Ham.	*Hamartigenia* (The Origin of Sin)
In Sym.	*In Symmachum* (Against Symmachus)
Perist.	*Peristephanon* (Crowns of the Martyrs)
Psych.	*Psychomachia*
Ps.	Psalms
Ps-Apoll. *Met. Psalm.*	Ps-Apollinaris, *Metaphrasis Psalmorum*
Quintilian	
Inst. Or.	*Institutio Oratoria* (The Orator's Education)
Rom.	Romans
Rufinus	
HE	*Historia Ecclesiastica* (History of the Church)
Sallust	
Bellum Jug.	*Bellum Jugurthinum* (Jugurthine War)
Cat.	*Catilina*
Seneca the Younger	
Ep.	*Epistles*
Socrates	
Hist. Eccl.	*Historia Ecclesiastica* (History of the Church)
Sophocles (Soph.)	
OC	*Oedipus Coloneus* (Oedipus at Colonus)
OT	*Oedipus Tyrannus* (Oedipus the King)
Phil.	*Philoctetes*
Sozomen	
HE	*Historia Ecclesiastica* (History of the Church)
Sulpicius Severus (Sulp. Sev.)	
Chro.	*Chronicle*
Ep.	*Epistles*
VM	*Vita Martini* (Life of Martin)
Tacitus	
Ann.	*Annals*
Tertullian (Tert.)	
Adv. Marc.	*Adversus Marcionem* (Against Marcion)
Apol.	*Apologia*

De cor.	*De corona militis* (On the Soldier's Crown)
De spect.	*De spectaculis* (On the Shows)
Thess.	Thessalonians
Varro	
Ling.	*De lingua latina* (On the Latin Language)

Introduction

'I've been on calendars, but I have never been on time'
Marilyn Monroe

In my view, there is no artwork that captures the modern sense of time as profoundly as Christian Marklay's installation, *The Clock* – first produced in 2010, and, since its opening, repeatedly staged in galleries around the world, to amazed reviews. It is, as Zadie Smith declared, 'sublime'.[1] *The Clock* is made up of around 12,000 short film and television clips that run on a 24-hour loop. In every single clip, you can see a watch or clock which shows the exact time at which you are watching *The Clock*. The synchronization is both funny and uncanny. If you start watching at 2.10, each of the short extracts contains a timepiece showing 2.10 – often several clips for the same minute. At 2.11, it is all 2.11 – and so on for twenty-four hours. At 6.00, a string of hatted men suggests a cocktail; tea is taken repeatedly between 4.00 and 4.30, tea-time; high noon looms and awaits its gunshots. The joy or frustration of interruption is replayed again and again with an extraordinary fascination. It is so easy to be hooked by the briefest narrative of suspense, to be caught by the excitement of a flash of racing, to imagine for whom the cute guy is getting dressed in his favourite shirt, to wonder if the gun to the head will be escaped.

Through these brief fragments of an infrastructure of time, the spectator becomes acutely aware of how often the affordances of time in cinema are themselves a structuring device of visual narrative. *The Clock* shows again and again cinema's love of the establishing shot – a man looks at his watch and then ahead expectantly; or the desperate race against time ('We only have twenty minutes, Jim'); or the build-up of stressful suspense (how long can he hang on to the rock?). What the theatre calls 'business' embodies and displays a sense of existing in time: the cigarette lit under a lamp-post;

---

[1] Smith (2011). Short clips are temporarily available on the internet.

the stamping feet of the waiting watchman; the stifled yawn of the bored beauty. Part of *The Clock*'s wit comes from unveiling the clichés of the cinematic enactment of time in and through the repetition of embodied gestures. You find yourself sharing Marklay's obsessiveness – watching for the time on the clock on the wall of the saloon rather than the fight in front of it, smashing the tables and spilling the drinks.

Yet, as you watch over a period of time, you may also become aware of the extraordinary skill of the editing of this installation – and not just by wondering at the sheer work of collecting so many clips with just that view of the time on a clock (it must have taken so much *time*). The soundtrack of one film drifts over the start of the next clip, linking them with a half-heard echo of overlapping themes; a door opens in one clip, only to lead into the set of another film; a running criminal from one film is chased by a policeman from another film, but at the same minute, precisely, in the narrative time of both films. This *Clock* certainly evidences its clock-maker. *The Clock* is a work of intricate beauty.

The spectator is made acutely aware of time in another sense too. You know at exactly what time you enter the installation and take your seat. You know what the time is, to the minute, as long as you stay. Watching the clock in a film is usually a sign of boredom. You are meant to lose yourself in the narrative, not glance at the time. Here you are riveted by watching the clock. Attentive to time. Of course at some point in the twenty-four hours you need to go – to go to the bathroom or leave. But when? Every minute counts: how much time in *The Clock* is the right time? When will you give up your seat to another? You see yourself in time, feel yourself embodied in time – we all have our body-clocks – and come face to face – through the face of the clock – with your own investment in the temporal calibration of your experience. You feel time passing, minute by minute.

This installation speaks to a uniquely modern sense of time, dependent on the social pervasiveness and accuracy of the clock. Consider, for a moment, as an equally modern but contrastive mirror for *The Clock*, the Superbowl as it appears on American television – another filmic display of clock-time. The programme is stretched out so that one hour (precisely) of game-time lasts for four hours; its final thirty seconds can last ten minutes. The game is fragmented into a series of plays, each repeated in slow motion and in real speed from different angles. The capitalist world of advertising invades the tension of the game – feeds vicariously off it – with commercial delays, where, with a different sense of counting, every minute also has a well-advertised price. (The cost per minute of an advertising slot is leaked avidly to the press). A fan always knows how much time there is on the game-clock, however long the show lasts. The referee could even announce a resetting of the game-clock, although the time of the show always just goes on, as it must. The Superbowl programme is the culmination of a long but specific history of the commercialization of time, its measured commodification. It is hard to think of parallels for such aggressive manipulation and regulation of time, minute by minute, in the spectacular shows of the past. Time was not always money. It was not always possible to watch time tick by like this.

It is a cliché of modernity's self-awareness that everything is getting faster, attention span getting shorter. *The Clock* performs this increasing fragmentation of time's flow, minute by minute. For a generation increasingly raised on the digital media, with the flash of pop-up ads or the apparently instantaneous communication of e-mail or Twitter, Warhol's promise of five minutes of fame may seem too long. This changing sense of time becomes especially salient when we reflect on how hard it is to imagine a life lived without an idea of minutes or seconds, as is the case throughout antiquity, for whom even the half-hour is a precariously utilized precision. How we measure out our lives – with coffee spoons or digital certainty – is integral to how we inhabit the world.

Indeed, I have started with Marklay's *The Clock* because it is a contemporary artwork that embodies in the most sophisticated and engaging way how a representation of time depends on a set of specifically modern ideas about time, practices of time, and aesthetics of time. Such a big claim could open into a book-length study in itself. But let me try to summarize very briefly what I mean.

In terms of modern aesthetics, first of all, we should note the classic modernist gesture of changing perspective so that what is usually in the background is brought to the fore: Marklay directs us towards the devices of *timing* by which a narrative is organized. *The Clock* shows how time is showed. *The Clock* epitomizes thus the formal self-reflexivity typical of modernist aesthetics. It achieves this display of time through a narrative of fragments. Since T. S. Eliot iconically declared 'These fragments I have shored against my ruin', modernist art has privileged not just the fragment but also the collocation of fragments in collage or – most recently – the remix or mash up.[2] Marklay's *The Clock* recuts and juxtaposes momentary, discrete extracts from films – but leaves the films on the cutting floor. This fragmentation changes how each extract can be appreciated. Because each

---

[2] 'The Wasteland' in Eliot (1920), with Varley-Winter (2018) for further bibliography.

moment is decontextualized, it is hard not to view them generically – through stereotypes, expectation, clichés – as I did above when I described 'the cute guy getting dressed in his favourite shirt' or 'the fight in the saloon' (I don't know what films the scenes come from: allusiveness is not the mode). Modernist aesthetics is obsessed with understanding modernity through repetition and its role in structuring social life: the most famous scene of Charlie Chaplin's *Modern Times* is the tramp desperately trying to stay in time with the production line's demands for repetition. The repetition of material modernity in its images of celebrity and commercial products are integral to the art of Andy Warhol (from Campbell's soup cans to the face of Marilyn Monroe); the repetitive scripts of social interaction are central to the language of the dramas of Harold Pinter or Samuel Beckett. The role of stereotypic filmic imagery in identity formation is brilliantly articulated in the art of Cindy Sherman. And we could add many other such examples. The repetition of clock-watching in watching *The Clock*, with its constant reframing of its scenes into less than a minute of anonymous celebrity, is deeply engaged with these aesthetic obsessions of modernity.

Film itself is a modern technology, with its 1,440 frames a minute, and its now digital capacities. Film changes the narration of time, how we see time. The technology of time, however, also alters the experience of time. There is a large-scale politics to this, of course.[3] The technical advances of clock-making are crucial for the history of seafaring, and hence trade and imperialism, for which western film, and Hollywood in particular, has played its role: how the West was won . . . and keeps winning the battle of culture.[4] Organizing time, synchronizing time locally, nationally and internationally, is no straightforward business, even when the technology should allow it. It is surprisingly recent – late nineteenth century – when an agreed national time, thanks to the railway system, was instituted in Britain and elsewhere; even later when international time was stabilized, Greece, proudly going alone, was still producing maps with Athens rather than Greenwich as the mean into the 1920s; the international date line was (re-) fixed only after the Second World War, changing the date, in a moment, on several islands.[5] *The Clock*, however, speaks more to how Western social life has been altered by the possibility of accurate time-keeping – which modern scholars agree is distinctive of modernity, though when and where

---

[3] See Clark (2019); Hartog (2011); Wilcox (1987). The bibliography of this introduction is indicative only: fuller bibliographies are in the following chapters.

[4] Sobel (1996) was trend-setting; see also the exceptional Galison (2003); also Ishibashi (2014).

[5] Rosenberg and Grafton (2010); Galison (2003); Schivelbusch (1986); Gay (2003), Ogle (2015) 20–46.

this modernity starts is debated and re-debated.[6] It has become a commonplace of the historiography of time that industrialization changes people's experience of time, demanding that time is measured and commodified in work, that work is thus defined by time (nine to five), or by units of production (the production line: time and efficiency studies), rather than by the necessary tasks of the agricultural year; and by increasingly small divisions of time.[7] An obsession with punctuality as a sign of good manners – a virtue even – is a modern politesse.[8] *The Clock* displays the degree to which modern social life – as represented in film – is regulated by the constant turn to the time. To be in the installation overlaps your own experience of the time of watching – minute by minute – a leisure time activity, with the representation of the pervasive need to watch the time. The installation – being in *The Clock* – is a necessarily reflexive experience of modern, clock-bound social time.

Ideas of time also alter as modernity progresses. The nineteenth century is the first era to recognize itself as a century.[9] Life expectancy in the twentieth century allows us to lament a young death at 65, something unimaginable in antiquity. Boredom as an expected element of work or childhood comes with industrialization. Only in modernity is human progress through technology or science an anticipation, an anticipation that seeds science fiction as well as social hopes.[10] Above all, since geology's scientific advances in the early nineteenth century and biology's theories of the mid-nineteenth century, time's abyss stretches back millions and millions of years and forward indefinitely – though the anthropocene may herald a more limited presence of humans within time.[11] Deep time is dizzying – and is explicitly set against Christian insistence on the beginning of time in creation and the end of time at the end of days, which postulates therefore a finite historical time span. Equally dizzying, however, are the smallest measurable units. To measure 10^{-21} of a second is an almost incomprehensible achievement.[12] Einstein is science's

---

[6] Sherman (1996); Dohrn-van Rossum (1996); Bartky (2000); Galison (2003); Glennie and Thrift (2009); Ishibashi (2014); Ogle (2015).

[7] Thompson (1967) is seminal; Le Goff (1980). See Nowotny (1989); Kern (2003) and (the equally seminal) Latour (1993).

[8] See Wolkenhauer (2019); Ker (2019): the Roman moral discourse of time did not include punctuality.

[9] Buckley (1966).

[10] Hartog (2020) 221 describes progress nicely as a secularization of perfection and perfectability (what Origen in *De principiis* calls 'our journey towards perfection').

[11] Gamble (2021); Buckland (2013); Rudwick (2005), (2008); Secord (1986) for the geology; Rees (2003) for the apocalyptic. Hartog (2011); Assmann (2013) (especially 131–208); Jordheim (2014) for the 'new regime of time'.

[12] On short time in antiquity see Miller and Symons eds. (2019).

most recognizable face because of his contribution to a new understanding of time, even if relativity is an understanding baffling to most, despite the massive sales of Stephen Hawking's *A Brief History of Time*. How humans see themselves in time depends on such grounding concepts, which change over time.

There are dozens of eye-opening books on this modern construction of time – which does not run at the same pace in all regions of the world or across all institutions or communities in any country.[13] Or, as Frederic Jameson puts it from his Marxist perspective, 'Each mode of production has its own temporalities.'[14] I have offered here no more than the briefest headlines of this fascinating and complex history, but I start with this modern artistic and intellectual reflection on the modernity of time because it is part of what has motivated this book. These attempts to locate and understand the rupture that defines modern time – along with the recognition of the continuities that make identifying such a rupture hard – have largely turned their back on what has a strong claim to be the first truly great transformation of thinking about time. This transformation is the Christian invention of time. The aim of this book is to describe and understand how notions and practices and experiences of time changed in late antiquity as the Roman empire became Christian, and how such a transformation transformed the representation of time in the literature of the era. There can be no adequate historiography of time that does not recognize this fundamental reorganization of Western thinking, institutionalization and experience concerning time.

Now, there are also many books about time in antiquity too, so many indeed that it has become a trope of their introductions not only to make such a statement, but also to note how many scholars have made such a statement before (time and time again). The few paragraphs that Aristotle dedicated to a theory of time – hugely influential paragraphs in the history of philosophy – have resulted in long books based on innumerable articles.[15] The history of the water-clock and sundial have been traced.[16] Augustine's brilliant discussion of time – his celebrated statement that he knows what time is until someone asks him is one of the few moments of antiquity to appear repeatedly in books on modernity – not only has

---

[13] Barak (2013); Wishnitzer (2015); Ogle (2015) 75–119; to add a (post-)colonial perspective to the works already cited. Banerjee (2006) smartly links temporal and monetary systems in colonialism.

[14] Jameson (2002) 79.

[15] Coope (2005) on Aristotle; Sorabji (1983) for the tradition, both with bibliographies.

[16] Allen (1996); Hannah (2009); Winter (2013); Talbert (2017); Jones (2019); and in a long history Dohrn-van Rossum (1996).

proved one of the most discussed arguments from late antiquity, but also has played a fundamental role in the by now extensive bibliography linking time and narrative.[17] The history of the calendar – like the history of astrology, often requiring obsessive attention to details of mathematical calculation – leaves a heritage on everyday living still, and cannot be told without Julius Caesar's interventions.[18] There are also discussions of time in various genres: historiography most intently, but also epic, tragedy and rhetoric.[19] More recently, we can see the beginnings of an interest in anachronism or 'queer time' in antiquity.[20] The whole history of classicism, indeed, the later reception of antiquity, depends on a genealogical view of how modernity is connected to the past – a construction of what it means to inhabit the time of now.[21] This too has its own historiography.

Yet it is striking that the history of the invention of Christian time – how Christianity's multiform development slowly changes the temporalities it inherited – has not been analysed from the multiple perspectives that such a large-scale cultural transformation needs, although the recognition that Christianity changed the understanding of time is readily acknowledged.[22] In part, this silence is a product of the institutionalization of the disciplines, so that classics and theology, ancient history and church history, are separated, institutionally and in practice, in modern universities (for all their shared backgrounds and past involvement).[23] It is still the case that most theologians and most classicists – even when both groups work on late antiquity – look anxiously (dismissively, longingly) at each other across the divide of their disciplinary expertises. In part, this silence is a reflex of the regular assertion that modernity secularized time – that such secularization is indeed a sign of its modernity.[24] Religion is the past to be left behind. The claim that modernity is the progress (in all senses) of secularism has been sharply dismantled by recent critics,[25] but its influence is evident in the unwillingness to consider the deep influence of religious thinking on the most basic concepts of time, despite the evident religious

---

[17] Ricoeur (1984); Kennedy (2013); Nightingale (2011); Pranger (2010); Allen ed. (2018).

[18] Rüpke (1995), (2006); Feeney (2007); Kosmin (2018); Stern (2001), (2012).

[19] Grethlein and Krebs eds. (2012); Grethlein (2013); Hartog (2011); Lianeri (2011), (2016); Bakker (2002); Purves (2019); Phillips (2020); Georgiadou and Oikonomopoulou eds. (2020).

[20] Atack (2020); Holmes (2020); Rood, Atack and Phillips (2020); Phillips (2020). Nagel and Wood (2010) is influential.

[21] The Postclassicisms Collective (2019).

[22] Hartog (2020: 83): 'There is for sure a Christian regime of historicity'. More commonly, it is totally ignored, as in Elias (1992).

[23] Conybeare and Goldhill eds. (2020).   [24] Davis (2008).

[25] Asad (2003); Taylor (2007) have been particularly influential; see Levey and Modood eds. (2009); Modood (2019); Calhoun, Juergensmeyer and van Antwerpen eds. (2011); Mahmood (2015).

roots of the week (as Zerubavel showed nearly forty years ago[26]) or the holiday or the idea of daily routine. In part, not looking towards the invention of Christian time is also a pragmatic if rather feeble response to the sheer scale of material and the complexity of the interlocking subjects that such a topic summons.

The first part of this book is an attempt thus to outline the Christian invention of time. The transformation that Christianity achieves engages both with the experience of time – the accepted structure of the seven-day week, the order of daily prayer, the festal calendar of Christmas and Easter – and with time's conceptualization – a new discourse of eternity, of life after death, of the triviality of the mundane, of waiting for the end of days. This Christian temporality was formed in and against the Jewish, Greek and Roman cultures in which it slowly developed. This, then, gives the founding question of the first section of the book: What were the institutions and languages which structured the experience and under-standing of time, and which Christianity inherited both from Greek and Roman cultures and from the Jewish tradition, and how did Christianity reshape such inheritances?

To answer this question the first section of the book contains ten essays on aspects of temporality. Each of these ten chapters takes a fundamental question of the discourse of temporality and explores how the traditions of Greco-Roman and Jewish culture are slowly transformed by the gradual dominance of Christianity. Each of these opening chapters is strictly an essay. They are short, with no pretension to comprehensive coverage, and each is designed to outline the parameters of what is a huge area of culture, rather than list all the relevant sources or give a full analysis of what are often heavily discussed and complex texts or conflicted institutions. They set out what I take to be the key questions of this history of temporality. The transformations of Christian time cannot be understood without this double address to both the pagan and the Jewish cultures in which it took shape (truly an 'entangled history'), and cannot be understood properly without an address to the conceptualization, experience and expression of time through the gradual development of Christian doctrine, power and institutionaliza-tion. Christian doctrine developed slowly and painfully and polemically (we should always talk of Christianities in late antiquity)[27] – and discussed aspects of time fervidly. This alteration of normative discourse required power to find social and cultural expression – through institutionalization,

---

[26] Zerubavel (1985), though there is much that is worryingly parochial in his evidence.
[27] See e.g. the exemplary Shaw (2011).

including the institutions of literary production. All this must be part of a history of temporality. Some readers may find the form of the essay too much of a provocation, some no doubt will find elements of superficiality in their areas of expertise. The aim, however, is to indicate how broad a cultural question the Christian invention of time is. I am fully aware that each of these chapters could be expanded to book length. The wager is that the scope of the foundational question that this organization of material allows to emerge justifies the essayistic treatment of each element of it.

The second section of the book has an equally large question: how did this transformation of temporality change the writing of late antiquity? To answer this question, this second section has five longer chapters, each of which looks in greater detail at specific authors and texts from late antiquity. These extended readings give me the chance to demonstrate how the questions outlined in the first section are embedded in the language and narratives that form the imaginary of the growing Christian community. Again, selection is necessary. There are chapters on Nonnus' *Paraphrase of the Gospel of John*, and on Nonnus' *Dionysiaca*; on Gregory of Nazianzus' poetry and prose, starting with his collection known as the *Aporrhēta*, 'Ineffable Matters', and concluding with a study of his sermon on Christmas Day 380. The final two chapters juxtapose Ambrose and Prudentius, who both wrote collections of hymns on the Christian day; and Sulpicius Severus and Orosius, who both wrote histories of the Christian world, and in the case of Sulpicius a very influential life of his master, St Martin of Tours. There are self-evidently many other authors who could have been included (originally considered chapters on Quintus of Smyrna and on Juvencus will appear elsewhere), but by this selection I cover Greek and Latin, prose and verse – and the most salient genres of epic, paraphrase, sermon, hymn, hagiography, and historiography, genres rarely brought together to produce a broad cultural picture. These two large-scale, interlinked questions – how the fundamental changes in Christian thinking about time are to be understood, and how these changes are embodied and embedded in the writing of late antiquity – structure this book.

An immediate caution is necessary. I have so far talked of Christianity, Greco-Roman culture, Judaism. Along with pagan and barbarian, these central terms of identification run the risk of concealing the fractured differences and competing claims that actually mark the transformation of the Roman world. Since at least Walter Bauer in the 1930s, it has been recognized not just that there were many different Christian groups from the beginnings of Christianity onwards, but also that any historiography that focuses on the opposition of orthodoxy and heresy is likely to rehearse the self-serving rhetoric of the later dominant church authorities that defined themselves as orthodox.[28] Whatever claims of universalism and truth we find, in each of the writers I discuss there is an evident rhetoric of conflict between pro- and anti-Nicaean Christians, between church authorities and charismatic individuals, between ascetic and civic projections of religion. Christianity remains Christianities. Similarly, while the Roman empire had an *enkuklios paideia*, a general course of education and culture, that started always from Homer and moved through a curriculum of reading to the institutions of rhetoric and philosophy, there are expressly debated and publicly enacted social and cultural differences both between Greece and Rome, and between different groups in the Greek-speaking East or Latin-speaking West (or the bilingual elites or the Latin-dominated army, say, which go between East and West). Tradition – to paraphrase Heidegger – is a rhetoric designed to present the past as self-evident – an ideological projection not just of what past is to be authorized but how the present finds its own genealogy within it. Again, as we will see, tradition is one of the most *contested* areas for each of the authors I discuss. Similar arguments could be made – and often have been made – for the language of pagans and barbarians. 'Pagan' and 'barbarian' are collective terms designed to conceal differences, to promote the values of Christian civilization as privileged and dominant: to simplify and polarize. These central terms of identification are all used as persuasive gestures of self-definition and need to be repeatedly pluralized and nuanced. The era of late antiquity is a time of transformation (as well as a transformation of time), and in it there are many shifting contingencies of self-positioning, networks of situated group formation, and traumatic explosions of hatred as well as religious conflicts and exclusions. The violence of supersessionism is integral to these narratives. The detailed readings of individual authors need to be extremely cautious about too clumsy claims of contextualization.

What is more, if there is no 'view from nowhere' within the exchanges of late antiquity, there is also no 'view from nowhere' in studying any aspect of the historiography of late antiquity, especially where religion is concerned.[29] The modern historiography of the Arian controversy, to take a specifically contentious example, could be said – conveniently and rhetorically – to be bookended by Cardinal Newman, the most celebrated

---

[28] Bauer (1971 (1934)).

[29] 'View from nowhere': see the seminal Haraway (1988). For how critics invented 'late antiquity' see Herzog (2002b); Vessey (1998); Rebenich (2009).

and controversial convert from Anglicanism in the nineteenth century, and Rowan Williams, leading academic but also Archbishop of Canterbury in the twentieth century.[30] It would be incautious not to reflect on how the intricate specificities of Newman's intense public and private engagement with what orthodoxy is, might affect both his research and his writing of such a history for a public readership, much as it would be unwise not to consider how Rowan Williams's intellectual credentials as a historian as well as a theologian, ameliorative public persona and religious liberal could inform his understanding of such an aggressive earlier Christian battle over the authorized truth. For both Newman and Williams, the Reformation – how it is understood historically and lived religiously – is also a fundamental moment of rupture to be negotiated, in their retrospective attention to an earlier conflict. For a historian writing about time in a pre-Reformation era, to disentangle what has become a dominant Protestant historiography within the university – a style of close reading, a view of change, a work ethic, an idea, above all, of what religion is – is an ongoing and pressing charge. In what follows, I will sometimes draw attention to striking and even dismaying examples where confessional commitments distort academic analysis. We should all also know that it is always harder to recognize one's own biases, one's own situatedness.[31] Working on antiquity is no excuse not to recognize the need to take account of the contemporary insistence on such anxious self-reflection. My minimal but requisite hope is that, with a self-consciousness that is not crippling, I will have properly respected all the authors, modern and ancient, I bring onto the page, and will have tried to understand why they have written what they have written and the way they have written it. To grasp what is at stake for yourself and for the figures you study is a basic requirement of criticism.

Although the second section of the book is focused on specific authors and specific genres, the ten essays of section one are each focused on a topic, and range widely over antiquity to show why and how this topic is integral and significant for the discourse of temporality. These ten topics are: (1) *God's Time* – how is the notion of a time of divinity – immortality – conceptualized? This chapter traces the shift from Homer's gods, without season, age or death on Olympus, through the new articulation of a doctrine of God's temporality around the time of the Council of Nicaea, to Augustine's attempt to imagine timelessness. (2) *The Time of Death* – how does the conceptualization of death inform a concept of temporality? Christianity aims to change fundamentally what death means for life, fame and human

---

[30] Newman (1833); Williams (1987).   [31] Goldhill (forthcoming b).

achievement – but with what effect on narratives? (3) *Telling Time*: the history of chronometry. How did time get told – measured and narrated – in antiquity and how does it relate to the history of modern technologies? What does the order of monastic time do to storytelling? Can a good monk have his own story? What, in short, is the relation between counting and recounting? (4) *Waiting*. Narrative – as Barthes recognized long ago[32] – requires its delays: ancient epics, indeed, are all constructed around delay, from Achilles' refusal to fight onwards. Yet Christianity insists that waiting for the Second Coming is the desired achievement, the crisis of now. This chapter therefore looks at the shifting patterns of hesitation, prophecy and fulfilment that structure ancient narratives – and how Christianity changes such dynamics. (5) *Time and Time Again*: this chapter considers how the role of *exempla*, integral to classical rhetoric and normativity, are redrafted by the logic of typology: repetition and normativity. What is the relation between time and linear narrative with both *exempla* and types? (6) *Making Time Visible*. Between memorial and memory, how does a culture make its past visible. What is historical time, what sort of memory or future does it project? What is pilgrimage as a culture of viewing the *topoi* of the past? (7) *At the Same Time*. Simultaneity is the problem of Einstein . . . What does it mean to say two events happened at the same time? This chapter explores how classical and Christian historiography use – differently – the notion of 'at the same time' to construct patterns of comprehension, affiliations of power, and a sense of world history. (8) *Timelessness and the Now*. This chapter looks at two competing notions and how their interactions are differently expressed over time. The first is 'timelessness' and its difference from eternity; the second is 'the now' – how the moment is conceptualized. Is time a series of nows? Or is it an unbroken continuity? What can it mean 'to live in the moment'? (9) *Life-times*. This chapter explores how a concept of time is integral to narratives of a life (life-writing/biography/hagiography). What strategies do Christian biographers adopt and adapt from the pagan and Jewish traditions to rewrite what it means to have a life, to live a Christian life-time? (10) *The Rape of Time*. This chapter considers rupture – the sudden – as a force not just in narrative but in ideology. It takes its title from a poem inscribed on a grand church in Constantinople where the patroness who paid for the building is described as having 'forced' or 'raped' time because her building surpasses Solomon's Temple. How, then, is change to be conceptualized? This is crucial for conversion, say: is it a sudden moment (as Kierkegaard theorizes, basing his account of the

---

[32] Barthes (1975).

sudden in Plato) or a development? Can one 'make a new start'? Inevitably, this turns not just to a philosophical idea of change but also a politics of revolution. How violent is the demand of supersessionism?

Although each essay can be read as a free-standing introduction, together they construct an argument. The opening chapter's grounding concern with the contrasting models of how divinities exist outside time – with all the complications of how a son of God can be accommodated within such timelessness – leads into a discussion of how such embedded thinking about the nature of immortality redrafts a set of attitudes towards human life and death, human achievement and what lasts. This consideration of what lasts in turn leads into a discussion of how the time of mortality can be calibrated, measured, organized and thereby lived. How is the time of mortality – a life-time – to be ordered, regulated? Crucial to the new Christian understanding of such a nexus of ideas is what waiting means – how looking forward affects the sense of the duration of time and how time is to be inhabited – and what life stories are to count. To understand what counts, the exemplary life is crucial. What, then, is the role of the examples of the past for the present and the practice of typology, which forces the present and past into models of each other? How do the cultures of Greece and Rome and then the culture of Christianity make such exemplarity visible, what memorials of the past, what practices of seeing the past are encouraged and institutionalized? How is a place in history made material? This interest in memorialization looks back significantly to the earlier discussion of what counts and what lasts in a human life. One fundamental question for such practices of historiography – one of the key genres of memorialization – is what it means to say that two events happened at the same time – and what such synchronicity is taken to indicate. Synchronicity as a problem is integral to the arguments about the Trinity that grounded the opening description of divine time. Thus, chapter 8 picks up the first chapter to see how timelessness as an idea is developed and how it relates to the contrasting claim to be 'in the moment', the complete absorption in a fragment of time, 'the now'. The final pair of chapters, in turn, develop the question of what is to count in a life-time to look first at the narrative genres of life-writing – how a life-story can be recounted – and, finally, the models of rupture and continuity that underlie the concern with timelessness and the moment – or the very passage of time. Together – and the cross-cutting questions here are evident – these ten essays argue that concepts underpinning temporality – timelessness, synchronicity, the moment, waiting, repetition, rupture – inform the ontological understanding of god and man, eschatology's transformation of history, the human sense of living in time and in the passing of time, how a life is ordered, valued, memorialized and narrated, and consequently the politics of how a group sees itself within history. And, most importantly, this opening section shows that the development of Christianity fundamentally changes how temporality is structured and therefore structures how humans place themselves in time and the practices of temporality: how life is (to be) lived. There is an ethics and politics embedded in the brackets around '(to be)': the often veiled dynamic between description and normative demands is what makes Christian temporality instrumental. This is the Christian invention of time.

These ten chapters set up the terms and debates in which the five close readings are framed. It is in this first section where many of the most well-known writers who write explicitly about time appear – philosophers such as Plato and Aristotle; chronographers such as Eusebius; poets such as Ovid, with his *Fasti* ('Calendrical Records'); theologians such as Augustine, whose profound discussion of time in the *Confessions* is a recurring touchstone for this book. These writers would appear in any list of the usual suspects for a book about ancient temporality, and they have due place in these opening essays. These essays, however, are thematic: there is no attempt to catalogue comprehensively by author Christian views on time ('Time in the New Testament', 'Time in Augustine', 'Time in Origen' and so on). The second section, by contrast, *does* focus on specific texts and authors. The selection of authors for the second section of the book was motivated by wanting to demonstrate how the Christian transformation of time was instrumental in writing which was not usually placed in the list of usual suspects, because it offers a less obviously or straightforwardly direct discussion of time within a philosophical or theological tradition. My argument is that it is in these texts where we see the impact of the change of thinking about time most saliently. It is here where we see the temporal imaginary in formation. For the Christian readers of late antiquity, I argue, an understanding of how the self inhabits time is forged through the experience of encountering these and other such explorations of the ethics, narratives and history of being in a Christian world. None of the primary texts studied in detail in this section are a regular part of the contemporary canon, and part of my polemic is to widen our notion of the literature of antiquity for classicists, theologians and historians. This is not just a question of bringing under-appreciated texts into the limelight, important though it is to contest and broaden what tradition means; it is also a way of re-visioning the more familiar texts by reframing them within this wider and differently charged narrative. Time also becomes differently visible in retrospect.

Although I use modernity, as in this introduction, as a means to explore contemporary critical self-placement and as a contrastive mirror for late antiquity – modernity from the Renaissance onwards – I do not focus purposively or in detail on texts after Nonnus in the fifth century. I do no more than allude to the Byzantine or Latin medieval continuation of these changing ideas and ideals of temporality – a topic which would require another book. Consequently, although chapter 14 on Prudentius and Ambrose looks specifically at how the religious day becomes organized, hour by hour, office by office, I do not discuss the long tradition of Books of Hours, the material history of the later church, its continuing practices of liturgy and ritual, the medieval monastery, and the effect of the Reformation, still continuing, on how modernity's temporality makes sense, although reading and thinking about such later changes has helped focus why the invention of Christian time in late antiquity is so important.[33] Nor do I have the languages to explore Islamic material, though I find it easy, even in translation, to be beguiled by the statement of Imam al-Shafi that the 103rd Sura, *al-Asr*, can act as a summary of the Qu'ran, in that, if this chapter alone had been given, everything else would follow. *Al-Asr* reads: 'By time, the human being is in loss, except those who believe, and do good deeds, and advise each other to truth, and recommend patience'. *Al-Asr* sharply asks, and gently but tellingly answers, what salve there could be for the loss that time brings to human life. One day, perhaps, I will have the tools and the opportunity to pursue the ideas of this book further, to make good my losses and deficits.

It will already have been evident, though it is worth being explicit here, if only to forestall a certain familiar type of review, that I have also made no attempt at providing the sort of full bibliography expected of a *Habilitationsschrift*, say, for each and every figure or subject that this book treats. I have tried to indicate, where it is most relevant to do so, from where I have learned in particular, what the most useful books and articles are, and where further annotation can be found by anyone who seeks it, but without feeling the need to record again the whole history of particular debates or critical approaches. Even so, the bibliography is nigh on fifty pages long, and it seemed better to lessen the scholarly annotation in order to ease the flow of reading. There are inevitably ignorances with a project

---

[33] I have had my go at ritual and liturgy in the nineteenth-century church in Goldhill (2016) 249–83, with further bibliography.

on this scale, but the scope of the footnoting is deliberate. This, too, follows the usual practice for the essay form. But I would not want this designed refusal to engage extensively in the narcissism of small differences, which is the bread and butter of scholarship, to obscure what I hope will be this book's contribution to the discipline. The book hopes to exemplify how classics, along with theology and history, can expand its perspective from its narrow, historically and ideologically determined canon of texts, to broaden the languages, range of material, and scope of questions it allows, in order to broach the large-scale questions with which the field's claim to be foundational needs to engage. Hence the book takes on texts in Greek, Latin and Hebrew across a full range of genres to explore as fully as it can a truly major question about the nature and impact of the culture of late antiquity.

The long inheritance of Christian time in the modern West is evident in multiple forms, from the way that the seven-day week feels like such a natural division of time, to the now half-heard church bells that ring every hour in my university town. But the focus of this book is not on such an inheritance, but rather on the transformation of time in late antiquity. The impact of this transformation is certainly long lasting and profound, but it is the complexities of understanding how such a gradual revolution was shaped that interest me, and, in particular, how the writings of late antiquity embed and embody such a new imaginary of how the self can inhabit time. This book, thus, is not about the philosophical or mathematical quiddity of time, nor is it a history of ancient books on the theory of time or the mathematics of calendars or the technology of clocks. It is about how the idea of time was re-shaped in and for the West, under pressure from a multiform religious agenda, that changed the experience as well as the understanding of time. This is a major cultural revolution whose history needs telling. This book is the first step towards such a history.

PART I

CHAPTER I

God's Time

There is an extraordinary moment in the history of the translation of the Bible, that opens a vista not just on to the cultural politics of translation but also on to the way that theories of time frame scripture's narratives of God.

At the beginning of chapter 2 of Genesis, after the account of the six days of creation, the Hebrew text reads (Gen. 2.2): 'and on the seventh (*ha'sh'vi'i*) day God finished His work which He had made; and He rested on the seventh day from all His work which He had made' – as the King James version has it. The authoritative Greek translation of the Hebrew, the Septuagint, reads: 'And on the sixth (ἕκτῃ) day God finished His work which He had made; and He rested on the seventh day from all His work which He had made'. How on earth can 'seventh' have been translated as 'sixth'? How can one of the most familiar sentences of the Bible, a sentence that explains the foundational ritual of the Sabbath, be renumbered in this way? What is the Greek doing to the Hebrew here?

Modern scholarship reminds us that there were other translations in circulation, and that even the fixity of the now established text of the Septuagint was hard won.[1] Nonetheless – or perhaps because of these different translations – the status of the Septuagint as the authoritative translation of the Hebrew Bible was proclaimed in antiquity as divinely guaranteed, and the Septuagint was thus used in ritual and in theological debate by both Jews and Christians for centuries – and quoted repeatedly by the Gospels and Paul as the Old Testament.[2] The different ways in

---

[1] Grafton and Williams (2006); Rajak (2009).

[2] Origen's *Hexapla* set three other Greek translations against the Septuagint, but even in this scholarly work, it was only the Septuagint he annotated with textual marks to distinguish where it differed from the Hebrew version, and both Rufinus comments that the Septuagint is 'ours', *nostra* (*HE* 6.16.4), and Epiphanius noted the Septuagint was the middle column utterly to refute those on either side (*de mens. et pond.* 535). See Grafton and Williams (2006) 94–5. On the status of translation, see Smelik (2013).

which this authority of the Septuagint is asserted is itself telling. Three Jewish writers, each writing in Greek, Aristeas, Philo and Josephus, lovingly tell the miraculous story of how Ptolemy Philadelphus, ruler of Egypt, commissioned the Septuagint in the third century BCE, and each has a different version of what is at stake in such a translation. Aristeas explains at greatest length the deep respect with which Ptolemy treated the Jews.[3] Ptolemy requests that seventy-two experts be sent from Jerusalem to Alexandria to complete the translation, commissioned on the grounds that no library could be thought comprehensive without these revelatory texts of Jewish law. Aristeas lists and describes the extravagant gifts the king sent to Eleazar, the priest of the temple at Jerusalem, along with his request for translators (an ecphrasis thoroughly Greek, and thoroughly unbiblical, in its expression and expectation of a sophisticated sense of realism: Aristeas' competitiveness is integrally Hellenized).[4] He describes the lavish feast with which the experts are received in the palace at Alexandria, and their individual, brilliant, summary answers to the king's seventy-two profound philosophical questions (Jews cleverer than Greeks . . .[5]). The Jews duly retreat to the island of Pharos, and in seventy-two days produce an agreed text, the absolute accuracy of which an audience of Jewish notables confirmed. The *Letter* of Aristeas is clearly designed, like other texts of the flourishing Alexandrian community of Jews, to project and promote the status of the Jews primarily in their own but also in others' eyes through the celebration and dissemination of the community's foundational texts. As ever, *translatio* is not just an exercise in rendition but a bid for cultural capital.

Philo was a leading figure of the Alexandrian Jewish community, who led an embassy to Caligula in Rome.[6] His numerous works read the Hebrew Bible through the lens of Platonic philosophy, often with an extensive allegorical apparatus. Philo declares that Ptolemy had developed an 'ardent passion' (ζῆλον καὶ πόθον, *Vit. Mos.* 2.31) for Jewish laws, and this motivated his commission of a translation. The seventy-two experts from Jerusalem retreat to Pharos because they want to avoid the confusion of Alexandria with its mix of different animals, peoples, sicknesses and

---

[3] On Aristeas, see Honigman (2007); Pearce (2007); Orlinsky (1975); Gruen (2008); Hatzimichali (2017).

[4] Pearce (2013); thanks to Max Leventhal for discussion. For later Jewish visuality, see Neis (2013); Levine (2012).

[5] On the importance of the question and answer to Aristeas see Adams (2020) 118–34. *Midrash Eikah Rabba* 1.1.6–13 collects stories in which Greek philosophers are outwitted by Jewish children.

[6] On Philo, see Niehoff (2018), with extensive further bibliography (also Runia (2012)); Lévy (1998); Niehoff (2001); Hadas-Lebel (2003); Adams (2020) 118–34, 139–46, 277–90.

noises: they seek a place of tranquillity and purity for their souls – the philosophical symbolism here is patent enough – and they find a place of elemental nature to write of the genesis of nature. The translation emerged 'as if they were inspired, not with each man producing his own scriptural exposition (προεφήτυον), but everyone using the same words and expressions, as if there was an invisible dictation to each of them' (*Vit. Mos.* 2.37). Philo writes a full paragraph explaining that the similarity of the Hebrew (Chaldaean) and Greek languages makes the process of translation close to swapping mathematical symbols of geometry: the Hebrew and Greek tongues are 'like sisters' running together 'in the purest spirit' (2.41) – without the variety and treachery usually associated with translation. This truly unconvincing claim – made with a flourish of linguistic science – reveals Philo's deep ideological investment in the harmony of Hellenism and Jewish culture.[7] So, he adds as proof, there is a festival every year in Alexandria to celebrate the day of the completion of the translation, shared by Jews and Greeks alike – an occasion publicly to perform this sisterhood. For Philo, the Septuagint is the sign of a fully Hellenized Judaism, and, he hopes, the harbinger of a Judaized Hellenism. *Translatio* – without addition or subtraction – is the mark of the idealized cultural hybridity Philo himself embodies.

Josephus straddles the boundaries between Jewish and Roman culture.[8] Written in Rome for a Greek-speaking audience by a Jew who was once the leader of the revolt against Rome, but who now lives in the imperial palace, his *Jewish Antiquities*, a paraphrase of the Jewish scriptures (as it has been called), is a text of translation as exchange – where the treachery of translation and the treachery of cultural transition and the treachery of political collaboration are never far apart, in the text's reception at least. Josephus copies into his history a very close and very full version of Aristeas' *Letter* (*AJ* 12.11–118), and in the Preface to the *Antiquities* (1.10–13) takes the example of the Septuagint as a model for making available Jewish sacred writing to a wider audience. For Josephus, self-serving as ever, the Septuagint is no more than one restricted predecessor of his own project, a predecessor that he claims disingenuously to have 'found' (εὖρον, 1.10), as if the liturgical text of the Alexandrian community (and far beyond) needed him to find it. Nonetheless, when he retells the story of Genesis,

---

[7] See Niehoff (2001), (2018) and for the connection of Homeric and Jewish scholarship Niehoff (2012); Honigman (2003); with crucial background in Nünlist (2009).

[8] On Josephus, see Cohen (1979); Mason (1998); Rajak (2002); Edmondson, Mason and Rives eds. (2005); Cohen and Schwartz eds. (2007); Goldhill and Morales eds. (2007) each with further bibliography.

God ceases labour on the *sixth* day, just as the Septuagint has it (*JA* 1.33). Josephus writes within an intellectual, historical and cultural tradition of empire: he offers his account of Judaism as 'worthy of serious attention' (ἀξίαν σπουδῆς, 1.5). The translation of the Septuagint is now subordinate to Josephus' translation of scripture into the privileged genre of classical history for a Greco-Roman audience.

The story of the Septuagint does not stop with Josephus' appropriation. Philo's story, with its imagery idealized to the point of allegory, rather than Aristeas' more prosaic account, may well be the source for the story that circulates later among Christians in particular, but also among Jews, that the seventy-two scholars were each put in separate rooms and nonetheless miraculously came up with the same version (the joke, of course, is that it would have been more of a miracle to get seventy-two scholars in the *same* room to agree) – a fourth version of what translation portends. Clement, Tertullian, Eusebius and Augustine all refer to this story, which is also given as a factual example (*ma'aseh*: 'it actually happened that . . .') in the Babylonian Talmud (Megillah 9a) during a discussion about where and how Greek, as the threatening language of the dominant culture, can be accepted into Jewish religious practice.[9] Jerome, translator supreme himself, who produced the Latin of the authorized version, tersely dismisses the tale as myth.[10] Augustine, however, uses the story of the translators' inspiration to prefer the Septuagint whenever it differs from the Hebrew text: he attributes the previous lack of a translation to Jewish 'religious scruple or envy' (*religione vel invidia*), a typical sign of how what was told first as a story of the importance of Jewish scripture becomes a supersessionist tool, complete with a bitter disdain.[11] The story of the Septuagint's inspired composition is repeatedly retold for over 700 years. As Christians make Greek (and then Latin) the language of scripture, the Septuagint and its authority as the word of God takes on an increased significance: it is a miracle that inevitably evokes the twinned issues of how conflict between cultures and languages is negotiated, and how authority is transmitted – and, eventually, how Christianity triumphed over (the Septuagint's) Jewishness.

So, granted this privileged status of the Septuagint, and the charged politics of translation, how should the extraordinary translation of 'the

---

[9] Eus. *Praep. Evang.* 13.12.2; Clem. *Strom.* 1.22.168; Tert. *Apol.* 18; Aug. *Civ. Dei* 18.42. On Tertullian, see Osborn (1997); on Clement, see Osborn (2005); Heath (2020); Thomson (2014).

[10] *Praef. In Pent.*; *Adv. Rufin.* 2.25. See Cain (2009); Chadwick (2001) 433–55 and especially Vessey (1993).

[11] *De doctr. christ.* 2.15 (22).

seventh day' as 'the sixth day' be understood? Rabbinical writing offers a remarkable answer to this question. *Midrash Rabbah Bereshit* (10.9), a compilation of midrashic and homiletic expositions on verses from Genesis, put together in the fifth century CE, explains that 'sixth day' was a deliberate mistranslation, made specifically for Ptolemy. (In the Talmud, b. Megillah 9a, there is a list of such selective and pointed manipulations in the Septuagint, a list, however, which offers no explanation of any of the omissions or changes.) The Hebrew text of Genesis states that God finished his work on the seventh day and rested on the seventh day. Does this not imply that God worked on the seventh day? *Midrash Rabbah Bereshit* 10.9 starts to explain away this problem with an analogy ('It is like a man striking his hammer on an anvil, raising it by day and bringing it down after nightfall'), but then offers this more revealing answer attributed to Shimon ben Yochai: Mortal man (*basar vadam*), who does not know his minutes, his times or his hours, must add from the profane to the sacred; but the Holy One, blessed be He, who knows His moments, His times and His hours, can enter by a hair's 'breadth'.

God and humans exist in time differently. Because of the condition of human knowledge, the division between the profane time of the week and the sacred time of the Sabbath needs an articulated division, which takes time: it is an addition ('must add'). The Sabbath begins before sundown to allow for the frailties of human knowing about time. For God, whose knowledge is perfect, there is no space – a hair's breadth – between beginning and ending. Thus he can indeed finish work and start rest on the seventh day without work having any duration on the seventh day. Ptolemy – and by extension the world of Greek learning – could not be expected to understand this subtle argument, and thus to avoid any vexing controversy, the text was changed. The translation is designed to simplify things for the Greeks.

Now, *Midrash Rabbah Bereshit* was compiled in the fifth century, more than 700 years after the Septuagint, and b. Megillah probably a century after that. It would be easy to see these rabbinical explanations as a *post factum* rationalization. The book of Jubilees, a Hellenistic rewriting of Genesis and parts of Exodus – a book obsessed with the ordering of time and the importance of the Sabbath – already had bluntly written (2.16), 'He finished all his work on the sixth day'.[12] *Jubilees*, a sign of the dynamic fluidity of engagement with scripture in the Hellenistic world, although it is cited by Christians for centuries and was important for the sect of the

---

[12] VanderKam's work, especially VanderKam (2018), is seminal. See also Reed (2015) with further bibliography; Kreps (2018).

Dead Sea Scrolls, was eventually excluded from the normative canon by religious authorities, except by both Jewish and Christian groups in Ethiopia. The rabbinical explanation is a rationalization that self-evidently emphasizes the excellence of rabbinical exegetical comprehension over and against Greek understanding.[13] Yet what is both crucial and paradigmatic is that the explanations depend on seeing *God's time* as different from human time, and for the rabbis this difference is based on how time can be known and measured – inhabited.

I have started this book with the story of creation (of course), for two interconnected reasons, then: first, because it establishes how the special nature of God's time becomes a fundamental issue in how temporality is conceptualized – this chapter's central question – and, secondly, because it demonstrates how to ask in this way how God's time is conceived immediately opens into issues of translation, cultural difference, transmission of and exclusions from authoritative tradition, the anthropology of religious institutions, principles of measurement, as well as ideas about divinity's ontological difference from humanity. That is, from the start, I can begin to explore the spreading interconnection of different questions and disciplinary frameworks (intellectual history, cultural history, theology, anthropology, and so on) that any issue concerning temporality will provoke – which is a grounding motivation – the intellectual excitement – for this project.

*

This chapter will end with Augustine's insistence, too, that God has a relation to time that is quite other from humans. But to appreciate how both these Jewish and Christian arguments from late antiquity are distinctively formed against a longer Greco-Roman cultural tradition, we need to look backwards to that tradition. Here, there is another inevitable starting point – namely, Homer and Hesiod, who, as Herodotus declared, 'gave the Greeks their gods'.[14] (We will come to Plato, Aristotle, Cicero and Lucretius, other starting points for the time of divinity, later in the book.) Homer and Hesiod are foundational texts for Greek culture, and, as Herodotus indicates, provide a conceptual frame for representing divinity

---

[13] On later rabbinic anxieties and confusions about the Septuagint as translation, see Smelik (2013), especially 298–320. On Talmud and classical culture see Heszer (1997); Schwartz (2001), (2009); Rubenstein (2003); Kalmin (2006); Lapin (2012); Boyarin (2009), (2015) – and Lieberman (1942), (1950).

[14] Her. 2.53: see T. Harrison (2000); Gould (2001) 359–77; Mikalson (2003); and more generally Eidinow, Kindt and Osborne eds. (2016).

that echoes throughout later Greek writing of all genres, and, differently, in Latin texts too. How, then, does divine time get shaped by these formative narratives?

Homer's *Odyssey* starts its narrative at a moment when Athene reminds Zeus that Odysseus is languishing on the island of Calypso.[15] The hero has been on Ogygia, the belly-button of the ocean, for seven years, forced to have sex with the goddess, and crying each day on the beach, longing for his own homeland. When Calypso tells him that Zeus has demanded he be allowed to leave, she starts a conversation in which she compares herself as a goddess to Penelope, Odysseus' wife. If he knew the sufferings he would face, she declares, he would stay with her and become immortal, rather than striving to return to his human life. Can Penelope match her in 'body and form' (5.211–12)? Odysseus agrees that Calypso is more impressive in 'form and stature' (5.217) because 'she is mortal but you are immortal and ageless' (5.218). Nonetheless, he wants and hopes all and every day (ἤματα πάντα, 5.219) to go to his *oikos* and see the day of his return (νόστιμον ἦμαρ, 5.220).

This brief conversation at the starting point of Odysseus' narrative of return establishes a basic and complexly layered opposition that runs through Homeric discourse on divine time.[16] Gods do not experience death (ἀθάνατος) and do not experience old age (ἀγήρως). Humans experience both. Thus, gods 'do not know the externality of the divisions of time',[17] but humans have wishes and expectations that last all day, and each day, on their limited journey towards an inevitable end. A human's life is marked out, as Hesiod's epic of the mundane emphasizes in its very title, by 'works and days', a calendar of timed activities with duration and senescence. It has its singularities ('the day of return'), but is framed by the household (*oikos*), whose continuity exists through the generations. The generations of humans, as Apollo famously puts it, are like leaves that fall to the ground and are replaced. The warrior who falls to the ground in death underlines that 'falling' is what humans do in their striving.[18] To the eyes of Apollo, the continuity of the generations is not the equivalent of the permanence of divine order precisely because divine deathlessness changes the experience of time. Humans cannot experience the continuity of the household over the generations, only project it, record it, hope for it; the unbroken, unfragmented time of gods is other. So, there are no seasons or

---

[15] Clay (1997) is excellent on this.
[16] Vidal-Naquet (1981) 69–94; Levy (1979); Clay (1981–2); Vernant (1959).
[17] Vernant (1991) 46; Bergren (1983).   [18] See Purves (2019) 37–66.

weather on Olympus (although there is sunrise and sunset). From Olympus, they observe human activities, and, in Homer at least, Zeus, the supreme god, never leaves Olympus. As Alex Purves smartly summarizes:

> Time on Olympus is of a different quality, and is also experienced in a different way, from time on the ground. Gods understand the movement of human time and use some of the same temporal markers as humans do (e.g. sunset) to determine it. However, they do not typically know what it means to *experience* time or – because they live for ever – what it means for time to have *length*. Another way of putting it would be to say that gods do not have their 'own' time. Instead, from a position outside time, they are able to observe and monitor the time of mortals.[19]

On Ogygia with Calypso, Odysseus cannot work and his days are without distinction (no 'works and days' for him).[20] Food, a human necessity, requires no effort of production.[21] There is no ritual calendar, not least because when a man lives with a divinity, rituals which express the normative articulated separation of man, beast and god, must be suspended. There are no sacrifices on Odysseus' journeying, except for the disastrous parodies of the Cyclops eating his men or his men killing the oxen of the Sun.[22] The moment of return is also a return to work as Odysseus starts his journey by building his raft: work, reasserting Odysseus' rejection of immortality, is an intervention in and against divine temporality. (Hephaestus, the god who does work – he manufactures wonderful artefacts – is also the one god to be thrown from Olympus by Zeus – he too falls – and who is humiliated in his bodily disability by the gods' laughter and by being cuckolded by Ares.) But even more tellingly, Odysseus and Calypso do not have any children despite having sex for seven years. 'A divinity's seed is never in vain' – except here. Odysseus' first moment of human recognition on Ithaca will be with his son, Telemachus; the first four books of the *Odyssey* have been taken up with the Telemachy, Telemachus' search to find out about his father and to demonstrate his incipient maturity. The father–son relationship grounds the patriarchal household, its genealogy.[23] Childlessness with Calypso is for Odysseus to stand outside the temporality of the household. With Calypso, unworking, in days without transformation or distinction, and without the children that mark the continuity of the human condition in the household,

---

[19] Purves (2019) 53.   [20] Segal (1994).   [21] Seminal is Vidal-Naquet (1981) 39–68.
[22] Vidal-Naquet (1981) ch. 1, with now the stimulating Stocking (2017).   [23] Goldhill (1991) 9–24.

Odysseus is outside the generations of man, like leaves – in a form of God's time.

The different relation of gods to time is marked by the alpha-privatives with which the standard Homeric vocabulary represents divine difference: gods are *a-thanatoi*, 'without death'; gods are 'without age', *a-gērōs*. This difference is intricately expressed in how work, bodies, food, sexual activity are represented. Gods are born and come of age, but they do not progress beyond the age their narrative allots them. (Christian and, differently, Muslim writers will determine these ideas as profoundly other to their own theologies.) Hermes is always youthful; Apollo a beardless young man. There are no old female divinities in the pantheon, however distinguished. There are certainly old men and old women in Homer, however, and the miseries of old age are a commonplace of archaic poetry, an expected and lamented restriction on human life. Humans die, but they are not by dying removed from time.[24] When Odysseus visits the underworld in Book II of the *Odyssey*, he sees the previous generations and his dead friends and relatives. They continue in a form of existence. They require blood from a sacrifice to speak; they are without substance and cannot be embraced ('three times though he tried'); but they are in time: changeless now, 'striding across the fields of asphodel', but not divine. When Achilles famously wishes that he could be the humblest man on earth rather than king of the dead, he marks not only the *Odyssey*'s committed preference for survival even at the expense of humiliation, but also the difference between the eternity of divine deathlessness and the eternity of human mortality.[25] *Dissatisfaction* distinguishes human existence in the underworld, whereas divine eternity, whatever its local arguments and violence, is marked by established ease. When Odysseus is to be told he will leave Calypso, he is found on the beach 'his sweet life (*aiōn*) was dripping away in grief for his return, because the goddess no longer (*ouketi*) pleased him' (5.152–3). 'No longer' may hint at a counterfactual narrative – there was a time when . . . – but more importantly his life-time is melting as he cries over his lost return, and expresses his dissatisfaction. The ideas are interlinked: the return to his human world implies the slipping of time which is the condition of human existence, and the dissatisfaction which is its mark.

Gods' time is not human time, then, in this early Greek writing. Can it be asked, then, how does gods' time come into being? Is there a creation

---

[24] See in general Falkner and de Luce eds. (1989).
[25] See Goldhill (1991) 72–91; Redfield (1975) 1–41; Nagy (1979); Griffin (1980); Edwards (1984); Lynn-George (1988).

story for gods' time? Rabbinical commentators declare that the question of what happened before the beginning of *bereshit*, 'In the beginning ...', should not be asked.[26] This may look like the pious exclusion of faith-challenging philosophical questioning, but is perhaps better seen as an indication of the senselessness of asking about a narrative, which must take time, before time has come into existence. As we will see later, the eternity of divinity in Christian thinking becomes a specific and deeply contested idea when the relationship between God the father and God the Son becomes an issue. (The word *aiōn*, which defines Odysseus' life-time, will become in later Greek, as we will see, a basic term for 'time' itself or 'eternity'; 'eternal', *aiōnios*, a term of ideological contestation.) But for archaic Greek writing, Hesiod provides the paradigmatic narrative of the genesis of divinity and the order of time. While the *Works and Days* divides human time into its repeated patterns of ritual, labour, and family continuity and disruption, the *Theogony* narrates the coming to being of the divine order: the structure of the divine world is narrated – framed in time – by virtue of a genealogy.[27] Genealogy, however, is also change, a narrative of a power relation between generations, which is expressed as violent hostility between the male figures of Ouranos, Cronos and Zeus, with the complicit trickery of the mothers, Gaia and Rheia. The tension which drives the *Theogony*, therefore, is: how can genealogy, with its assumption that the son will replace the father, produce the stability of unchanging, unconflictual order? As Jean-Pierre Vernant's celebrated analysis explains, after Zeus binds his own father and fights successfully to take up the kingship of heaven and earth, he first swallows Metis (his first wife, as she is described [886]). Metis, the goddess who embodies cunning intelligence, the attribute necessary for the weaker (the son) to outwit and overthrow the stronger (the father), is now embedded in Zeus; he possesses and controls the very quality necessary to challenge him: he is to be known as *mētieta*, 'metis-ized'. ('Cunning', or 'all wise', the usual translations, conceal the full transformative sense of this adjective, applied only to Zeus: he has *incorporated mētis*). This marriage – or total possession – produces Athene, born from the head of Zeus, nourished in his body. Athene is, as Aeschylus puts it, 'totally of the father' (*Eum.* 738). Now filiation is without threat to the father, genealogy without the danger of disruption. With *mētis* embodied in Zeus, power is established without the possibility of being overthrown. Second, however, Zeus marries Themis, the goddess of fixity and regulation – and their first

---

[26] See Schäfer (2008); Niehoff (2005); Alexander (1992a), each with further bibliography.
[27] See Vernant (1965), (1989); Fowler (1998); Clay (2003) with extensive further bibliography.

children are the Horai, the Seasons or the Hours – regulated time. Regulated time is born of the union of rule and order. Legal Order, Justice and Peace come from the same union, figures who 'regulate (ὡρεύουσι) daily life (*erga*) for human beings' (903). The etymological play between 'Seasons' (ὧραι) and 'regulate' or 'have concern for' (ὡρεύουσι) is pointed: as social order requires regulation, so too does time. Fixed time is the result of the end of the genealogical violence of the *Theogony*. As individual gods may be born and grow to a particular state which becomes unchanging, as the race of gods too can be said to be 'always existing' at the start of a poem that narrates their coming into being (αἰὲν ἐόντων, 21), so the structure of the divine world and with it the order of time can be narrated as coming into being *for all time*: structure expressed through genealogy. The apparent paradox of the coming into being of the everlasting receives narrative resolution in Hesiod's genealogical myths. The agricultural time of the seasons, the necessary condition of the works and days, is the product of the establishment of gods' time.

Humans, then, come into existence only after the creation of the time of the gods, a genealogy that grounds the opposition of divine and mortal. Yet in Homer, Hesiod and in other archaic narratives the boundary that such an opposition constructs is also explored, played with and contested in fascinating ways. Odysseus meets Heracles in Hades, or rather he meets the *eidōlon*, the 'image', or 'wraith' of Heracles (11.602). Heracles himself, explains Odysseus, enjoys feasts with the deathless gods and has Hebe ('Prime of Life') as his wife. These lines have caused anxiety since antiquity, and were declared to be an interpolation early on. It is indeed a surprise that an *eidōlon* of Heracles can be in Hades (where he talks and looks menacing) while the figure of Heracles himself on Olympus can be married and enjoy divine celebrations. But there is a parallel not regularly cited here. Helen of Troy, according first to Stesichorus and then Euripides, did not go to Troy, but only her *eidōlon*: she herself remained in Egypt.[28] She too is a child of Zeus and destined for a special afterlife. But more importantly Helen too is a figure who confuses the standard normative oppositions, in her case, of gender.[29] Helen commits adultery with Paris, but in the *Odyssey* is again living with her husband Menelaus, and, when Telemachus arrives in Book 4, the couple are celebrating the wedding of their daughter to Neoptolemus, Achilles' son (and a second wedding of Menelaus' son from a slave). In the scenes that follow, Helen tells a story in

---

[28] See Woodbury (1967), and, more critically, Bassi (1993).
[29] See Bergren (1981); Katz (1991); Wohl (1993); Goldhill (1994); Felson Rubin (1994); Doherty (1995); Zeitlin (1996) 19–52; Worman (2001).

which in Troy she longs to return to Greece and actively helps the Greek mission against Troy, but this story is juxtaposed to Menelaus' tale in which Helen actively if ineffectually works to scupper the Greeks in the Trojan horse. The contrast of the two stories has prompted heated discussion about the doubleness as well as the duplicity of Helen since antiquity.[30] The story of the *eidōlon* and the real Helen as two figures in the same time-frame embodies this doubleness and duplicity. Heracles, in turn, is the paradigmatic human figure who transcends the expectation of mortality and becomes divine. As such, he is a contrastive limit case repeatedly in Pindar's poetry of praise, to set against the limits of achievement humans can achieve.[31] His existence in the *Odyssey* as an *eidōlon* in Hades and as a divine figure on Olympus, does not need to be explained as the confusion of a poor poet in the past, so often the last resort of self-aggrandizing modern textual criticism. Perhaps it is better to see it as the embodiment of the anomaly that is Heracles, the figure who precisely confuses the expectations of human time and gods' time.

In the underworld, Odysseus also describes Castor and Polydeuces (11.300–4), the brothers of Helen. They have an honour (*timē*) 'equal to the gods' in that on alternate days they are dead in Hades and alive on earth. The unparalleled repeated transition between life and death is marked by an unparalleled adjective, ἑτερήμεροι, 'alternate day-ed' – 'One time they live, another time they are dead – alternate day-ed' (303–4). Philo, much later in Alexandria, and obsessed, as we saw, with rewriting the stories of the past, is the only ancient author we know to have picked up on this strange word significantly. He used it to write a beautiful paragraph about the life of a man who pursues virtue (*De somn.* 150–6). Such a life of *askēsis*, writes Philo, is 'alternate day-ed, as someone said' – self-consciously marking the odd word as a specific allusion, without a named author, asking for our complicity – 'one time alive and awake, another time dead and asleep' – a phrase which follows Homer's syntax and vocabulary to enforce the connection. The wise inhabit Olympus, as it were, he goes on, constantly rising in and through learning, the bad live in the pits of Hades, proceeding 'from swaddling clothes to old age, familiar with decay' (151). Typically for Philo, time itself is moralized, as if only the bad age and weaken. So those who strive for virtue live as if on a ladder (the image is designedly Platonic), trying to go up, but often slipping back down – an allegorical exposition of Homer's adjective. Polydeuces and Castor, shifting between life and death, become figures to express the daily struggle in

---

[30] Goldhill (1988).   [31] See Currie (2005).

a man's life between the higher aims of virtue and the lower desires. The unique word in Homer, describing a unique condition of mortality, in Philo becomes a model for all human striving.

Homer's concise image of Castor and Polydeuces is expressed more fully and in what becomes its most familiar version, however, in Pindar's ode, *Nemean* 10, where Zeus offers Polydeuces a choice.[32] As with Odysseus, a profoundly alternative narrative is briefly imagined but dismissed. Since he is the son of Zeus, Polydeuces may, if he wishes, avoid 'death and hated old age' (83–4) and live on Olympus with the gods; or he may, out of love for his mortal brother, choose that they both live 'half a life beneath the earth, half in the golden halls of heaven' (87–8). Polydeuces chooses to share (im)mortality with his brother. Again, as with Helen and Heracles, when the boundary between human and divine continuity is crossed, the story takes the form of a double embodiment and a double experience of time. Within the standard matrix of the contrast between death and deathlessness, ageing and agelessness, and its concomitant horror of decay, there is an imagining of another relationship to time, where two brothers can share both the divine and human experience of time.

Such doubleness is modelled in the divine world by Persephone. The daughter of Demeter, picking flowers, is raped by Hades, who takes her down to the underworld. Demeter wanders the world in grief looking for her child, with the consequence that the agriculture necessary for human life fails. When Persephone is found, her return to the world above is dependent on whether she has eaten pomegranate seeds in the underworld. Because she has been forced to eat, she is compelled to spend six months of the year in the underworld and six months in the world above (raped, forced to eat, sent to earth and back . . .). This narrative – most familiar from the Homeric *Hymn to Demeter*[33] – gives a mythic explanation of the seasonal disappearance and reappearance of crops. But it also links human daily life to the time of the gods, and sets a half-life in the world of the dead and the world of the living at the heart of the divine pantheon.

The sexiest and funniest version of the difference between gods' time and human time is the Homeric *Hymn to Aphrodite*.[34] This hexameter poem tells how Aphrodite was made to fall in lust by Zeus in order to stop

---

[32] Young (1993). The relation of this poem to Theocritus 22, as well as Homer, is much discussed: see e.g. Hunter (1996) 65–76; Sens (1992).

[33] Foley ed. (1993); Richardson (1978); Richardson (2011); Clay (1989); and for its extensive afterlife, Hinds (1987).

[34] Clay (1989) is the best guide to the Homeric Hymns; see also Faulkner ed. (2011); and on the *Hymn to Aphrodite* Smith (1981); Bergren (1989); Schein (2008); Faulkner (2008a).

her constant boasting of how she had made the gods desire humans; and how thus she is led to seduce Anchises on a mountain top above Troy and becomes pregnant with Aeneas – starting the family line that will lead (eventually) to the Roman empire. The poem begins by asking the Muse to sing of the *erga*, 'the works', of Aphrodite – a phrase which can imply the divinity's outstanding achievements, but which is also a common expression for 'sex', which is the deed with which Aphrodite as the goddess of sexual desire is most associated; and the central act of the poem will be the conception of a hero. Aphrodite's power is said to conquer every being, gods, humans and beasts alike, an announcement immediately qualified by a long account of three divinities who escape her influence, Athene, Artemis and Hestia. They are all *parthenoi*, a term usually translated as 'virgins'. This vocabulary does not merely indicate a sexual status but immediately establishes a contrast not just with Aphrodite and her victims – polytheism is a system where divinities form a network of power, attributes and influences – but also with humans. In early Greek cultural values, to be a *parthenos* is a transitional state, which is to be passed through as quickly as possible. The human virgin is imaged as a dangerous wild animal who must be tamed by the yoke of marriage, and, by the fifth century BCE, medical treatises, such as *Peri Parthenion*, have formalized prolonged virginity as a potentially lethal condition with dismaying psychological as well as physical symptoms.[35] The household with its patriarchal genealogy provides the frame of human life, and this integrally organizes the role of the *parthenos* within the exchange of marriage and the expectation of the production of children. The lengthy description of the three goddesses as permanent 'virgins' constructs a powerful contrast between divine and human experiences of time, ageing, change – before a story that will powerfully reassert the necessity of humans staying in human time, however close they get to a goddess.

As befits a hymn to the laughter-loving goddess of desire, there is a good deal of delightfully witty writing in the poem – Aphrodite's denial at first that she is a goddess, although she has appeared on the mountain top dressed as a virgin on her wedding day; her modest downwards glance as she goes to his bed, imitating her frailer victims as she herself becomes a fool for love; Anchises' later admission that he had his doubts about her; the poem's disrobing of Aphrodite, a reverse arming scene – do not all seductive scenes make the reader dangerously complicit with the snares and tricks of desire? – and, above all, the final sting where Aphrodite threatens

---

[35] See King (1998).

to get Zeus to zap Anchises with a thunderbolt if he tells anyone about their tryst (how, then, do we know the story?). As readers, we participate in the world of Aphrodite's sexy games, her *erga*. But it is in Aphrodite's final speech to Anchises that the difference between humans and gods becomes most fully articulated. As Jenny Strauss Clay has most cogently argued, the goddess mobilizes the examples of Ganymede and Tithonus to explain why she will *not* make Anchises immortal, 'an illogical and specious argument', even if perhaps 'rhetorically effective'.[36] Ganymede was taken to heaven, snatched by Zeus, and retains his youth and is immortal, if never more than a boy. Tithonus was given immortality at the request of his lover, Dawn, but, without the gift of agelessness, becomes increasingly withered and decrepit, until she locks him in a chamber in disgust. The figure who divides time into day and night, Dawn, is tricked by the time-bound decay of her human object of desire. The baby Aeneas is to be given to the wood Nymphs to be brought up, instructs Aphrodite in her childcare arrangements, explaining that the Nymphs do not grow old, but will die: they are liminal, 'neither divine nor human' (259). The Nymphs will bring up Aeneas, a 'young shoot' (*thalos*, 278), who, like the leaves and like the Nymphs, will in his turn fall in death. Anchises, the human who will age and die, is thus contrasted with figures who are immortal and unchanging as the object of desire (Ganymede), who suffer from ageing but are immortal (Tithonus), and who are un-ageing but will die (Nymphs). Anchises is told of the variations of agelessness and deathlessness, but is doomed to experience neither. As the angry and humiliated Aphrodite sums up, 'If you could live as you are in body and looks, you could be called my husband', but as things are, 'destructive, burdensome old age' will overwhelm you, old age which 'the gods hate' (241–6). Aphrodite leaves, and the poem ends without Anchises being able to reply to her farewell, silenced before her physical and conceptual separation. Divinity and humanity are certainly intimate in this poem – for the last time, if Jenny Strauss Clay's bold analysis is correct[37] – but the feints and laughs of the narrative leave the gap between human time and gods' time unbreached, except in the poem's own imaginary of mythic figures, manipulated by Aphrodite to enforce a separation between herself and the lover she has taken. Odysseus refuses the opportunity of immortality, to stay for ever in gods' time, in order to start his journey back towards his *oikos*, Anchises is

---

[36] Clay (1989) 190. See also King (1986).
[37] Building on van der Ben (1980) but doubted by Faulkner (2008a) 3–18 and (2008b); Thalmann (1991) 146. See in general Murnaghan (1992); Pucci (2000).

designedly not made the offer. When gods enter human time, the confusion of temporal order erupts: here, living with the consequences of sex ('like a virgin'), the immortal is pregnant with a human child (for nine months?) who will, like his father, grow old and die, after founding an immortal city. The discourse of temporality struggles to keep gods 'outside (human) time'.

When we turn again to the narratives of Christianity in late antiquity – after Plato, Aristotle, Cicero and Lucretius, amongst others – we will come back to how God's time becomes a burning question posed by the birth of a God, the sexuality of a human in relation to a divine birth, and how the coming of age and death of a divine figure can be comprehended. But there will be very little flirtation, amused mockery and ribald laughter in such violent and fearful theological arguments. Indeed, seriousness was never more deadly.

*

Homer and Hesiod remained a constant presence in Greek writing and education, and their representation of divine time thus continues to be instrumental in the Greek imaginary. This continuity constitutes a vexing problem for Christian intellectuals in the developing church. The leading Christian churchmen of the fourth century, Gregory of Nyssa, his brother Basil of Caesarea, and their close friend, Gregory of Nazianzus, stridently upheld the value of the tradition of Greek learning which privileged Homer as the foundational poet of such a tradition.[38] This was no straightforward position for them to take. On the one hand, their stance set them against other Christians, 'who despise learning (*paideusin diaptuein*)' and especially learning from non-Christian traditions (*exōthen*), as 'treacherous and dangerous and keeping us far from God'.[39] Such traditional education, however, as Gregory of Nazianzus insisted – and embodied in his own work – must be regarded as 'the first of good things': Gregory's letters are dotted with knowing quotations from Homer and other writers of the classical tradition;[40] and Basil both defends the study of such writers and demonstrates the moral value of Homer for Christians. On the other hand,

---

[38] See e.g. Elm (2012); Limberis (2011); Daley (2008); Børtnes and Hägg eds. (2006); Rousseau (1994); Schwab (2012); and more generally Bowersock (1990); Eshleman (2012); Jones (2014); MacMullen (1997), and on Latin Kahlos (2007). The discussion of Homer was already central to Origen's response to Celsus, see Hunter (2021), which may go back to the Gospel of John, see Van Kooten (2021).

[39] *Funeral Speech for Basil* 43.11.

[40] Storin (2019) is ground-breaking for Gregory. McLynn (2015) is good on the politics the letters perform.

Julian the Apostate, who also deeply valued the so-called pagan past, and resisted the rise of Christianity, attempted to enact a law which banned Christians from teaching such traditional material on the grounds that for a Christian to teach Homer either made them teach it falsely or made them hypocrites with regard to their own beliefs. This group's commitment to *paideia* was thus fully part of a politics of culture – both a politics between Christians, and a politics between Christians and Julian and his supporters.[41] Each was also deeply involved with theology, of course, and their writings indicate vividly how contentious and even violent the issue of God's time could become for the Christians of the fourth century, as Christology – the nature of Jesus Christ within the Trinity; how to calibrate the divinity and humanity of the Son – became a battleground of belief and authority, within these conflicts over the direction of the culture of empire.

Gregory of Nyssa paints a wonderful portrait of just how pervasive arguments over Christology had become, a portrait tinged with satire and his own polemical agenda, for sure:[42]

> The whole city is occupied with such discussions: the alleyways, the market-places, the broad avenues and city streets; the clothes sellers, the money-changers, the food vendors. If you ask for change, they philosophize about the Begotten and the Unbegotten. If you ask the price of bread, the answer is 'The Father is greater, the Son inferior'. If you ask 'Is the bath ready?', he will answer 'The Son was created from nothing'.[43]

Such religious fighting talk produced civil discord – a law was even passed trying to prevent bath attendants from discussing such matters – and Gregory of Nazianzus sniffily makes his own highly rhetorical plea to keep such potentially violent debates among the elite: 'Serious theology (*to peri theou philosophein*) is not for everyone, I tell you, not for everyone . . . Is any of the crowd (to participate), unfit as they are for such sublimity and contemplation? Utterly unhallowed? Let him not come near; it is not safe'.[44] But between Alexandria, Antioch, Rome and Constantinople, different ecclesiastical authorities, with different groups of supporters, were struggling not only over belief but also over power, authority and

---

[41] See the excellent Elm (2012); reading Homer was already a source of contention for Origen on Celsus, see Hunter (2021).

[42] As Williams (2017) 16–22 reminds us forcefully.

[43] Gregory of Nyssa, *On the Divinity of the Son and the Holy Spirit*, PG 46 557; on public disputation in general see Lim (1995).

[44] *Or.* 27.3. Gregory is also quite capable of arguing for the universal, easy message of Christianity, when it suits him politically: see *Or.* 4.73. Thanks to Lea Niccolai for discussions on this.

status: a bishopric could enrich, excommunication and confiscation of property crush a family.[45] The power games of empire at a local and wider level were intricately intertwined with the church's contests over doctrine. Gregory of Nazianzus, burned too often, tellingly described his fellow bishops – he even imagines them offering it as a self-description – as 'wicked umpires of ambition and ignorant judges of politics'.[46]

Gregory of Nyssa offers an exemplary case of how God's time could become a contested question.[47] In his treatise *Against Eunomios* he attacks Eunomios for arguing – as Arius had before him – that the Son had not always been in existence, but by virtue of being begotten was secondary to the Father. The watchword of the Arians, according at least to their enemies, was: 'There was a time (*pote*) when he was not', ἦν ποτε ὅτε οὐκ ἦν. The creed proclaimed at the Council of Nicaea specifies this statement, along with the parallel idea that 'Before he was begotten, he was not', as two of the heretical claims that lead to immediate anathematization.[48] To make his case, Gregory constructs an elegant contrast between God's time and the mundane time of humans, which opens out into a full theology of time and man's agency – and also reveals his reading in Neo-Platonic philosophy, the intellectual project from outside Christianity that nonetheless intimately informs Christian thinking. Human time is marked by its 'extension', *diastēma*, as is human space. That which is created is defined by such 'extension'. Extension can be divided and measured. But unlike this created life, 'the life of God is entirely separate from temporal measurement ... Time (χρόνος) is a characteristic of created reality only.'[49] As Gregory writes: 'That life (God's) is not in time, but time came from it'[50] – an argument we will shortly see developed further by Augustine. So if the Son was secondary to the Father, it would be like putting two sticks, a shorter and longer one, next to each other. However close they were, they would still embody and make visible measurable distance. Such visible measurement is alien to the eternity (*aidiotēs*) and ungraspability of the infinity of the divine.[51] Human

[45] Rapp (2005).
[46] Gregory of Nazianzus, *Or.* 42.22. On his iambic poetry of complaint see Hawkins (2014) 142–80.
[47] On Gregory of Nyssa, see Coakley ed. (2003); Ludlow (2007); Marmodoro and McLynn eds. (2018), each with further bibliography; and specifically and with further bibliography Balás (1976); Plass (1980); Boersma (2012). On Maximus the Confessor's adoption of Cappadocian theory see Mitralexis (2016).
[48] Good introductions in Young (2010); Ayres (2004); Behr (2004) – and the earlier studies of Hanson (1988) and Simonetti (1975). On Basil's attack on Eunomius, especially in the context of his pedagogy, see Rousseau (1994) 93–133.
[49] Boersma (2012) 583.   [50] *Contra Eun.* 1.   [51] Balás (1976) 151–3.

beings, finite as they are, cannot comprehend the infinity of God ('to tell of God is impossible . . . but to know him even more impossible', *phrasai men adunaton, noēsai de adunatoteron, Or.* 28.4). As humans, 'we are tethered to this temporal existence with its diastemic character', 'inescapably temporal'.[52] Humans can strive to ascend towards God, to transcend temporality, but this process is unending, constantly marked by the unfulfillable desire for the divine. God's time is otherwise.

God's creation of time, rather than any existence in it, explains Gregory's reading of the Easter story, where he analyses the darkness at the occasion of the crucifixion as a night dividing the day into two days. Christ rules the temporal order, Gregory writes, so that 'his works should not necessarily be forced to fit set measures of time (τοῦ χρόνου μέτροις), but that the measures of time (τοῦ χρόνου τὰ μέτρα) should be newly contrived for what his works required'.[53] God is not to be 'bound by time (τῷ χρόνῳ) for his actions, but to create time (τὸν χρόνον) to fit his actions'.[54] For Gregory, God as the author of time can recreate the pattern of time to his will. Easter is a demonstration of God's transcendence of time.

Gregory of Nazianzus, who will be the subject of chapter 13, also writes against Eunomius. His powerful, intense orations – which gave him the honorific sobriquet of 'The Theologian' in the Byzantine era – offer an aggressive theoretical rejoinder to his opponents. For Gregory of Nazianzus, as humans, we are locked into a paradox of language when it comes to God's time. 'When we wish to express what is above time', he writes, 'we cannot avoid the indication of time (*chronikēn emphasin*)'.[55] He explains what he means: 'For such expressions as "when" and "before" and "after" and "from the beginning" are not timeless (*achrona*), however much we force their meaning.'[56] The crucial terms of the debate about the priority of God the Father, or even what the first line of the Gospel of John might mean, are already flawed by their expression in a language which cannot escape its temporality. Thus, he concludes, 'It will be necessary for us to adopt the standard of Eternity, that interval (*diastēma*) which extends through all things above and beyond time, and which is not divided or measured by any movement, nor by the revolution of the sun, as time is measured.' Like Gregory of Nyssa, Gregory of Nazianzus resists the measurements of time. The notion of *diastēma*, 'extension/interval', is now applied not to human time but to eternity

---

[52] Boersma (2012) 589.   [53] Boersma (2012) 595 quoting *De trid. spat.* 290 Jaeger.
[54] Boersma (2012) 595 quoting *De trid. Spat.* 290 Jaeger.   [55] *Or.* 29.3.   [56] *Or.* 29.3.

that transcends time – as Gregory himself strives to transcend his own inhabitation of human language, and performs its impossibility. In a wonderfully striking turn of phrase, he dismisses his opponents as *philochronoi*, 'lovers of time'.[57] Anyone who resists the understanding of God's time that is encoded in the Nicene creed is marred by their corrupting desire for mundane, limited, worldly temporality. Like Achilles in the underworld, they long for life, any life. The true Christian must be in love with eternity.

The Council of Chalcedon in 451, necessary because 'Nicaea bequeathed both terminological confusion and mutual recrimination', was a major political attempt to regulate such debates.[58] The council set out to define what faith is to hold – and it was also a heated drama of infighting between the barons of the church, between the Western and Eastern Churches (the Latin prelates were hampered by their need for translators for the debates in Greek); even the official record of previous council discussions was fiercely contested. At one level, it was a success, not least in still being held authoritative by many Christians including the Catholic Church. It established the Nicene tradition – 'the growing narrativization of the idea of 'Nicaea''[59] – that Basil and both Gregories supported, and, most instrumentally, Cyril defended so aggressively.[60] At another level, not only did violent contentiousness continue for some decades, especially, say, in Jerusalem, but also, over a longer perspective, 'the divisions stemming from Chalcedon remain unresolved'.[61] The history of the build-up to Chalcedon, the business of its sessions, and its consequences, make for an extremely intricate story that has been intensely discussed by scholars. But for our purposes here, and against its complex backstory, it is crucial that the council resolved as a statement of faith that the Son was 'begotten from the Father before the ages in respect of the Godhead' and that 'this birth in time neither subtracted anything from his divine and eternal birth nor added anything to it'. God the Son and God the Father are co-eternal. The Word that was with God at the beginning, as John announces, is Christ, the Son. Chalcedon's statement of the Incarnation continued to be

---

[57] *Or.* 29.5. The concept of *diastēma* in Gregory of Nyssa is much discussed: see Verghese (1976); Peroli (1997) 123–5; and, from a Derridean perspective, Douglass (2005).

[58] Smith (2018) 12. For the texts see the useful Price and Gaddis (2007).

[59] Smith (2018) 199. For the slow development of Nicaean orthodoxy see also Ayres (2004); Behr (2004).

[60] On Cyril, a peculiarly nasty political operator for a saint, see also Wessel (2004) with further bibliography.

[61] Price and Gaddis (2007) 56. See Need (2008); Price and Whitby eds. (2009); Young (2010); and especially Smith (2018).

resisted by some and fought over, but remained the authoritative declaration of faith. For Christians – even in opposition – the paradox – or
mystery – of the co-eternity of the Son and the Father now defines God's
time.

*

Between the era of Basil and the Gregories in the Greek East, and the
Council of Chalcedon (with its Latin- and Greek-speaking prelates,
endorsing Leo's Tome, and the tradition such endorsement sought to
establish), Augustine in Africa made his struggle to understand the temporality of God an integral argument in his exploration of self-
understanding in the *Confessions*. In Augustine's hands, God's time
becomes a complex and compelling theological idea, the foundation of
Western theology to come. Book 11 of the *Confessions* constitutes the most
profound reflection on time and God's time in antiquity – it is a text we
will come back to repeatedly – and in its very depth and rhetorical
brilliance, it shows up the limitations of the discussions of the Greek
theologians, and the politicized silence of the records of the Council of
Chalcedon. Augustine is where my discussion of God's time needs to
conclude.[62]

In Book 11 of the *Confessions*, Augustine asks how God created the
heaven and the earth. It was not made by any process that can be
understood by any analogy to human artifice; rather, *dixisti et facta sunt*,
'you spoke and they were made' (11.5). Augustine immediately asks, 'But
how did you speak?', *sed quomodo dixisti?* (11.6), and this question opens an
extended discussion of the relation between God's language and time.
God's words of creation, argues Augustine, cannot have been like a voice
from heaven that utters a sentence heard by men and that takes time to be
said, syllable by syllable, sound against and out of silence. For those words
were for the moment (*ad tempus*) and for the external ear, whereas the
internal ear is turned towards an everlasting word, a word *in* silence.[63] This
is something quite different, something quite different, declares Augustine
(*aliud est longe, longe aliud est*, 11.6) – a repetition that is not just emphatic
but dramatically performs its extension in time (*longe, longe*),
a demonstration of what Augustine is talking about. For if the act of
creation required words that sounded and took time to say, there would

---

[62] A huge bibliography could be given. I will single out here Ricoeur (1984); Kennedy (2013);
Nightingale (2011); Conybeare (2016); Clark (1993); Kelly (2004); Pranger (2010) especially 35–54;
Wetzel (1992) 17–44.

[63] Polk (1991) 64–5 – an article also good on the relation of Augustine to Husserl.

need to be pre-existent before the creation of the material world a material form to make the sound. God's voice must be something else. God's word 'abides forever', *manet in aeternum*. This phrase quotes Isaiah 40.8, a verse that begins 'The grass withers, the flower fades, but ...', thus setting the permanence of divine utterance against the time-honoured analogy for human fleetingness. For Augustine, more precisely, however, God's word of creation must *always already* be spoken – not as successive syllables even, because otherwise there would be 'no true eternity, no true immortality', *nec vera aeternitas, nec vera immortalitas*. Instead, there would be time and change, *tempus et mutatio*. That is, because God is immortal and his existence is eternal, he stands outside time and change.[64] Therefore his language cannot be in time. 'Abiding forever', God's word too must be changeless and without the expressivity that is formed in time.

Augustine declares he knows this (*novi*, 11.7), this wonder of creation; indeed, everyone who is grateful for the truth, knows this; 'we know this, lord, we know this' (*novimus, domine novimus*), he writes, upping the ante of our complicit agreement. But he is quickly thrown back into *aporia*. If God's word is eternal and nothing is made but by God's saying, how come everything is not made at the same time? ('I don't know how to express it ...', confesses Augustine.) More strikingly, the Gospels themselves are an example of the Word made Flesh. God speaks through the flesh – *per carnem ait* – and the written word 'sounds in the outer ears of humans', but only so that it should be believed and sought inwardly and found in eternal truth. The Gospels are not simply the unmediated word of God (and thus true), as so many later religious writers will claim. The Gospels have to exist in human language, time-bound, and thus heard through the frailty of human bodies – which is an epistemological as much as a physical condition. The Gospels, insists Augustine, have rather to be heard inwardly, in silence, sought for and found in a space beyond the materiality of their letters, if their eternal truth – God's truth – is to be appreciated. 'The pressure of eternity on the present ... is all pervasive'.[65] God's existence outside time is a barrier to the human comprehension of the human form of the Gospels, God's words.

The language of the Gospel of John in particular reverberates through these passages. John's beginning, an aggressive rewrite of the beginning of Genesis, changes how we hear the idea of God's language. For Augustine

---

[64] For the crucial place of eternity in Augustine's thinking see Meijering (1979) with the commentary of Wetzel (1992) 26–44; also Polk (1991) 77 'temporality has its source in eternity'.
[65] Pranger (2010) 40.

now redrafts the creation story again (11.9): *In hoc principio deus, fecisti caelum et terram in verbo tuo, in filio tuo, in virtute tua, in sapientia tua, in veritate tua miro modo dicens et miro modo faciens, quis comprehendet? Quis enarrabit?* This is an exemplary sentence of Augustine's remarkable expository technique. It begins with what appears a recapitulation: 'In this beginning God, you made heaven and earth' – a recapitulation of the previous, now explained, starting point of God's creation, and a recapitulation of the opening verse of Genesis. But it is immediately qualified through the opening of John's Gospel: *in verbo tuo, in filio tuo*, 'in your word, in your son'. Now, if not throughout the previous exegesis, the word *verbum* ('w/Word') is invested with its full theological implications. The beginning of the Gospel of John announces that *in principio erat verbum*, 'in the beginning was the Word': it is this Word which is now evoked with *in verbo tuo* which implies both the eternal utterance of creation and the *verbum* of John, the Word that is God and with God – as the next phrase *in filio tuo*, 'in your Son' requires: the Son is already present (*apud*) at the creation, co-eternal with the Father. There may be a nativity story for the incarnated Jesus in the Gospels and a tale of death, but the proclaimed co-eternity of God the Son is a constant rejection of the possibility of seeing such stories as similar to Hesiod's theogonic genealogical narratives. But Augustine extends his incantatory list first with *in virtute tua*, 'in your virtue': for Augustine, human virtues are a complex issue, 'splendid vices', unless referred to God.[66] God's virtue is other, a mark of perfection, not a striving to control vice. *In sapientia tua*, 'in your wisdom', recalls not just the Psalms (104.24: 'you made all things in wisdom'), but also the *sophia* of God, 'the Holy Wisdom' which like *Verbum* can be an expression for God the Son – itself an appropriation of the mystical value of *sophia* in Hellenistic philosophy as an abstract expression for how humans can strive towards the perfection of God ('He is also called 'Wisdom,' insofar as He is the knowledge of things divine and human' (Gregory of Nazianzus, *Or.* 30.20)). *In veritate tua*, 'in your truth', recalls here specifically the importance of truth as the embodiment of God's word, the eternal and unchanging value of God's utterance of creation. These five phrases, each theologically richly layered, lead to Augustine's exclamatory *miro modo dicens et miro modo faciens*, 'speaking in a wonderful way, making in a wonderful way', which appears to capture in awe the opening *dixisti et facta sunt*, 'you spoke and they were made'. But this rhetorically ringing sentence of majestic worship has the rug pulled out

---

[66] Wetzel (1992) is superb on the issue of virtue in Augustine.

from under it with the final questions: *quis comprehendet. Quis enarrabit?* 'Who can understand it? Who will give an account in words?' The very thought fills Augustine with 'terror and burning love'. The extraordinary journey in this one utterance travels from recapitulation through theologically committed reverence in a rhetorically charged and emotive expressivity, to a sudden reversal of baffled awe. Augustine is determined that the sheer unthinkability of God's time will remain at the forefront of his exposition, and the very structure of his sentence leads the reader towards the same blunt question: *quis comprehendet?*

Augustine's baffled question opens a further long exposition of God's time. He begins with the problem that we have already seen in the rabbinical commentaries – the people who ask what God was doing before the act of creation. He contrasts starkly the flitting hearts of humans living in a world of change, on the one hand, when even a long time is made up of no more than a succession of moments, and, on the other, the eternity of God, where 'nothing is transient and the whole is present'. He cites but rejects the joke that what God was doing before Creation was 'preparing hell for people who ask about profundities'. But these brief comments are no more than ground-clearing gestures before the crushing culminatory argument. God is the creator and establisher (*auctor et conditor*) of the centuries; he is the 'causal force' (*operator*) of all times; in short, God is the author of temporality: *id ipsum enim tempus tu feceras*, 'For you have made time itself'. Therefore 'time could not pass before you made time'. To ask what happened before creation makes no sense. *Non enim erat tunc, ubi non erat tempus*, 'For there was no then, where/when there was no time'.

Augustine strives to get this difficult point across. 'It is not in time (*tempore*) that you precede time (*tempora*)', he continues. Your years neither come nor go, where our years come and go in succession. But your years are 'one day' and your day is 'not any or every day but Today'; because there is no yesterday or tomorrow for you: 'your Today is eternity', *hodiernus tuus aeternus*. Thus, concludes Augustine, 'You created all times, and you are before all times; nor was there any time when time did not exist'. Augustine inevitably struggles with – or performs the paradox of – expressing timelessness in a language marked continually by its temporality, its tenses and duration. His attempt to express how God's present is eternal is even more fraught in Latin without the inverted commas and capital letters that translators use to contrast 'your "years"' with 'our years', or human 'today' with God's 'Today'. Augustine is challenging his readers to try to imagine from within time how God is timeless, to experience the conceptual crisis that such a challenge provides. How can existence be thought without time?

It is this argument about God's existence outside time that leads to Augustine's question, so repeatedly quoted by later writers on temporality, 'What, then, is time?'. And its stunning first response, 'Provided no-one asks me, I know ...', another dramatic, paradoxical swerve between apparent sanguinity and bafflement. We will return to his dense and brilliant attempt to answer this question later in this book, when we turn to the experience of the self in time and the role of memory. But I am hoping that the self-evidently rhetorical and delimited contrast I have constructed between Gods' time in archaic Greek writing and God's time in the rabbis, the Cappadocians, and Augustine is nonetheless useful enough in bringing one conclusion to the fore. The contrast between the archaic Greek past and the Greek, Hebrew and Latin writers of late antiquity is not best focused on the contrast between the immortality of God(s) and the mortality of humans, nor on the ageing or agelessness of God(s) in contradistinction to humans. Rather, in archaic poetry, gods are conceived as transcending time both by not experiencing its vicissitudes, and also by enjoying its opportunities – drinking, sex, watching, arguing. Humans can imagine and represent interstitial conditions between the condition of a god and the condition of a human, and also narrate a translation from the human to the divine – while also arguing for the unbreachable boundary between mortal and immortal (and gods never become human, permanently, even if they look like them on occasion for nefarious or dramatic purposes). It is easy for Greek gods to be represented thus as having sex with humans, producing creatures who straddle the boundaries of human and divine, and, as in the Homeric *Hymn to Aphrodite*, as engaging with humans in a resolutely temporal narrative of sex, pregnancy, five years of childcare and a demand for silence about a sexual peccadillo in the past. In the conceptualizations of late antiquity, however, at least for the rabbis, the Cappadocian Fathers and Augustine, the question is how to imagine God as timeless, how to imagine a divine existence without time, and the agency of a timeless God in the history of humans. There is now a theology of time.

Now, in one case, of course, we have looked at hexameter poetry within an epic tradition, and, in the other, prose from within a tradition of religious commentary – across a gap of perhaps 1,200 years, a gap filled with other writing about God's time(s), which we will discuss in coming chapters. Nor, for the moment, have I offered anything by way of a cultural history for either set of texts, as I would normally demand of myself. The differences in usage between Homer and Augustine will not be smoothed over by calling Homer 'the bible of the Greeks' (a phrase so many have used). Nor is it enough to point out that both sets of texts are privileged genres within their communities of production. Yet the contrast between Homer and late antiquity, for all its demonstrable insufficiencies, also helps us see something that is often occluded in histories of writing from antiquity but which is a fundamental tenet of this book. In late antiquity, unlike most earlier contexts, time becomes subject to a thoroughgoing theological framing, which changes the conceptualization of divinity (or is it that the change of the conceptualization of divinity changes the comprehension of time?): God's time changes; and with it, the forms of narrative also change. As Lactantius writes (*Inst.* 4.8), 'No-one hearing the expression "son of God" should conceive such evil in his mind as to think that God procreated as a result of marriage and intercourse with some woman.' No-one should hear in Christian expressivity any possibility of familiar so-called pagan genealogical myths of divine coming to be (such as the Homeric *Hymn to Aphrodite*). For all the complexities of incarnational theology, divine coming to be cannot be a Christian narrative. As the word 'word' takes on a new, profound and contested range of meanings, which challenge the inherited relation between the language of God and its nature in time, so too the phrase 'son of God' must be parsed afresh and re-narrated, within a frame that rejects the genealogical, time-bound implications of the word 'son', who is now 'coeval with the father'. This history of change helps reveal the specificity of the strategies of the representation of divinity and time in *both* archaic *and* late antique discourse. One starting question for this book, then, will be this: how do the narratives of late antiquity change when the concept of God's time changes? How can *timelessness* and narrative rhyme?

CHAPTER 2

The Time of Death

How eternity grounds early Greek thinking about death and human achievement is articulated vividly by Plato, especially in his response to Homer's paradigms.

Plato has a profoundly ambivalent relationship with Homer. Homer he calls the '*most divine*' of poets, and (thus) Homer is also the *most dangerous* in his seductive, emotional lure away from the philosophical rigour of true knowledge. The threatening normativity of the poets, in Plato's construction, requires strict censorship. Yet the pleasure we all feel in the performance of Homer – or tragedy – constantly shadows Plato's dismissals.[1] It is not always in his most explicit engagement with epic poetry, however, that Plato reveals Homer's influence on his thinking, even and especially in the *Republic* whose establishment for all time bans the epic poets.[2] When he comes in the *Laws* to discuss the regulations for family life, for example, he takes it for granted that the human race, by its very nature, pursues immortality: it is implanted in humans. 'The desire to prove oneself famous', writes Plato (721b–c), 'as opposed to lying in a nameless grave', embodies this 'pursuit of immortality'. Plato here invokes Homer's epics, which provide the paradigmatic model for the pursuit of glory, a model which establishes the contrast between the anonymity of the unmarked grave and the fame that epic poetry itself provides – to be sung on the lips of men, for time to come. Yet Plato, with his typical intellectual expansiveness, immediately links this human desire for immortality with an understanding of time itself. The Athenian Stranger declares (721c3–7):

γένος οὖν ἀνθρώπων ἐστίν τι συμφυὲς τοῦ παντὸς χρόνου, ὃ διὰ τέλους αὐτῷ συνέπεται καὶ συνέψεται, τούτῳ τῷ τρόπῳ ἀθάνατον ὄν, τῷ παῖδας

---

[1] See Rosen (1988) especially 1–26; Ferrari (1989); Gould (1990), especially 4–69; Nightingale (1995) 60–7; Destrée and Hermann eds. (2011), particularly Most (2011).
[2] Hobbs (2000).

παίδων καταλειπόμενον, ταὐτὸν καὶ ἓν ὂν ἀεί, γενέσει τῆς ἀθανασίας μεταληφέναι.

The race of humans is thus in a way organically connected with the totality of time. It accompanies it and will accompany it through to the end, being in this way immortal, in that it leaves behind children of its children, being one and the same each time anew, and it is in procreation that the human race takes part in immortality.

The phrase 'the human race' (*genos anthrōpōn*) is not as common in Greek as it is in English, nor is it neutral. ('The race of women' is perhaps a more familiar ideologically loaded expression.[3]) This language implicitly contrasts humans with gods and animals, in the standard tripartite systematization of the world that underlies Greek cosmography. Neither animals nor gods can desire immortality. The human race, the Athenian Stranger states, is 'organically connected with the totality of time', an extraordinarily dense idea which his paragraph goes on to unpack. The human race goes hand in hand with time, as it were, and will continue to do so. Unlike stories or science's predictions, now all too pressing, which imagine a time in the past or a time in the future without humans, here humans and time go together 'through to the end (*telos*)', a distant teleology. The human race is thus 'deathless' (*athanatos*). The term that defines the gods and their time is now appropriated to humans. How are humans immortal? By having children, and children's children – each generation the same, each new time (as we saw with the *Odyssey*). In this perspective, there are no distinctive generations, stronger or weaker, no golden or silver ages. Rather – the conclusion – the human race partakes in immortality through procreation (*genesei* – echoing the opening *genos*: the being of the race is etymologically connected to its coming to be). For Plato, the tension remains between the desire for glory, not to be nameless, and the sheer sameness and continuity of the generations. Apollo, we recall, in the *Iliad* compared human beings to leaves that fall, insignificant in their repeated fleetingness (21.461–7). Glaucus, as he prepares to fight with Sarpedon, uses the same analogy, seeing himself for a moment, before risking his life, in the sobering perspective of divine indifference to human (in)distinctiveness – the contingency of his memorialization. But Glaucus, who specifically names the season of the leaves as a 'generation' (*geneē*, 6.146), goes on to declare his proud genealogy, asserting a distinctive line of the generations and his place in this history. To construct such a line of memory, to take up one's place in such a fight for glory, is necessary

---

[3] Loraux (1981a) 75–117.

precisely because humans die. This is the proclaimed logic of the Homeric warrior.[4] Plato takes up the inherited compulsion of epic, its expressivity, and reformulates it as a general, philosophically dense idea about humankind. What it means to live in time is to struggle to transcend the insignificance of the fleetingness of the time-bound.

Is death the defining moment of glory, then? Is this what Plato suggests in the contrast between the unmarked grave and becoming famous? The Iliadic figure of Achilles, at least as he has been understood by modern readers, might suggest that the answer to such questions is simply yes.[5] (As Solon says to Croesus in Herodotus, famously and smugly: 'Look to the end'.)[6] Achilles, whose body is as close to divine as a human body can be, is offered a choice of a long life of comfort but a lack of recognition, or a short life of immortal glory. He chooses immortal glory and thus a necessary young death. Death is the price of fame, a 'tragic consciousness of the untimely' his condition of living.[7] 'Let me die immediately', Achilles declares to his mother, as he reiterates his commitment to his self-defining wager. The precise phrase *aphthiton kleos*, 'immortal glory', occurs only once in the *Iliad* but (self-performatively) its afterlife has been long.[8] The term *aphthiton* is another alpha-privative, 'unwasting', 'not decaying' – and we have already seen a distant but pointed echo of this language in Philo's arresting expression for the humans who do not strive for excellence as 'familiar with decay' (*phtharsis*). Indeed, in Homeric discourse, the very term *chronos*, 'time', implies not so much a general concept of time, as the deadness of mere passing, the threat of what Hermann Fränkel influentially termed 'empty time', an indistinct passage of the days, which *kleos* disrupts with a claim to the significance of moments of glory that will last in memory and record.[9] So, Egbert Bakker writes:

> The antithetical relation between *khronos* and *kleos* sheds light on the instances of *khronos* that do occur in Homer. These denote Fränkel's 'empty time' only when we view *khronos* in our perspective of temporality. But in connection with *kleos*, the time in which 'nothing happens' becomes precisely *khronos* as factor that is averse to *kleos*, the dimension in which people just age and can do nothing to make up for it.[10]

---

[4] Seminal discussion in Redfield (1975); Nagy (1979).

[5] From a large bibliography see Redfield (1975), (1979); Bergren (1980); Lynn-George (1988); Goldhill (1991): 69–108; Li (2018). For a different frame see Butler (2009), looking back to Weil (1939), on which see Lindheim and Morales eds. (2015).

[6] Her. 1.32.9.   [7] Phillips (2020) 42.

[8] As well as works cited in previous footnote see Finkelberg (1986), (2007); Edwards (1988); Anderson (1981).

[9] Fränkel (1960) 1–22.   [10] Bakker (2002) 28.

In Homer, *chronos* is brought into visibility 'only at moments in which heroic action or the progression of the narrative is stalled, and so the possibility to build *kleos* temporarily blocked'.[11] The hero who risks his life in battle for the glory of victory is the inverse of Odysseus hidden on Calypso's island, where his 'sort of gods' time' of unchanging, workless inactivity makes him so unknown that his son must travel the world to search for his *kleos*. While every warrior risks his life for glory, Achilles willingly exchanges the possibility of survival for the certainty of fame. His 'undecaying glory' is the iconic opposite of the 'wasting' of Odysseus' house at the hands of the suitors, in the absence of the master; or Penelope who 'wastes' her life (*aiōn*) longing for her absent husband. The term *aiōn* 'a life-time', 'period', will become a key term in the theories of time in late antiquity, as we will see: she wastes her *aiōn* here, however, in direct parallel to her husband's wasted *aiōn* on Calypso's island, as we saw in chapter 1. Or, indeed, the fate of 'humans, like leaves' which, in Apollo's words, is to 'decay without spirit', *phthinuthousin akērioi*. The implications of this adjective *akērioi* are made clearer by Menelaus. When his fellow Greek princes hesitate to face Hector in single combat, he accuses them of sitting *akērioi aklees*, 'without spirit without glory' (7.100): their failure to demonstrate the stomach for a fight is a self-fulfilling indication of their future lack of reputation. To be 'spiritless' is the psychological barrier to the achievement of fame; it restricts a person to inhabit 'empty time'. The leaves that fall, the men who live like leaves, are spiritless and thus without distinction or lasting reputation. *Akērios* in these lines is normally understood as 'without spirit' (*a-kēr*); the same letters can also be understood, however, as 'not touched by fate' (*a-kēr*). Hesiod in the *Works and Days* talks of 'days that are *akērioi*', 'not touched by fate', which he specifies as days on which nothing happens. The 'spiritless' heroes in their cowardice threaten to make the challenge of Hector into 'a day untouched by fate'. Their delay of action risks stretching the moment of potential glory into 'empty time' – as Achilles' seventeen-book withdrawal from battle is a threat to return to the undistinguished life of comfort he has previously rejected. But the leaves, do they decay 'without spirit' or 'untouched by fate'? With either of these senses for *akērioi* – and both are possible – decay and a lack of distinctive, lasting identity are mutually implicative. It is against such a threat that the Homeric hero, and Achilles above all, pursues glory, the glory that defeats such (empty) time.

---

[11] Bakker (2002) 28.

From the paradigm of Achilles, then, it might seem indeed that the pursuit of glory (*kleos*) is intimately connected with the moment of death. As Achilles' 'undecaying fame' is won at the expense of a necessarily short life, so each hero risks his own life in battle to transcend the time-bound human condition. The barrow, tomb or memorial is the physical instantiation of the fame of the dead hero, the *sēma* that marks distinctiveness. Epic is the very performance of such fame, creating the *aei* ('always') of continuity through the bard's singing, **aeidein**. The necessity of death grounds the experience of time.

But things are not quite so simple. Helen, the daughter of Zeus, as she tries to persuade Hector to sit and rest from war, declares memorably – and, it turns out, rightly – that she and Paris will be *aoidimoi* 'the subject of epic', 'celebrated in song', for generations to come (*Il.* 6.358). Helen's death is not mentioned in Homer. In later literature – Euripides is an especially pertinent example – there are stories that she is turned into a star, or that she goes to the Isles of the Blessed (with Menelaus, in Lucian's amused version of literary afterlife). These Euripidean stories all occur in speeches from a *deus ex machina*, where the dramatist is particularly creative in his use of imagined future mythic narratives. His inventiveness and the multiplicity of stories indicate that there appears to be no single privileged or expected account of her end. But Helen's anticipation of her fame is not dependent on her death. Penelope is the only woman in the Homeric poems who is described with the specific language of *kleos* (*Od.* 24.192–202).[12] Her death is nowhere depicted. Like Helen, her death is not integral to her story. In the underworld, Agamemnon praises Penelope in comparison to his own wife, Clytemnestra, whose death, told or suppressed, is crucial to her narrative; but for Penelope too, her fame depends not just on Odysseus coming home to find her chaste, but also on her survival of the 'dead time' of his absence. Odysseus' death, in turn, is certainly projected in the *Odyssey* through the mysterious prophecy of Teiresias that 'death will come from the sea', a far-off event (11.97–137). But Odysseus' *kleos* depends on his survival not his death; he returns from the underworld, where the celebration of his fame is asserted against the 'most shameful death' of Agamemnon, a death that threatens the king's very *kleos*, and against the glory of Achilles. Achilles in the *Odyssey* is made even to wish to give up all his post-death glory in order to live humbly on earth, a reverse of his defining Iliadic choice.[13] Significantly, Odysseus is the only Homeric hero

---

[12] See Katz (1991); Foley (1995); Mueller (2007).
[13] See Edwards (1984); Pucci (1987); Peradotto (1990); Goldhill (1991), each with earlier bibliography.

who announces his own fame with the explicit language of *kleos* and in the present: 'my *kleos* reaches heaven', he declares to Alkinoos, king of the Phaeacians (9.20). Over and against Achilles in the *Iliad*, the *Odyssey* competitively recalibrates the temporal dynamics of *kleos*.

Indeed, the *Odyssey* – to a point of idealization – revels in the generations and the continuity of the family, the passing of life that motivates the Iliadic Apollo to dismiss the significance of human striving. When the families of the Suitors seek revenge for the death of their children, they are faced by Laertes, Odysseus and Telemachus. Laertes, although he had fainted away in weakness, when Odysseus had confronted him, bowed with age, in his orchard, powerfully throws the first spear and kills the father of Antinous, the first suitor killed by Odysseus in the hall, a significantly generational pattern of killing. Before this, he boasts 'what a day this is! My son and my grandson are competing in valour!' (*Od.* 24.514–15). The three generations of single sons, itself a wonder of genea-logical survival, all appear to be of fighting age at the same time.[14] Odysseus, like Penelope, has his body made prime and beautiful again by Athena; Telemachus has proved himself in the battle of the hall to have achieved the manhood that was his quest from Book 1; the household property has passed from Laertes to Odysseus without him passing on. In this closing, iconic image of Odysseus' family in the *Odyssey*, time's inevitable trajectory is bent so that, unlike the decaying, falling leaves, the three generations of men can fight, all in the prime of life, side by side. It may well be, as Aristotle insists (*Phys.* 221b), that 'time in and by itself is the cause of decay (*phtharsis*)', but the *Odyssey* imagines the family of Odysseus raised above such necessity. Homeric epic can also imagine, for a moment, the time of the generations of men reordered, a different way for men to transcend the exigencies of time.

*

If there is one word which goes right to the heart of this time of death, it is *kairos*.[15] If *chronos* signals the empty time of duration, *kairos* marks a crisis, the killing time. To be struck *kairiōs* is to be delivered a mortal blow: it is your time. The recognition of the time of *kairos* is to know you are at a turning point, and the time is ripe; *kairos* is often translated as the 'right

---

[14] Goldhill (2010).

[15] Trédé-Boulmer (2015) is essential and goes beyond e.g. Wilson (1980), (1981). See also Arrighetti (2006) 90–101 and Dickson (2019) (on Homer). More speculatively, Gallet (1990) picks up on Onians (1988) 343–9; Andrews (2020); Cassin (2014) takes the term in a theoretical direction.

time', but is better understood as a 'decisive time', a moment when a decision will make all the difference, a turning point in how the future will unfurl. Oedipus had to reach the crossroads at just the moment to meet his father, a moment when a decision and a cutting make all the difference to Oedipus' future. Before historiography made a topos out of the opposition of *chronos* and *kairos*,[16] 'duration' and 'occasion', and rhetoric discovered it as a term of theory, tragedy became the genre above all that insists on the story of *kairos*, the one moment that marks you for death, or, worse, suffering – the recognition, which is always time turning back on itself, seeing oneself again, and thus otherwise, that 'All too late, I learned in full . . .'.

Clytemnestra, whose death at the hands of her son is silenced in the *Odyssey*, screams her story aloud, centre-stage, in Aeschylus' *Oresteia*. When she appears at the doorway of the palace, over the concealed body of her husband, to describe his murder at her hands, with a sexualized and shockingly ritualized description of his blood spurting over her, she announces that 'I have said so many things before to suit the occasion (*kairiōs*) but I will not be ashamed to say the opposite' (*Aga.* 1322–3). She has lied to the citizens, manipulated her husband, tricked and deceived her way to this moment: she sums up the power of her speech simply: 'to have spoken to the moment'. But even this apparently transparent declaration reveals the buried, dangerous life of words in Aeschylus' dramatic vision. The off-stage death cry of Agamemnon that she is now recalling with the description of his murder, was the extraordinary verse: 'Aaargh, I have been struck a killing (*kairian*) blow inside' (*Aga.* 1343). As she walked to her death, a moment she knew she could not avoid, Cassandra had hoped for *kairian plēgēn* (1292) 'a timely/killing blow' – a prophecy, it seems, of Agamemnon's death too. Clytemnestra's speech 'to the moment' (*kairiōs*) was 'lethal' (*kairios*). Her clear speech now echoes with the killing blow she made her husband bellow out loud.

Every tragedy dramatizes the terror of such tipping points, what Aristotle calls *anagnōrisis*, the sudden recalibration of past knowing, or *peripeteia*, the disastrous fall, faster and heavier than leaves. But in an era which made causation an obsessional intellectual interest, the question of how a tipping point is approached becomes insistent. For Oedipus to reach the time and place of the crossroads, what were the determining factors? Was it the drunken man at the feast who taunted him for his unknown birth? Was it the shepherd's gesture of pity? The decision to expose him

---

[16] See chapter 7 below, p. 138.

rather than put a knife to his throat? His own decision to go to Delphi and by that road? Was it all part of Apollo's plan? Or was there a destabilizing narrative of chance at work? Is *tuchē*, a word the play returns to repeatedly, to be understood as fate or luck?[17] Tragedy's dramatization of the killing time, *kairos*, turns *chronos* away from 'empty time', 'dead time', into a dangerous destabilizing expanse of uncertain knowledge, concealed motivation, and hidden connections.

Aeschylus dramatizes this buried life of *chronos* with remarkable intensity, especially in the *Oresteia*. The death cry of Agamemnon is approached with gathering darkness. Clytemnestra prepares for his return with increasingly double-edged speeches of anticipation and the chorus rehearse the intricate narratives of the past, its errors and demands, and struggle to explain the desired normative pattern of justice – a pattern in which Agamemnon's death will be placed. The connections in this narrative are terrifying and conflictual. Zeus, the chorus assert, sends an omen of eagles to direct the expedition to Troy, but this very omen also drives Artemis to stay the fleet. Calchas, the prophet who interprets the oracle, describes the eagles as 'sacrificing (*thuomenoisin*) a cowering hare' (*Aga.* 136), and this leads him to fear that Artemis will demand 'another sacrifice (*thusian*)' (*Aga.* 150) – as if the metaphor of 'sacrifice' *results in* the violent sacrifice of Iphigeneia, which Clytemnestra will cite as part of her motivation for murdering her husband. This uncertain interconnectivity of events, the doubt about how actions or words will prove to have been causal, produces in the chorus – and the audience? – a miasmic fear and desire for stable knowledge.[18]

The chorus' frightened and desperate want of knowledge culminates in their response to Agamemnon's death cry. In a strikingly unique piece of stagecraft, Aeschylus has each member of the chorus speak in his individual voice. The crisis of Agamemnon's scream divides the collectivity into a fragmented chaos of inadequate decision-making. The chorus' lack of knowledge stops them from acting. As one chorus member summarizes (13.56–7): χρονίζομεν γάρ, οἱ δὲ τῆς μέλλους κλέος | πέδοι πατοῦντες οὐ καθεύδουσιν χερί, 'Yes, we are wasting time (*chronizomen*); they are not asleep in their action; they are stamping underfoot the renown of delay'. In the 'real time' of the staged drama, the chorus member marks what is happening: they are, like the Greek princes at Troy in Menelaus' reproach, stuck in 'empty time' – *chronizomen* – refusing to act. As his colleague had said, but without taking any action (1353): 'Now is the perfect time *not* to

---

[17] Pucci (1992).   [18] Goldhill (1984) traces this in full.

hesitate to act' (or 'to delay' (*mellein*)). Their opponents meanwhile stamp under foot *tēs mellous kleos*, a typically impossibly complex Aeschylean expression. *Kleos* can mean 'report' or 'rumour', as well as 'renown', 'fame', and here, therefore, there is an implication that what is being destroyed is also a *story* of delay. The chorus is desperate for information from the palace: what they won't get, it is implied, is a tale of delay. In the *Odyssey*, when Odysseus has killed the suitors in the palace, he arranges for a fake wedding song to be performed so that no *kleos* of the slaughter will get out (*Od.* 23.133–40). No such *kleos* of delay will emerge, the chorus fears, here and now. It may even recall their own earlier blithe misogyny, when, conclusively but with misrecognition, they commented on Clytemnestra's story of the fall of Troy (*Aga.* 486–7): 'swift to die, a tale (*kleos*) told by a woman is wiped out'. *Kleos* that should aim to be undecaying, perishes quickly, like an Achilles but without glory, when it is spoken by a woman. Their misplaced dismissal of Clytemnestra is now repeated, as they fail even to understand what has happened in the palace and once again hope for a story to guide them. But *kleos* is also the disruption of and triumph over empty *chronos*: their delay (*chronizomen*) seems to invoke the language of *kleos*, predicated on the action they cannot undertake. *Mellous*, however, is a very rare and strange word, and may also imply what is intended or likely to happen (it is an abstract noun from *mellein*, and *to mellon* becomes the normal Attic for 'the future', though I suspect it never completely loses to abstraction the sense of a person expecting or waiting for something to happen).[19] Those in the palace – actually Clytemnestra – are crushing also 'the renown' of what might be expected or intended – both Agamemnon's *kleos* damaged by his disgraceful death, and their own *kleos*, disbarred by their inactivity. Between the *kairios* blow to Agamemnon by Clytemnestra and her announcement of her previously *kairios* speech, the chorus' fragmented, desperate interchange dramatizes *chronos* as a space of delay and, above all, ignorance that blocks action. To get the right time involves the confidence of right knowledge. The 'right time' makes human decision-making not just a temporal but also an epistemological issue.

---

[19] Pindar, *Olympian* 2.56, one of the most vexed lines of Pindar, with its incomplete conditional clause, begins 'If . . . someone knows the future . . . (*to mellon*)', an 'if' that is incomplete if any is. For the development of the notion of the future in Greek historiography see the essays collected in Lianeri ed. (2016), with the background of Hölscher (1999); Kosselleck (2004). On expectation and the future, see on Thucydides 1.138.2 (Themistocles' ability to estimate what would happen), Greenwood (2016) 87–91 – with further indicative examples.

The figure in the *Oresteia* who seems to transcend – and thus most strikingly to emphasize – how '*kairos* defines the human approach to, and separation from, divine power as a mode of apprehending the fluctuations of human temporality',[20] is Cassandra. In stark contrast to the faffing chorus, Cassandra has perfect knowledge of the past and future, a gift from divine power (double-edged, of course, in that she can convince no human of her certain understanding). She alone has the language and knowledge to express the interconnectivity of events and to determine the *prōtarchon atēn*, the 'first and originary transgression', a moment that the human subjects of tragedy so often seek in their counterfactual longings or explanatory exculpations. But the paradox of Cassandra's certain knowledge is that she cannot affect the unfurling of the story. Perfect knowledge of the future means no free will.[21] The chorus imagine that if only they had better knowledge, they would know how to act to make a difference to the future. Cassandra, who does know, can do no more than assert the inevitability of the future, her speech a devastating superfluity. So she tells the wondering chorus, as she walks, as calm as a sacrificial ox, to her death (1299): 'There is no escape; my hosts, no more time (*chronon*)'. The chorus comment, 'The last moment of time at least is privileged'. Their temporizing sop to hope is bluntly rejected: 'That day is here and now'. Cassandra's perfect knowledge means she alone of the human characters in the *Oresteia* understands exactly and without recourse when the *kairos* has arrived. Her refusal of the desire of the god of prophecy, Apollo, leaves her, thanks to her gift of complete foreknowledge, without recourse to counterfactuals or desire, inhabiting time without surprise or alteration. Is this not Cassandra's horrible paradox, to escape from the human experience of time and yet knowingly to face mortality, her time of death?

Prometheus Bound – which in its current form may not be the work of Aeschylus, or not entirely of Aeschylus[22] – offers the most dramatically bizarre and yet lastingly influential engagement with these tensions between action, duration and intervention. Till Samuel Beckett's works, it is the only play I know to have a central character bound unmoving – incapable of action – through the whole drama. Prometheus is tied to a rock and is to be assaulted daily by an eagle for 10,000 years. This is also the only extant Greek play which also has no human, social scenario – the main actors are all divine; only Io, a woman turned into a wandering cow, is not an immortal; the setting is a mountain range in the Caucasus. The plot of the play, however, depends, as with Cassandra, on a paradox of

---

[20] Fitzgerald (1987) 10.   [21] Goldhill (1986) 23–8.   [22] Griffith (1977) remains the best discussion.

knowledge and the passing of time, but now, crucially, without facing a time of death. Prometheus, who gives humans the knowledge they need to progress – not just fire – knows a secret. The secret is that whoever sires a child with the nymph Thetis will produce a son greater than his father. This secret links the play both to our discussion of the genealogy of the gods, where the threat to the stable establishment of Zeus was precisely such a continuation of violent intergenerational succession; and to Achilles, who is the prophesized son of Thetis, and whose immortality will be of fame, and thus no threat to Zeus: the network of myths makes time visible. Prometheus knows that his knowledge can thus destroy or maintain the rule of Zeus. Zeus has bound him to the rock. Zeus will not set him free until he reveals the secret; Prometheus will not reveal the secret until he is set free. It is a standoff, made a paradox by the fact Prometheus has foreknowledge of the future. So how long should the standoff last? When will there be a *kairos* to shift the actionless action? What could change such stasis? By making a drama out of gods' time, the play becomes a form of *chronos*, a continuum unbroken by the prospect of recognition or *peripeteia* or *kleos*. Without the intervention of death, even constant suffering makes for poor storytelling (as we will see, both the everlasting joy of heaven and the everlasting suffering of hell are hard to turn into narrative, which loves its endings). The play can be saved only by the prediction of a saviour, Heracles, in a further play, who will break the standoff by releasing (*luein*) Prometheus and thus find a resolution (*lusis*) for the plot. *Prometheus Bound* – in the time of the play – asks us to imagine Prometheus' experience of time stretching unendingly onwards. It dramatizes the 'empty time' of *chronos* as waiting and suffering.

With Cassandra and Prometheus, tragedy explores the paradoxes that arise from the tension between *chronos* and *kairos*, when, under unique circumstances, ignorance of what is to happen is removed. It is a dramatization that also highlights the normality of human uncertainty and its consequences. When ignorance of the future is the condition of humans, it is only 'too late' that full knowledge can be recognized, only ironically that Oedipus can claim to know. Misrecognizing how *chronos* relates to *kairos*, how the unfurling of events will produce a killing time of crisis, is tragedy's story. When Christian texts turn to make *kairos* a term of theological necessity, it comes trailing clouds of tragic conceptual anxiety about how the moment can be seized or lost, how the killing time defines a life. How, then, should a Christian face the end?

*

Plato has Socrates declare that true philosophers are engaged in a life-long 'training for death' (*Phaedo* 67e). In classical Athens, every citizen trained for military service, and risked dying in battle for the state, which in civic discourse becomes the epitaph of 'having proved yourself a good man', *agathos genomenos*: a man's goodness itself becomes linked to such a death in war, and the model of Achilles and the Homeric heroes is never far from the imaginary of such commitment.[23] Even – especially – Socrates fought for his city, fulfilling the duty that Aristotle sees as a requirement of democracy, to bear arms for the state. But Plato's idea takes preparation for death in a new direction, which redefines how life is to be spent and valued. 'The unexamined life is not worth living': a life without philosophy has lost its worth. But, further, Plato has Socrates insist first that the soul is immortal – which makes death, he concludes, unfrightening – and, second, that the philosopher's soul in particular, purified as it is, will rise into a bliss of contemplation that will last.[24] This idealism – 'the prize is noble, and the hope is great' (*Phaedo* 114c) – is surrounded, however, by hesitations. Socrates has to argue against the doubts of his friends, and turns to a myth of the underworld to persuade them. Their grief continues, however. It is not clear that the immortality of the soul would have been anything other than a very outlandish notion to most of his contemporaries, as odd as the Pythagorean claim of the transmigration of souls – unlike the familiar and institutionalized mystery cults' promise of the continuation of feasting and fun for their initiates, or Homer's picture of bodily ghosts waiting for blood to talk, or the lasting power of the heroes of cult, including the Athenians who fought and died at Marathon. More than four hundred years later, Lucian, smart-ass as ever, portrays an underworld where Socrates, far from a soul in contemplation, is still nagging away annoyingly at anyone he meets (*Ver. Hist.* 2.17). But however much the philosophical claim of the immortality of the soul has one historical, intellectual trajectory, Plato's portrayal of how Socrates faced his time of death embodies the philosopher's preparation in a calm, reflective and even witty response to the end of life. For Socrates, the *kairos*

---

[23] Loraux (1981b).
[24] Long (2019) is a useful introduction to this extensively discussed topic. See also Sedley (1989), (2009); Castagnoli (2019b) all with further bibliography.

of his execution becomes a willed transition, a fulfilment of his role as a philosopher, much as Achilles' acceptance of his own young death fulfils his role as hero. This image of Socrates at the time of death has been powerfully instrumental in the Western imagination.[25] Socrates becomes a figure easy to assimilate to Christian normative history because Plato's representation of his death seems to anticipate the moral certainty and spectacular self-possession of the hagiographic picturing of the martyrdom of saints.[26] It is one way to face the end: a fulfilled martyr to the truth.

If Socrates provides one necessary strand in the back history of the representation of martyrdom, the self-representation of the Roman adaptation of Hellenistic schools of philosophy and especially Stoicism, is also integral. Stoicism became the philosophical *lingua franca* of the educated elite of the Roman empire, and its promotion of fearlessness at the prospect of the end of life happily adapted and theorized the Roman penchant for bravery in the face of death, a penchant which also made gladiators stars of the games' spectacles of violence.[27] It is not by chance that Seneca, whose vein-opening suicide in the bath became an icon of self-control, exchanges letters with St Paul, at least in the apocryphal tradition: in the Christian imaginary, Seneca's paraded calm at the prospect of death and his philosophy of self-examination makes him a worthy interlocutor for the apostle.[28]

This fearless calm becomes the trademark of the Christian martyr. Prudentius' fourth-century collection of poems known as the *Peristephanon* ('Crowns of Martyrdom') adopts the metrical forms of the Latin poetic tradition to tell the stories of Christian martyrs at the point of their martyrdoms.[29] For these martyrs, there is no preparation for death but their Christianity itself. Eulalia is a young woman who is kept at home by her parents, but she escapes to town in order to refuse to sacrifice and thereby to bear witness to her Christianity.[30] As with the other martyrs Prudentius depicts, there is no representation of any pattern of reasoning that leads to the martyrdom, not even the sort of motivating dream that Perpetua experiences. The *Peristephanon* is a text for the Christian reader – mediated through Prudentius' self-representation as worshipper – to contemplate intense

---

[25] In general, Trapp ed. (2007); Judson and Karasmanis eds. (2006); Moore ed. (2019); more focusedly, Lane (2001); Nehemas (1998); Edwards (2007).

[26] See Frede (2006); Edwards (2007) and Franek (2016) with bibliography to a long history of debate.

[27] See e.g. Hopkins (1983); Barton (1993); Wiedemann (1992); Kyle (1998); Gunderson (1996); Edwards (2007) – all of which pay respects to Veyne (1976).

[28] See in general, Griffin (1976); Wilson (2014); and especially Edwards (2014).

[29] See Palmer (1989); Malamud (1989); Roberts (1993); Mastrangelo (2008) and generally Castelli (2004).

[30] Goldhill (1999).

suffering as an *imitatio Christi*, and as an act of public commitment to Christianity's truth against fierce and violent opposition. The martyr transcends torture – and it is crucial that every story is a story not just of pain or death but of torture – through the promise of immortality to come, and demonstrates this transcendence with a verbal performativity, a wit, that is the culmination also of Prudentius' self-display of verbal brilliance as his worship. 'Turn me over; I am done on this side of my body', quips Lawrence as he is burnt to death on the barbecue.[31] Such stories of the martyrs not only make the time of death a defining event in life – death is very much experienced – but also change the relation of the Christian to time. Salvation is a promise of immortality, to join the *sancti* in heaven; how death – as the ultimate moment of witnessing to Christ – is faced, is determinative of the time to come. Death as a process of transition is profoundly moralized. Glory, that motivation of Homer's heroes, is now an attribute of God, and God alone. To die for God is to witness *his* glory. Humans, like the cherubim and seraphim, declare the glory of God, in the highest. To read Prudentius piously is to understand both his poetry and the stories of the martyrs as enacting service to the glory of God, remediating the fame of the poet and the fame of the hero in a different direction.

Repentance – a form of death and rebirth in the language of Christianity – changes everything: it is to be a complete and thoroughgoing realignment of the soul and the life. So – to take one of very many possible examples – St Mary of Egypt ran away from home at age 12 (the same transgressive choice that Eulalia makes), but does so to become a prostitute in Alexandria. For seventeen years she lives her dissolute life, until she has a remarkable conversion in Jerusalem. From this moment, she retreats to the desert, taking only three loaves of bread, where she lives for thirty years alone as an ascetic. When she is finally sighted by Saint Zosimas, she is naked and appears barely human – her wasting away is transformed into transcendent ascetic piety – and dies on the very day he gives her communion. Her new life is a remove from all social interaction; the thirty years are described only as a continuum of her turn towards God.[32] The experience of *chronos* as an indistinct time that negates the pursuit of *kleos* has become the very fame of the saint. So Simeon Stylites becomes a spectacle of holiness by displaying himself on a pillar for year after year, where the mundanity of his experience of time in public is articulated in his hagiography only by increased suffering.[33] The promise of immortality after the transition of death changes how human life can and should be experienced.

---

[31] Conybeare (2002).   [32] See Burrus (2004) 147–54, (2019) 93–117.   [33] Burrus (2019) 123–34.

One history of Christianity – or, better, Christianities – would trace the repeated tension between the desire for a removal from society into a *chronos* of contemplation and worship, and the inevitable engagement in the messy politics of a church that is central to the power of worldly empires. Christianity's projection of temporality from the creation of time itself to the end of days not only changes how the time of death is to be conceptualized, but also thus redefines the significance of the mundane – daily time.

This contrast between the long Greek tradition of memorializing, representing, conceptualizing the time of death and the Christian reconstruction of such a tradition is on display with particularly insistent architecture in the juxtaposition of Books 7 and 8 of the *Palatine Anthology* – a juxtaposition almost certainly established only in tenth-century Byzantium, when the anthology in its current book form was put together.[34] Book 7 contains 748 epitaphs from across several hundred years. Some record famous warriors who died for their countries in poems that became famous memorials ('Stranger, tell the Spartans that here we lie, obedient to their laws', *AP* 7.249). Many record deaths abroad of merchants or those who, shipwrecked, left empty tombs. There are poems for cats, locusts, pets of all sorts. Others are composed for literary figures or for poets or civic figures of the past. It is significantly difficult to distinguish a hard and fast line between records from actual tombs and literary imitations of the form, as both testify to how death and memorial are to be conceptualized. Many assume that a stranger is passing by and must be addressed: that is, the tomb does not speak to a social community or a family, but to an imagined other, who like us as readers, scan the inscription and move on. There are few or no sequences of poems in the collection. There are some epigrams that recognize paradoxes or bizarre deaths, particularly of brides and grooms, but little recognition of noble deaths, and, above all, no acknowledgement of an afterlife or that the values of a lived life may be directed towards a life after death. Rather, the briefness of the form of the epigram turns the recognition of the fleetingness of life into a repeated display of the imagined observer's sophisticated awareness of the epigrammatist's sophisticated turn of perspective.

So, Simonides, from the fifth century BCE, writes a four-line epitaph for Pythonax and his brother who both died before they saw the prime of life (*AP* 7.300). Their father Megaristeus set up this memorial to the dead, records Simonides, 'a memorial immortal to his mortal sons a gift',

34 Cameron (1993).

μνῆμα ... ἀθάνατον θνητοῖς παισὶ χαριζόμενος. The echoes of Achilles' choice are emphasized by the pointed juxtaposition of 'immortal' and 'mortal'. The poem's verbal dexterity, by demanding the reader's recognition and reflection, enacts the injunction to each passer-by to stop in time before the tomb – a stopping which by reading the inscription becomes the act of performed memorial. By contrast, Aceratus the Grammarian (perhaps from the first century CE), writes a particularly self-conscious version of an epitaph for the Homeric Hector (*AP* 7.138):

> Ἕκτορ, Ὁμηρείῃσιν ἀεὶ βεβοημένε βίβλοις,
> θειοδόμου τείχευς ἕρκος ἐρυμνότατον,[35] (nb text Gow)
> ἐν σοὶ Μαιονίδης ἀνεπαύσατο. σοῦ δὲ θανόντος
> Ἕκτορ, ἐσιγήθη καὶ σελὶς Ἰλιάδος.

> Hector, constantly bruited in the Homeric books,
> Strongest bulwark of the god-built wall,
> With you Homer made his end. When you died,
> Hector, so too were silenced the pages of the *Iliad*.

There is no doubt that this epitaph appeals to a literary audience: the written text of the *Iliad* is its world. Hector, who shouts a lot in the *Iliad*, is always 'being shouted about' in Homer's books (the 'always' of immortal glory is here, more bathetically, Hector's repeated, loud presence in the epic). But Homer, here referred to with the patronymic which imagines for him a descent from Orpheus, made his end – both rested and stopped – 'with you'. The riddle of this striking expression is immediately explained: when Hector died the *Iliad* fell silent. The *Iliad* indeed ends with the funeral of Hector, but the epigram nicely contrasts the 'shouting' of Hector with the 'silence' of the end. Homer, Hector and the *Iliad* share a life of sound and an end in silence. The time of the poem and the time of its dead hero and the time of the poet are overlapped in this self-reflexive redrafting of Homeric temporality into the discourse of narratology *avant la lettre*. The time of death is the opportunity for a literary critical *jeu d'ésprit*.

Book 8 of the *Palatine Anthology* provides a remarkable contrast with the displaced and deracinated tombstones of Book 7.[36] It is a volume by a single author, Gregory of Nazianzus. The collection may well have been ordered by him, as its poems talk explicitly about poems being set

---

[35] Page (1981) 3, very aggressively, prefers the comparative to the superlative here.
[36] The following discussion follows closely Goldhill (2020) and especially Goldhill and Greensmith (2020), where a full bibliography can be found.

in sequence, and about the number of poems in such sequences. The book is indeed structured overall as a diptych, with, first, 165 poems on his family and friends from his Christian community, and, second, 89 poems attacking those who desecrate Christian tombs. As Homer constructs the anti-funeral in the desecration of the corpse as a contrast to the hero's search for the glory of a noble death,[37] so here Gregory contrasts the anti-sepulchral festivities and desecration of the grave with the celebrated lives of his Christian heroes. The book is organized around this juxtaposition.

Gregory's poems are educated, sophisticated verse, replete with the *paideusis* we would expect from his self-representation elsewhere, as we saw in chapter 1, but they echo the epitaphs of the past to redraft the form into a stridently Christian discourse. The first 165 poems construct a portrayal of a holy Christian family within an extended Christian community. He writes a dozen poems, he announces, for his friend and mentor, Basil of Caesarea; then a dozen for his father; some 53 for his mother; a good handful for his brother; others for close friends and religious colleagues. Each individual is evaluated within this community, for their Christian virtues and their contribution to the community. An image of an ideal Christian family, a holy family, is projected. Gregory sets himself at the very centre of this network (and even writes epitaphs for himself). This idealism is also a polemic. Gregory rejects the sexual life of the family for himself in a commitment to chastity, but lives within this community fully. This is a considered and specific position in the ongoing fights over how a Christian should live, rejecting the strict asceticism of a desert saint, but also rejecting marriage for himself.[38] Readers are not addressed as passing strangers but as potential intimates, observers of a community they can aspire to join.

The values which Gregory praises in these epitaphs offer a different moral framework from his classical predecessors. His father's 'winged soul', he announces, 'is with God' (*lachen theon*, AP 8.12.5); it is 'near the Holy Trinity' (*AP* 8.14.4); 'raised up' (*AP* 8.20.2). This triumphal discourse of transformation to heaven is strikingly contrasting with the descent down the 'iron road of Hades' (*AP* 7.412.8) which is the only end imagined in Book 7. This transcendence is directly linked to his father's life in the church. Gregory emphasizes his father's contribution as a bishop, a shepherd to the people, and his late but wonderful conversion from idolatry to Christianity, and the family of priests he has sired. (Gregory is never shy of pointing out how his own marvellous achievements redound

---

[37] Segal (1971); Redfield (1975).   [38] Elm (2012); Gager (1975).

to his father.) The Christian life praised depends on a good life leading to a death that leads to ascension to the house of God.

Gregory most intensely honours his mother, Nonna, with 'so many epigrams', and dramatizes the time of her death in a remarkable sequence of varied responses, and the very variations indicate the ongoing construction of the discourse of death. ('Death is not an event in life, we do not live to experience death', as Wittgenstein wrote,[39] but, as he did not continue, it is constantly imagined and ideologized through an increasingly intense Christian gaze at the moment of the death of others.) Each epitaph here emphasizes that she died in old age in the church praying, and each promotes a version of Christian virtue in a woman. 'One woman is famous (*kleinē*) for household tasks, another for grace and chastity, another for piety and the pains of the flesh, with tears, prayers, and charitable actions. Nonna is celebrated (*aoidimos*) for everything. If it is right to call this the end, she died praying' (*AP* 8.31). This epitaph lists the virtues of a woman. If household tasks, grace and chastity have a long history of praise, the addition of 'pains of the flesh' along with 'tears, prayer and charitable actions' reframes the tradition in a specifically Christian discourse of suffering and public service. Nonna for all these virtues is *aoidimos*, 'to be sung of' – the word Helen had used of herself in the *Iliad* – and the contrast with the tradition of epic understanding of a woman's worth is pointed by such vocabulary. The paradox is that the genre of epitaph is integrally the performance of memorial as a performance of fame. Nonna is *aoidimos*, *kleinē*, the object of praise. The very form Gregory adopts inevitably enacts the tension between humility and pride in the exemplary. She died praying, he states at the end of this epitaph (and many others), but immediately glosses this with 'if it is right to call this an end/a death (*teleutēn*)'. Death is indeed no end. As the next epitaph announces, in the voice of Nonna herself, she has gone to *zōēn ... ouraniēn*, 'a heavenly life'. The time of death has become the beginning of a life in heaven, a life in light – if 'to see the light' is a common expression for life as opposed to death in earlier Greek, now death is a transfer into light rather than darkness, a new life. And her very death in church praying is, Gregory insists, the happy *result* of her prayers, tears and vigils through her life. In this long sequence of poems on his mother, Gregory is redefining the traditional language of memorial so that the good Christian's life on earth produces the death that leads to a life in light in heaven. Life on earth, in turn, is praised as crying, suffering, praying, and, for some, producing pious children; a rejection of the physical in the

[39] Wittgenstein (1922) 6.4311.

name of the spiritual, and thus a rejection of the standard marks of social esteem in wealth and material benefit. The genre of epitaph in Gregory's hands redrafts how the ordinary time of life is to be evaluated. The 'god-given gleam' that transforms striving into a triumph of *kleos* – the model for men's competitive commitment to a life of action from Homer to Pindar and onwards – is transformed now into praise for an existence where a relation to God is lived through a refusal of the pursuit of *kleos* in this life in the hope and expectation of the true life to come. The immortality of fame is no longer the aim or promise, but an immortal life.

Achilles' renewed determination to die young to win immortal fame, and his Odyssean desire to be the humblest man on earth rather than king over the dead, should be equally impossible thoughts for the Christians of late antiquity. The martyr's death is for the glory of god, not the glory of a human (though the martyr certainly becomes an exemplar of religious commitment, a life to be lived up to, a life to be 'embraced with the arms of the mind'[40]). A longing for life on earth is a denial of the very promise of Christianity. This does not stop Homer continuing to be read, studied, loved … Yet what this Christian understanding of the time of death intro-duces is a repeated tension in the experience of time and thus the narratives of a life, a tension between removal into contemplation, mortification of the flesh, prayer, as expressions of a relation to God, and engagement with the church and the material world as an expression of mission. What the end is going to be changes how stories are told. The anticipation of the end of days, we will see shortly, is one motivation for complete sexual abstinence, the attempt to stop the generations of men, like leaves, falling – or to rethink entirely how humanity is 'organically connected to the totality of time', as Plato put it. How the time of death is thought to make sense of the human experience of mundane time becomes a new and vexing question in the Christianity of late antiquity. The transformation of the time of God comes with a change in the understanding of the end that is death and with it a change in the wagers of human achievement: it changes how a human is to inhabit time. The possibility of a martyr's iconic death – a killing time that transcends love of earthly life – is formed in relation to a saint's continuous *chronos*, an unworldly continuity, gazing at and longing for a promised eternity: both are transformative ideals of an end, for the labourers of the church on earth to contemplate. A second foundational question, then, for this book is: how do these shifting conceptualizations of the end change the narratives of a life, and transform the horizons of how time is to be inhabited?

---

[40] Augustine, *Sermo Denis* 14.3; see Perkins (1995); Clark (1999); Castelli (2004); Kelley (2006).

CHAPTER 3

Telling Time

The messianic religions of late antiquity are obsessed with getting the time right, to know the right time – both at the level of daily, weekly, monthly or annual rituals, and at the level of world history. Where the prophet Cassandra can say for herself, 'the day has come', the Gospels will insist that everyone must be anxiously aware that 'the hour is coming and is now here'. The desire to be certain about one's place in time produces an extended, competitive and argumentative scholarly literature, which is never simply about the correct calibration of time. Rabbinical writing, first of all, is exemplary of these temporal obsessions.

'There is no early and late in the Torah'. This rabbinical principle is a response to when the Pentateuch seems to narrate events out of chronological order. An immediate example: Leviticus 24 lists legal regulations, and describes the punishment of a man who has blasphemed – a stoning that takes place while the Israelites are travelling in the desert, after the revelation on Mount Sinai. But Leviticus 25 begins, 'The Lord spoke to Moses on Sinai, saying . . .' – an intervention that must have taken place earlier than chapter 24. There is no inconsistency here, it is argued, because 'there is no early and late in the Torah': linear temporality is an unnecessary expectation. For rabbinical biblical exegesis, the principle of 'no early and late' not only allows cross-referencing between passages from within apparently different temporal frameworks, but also produces a creative reading strategy that recognizes but recalibrates such temporal disruptions in the name of the unity and permanence of the Torah. The Torah exists, it is argued, before the giving of the Torah on Sinai: Noah, it is pointed out, takes into the ark 'pure and impure animals', a distinction established, it might be thought, only by the text of the Torah. The Torah pre-exists the revelation to Moses, as the embodiment of God's wisdom, which is the 'paradigm' or 'archetype' – the blueprint – of creation itself (Philo already argues this, separately from the rabbinical tradition).[1] The constant and divided rabbinic arguments about what the text of the Torah signifies is grounded in the 'always already' of its continuing significance.

The challenge to linear chronology and its causality is enacted in the further rabbinical principle known as *bererah* – legal retroactivity. This problematic idea has been superbly analysed most recently by Lynn Kaye. With *bererah*, 'different temporal configurations of events are presented as coexistent';[2] its aim is to replace previous doubt with ritual certainty through what can be called a legal fiction about time. So Kaye's paradigmatic example concerns the establishment of a legal residence to allow movement on the Sabbath. An observant Jew cannot walk more than 2,000 cubits from his house on the Sabbath. But to allow someone to hear a visiting lecturer, say, who will be further from the home, a temporary legal residence may be appointed to allow a longer walk. But if you do not know on Friday night, when the Sabbath starts, which direction you will be travelling, you cannot determine the temporary residence as you are required to do; but if you do subsequently discover the destination on Saturday, then, the Talmud asserts, the decision on Saturday will reveal what the intention on Friday really had been – although evidently the very problem is predicated precisely on *not* knowing the intention when the Sabbath starts. The principle of *bererah* thus allows 'a later action to *have actually taken place earlier*, replacing the ambivalence with certainty'.[3] A logically challenging case of reverse causality . . .[4]

Now, when and how this principle is invoked by the rabbis is a complicated matter of legal nicety, which could detain us here for some time.[5] The vista it opens onto the Talmud's temporality, however, is instructive. In a perhaps less worrying and more standard scenario, the Talmud has no anxiety about imagining conversations between rabbis who lived centuries apart as a way of dramatizing differences of opinion. Linear chronology is no bar to interaction – without any of the complex heroic, dream-time or necromantic apparatus typical of Greek and Roman moments of meeting figures of the past. So, too, when it comes to a discussion of the interpretation of dreams, it is declared with a certain psychological plausibility that the interpretation 'protentively confers reality upon the dream, and as a consequence, the dream shapes reality': a double

---

[1] Philo, *Opif.* 15–25: see van Winden (1983).   [2] Kaye (2018) 112.   [3] Kaye (2018) 3.
[4] And a much more interesting and salient case than the examples dreamt up in Dummett (1954), (1964). A better case for Dummett might be the laughter which retrospectively turns a comment, not intended to be humorous, into a joke.
[5] Kaye (2018) of course handles these too.

movement of retrospective and proleptic making of truth.[6] Most saliently for me here, however, is how this overturning of linear temporality informs the Talmud's style of history writing. From Thucydides' example onwards, historiography – to speak *grosso modo* – expects the unfolding of a narrative in a linear and causal patterning, where digressions are self-consciously marked as such. This is not, however, how the Talmud imagines the interlinking of events. Within a broad patterning of divine retribution, each moment in the national story is morselized and becomes the fulfilment and exemplar of biblical verses, which have always already encompassed these stories to come. Digressions are so prevalent that they form the normative style of interconnection. The record and memorialization of the turning points of national history turn out to be stories about good or, usually, bad halachic decisions (decisions on religious law and behaviour), where it rarely matters which emperor is the oppressor or in what order the military or political events are dated. Indeed, major disasters are represented providentially as paradigms of each other, a repetition enforced by determining the same date in the calendar for these overlaid occasions: the first and second temple were both destroyed on the ninth of Av; the Greeks desecrated the altar on 25 Kislev which was also the date of its rededication. In these scripturally grounded, theologically underpinned narratives, repetition is explanatory and causal. (Occasional such comments from earlier in the Greco-Roman tradition, such as Timaeus' claim that Rome and Carthage were founded in the same year, are exceptions rather than a discursive norm, and, as we will see in chapter 7, simultaneity is a *problem* for historiography.) Furthermore, such events designedly form the aetiology of a festival remembrance. The exodus from Egypt is announced to be the future festival of Passover before it has happened; the occasion of the translation of the Torah into Greek, when the world went dark for three days, is marked out for annual memorial. In this way, singularities of the national story are transformed into the regulation of the religious calendar's repetition. As has been often stated and much debated after Yerushalmi's seminal book, *Zakhor*, the Talmud does not write or engage with history, but with memory and, above all, religious regulation.[7] Nor are matters different in other rabbinical writings: with the striking

---

[6] Wolfson (2011) 147. Wolfson (2006, 2015) fascinatingly takes forward such issues into later Jewish mysticism.

[7] Yerushalmi (1982): the contrast with Jewish texts in Greek, for whom historiography is the 'most important Greek genre' (Adams (2020) 299), is stark; see also Funkenstein (1989); Gafni (1996); Neusner (2004) and the works cited below in chapter 6, n. 1. I have discussed this in Goldhill (2020) 194–235.

exception of the first-century CE *Seder Olam*, a unique and polemical book of 'exegetical chronography', and in contrast with Hellenistic Jewish writers, including Josephus, the rabbis show no historiographical interest.[8]

In this sense at least, according to the rabbinical world-view, the insistence on halachic life – a society ruled by the regulation and debate of religious law – is a *timeless* project. The historical time-frame does not affect the halachic issue. The Talmud has no difficulty in imagining that the patriarchs, Abraham, Isaac and Jacob went to *yeshivah* (rabbinic schools), and argued halachic issues, although one might have thought that they were the first and only Jews in their own time. This is not so much an anachronism, as an assumption of sameness across time. When the Torah says of king Solomon, 'I built houses, I planted vineyards, I made gardens and orchards, I planted in them trees of all kinds of fruit', *Midrash Kohelet Rabbah* (2.4–8) analyses it to say: 'buildings, this means synagogues and schoolhouses; vineyards: this means rows of scholars who sit like rows of vines; gardens and orchards: this means the great *mishnayot* (religious regulations of the rabbis); trees of all kinds of fruit: this means the Talmud'. The world is to be viewed now and forever through the lens of rabbinic learning and practice. This all-embracing view may well be formed by the stringent politicized concern of the rabbis for the boundaries of their own world, a reflex of the porous and fragile boundaries of the Jewish community after the destruction of the Temple and the loss of the possibility of self-determination.[9] Certainly the destruction of the Temple marks a divide in practice, with the consequent allowance that cultic and domestic regulations can change over time; and the broad generations of the earlier and later rabbis, the *tannaim* and *ammoraim*, are recognized as having different authority.[10] Yet it is integral to rabbinic writing that the interconnection or separation of stories is generally not expressed along a temporal axis, but through a network of interrelated normative religious regulations. The narrative form of the Talmud is not articulated according to a linear temporal succession. Contingency – most directly theorized in 'emergency laws' – can indicate a specific moment in time, but there is no sense of 'olden days', as with Roman writers, or that different eras have different lineaments. Anachronism is not a pressing issue because there are not two different times to compare.[11]

---

[8] See Milikowsky (2014) from whom the phrase 'exegetical chronography' is taken; Stern (2001), (2012); Bickerman (1968).

[9] Boyarin (2015); Rubenstein (2003); Schwartz (2001); Kalmin (2006); Secunda (2013); Lapin (2012).

[10] On the role of tradition within the Talmud, Vidas (2014) is a fascinating critique of the standard view upheld by Rubenstein (2003); Halivni (2005); Friedman (2010).

[11] Rood, Atack and Phillips (2020) – of course, I am tempted to add, sadly – despite their title include neither Jewish nor Christian sources in their antiquity.

Yet there is in the Talmud – at the same time – a fascination with telling the time. Any mechanical technology for telling the time is barely mentioned in the Talmud (and unlikely to have been regularly available to the students of the Talmud). Yet the need for rituals to be observed at the correct time motivates many long discussions. The beginning of the Tractate b. *Berakhot* opens with a discussion of the timing for reciting the *shema*, the fundamental liturgical prayer of Judaism. The *shema* says it should be recited 'when you lie down and when you rise up' – evening and morning. From what time, then, in the evening, and until what time, may it be said? The answer has two forms. The first is a parallel: 'From the time when the priests enter to eat the *teruma*', that is, the time of evening's start is marked by another ritual, the entrance of the *cohenim* into the temple to eat the fat of the evening sacrifice. The second answer, however, is 'Until the end of the first watch' – a diacritical temporal marker. The next pages consequently open an intricate double argument, first about the parallel – can this be adequately defined also as 'when the stars appear'? Is it the same time as 'when a poor man eats bread', mentioned elsewhere in such debates? Or when people go to eat their Shabbat bread? – and second about the watch: how many watches of the night are there, three or four? From what biblical verses can the case for either number be made? If one text (Judges 7.19) says that Gideon 'came to the edge of the camp at the beginning of the middle watch', it proves there must be three watches, as 'middle' implies one before and one after; but if two verses of a Psalm (Ps. 119.62/119.148) indicate that David rose at midnight (to study, of course) after two watches, surely there must be four watches (two before and two after the midpoint). With a certain intellectual flair, the text revels in inconclusively reconciling these apparent contradictions.

This extended argument is exemplary for at least three interrelated reasons: first for the network of different potential ways of marking time – different ways of experiencing time as measurement, or seeing oneself located in the right time as a social and religious positionality; second, for its use of biblical verses to explore how to divide the night – for making time a question of scriptural authority, thus requiring the expertise of rabbinical interpretation; thirdly, for its insistence (and performance) that telling the time is an epistemological crux. To tell the time requires knowing the time. So, when a verse of the Bible reads 'Thus saith the Lord, "About midnight, I will go out into the midst of Egypt"' (Ex. 11.4), it is immediately pointed out that it cannot be God who says 'about midnight' as God certainly knows time (as we have seen). Does this, then, mean that Moses did not know exactly when midnight was? Rather, Moses calculated that if Pharaoh's astrologers made a mistake about when exactly midnight was, they would say that Moses was a liar (it is added elsewhere, and they would thus doubt God). So to say 'about midnight' is a motivated inaccuracy – a knowing refusal to say 'I know' – to avoid getting entangled in a mess of false expectations and accusations.

A witness called before the court, however, is granted the leniency of human failing. 'If one says "at the second hour", and the other says "at the third hour" their testimony stands' (m *Sanhedrin* 5.3) A contradiction of an hour between two witnesses is to be ignored. But 'If one says "at the third hour", and the other says "at the fifth", their testimony is rejected. Rabbi Yehudah says it stands'. Two hours, it seems, is too large a discrepancy, though there is a countervailing opinion. But, the mishnah continues, 'If one says "at the fifth hour", and the other says "at the seventh", their testimony is rejected since at the fifth hour the sun is in the east while at the seventh hour the sun is in the west'. When there is a substantial, astronomical reason to make the difference between witnesses less likely, that difference must be accounted for. The debate that follows from this initial statement of the issue depends not just on human inability to tell the time accurately – and thus challenging their role as witnesses, where authoritative knowledge, precisely, is at stake – but also on the frameworks of measurement. The physical difference of the sun being in the east rather than the west makes it less plausible that time could be confused between two witnesses. How tightly time can be measured and thus form part of regulation becomes a pressing concern for religious authority. This stretches to the calendar itself, where the recognition of the precise date of the new moon depends on the witnesses who have seen it, and where the difference between the lunar and solar calendars requires the repeated recalibration of intercalated months to keep the festivals in their proper seasons. Telling time is also the announcement of the proper times of the religious calendar by the authorities. Indeed, for the rabbinical communities of late antiquity, as we will see, the regulation and recognition of their community in and against the dominant structures of imperial authority is maintained by the regularization and enforcement of their lunar calendar against the Julian calendar, the time of empire: telling time can become an act of community-building as resistance.[12]

The measurement of time, the ordering of the calendar, and the narratives of time are three pillars of the following discussions, but it is the connection between them that provides my focus: how telling time and

---

[12] See Stern (2001), (2012); Feeney (2007); Kosmin (2018).

telling stories are mutually implicative processes. Between counting, accounting, recounting (or *zählen* and *erzählen*), how are the ordering of narrative and the ordering of time to be interrelated?

*

Measuring time is obsessive. As is the historiography of time's technologies. Histories of time are rarely brief.[13]

There are hundreds of thousands of baked clay tablets written in Akkadian and Sumerian from the Assyrian kingdom, which reached its zenith in the eighth and seventh centuries BCE (and remained an object of fascination for both the Israelite and Greek communities that remembered the imperial figure of Senacherib in particular). These tablets record in cuneiform script, amongst other lists and inscriptions, scholarly calculations of astronomical time. The educated elite of the court created 'a highly competitive atmosphere which drove intellectual innovation'.[14] Their aim appears to have been to determine through celestial omens the gods' will, and thus it was particularly important to have accurate predictions of the times of solar and lunar eclipses, always especially dangerous moments. Much Assyrian scholarship depended on their smart neighbours, the Babylonians, against whom the Assyrian kings even organized library raids. In the Assyrian *Epic of Creation*, however, the regularity of astronomical measurement comes from divine creation. The god Marduk creates time before the natural world or humankind. He sets the heavenly bodies in place to follow a temporal pattern and thus mark out time ('Time comes into being with the heavens', Plato wrote, many years later).[15] 'Go forth every month', Marduk says to the moon,

> without fail as a crescent disc, at the beginning of the month, to wax over the land. You shall shine with horns to mark out six days; on the seventh the disc shall be half. On the fifteenth day you shall always be in opposition, at the mid-point of each month ... on the ... thirtieth day you shall be in conjunction with the sun again. (Tablet 5.16–22)

The calendar here is evidently made up of twelve thirty-day months. The actual lunar year has fewer days, of course; the solar calendar more.

---

[13] From a huge bibliography on modern time(s), let me single out Galison (2003); Wilcox (1987); Sherman (1996); Gell (1992); Glennie and Thrift (2009); Bender and Welberry eds. (1991); Koselleck (2004); Birth (2012); Buckley (1966); Clark (2019); Currie (2007); Landes (1983); Schlögel (2016).

[14] Robson (2004) 46 – with further bibliography to which Haubold, Steele and Stevens eds. (2019) is an important addition. The following paragraph follows Robson's analysis. Rochberg (2020) looks at the influence of Babylonian work on Hellenistic astronomy.

[15] *Tim.* 38b6 discussed below pp. 165–9.

Reconciling these different frameworks involves a mass of continuing scholarly activity of recalculation and intercalation. Astronomy provides one technical system for measuring time (and we will return to its Christian significance in chapters 12 and 13). The omens this system produces provide an interface between the technology of time-telling and the structure of royal power, and the performance of the rituals of power, which also make time visible. The Assyrians, however, also kept *limmu* lists, lists of court officials, and lists of kings, year by year, which construct a human deep history back into the second millennium BCE. Such genealogies, fictional and historical, express, as ever, the self-representation of contemporary rule along with the narratives of succession that make sense of dynastic transition – that develop, that is, the discourse of authority. Telling time is deeply involved with the statements of power.

In classical Athens from the fifth century BCE, the technology of water-clocks, which dated back many centuries in other cultures, was adapted with loop-back systems to regulate flow, and utilized with greatest public awareness in the courtroom to limit the time of speeches by defendants and prosecutors. The clock became the sign of the court and its legal process. (Plato, typically, has the conversation-hogging Socrates – a notable failure in court – contrast the restricted flow of the water-clock with the unrestricted to and fro of true philosophical debate (*Theaet.* 172d).) The courtrooms of Athens embodied the essential ideological tenet of democracy that guaranteed equality before the law for all citizens, and the clocks guaranteed equality of time between competing sides within the legal process. Danielle Allen generalizes the point nicely: 'Within the polis, therefore, time as measurement primarily was a mark of citizenship and privilege in Athens, separating politically powerful from powerless'.[16] Regulating time is regulating the exercise of the public spectacles of political power.

Rome, too, used the water-clock in the courtroom but its non-democratic politics uses the technology in a tellingly different manner. The court decided the amount of time allowed to each speaker, and under the principate, looking back to the Republic, the control of the time of speech became a contentious issue of political freedom: *libera tempora* overlapped the liberal use of time with the time of political liberty. Pliny in particular dramatizes 'the perceived boundary between the rapid present of the principate and the slow past of the republic' in what he excoriates as 'a conspiracy of haste among both speakers and judges':[17] 'We fast track our

---

[16] Allen (1996) 163.   [17] Ker (2009) 300, 290. See also Riggsby (2009) (and Riggsby (2003)).

cases', he laments in a letter, 'using clepsydrae less numerous than the number of *days* it once took for cases to be set forth', *paucioribus clepsydris praecipitamus causas quam diebus explicari solebant* (*Ep.* 6.2.6). Yet Pliny himself is also capable of proudly boasting how in Trajan's court he was granted extra *clepsydrae* to make his case (*Ep.* 2.11) – reflecting the *pulchrum et antiquum* 'noble and ancient practice' of the time of the Republic (a simplification that Cicero's description of being ordered to speak for only half an hour might belie). Pliny also praises his own performance as a judge because he grants speakers the time they require (*Ep.* 6.2.7–8). Pliny also shows, that is, how amid the complaints about *tempora*, he earnestly manipulates the possibilities of (self-)representation to project the possibility of good governance. The power of regulating time is negotiated as well as exercised.

This connection between political control and the exercise of time is a constant in the historiography of time. 'Neither political Space nor political Time are natural resources. They are ideologically constructed instruments of power.'[18] As the inscription on the clock-tower in Adana in Turkey, built in the late Ottoman empire, declares with extraordinary explicitness: 'It seems like a clock striking but in fact it is the government calling.'[19] Social systems, as Niklas Luhmann argued, are intimately connected with their temporal horizons, their comprehension of *Weltzeit*.[20] Or, to use François Hartog's productive expression, regimes of historicity, which determine the self-awareness of a person through the interrelations of past, present and future, depend on theories of time. This is nowhere clearer than in the eighteenth and nineteenth centuries, where, as Reinhart Koselleck's seminal analysis of the 'semantics of historical time' asserts, a new dominance of historical self-awareness and a new professionalization of the discipline of history become prevalent: political regimes both write the past in self-justifying, teleological narratives of their rise to power over time, and display the past in similarly performative material and ritual forms, from museums to coronations.[21] As Chris Clark puts it, power bends time to its own narrative – 'regime chronopolitics', each with its own 'distinctive temporal signature'.[22] The Victorian fantasy of Merrie Old England, or Walter Scott's medievalism, or the German-speaking states' obsession with their Teutonic past, create an imagination of the past that informs the political dreams of the future, and explains the political engagement of the present. It is not by chance that the very

[18] Fabian (1983) 144.   [19] Wishnitzer (2015) 149.   [20] Luhmann (1975), (1982).
[21] Koselleck (2004).   [22] Clark (2019) 16, 211.

concept of heritage takes its now instrumental cultural and political shape only in the nineteenth century.[23] Periodization takes on a new ideological force in such nationalist narratives – terms endemic to this book such as 'Hellenistic' or even 'late antiquity' are intellectually loaded constructions of this era[24] – and a heightened consciousness of 'living in the present moment', in 'modernity', is symptomatic of the time's self-consciousness and rhetoric. Industrialization is a crucial part of this process of the ordering of time. A national time – time should be measured on the same scale across the jurisdiction of a nation state – is introduced because of the need of harmonizing railway timetables, a system dependent on the accuracy of timepieces as well as the state's will and power to enforce it.[25] With the transformations of transport, which reduces distances by speed, time itself seems to hurry, or at least the experience of travel brings a vivid sense of acceleration, the feeling of time, as well as the landscape, rushing by. Work and days are reorganized by industrial capitalism's insistence that time is money, and 'time-discipline' and 'time-thrift' become categories instrumental in defining work through efficiency and productivity in a set time rather than through tasks completed according to the order of time. As Wordsworth wrote, from his Romantic longing for a different artisanal understanding of a relation to work and the natural world, labouring men and women are now overseen by 'Stewards of our labour, watchful men | And skilful in the usury of time . . .'.[26] In this 'usury of time' – wonderful phrase – punctuality is a virtue, wasting time a grievous error. By contrast, the first state prison in England, with its forced inactivity of timed incarceration, is also a nineteenth-century introduction. Now punishment is – for the first time? – to be measured by time, time 'inside', time judged to fit the crime: prisoners will henceforth have to 'do time'.

Through the nineteenth century, the organization of historical time into a universal developmental model grounds political movements, investing terms such as 'revolution', 'progress', 'the state' with new force. Marxism's passion for explaining the past in service of a normative future is only the most obvious example where a theory of work and industrialism's time-keeping explains how the experience of everyday time is embedded in a model of the necessary unfurling of a universal development of

---

[23] Swenson (2013).

[24] See e.g. Momigliano (1994) on Droysen; Lianeri ed. (2011); Herzog (2002a) on 'Late Antiquity'. See also Bowersock, Brown and Grabar (1999); Rousseau ed. (2009); Brown (1998).

[25] See Galison (2003); Wilcox (1987); Sherman (1996); Birth (2012).

[26] Wordsworth, *Prelude* 5.378–9. See Sennett (2008), (2005).

politico-historical time.[27] This politicized notion of a universal time –
'Godless, continuous, empty and homogenous'[28] – continues to ground
the narratives of political dominance.[29] (Chakrabarty may be right to
emphasize 'godless', but the contrast, easy to oversimplify, between
Catholic ideals of church tradition and Protestant notions of reformation
also underlie this historiography, especially in Western Protestant university environments, as is demonstrated by the ease with which the so-called
Protestant work ethic is assimilated to such economic models of modernity.) Geology, however, was the discipline that, with greatest immediate
shock value, revolutionized Victorian comprehension of deep time.[30]
Geology produced a model of the long history of the material world
which challenged in particular the possibility of traditional Christian
chronology, Protestant or Catholic, with its commitment to the limited
span of time in its eschatological vision. In its invention of stratigraphy,
geology also offered a way of conceptualizing, representing and articulating
the past, which, in combination with later evolutionary science, made cyclical
or non-dynamic natural history harder to maintain. 'Revolutions still more
remote appeared in the distance of this extraordinary perspective (of geology)',
as one challenged scientist wrote, striving to capture his sense of wonder, 'The
mind seemed to grow giddy by looking so far into the abyss of time.'[31] As with
Darwin's vision of the infinite number of generations each contributing to the
incremental work of evolution, so the abyss of geological time, stretching,
unendingly, into the past and the future, terrified, excited and overwhelmed
imaginations in the nineteenth century – and both geological and evolutionary biological theory profoundly influenced the narrative of the novel, for
example, as scholars, extending the revelatory work of Gillian Beer, have
outlined tellingly. 'Comte, Hegel, Darwin, Spencer and Marx shared the
idea that philosophies, nations, social systems, or living forms become what
they are as a result of progressive transformation in time, that any present form
contains vestiges of all that has gone before.'[32] For some, the very word vestiges
will recall Robert Chambers' *Vestiges of the Natural History of Creation*, one of
the nineteenth century's most provocative books in such a tradition.[33] In

---

[27] See e.g. Buckley (1966); Burrow (1981); Bowler (1989); Fritzsche (2004); Hesketh (2011);
Koditscheck (2011).
[28] Chakrabarty (1997) 36; see also the other essays in Lowe and Lloyd eds. (1997); and the trenchant
Davis (2008).
[29] Chakrabarty (2000); Jameson (2002); Banerjee (2006).
[30] Lyell (1837): see Rudwick (1992), (2005), (2008); Secord (1986).
[31] Playfair (1805) *Transactions of the Royal Society of Edinburgh* 5: 73.
[32] Kern (2003) 51. See Beer (1983) and e.g. Shuttleworth (1984); Buckland (2013).
[33] See the wonderful Secord (2000).

short, between historiography, science, literature, theology and political theory, how both the deep time of the past and the everyday time of social life are to be comprehended and narrated has become a battleground of competing theories of time and forms of narrative. Telling time is integral to these discourses of authority and power, and telling time is instrumental in how these discourses changed everyday life, its comprehension and exercise, in the nineteenth century and beyond. Who is to control the telling of time? When Philip Gosse, fundamentalist Christian and biologist, was laughed into public humiliation for suggesting that God placed fossils on earth as fossils, he became a lightning rod for the profound differences between religious and non-religious understandings of time.[34] Can we write a history of telling time, then, without turning back to God's time?

One starting point for such a history might be Aristotle. In Book 4 of his *Physics*, Aristotle gives a compressed account of time that has provoked many responses over the centuries.[35] He begins with three paradoxes about the very existence of time, for which he surprisingly does not offer solutions. First, he asks: if time is made up of the past, which has been but is not, and the future which will be, but is not, how can time participate in being, since it is made up thus of non-being – what is not. Second, if something is divisible, its parts must exist. But time is divided into the past, which is not; and the future, which is not, and cannot be made up of just 'nows'. Third, he asks: is the now always different or always the same? The now cannot be different, he argues. If the current time (now) is three o'clock, when will the instant be when this now will have ceased to be? It cannot be at three o'clock, for that is when the moment exists. It cannot be the very next instant because two instants cannot be next to each other. If there is a gap between the first and second instant, the gap itself can always be divided. So if one were to say, it ceases to exist at three o'clock plus one second, what about three o'clock plus half a second? Does the now exist then? If it does, the instant was not three o'clock; if it does not exist, it did not cease at three o'clock plus one second. And so on in infinite regress. But if the two instants are without division, there is no moment when the first ceases to exist and the second exists; events of the past would be, impossibly, co-existent with events in the present.

---

[34] Brilliantly told in Gosse (1907).
[35] See Annas (1975); the seminal Sorabji (1983); the fully detailed Coope (2005); and, for modern response e.g. Mellor (1981), each with extended further bibliography. Garcia (2014) 177–88 shows incisively why the positions of McTaggart, Broad, Merrick and others back to Kant are destined to 'lead to impasses' (182).

These three paradoxes have fascinated modern philosophers of time, and they have argued extensively and intently about Aristotle's possible solutions, and traced the solutions to such problems of the instant and of change that are to be found in other ancient and medieval thinkers – and we have already begun to see how engrossed Augustine was in a similar array of concerns about the disappearing presentness of human time, already lost to memory and anticipating the future, as it is experienced. Aristotle goes on to argue that change and mind are preconditions for the existence of time, and to define time as a 'number of motion with respect to the before and after', ὁ χρόνος ἀριθμὸς ἐστι κινήσεως κατὰ τὸ πρότερον καὶ ὕστερον (220a25): change is essential to the recognition of time, as is, it seems, an ensouled being to recognize it. Whether 'number' necessarily implies a quantity that can be counted has been debated by philosophers, as has almost every aspect of Aristotle's dense few pages. Yet for my purposes here, rather than rehearsing the book-length debates about the precise sense and implications of Aristotle's argument, it is more important to note that Aristotle offers a theoretical account within a model of physics that does not refer to divine causation, nor insist on calendrical requirements, nor even debate at length about what it is to inhabit time; rather, he tries to argue about time itself. Telling time, for Aristotle, is a tightly focused philosophical project. Far from an engagement with the spectacles of politics, law and regulation, Aristotle's mode of telling time is resolutely abstract, rebarbative in its intellectual austerity, and, in contrast with Hartog's regimes of historicity and Clark's historicity of regimes, searching, it seems, for a universal – a timeless – definition of time.

This Aristotelian project also has a long heritage. Were we to return to the nineteenth century, we could happily follow Peter Galison's detailed studies of Poincaré, for example, whose work on simultaneity was fundamental in the lead-up to Einstein's publication of four seminal papers on time in 1905.[36] (Richard Sorabji, with a certain delight in his own unusual lack of restraint, declares that *only* Einstein has made a contribution to the theory of time of comparable ground-breaking novelty to Aristotle.)[37] We might also enjoy the tales of the experimental failures to capture simultaneity in signal transmission across the English Channel, which, like so many other Anglo-French collaborations, broke down into mutual accusations of incompetence. The international adoption of Greenwich Mean Time was seen in France as a national insult and resisted for some years – another example of the unfortunate interface of the science of time and nationalist

---

[36] Galison (2003).   [37] Sorabji (1983).

ideology. Such a history of telling time would also follow the especially rich recent bibliography on the development of the technology of timepieces, from the long, experimental struggle to measure longitude accurately – a project which was certainly neither austere nor abstracted from the politics of money-making and empire – through the equally fascinating story of clock-making, to today's removal of time-telling from any relation to the heavens – atomic clocks use the electromagnetic transitions in the hyperfine structure of the caesium-133 atom, accurate, it is boasted, to the equivalent of the loss of one second in 20 million years.[38] This history would note how for many centuries it was barely necessary or possible to measure time in daily life with more precision than an hour (a measure that in itself shifted in length through the year as it usually represented a twelfth of a day or the twelfth of a night, which thus expanded and contracted according to the season); how clocks and then watches introduced both more accurate and more personally owned expressions of precise time-keeping (the stories of Pepys, walking the streets of London to test his new watch's accuracy, are paradigmatic of how behaviour was changed by this technology); to the most recent nanotechnology that claims to measure time in extraordinarily small divisions, that are barely imaginable. When measurement claims to reach 247×10^{-21} of a second (247 zeptoseconds), is this still telling time? Under such circumstances we might be reminded of where Aristotle started his reflections, that time exists, 'barely, obscurely'.

It might seem, therefore – and this has often been argued – that from the Enlightenment onwards, modernity is thus distinguished by an increasing secularization of time. Isaac Newton is one of the highest high priests of Enlightenment science, but he was also a committed Christian who 'undertook an extraordinary programme of creative theological research, whose expansiveness, originality and radicalism was matched by only a handful of contemporaries' (including spending many of his scientific hours calculating the dimensions of the Temple of Jerusalem).[39] When it comes to Time, however, Newton writes what could stand as the credo of such secular belief: 'Absolute, true and mathematical time, of itself and from its own nature, flows uniformly, without regard to anything external'.[40] So, it has been claimed, at the turning point of the beginning of the nineteenth century, 'the linear, rather than the tabular historical

---

[38] See e.g. Landes (1983); Wilcox (1987); Sherman (1996); Gay (2003); Glennie and Thrift (2009); for seafaring as well as railways as a major influence see Ishibashi (2014); for the politics of national time see Bartky (2000); for the Ottoman empire, the fascinating Wishnitzer (2015); on Egypt, Barak (2013); for Greenwich time in England see Rooney and Nye (2009).

[39] Iliffe (2013); and see Iliffe (2016) for the full story.   [40] Newton (1846) 77.

timeline, the visual representation of time so familiar to us to be unre-markable, became a hegemonic temporal metaphor'.[41] Although quantum physics in the wake of Einstein came to change the game at least for science, it still appears that measuring time is generally treated as a regular physical event; and, in social life, time is invested with a mundane expectation of linear regularity; across the world, time is time is time. We glance at our watch or phone, and move on ...

But things are not quite so simple, and not only because of quantum physics' destabilization of the basic ideas of simultaneity, 'uniform flow', and uncertainty. Virginia Woolf, herself a high priestess of modernism and its narratives, captures one factor. 'An hour, once it lodges in the queer element of the human spirit, may be stretched to fifty or a hundred times its clock length; on the other hand, an hour may be accurately represented on the timepiece of the mind by one second.'[42] The experience of time resists the regularity of measurement: boredom stays time; excitement accelerates it. For Woolf, time is not in the province of the regular, but the queer.[43] Her own style of writing – like that of James Joyce – attempts to capture this 'queer element' of human consciousness, but such writing is inconceivable without its contextualization in modernist psychology, par-ticularly Sigmund Freud, and modernist theories of time, especially Henri Bergson, as was recognized at the time. Wyndham Lewis declared: 'Mr Joyce is very strictly of the school of Bergson-Einstein-Stein-Proust. He is of the great time-school they represent.'[44] He could have added Freud. Freud, translated into English by Woolf's friend, James Strachey, and published by her husband, Leonard Woolf, always insisted on the indes-tructability of the content of the unconscious. What is repressed will return (*Wiederkehr des Verdrängten*). It is essential to Freud's view of temporality and causation that these lasting experiences, memory-traces, impressions are revised and refitted to new circumstances, a process he termed *Nachträglichkeit*, 'deferred revision'. The stratification of the mind – and the archaeological metaphor is not arbitrary – keeps the past in the present, and, much as the past impacts on the psychology of the present, so too through *Nachträglichkeit* the present changes the past. The time of the mind is not linear, even and especially when considering sexual

---

[41] West-Pavlov (2013) 68.   [42] Woolf (1928) 98.

[43] The idea of 'queer time' has been especially taken up in medieval and early modern studies recently: see Dinshaw (2012); Freccero (2006). As with antiquity, the current discussions of 'queer time' in medieval Europe do not discuss how straight time seems to become culturally hegemonic only in the Enlightenment and its aftermath.

[44] Lewis (1927) 89.

development. The compulsion to repeat (*Wiederholungszwang*) is the most telling symptom of the failing repression of the past in the present.[45] Both Woolf's *Mrs Dalloway* and James Joyce's *Ulysses* are novels that take place in a single day and have thus a unitary temporal structure (the repeated ringing of Big Ben in *Mrs Dalloway* sounds out the clock of linear time), but each seeks to represent the multiform, non-linear, layered consciousness of its hero/ine. Each plays out the tension between the forward drive of the book's telling and the fragmented, echoing, interconnecting making of sense – between the linear progress of the narrative and a multi-directional, fissiparous experience of narrative. For both Joyce and Woolf, the time of the mind cannot be adequately fitted to the time of the clock.

Virginia Woolf hand-set 'The Waste Land' by T. S. Eliot. The career of Eliot – another of the high priests of modernism – could be marked out between the bleakest version of time's arrow in 'Fragment of an Agon':

> Birth, and copulation, and death.
> That's all, that's all, that's all, that's all,
> Birth, and copulation, and death.

and the famous opening of the 'Four Quartets' that draws not just on these modernist psychologies and theories of time, but also on Augustine:

> Time present and time past
> Are both perhaps present in time future,
> And time future contained in time past.
> If all time is eternally present
> All time is unredeemable.
> What might have been is an abstraction
> Remaining a perpetual possibility
> Only in a world of speculation.
> What might have been and what has been
> Point to one end, which is always present.
> Footfalls echo in the memory
> Down the passage which we did not take
> Towards the door we never opened
> Into the rose-garden. My words echo
> Thus, in your mind.

In between these two poems, Eliot had converted to Anglo-Catholicism, which Woolf wrongly predicted he would quickly drop like the bones from

---

[45] See, from within the hothouse of psychoanalytic literature, Laplanche (1999a), (1999b), (1999c), (2017), better than the deeply disappointing Green (2002).

an eaten herring.[46] Eliot resists the idea of an eternal present in the name of redemption, and the echoes of possibility ('perhaps', the performed hesitation in theorizing), set against the inevitability of the 'one end', redeemed by his religion. Eliot, too, insists on the echo in the mind as the experience of consciousness ('footfalls echo in the memory') and turns it back on his readers: 'My words echo, thus, in your mind', insisting, like Augustine that reading, the experience of language – *logos* – is integral to the understanding of time. Here, for the modernist Eliot, his religious hope as much as his sense of consciousness stand against the simple directional flow of time's arrow. The triumph of secularization will have to wait.

Digressions knowingly disrupt and draw attention to the interconnection of narrative and time. ('Let us talk of something else, for a while . . .'). I have taken this briefest of detours through more recent centuries partly because some of the very best theoreticians of time have worked on this era and the mobilization of their insights cannot take place without such contextualization. Partly, when considering time, of all subjects, it would be crass to think that the study of antiquity can take place as a view from nowhere, no time – an appeal to disinterested objectivity that will be discussed in the coming chapters: I write from within a modern set of assumptions about time that I hope, over time, to be able to make more visible. How time is told changes over the centuries, and how such technologies and theorizations change narrative form is also revealed vividly by this later history. It is a history of appropriation and changes of self-understanding, as well as technology or philosophy, that lets us see what might be meant when we privilege any ancient ideas as 'influential' – and thus to appreciate more precisely our own stake in such a history. We see what is specific to both, when the ancient and modern worlds are mirrored in each other, across the divide of the Christian invention of time. That is why such a digression as this is integral to my narrative.

*

The Roman day had its order, certainly. Martial, a poet with an extraordinary eye for the disruptive in the mundane, captures this potential for the compulsion of routine with characteristic vividness (4.8.1–7):

> The first and second hours wear out the clients,
> the third puts to work the hoarse lawyers,

---

[46] 'Drop Christianity with his wife, as one might empty fishbones after the herring'. Letter to Francis Birrell, 3 September 1933, cited in Cooper (1995) 200 n. 3.

Rome stretches her various labours till the fifth hour,
the sixth will give rest to the tired, the seventh will bring an end,
the eighth to ninth is enough for oily wrestling matches,
the ninth commands us to pile up cushions and crush them,
the tenth is the hour for my little poetry books . . .

The poem goes on to explain that it is only late, and at a decent party, that his poetry should come to the eyes of the emperor. The roll-call of the hours of the day – the clichés of mundane expectation – is a build up to the moment of relaxation when Martial's voice can dare be heard. The day is but a prelude to a moment of the night. As so often, Martial's description of the city-life of Rome and its power dynamics is also a self-serving search for his own place on the map, in this case, the map of time.[47]

Rome, too, had its timepieces to help with such a map. Augustus set up a huge sundial in the forum, his *horologeum*, an expression of power over space and time: 'the emperor measured time in such a way to show his natural destined part in its progress'.[48] Petronius imagines that Trimalchio had a water-clock with a trumpeter (real or mechanical? Vitruvius explains the technique at II.II) to sound out the hours, a sign of his vulgar extravagance, and the sort of innovation that makes Pliny the Elder grumpy. Astrologers calculated times through the maps of the heavens with great precision (and threat) and it seems that the most complex chronological machine as yet discovered from antiquity, the Antikythera mechanism, was designed for such astrological calculations.[49] Yet in later antiquity, Christianity, at least Christianity in its more extreme forms of monasticism, radically altered how the hours of the day were calibrated and experienced.

Anthony retreated to the Egyptian desert to live a solitary life of a hermit, and others followed his example – though not always to his extreme of fasting, prayer and solitude: reading the life of Anthony was a tipping point in Augustine's conversion, but he never saw the need to find a cave to live in; Jerome took his staff and library with him on his retreat in the Holy Land. Pachomius, however, starting around 318, organized a group of these individual ascetics into a novel community at Tabennisi in Egypt. The very term monastery, with its root, *monos*, 'alone', marks the paradox of the community of solitaries.[50] By the time

---

[47] See Fitzgerald (2007); Rimell (2008); and Wolkenhauer (2019), with Ker (2019).
[48] Allen (1996) 165. In relation to Paul, see Nasrallah (2019) 199–223.
[49] Jones (2017) sums it up. Talbert (2017) takes us from public grandeur to the hand-held. See Hannah (2020a) and (2020b) for short summaries. For the fullest account of ancient timepieces, with pictures, see Winter (2013).
[50] See Rousseau (1985), with the background of Rousseau (1978).

of his death in 348, there were eight other groups of what became known as 'cenobitic' communities ('communities of shared life'), and in the next generation the number of monks involved in such institutional living swelled into the thousands. Pachomius was the first to write a code for such monastic living and he organizes the day around the liturgy – regular prayers at regular hours – and expects the times in between to be spent on manual labour and spiritual reflection. Although the flight to the desert constituted a sort of 'anti-city' (as Peter Brown memorably describes it), the cenobitic monastery was also productive, a highly organized factory of manual and spiritual labour.[51] Indeed, for the families of Basil and Gregory – Basil was instrumental in the foundation of eastern monasticism[52] – such religious institutions were based on their existing wealthy estates and consciously avoided the extreme asceticism of an Anthony. The life of the monastery remained for many centuries a model of Christian time.

No figure is more salient in this history than John Cassian, not least because he was bilingual in Latin and Greek, and brought the principle of monastic regulation in the style of Pachomius' Rule to the West.[53] He visited monasteries in Egypt, and returned to export Pachomius' paradigm. Cassian's *Institutes* are full of instructions for the precise, repeated and zealously regulated use of time, including the night. Here is a typical rule for the monk entrusted with waking the others for night-time prayer (*Inst.* 2.17):

> But he who has been entrusted with the office of summoning the religious assembly and with the care of the service should not presume to rouse the brethren for their daily vigils irregularly, as he pleases, or as he may wake up in the night, or as the accident of his own sleep or sleeplessness may incline him. But, although daily habit may constrain him to wake at the usual hour, yet by often and anxiously ascertaining by the course of the stars the right hour for service, he should summon them to the office of prayer, lest he be found careless in one of two ways: either if, overcome with sleep, he lets the proper hour of the night go by, or if, wanting to go to bed and impatient for his sleep, he anticipates it, and so may be thought to have secured is own repose instead of attending to the spiritual office and the rest of all the others.

Time is to be regulated assiduously; daily habit retrained; precision required. Cassian even tells us at what time of day a monk might be

---

[51] Brown (1988). See also Gaca (2009).   [52] See Rousseau (1985) 190–232.
[53] Rousseau (1978); Markus (1990) 181–8; Stewart (1998); Driver (2013) (with full bibliography), all of whom cite Chadwick (1950).

afflicted by depression (*akēdia*, usually translated as spiritual 'sloth') – especially the fifth to sixth hour, it seems.

Benedict's Rule is perhaps the most influential monastic regulation over the centuries. It is influenced by Basil's Rule, but has a far more rigorous expression of how time is to be controlled.[54] Seven compulsory religious services at specific junctures (the seven so-called Canonical Hours) are supplemented by night-time vigils; meal times are to be taken at specific times in different seasons; the length of daily work is stipulated. In a way that is unparalleled in antiquity, the monastic rules insist that every hour counts and must be counted; that repetitive, stringent observance of such rules of time are integral to spiritual salvation; that such observance is indeed an 'office', an imposed duty, willingly upheld. In chapter 14, we will see how this institutionalization becomes part of internal life. The rule of the monastery provides a template in which the distinction either of *kairos* and of *chronos*, or of work and leisure, does not apply: all time is regulated to direct humans towards the worship of God.

Are there ways in which the industrious factory that is the monastery looks forward to the industrial factory of the nineteenth century?[55] The daily time that Christianity established makes the logic of my earlier brief digression on the modern history of time-telling, I hope, clearer. We will return to calendars, eras and the end of days before long: Christianity insists that time is unthinkable outside religious history and the Christian year (God's time).[56] But for the mundane telling of the time of day Christianity bequeathed a newly demanding idea of time that is not just regular, regulated, ritualized, but also stridently repetitive and compulsive. Looking to God re-orders how life is seen, and the technology of temporal control is integral to how God's time and human life are shaped. Unlike the violent and bloody tales of saints in pursuit of the crown of martyrdom, and unlike the Talmud, which images the necessity of a constant questioning and exploration of what correct *halacha* is, the monastic rule idealizes the tautological narrative of regulations obediently followed. We might ask if this extreme religious ordering of time leaves any narrative that is not of transgression or temptation. Is not the best monk a man without a story? Is not the desire for God's

---

[54] See Zerubavel (1981) 33–69; Adam (1995) 64–6 who is typical in taking the *regula* teleologically as the model of the modern schedule, without regard for its religious foundations.

[55] Weber (1981) [1923] 365.

[56] Keble's *The Christian Year* was one of the nineteenth-century's bestselling books, which capitalized on the link between daily devotion and reading – instituting the domestic performance of Christian time: see Tennyson (1981); Blair (2012); Lysack (2019).

timelessness – or even eternity – a desire for no more narrative? A man not so much without qualities, as without a story. Or, rather, does the necessity of temptation, which the Fall makes inevitable, mean that we know that such a story without conflict, without a hesitation, without a crisis, is impossible to tell?

CHAPTER 4

Waiting

One answer to the questions posed at the end of the last chapter is to recognize that Christianity gives new and powerful and transformative impetus to the question, 'What are you waiting for?'. That is this chapter's subject.

When the disguised Odysseus, after twenty years away, finally sees his father, the old man is in his orchard, and the pathetic recognition scene turns on the recollection of how Laertes had once named these trees for his son.[1] Leaves like the things of man ... The recognition that will guarantee the full genealogical significance of Odysseus' constantly repeated patronymic, 'son of Laertes', details another act of naming, naming tied to the long-term promise of husbandry that always plants for the future. To 'cut down the olive trees', a constant threat of ancient warfare, still being replayed, is not just to damage the resources of another state but to commit a brutal act that aims to wipe out the possibility of a future for the enemy.

A passage in the Talmud makes the ideology of planting explicit (*Ta'anit* 23a):

> One day, Ḥoni was walking along the road when he saw a certain man planting a carob tree. Ḥoni said to him: This tree, after how many years will it bear fruit? The man said to him: It will not produce fruit until seventy years have passed. Ḥoni said to him: Is it obvious to you that you will live seventy years, that you expect to benefit from this tree? The man said to Ḥoni: I myself found a world full of carob trees. Just as my ancestors planted for me, I too am planting for my descendants.

Ḥoni is a celebrated figure in rabbinical mythology whose magical prayers were strong enough to bring rain in drought. Yet here – a familiar trope of traditional tales of traditional wisdom – he learns from a simple man, tied to the soil and the generations of men, about how to think of

---

[1] Henderson (1997).

waiting for the future. The story takes on a more pointed polemic when it comes to the Messiah. In *Midrash Avot Derabbi Natan*, a commentary on *Pirkei Avot*, the *Ethics of the Fathers*, a text which begins with a genealogy of transmitted wisdom from the revelation on Mount Sinai, there is this truly amazing injunction (31):

> If you had a sapling in your hand and were told that the Messiah had come, first plant the sapling (and then go out to greet the Messiah).

The announcement that the Messiah has come should not stop you planting for the future. There is a pragmatic caution when it comes to eschatology. After waiting so long, and so many false messiahs, better to guarantee the planting for the future before going to greet the figure who will bring the end of things as they are. Anticipating, expecting, waiting for the end is what humans do.

There is a wry recognition of the mix of hope and disappointment in this pronouncement of *Midrash Avot Derabbi Nathan*. Back in 1961, the British comedian Peter Cook, part of *Beyond the Fringe*, wrote a celebrated satirical sketch about a millennial group awaiting the end of the world. When the predicted moment of destruction passed uneventfully (again), the dialogue concluded with the hilariously bathetic line, 'Not quite the conflagration I'd been banking on'. Anticipating the end of days, however, was a deadly serious concern of early Christianity, a concern which changed how it was thought possible to engage in everyday life and its continuity. For many today inhabiting the anthropocene, there is an anxiety of similar proportions, though with a very different hope: if destruction is predicted, how, then, is daily life to be lived? There is a pressing politics in awaiting the end.

The most read epics of antiquity are structured around delay – about willed, forced or misguided waiting. In the *Iliad*, Achilles spends seventeen books refusing to fight because of the slight to his honour. The death of Patroclus changes the motivation of his return – it is loss of a different calibre that drives him to re-commit to his own inevitable young death – but the structure of delay and fulfilment is integral to the moral compass of the work: the foundational question of what is it worth dying for is given narratological form in the narrative of refusal and re-commitment. The *Iliad*, however, anticipates but never represents the death of Achilles. For that, the readers too must wait. In the *Odyssey*, Odysseus is delayed for ten years from making his return. Seven years with Calypso, concealed; one year with Circe: both wish to make him their husband, to extend his stay in a permanent domesticity that fractures and distorts the home-life to which he wishes to return. With the deceitful and dangerous Circe,

Odysseus has to be reminded by his crew to leave, as if for once the hero has found his desire to travel back obscured. Even when he reaches Ithaca, kills the suitors and returns to Penelope's bed, he immediately announces that the end has not yet been attained and he has another journey to make, a journey to an unmappable place that does not know the sea, which Teiresias has indicated is a necessary destination before he can reach home properly. Each step on Ithaca marks thus the 'not yet' of return, each threshold is a temporary point of passage. Odysseus' unending delay makes Penelope the archetype of waiting (the sort of time which Fränkel saw as mere *chronos*). Weaving and unweaving the shroud of her father-in-law is a potent symbol of this stasis: awaiting Laertes' death; awaiting the return of her husband; doing and undoing, to stay still. The discovery of her complicity with delay is one catalyst for the plot to reach its end. For both the *Iliad* and the *Odyssey*, however, full of many deaths as well as the representation of the underworld, the end that is the death of the hero is deferred, and marks the ends for which they struggle as temporary and tear-stained: Odysseus will not manage to reach and stay at home; Achilles wishes to die because he has lost Patroclus. This melancholic structure invests delay with the double force of the 'not yet', not yet the defining victory, not yet the tragedy of heroic loss.

The two epics that redefine this 'not yet' most strikingly are Quintus of Smyrna's *Posthomerica* and Lucan's *Pharsalia*.[2] For Lucan, the politics of deferral is a resistance to the principate. The victory of Caesar, the triumph of Octavian, bring the disaster, as Lucan sees it, of imperial power. As long as the poet can keep Caesar from crossing the Rubicon, the advent of imperial tyranny is kept at an imagined distance. The horror of civil war, mapped in the violent distortions of the language of fraternal equality and fraternal (self-)destruction, is itself even a terrible bulwark against the political closure that is inevitable but hated. Lucan's poetry is a tortured resistance to its own end.

The *Posthomerica* begins with a 'when' – a connective back to the end of the *Iliad* – and ends with a storm, a narrative device that so often marks the crucial turning point at the *start* of an epic, and an anticipation of the *Odyssey* to come.[3] Quintus' epic is located between Homer's two epics, as a continuation and a connection. There are two narrative vectors in the *Posthomerica*, however, that pull in opposite directions. On the one hand, there is the parade of the stories and destructions of major heroes: Penthesileia, Memnon,

---

[2] On Quintus see Greensmith (2020); Maciver (2012); on Lucan see Henderson (1987); Masters (1992).
[3] Greensmith (2020); Maciver (2012); Bär (2007).

Achilles, Ajax. Then Neoptolemus arrives. Paris is killed by Philoctetes. In Book 11, both sides fight to a standstill, and that leads to the opening of 12: 'there was no end of war'. This vector is the drive forwards of the battle narrative. On the other hand, there is a second vector which also culminates in the fight to a standstill in Book 11. This is made up of the continual delays to this drive. There are more funeral scenes in the *Posthomerica* than in the *Iliad* and *Odyssey* combined, as the fighting repeatedly grinds to an exhausted halt for the collecting and burial of the dead. The heroes on both sides, especially the Trojans, lament the length of the war. Enthusiasm and terror oscillate, leaving neither side in a dominant position. These thematics of delay are mirrored at the narratological level by the constant turn to simile, epitaph, ecphrasis as a retardation of the action. This second vector makes repetitive gestures of delay the defining movement of the narrative. The poetics of the interval are defined thus by these competing vectors: on the one hand, the impulse of battle narrative with its structure of winners and losers, violent triumph and abject defeat, which cannot however reach the climactic fulfilment of either sacking the ships or sacking the city (and even when the city is sacked, once simply fighting is rejected, the narrative goes on to the further destruction of the storm and . . . so on); on the other, the repeated move away from the battlefield in similes and other gestures of delay – to the extent that by Book 12 it has become clear that the impulse of battle is a stalemate: but no end. As Emma Greensmith has brilliantly shown, the interval summons a poetics of stasis.[4] We are left waiting for the sack of Troy. As the prologue of Book 1 programmatically announces, ἔμιμνον . . . ἔμιμνον, 'they were waiting' . . . 'they were waiting'. The interval is the place where we wait.

Virgil's Aeneas may be *pius*, 'dutiful', but he is also *invitus*, 'unwilling': he does not desire to leave Dido in Book 4 of the *Aeneid*. In the *Aeneid*, written after historiography has recalibrated the writing of time, the foundational is now necessarily political and imperial.[5] The story of Rome's beginnings is a story *for* the powers that be. The vision in the underworld is a prediction of history to come, a teleology of power. Much as Hesiod does, Virgil uses the narration of the past to explain how things must be what they are. It is a prospect that promises its own timelessness: *His ego nec metas rerum nec tempora pono; imperium sine fine dedi* – Jupiter promises 'For the Romans I place no boundaries on things, no time: power without limit I grant' (*Aen.* 1.278–9).[6] It is against such a promise of lasting empire that Aeneas' crushed desire to stay with Dido is articulated as

---

[4] Greensmith (2020).   [5] Hardie (1986).
[6] For the Christian reappropriation of this phrase, see e.g. Prud. *In Sym.* 1.542; 2.541.

a necessary step towards the foundation of Rome. Delay here is a temporary and temporalizing block on the inexorable passage of history.

Yet not only does there remain a provocative openness about the ending of the *Aeneid* that, as Philip Hardie in particular has explored, prompted later Latin poets into varied theoretically and ideologically laden gestures of closure, but also it is against this foundational tradition embodied in the *Aeneid* that Ovid in his *Metamorphoses*, wilfully, knowingly and brilliantly, sets his epic's swirls of narratives of change, a poem without end (*perpetuum . . . carmen*, *Met.* 1.4), where even and especially the origins (*coeptis*, *Met.* 1.2) are ceaselessly open to transformation (*mutastis*, *Met.* 1.2).[7] Against the empire without end, Ovid constructs the subversive poem of change without end. The prologue promises a story *ad mea . . . tempora*, 'up to my times', but the epilogue promises that what will last for ever (*perennis, indelibilis*) is the poem itself and with it Ovid's fame: *vivam* is its last word: 'I will live'. Stories of identity, power, the past, in Ovid's hands change and threaten to keep changing beyond the ideological projection and regulatory power of imperial continuity. What is lasting across time is not empire and its values, but a poet's renown and his poetry of transformation.

Much more, of course, could be said about the dynamics of delay in each epic here and in other epics too, but enough has been said about waiting to offer immediately four provisional conclusions. First, epic narrative displays and manipulates its own temporal unfurling, as an integral function of the thematics of each poem. Narratives of action do love temporary blockages to achievement – struggle, difficult tasks, terrifying enemies on the way to a goal. But epic is particularly concerned to make the hesitations and refusals of delay significant thematic elements of the work. Second, each epic develops a narrative that is aimed from the start at an ending (*tantae molis erat . . . Romanam condere gentem*, 'it was an endeavour of so great a scale to found the Roman nation', *Aen.* 1.33). Perhaps what distinguishes all narrative, as opposed to the mere span of a life, is what Frank Kermode calls 'a sense of an ending'.[8] This should be understood most generally, as Viktor Shlovsky influentially asserted, as the ability of literary narrative to manipulate time.[9] But all the epics discussed in this chapter project their stories beyond the formal closure of the narrative, towards the coming death of the hero, the fall of a city, further travelling, or future

---

[7] Hardie (1993); Hardie, Barchiesi and Hinds eds. (1999); Brown (2005); Feldherr (2010). For the tradition see Barkan (1986).
[8] Kermode (1966).   [9] Lemon and Reis (1965).

foundation. The tension between the epic's formal narrative closure and such imaginative open-endedness invests the relation between delay and fulfilment with a disruptive question about what ending or endings count and how. The relation between *arma* and *virum*, announced (with *-que*) as the subject of the epic in line 1, is a question simply not resolved – or certainly not resolved simply or conclusively – in and by the final scene of the *Aeneid*. Third, this sense of an ending embodies the politics of epic's foundational desire. 'What are you waiting for?' becomes a politicized question. So – to come to an unfinished story of now – Martin Luther King knew that the advice to wait for a change to come was a poisonous encouragement of complicit delay: 'time itself becomes an ally of the forces of social stagnation'.[10] What is it worth waiting for and at what cost – Phoenix's challenge to the recalcitrant Achilles – is a question echoing through the revolutions of late antiquity. Fourth, the reader's pleasure is thoroughly engaged in this structure of delay and fulfilment: what Roland Barthes famously called 'the hermeneutic code' – the enigma that needs explication or solving in and by the narrative.[11] Reading for the plot, as Peter Brooks puts it, involves the anticipation of retrospection.[12] By the end we know we will have re-evaluated the clues when the murderer is revealed, and we read from the start with such expectation. Even more so when we re-read critically. Waiting, in this sense, is endemic to reading. (What *does* happen next?) The engagement of expectation, however, is not only pleasurable but also normative, as the progress towards fulfilment follows or plays with the tropes of genre, the dynamics of cliché and surprise, culturally dominant patterns of behaviour, feeling and social interaction. Stories train their readers, however recalcitrant, to inhabit the world. Waiting for the end of a story enacts a form of cultural assimilation.

*

Christianity makes delay and fulfilment a defining structure of religious narrative. 'Do not think I have come to abolish (καταλῦσαι) the Law or the prophets', says Jesus in the Gospel of Matthew (5.17), 'I have not come to abolish them but to fulfill (πληρῶσαι) them'. The repeated quotation of the Septuagint in the Gospels sounds out a regular tolling, reinforcing the logic that makes Jesus' story the deferred completion of the prophecies of

---

[10] King (1964) – the remarkable 'Letter from a Birmingham Jail', dated 16 April 1963, a powerful statement of what is (un)timeliness, specifically in response to a Christian call simply to wait for change.

[11] Barthes (1975).   [12] Brooks (1992).

the Hebrew Bible. The Pentateuch is not an expressly apocalyptic or eschatological text, but the prophets have many passages that image the end of days in terrifying and beautiful poetry. Often the comparison is between the promise of the eternity of God's salvation and the ephemerality of the time of man's achievement (Isaiah 51.6):

> Lift up your eyes to the heavens
>> And look upon the earth beneath;
> For the heavens will vanish like smoke
>> The earth will wax old like a garment,
> And they that dwell therein will die in like manner;
> But my salvation will be for ever,
>> And my deliverance will never be ended.

The physical world's destruction and the death of the humans inhabiting it are imagined as a majestic foil to proclaim the eternity of God's power. Such generalizing rhetoric of doom seeded a profusion of later apocalyptic literature; much as Ezekiel's vision of the valley of dead bones brought back to life is instrumental in the later eschatological proclamations of the Resurrection of the Dead. But the book of Daniel contains a more specific politicized vision of the empires of the world passing away. In chapter 7, his prophecy – an interpretation of the dream of Nebuchadnezzar – describes 'four great beasts', which are 'four kings who shall arise out of the earth'. In contrast to the Son of Man whose 'power is an everlasting (αἰώνιος) power, which will not pass away, and his kingdom will not be destroyed' (7.14), these four kings betoken empires which will defeat each other until the fourth king, 'which will exceed all the kingdoms' (7.23), comes to devour the whole world and destroy it. This terrifying king 'will think to change laws and times (καιρούς)'. The threat of his violent anarchy reaches to time itself – either the feasts and ceremonials of worship, or, even more grandly, to the ordering of months and seasons, the regularity of the cosmos. In a deeply obscure phrase, this kingdom is said to last ἕως καιροῦ καὶ καιρῶν καὶ ἥμισυ καιροῦ, which is translated (obscurely) in the King James Bible as 'until a time and times and the dividing of time'. This phrase has been taken to mean – somewhat bathetically – 'three and a half years', or, more grandly, three eras of different lengths; as well as various other calculations for the length of the rule of this monstrous authority. The threat of the 'change of times', however, seems to anticipate this phrase, which divides time and times into semantic confusion – an anarchy that prevents secure anticipation of the precise time of the end. Daniel's vision is a politicized history of the unfurling of time as the providential passage of four empires.

The book of Daniel, tellingly, is explicitly named and quoted in the Gospel of Matthew (and quoted in Mark 13.14), precisely when Jesus foretells the destruction of the Temple of Jerusalem which opens out his discourse on when the 'end of days', or 'the closing of the age' (συντελείας τοῦ αἰῶνος, 24.3) will come. 'When you see "the abomination that is desolation", as spoken of by Daniel the prophet, in the holy place (let the reader understand), then . . .' (Mat. 24.15–16). This is a carefully layered prophetic expression. They are to 'see' what the vision of Daniel has foretold, 'the abomination that is desolation' (Dan. 12.11, cf. 9.27; 11.31). Daniel's vision, exiled in Babylon, seems to allude to the occupation of Jerusalem by Antiochus IV Epiphanes of the Seleucid kingdom, which threatened to destroy cultic worship in the Temple – a desecration and a regime against which the Maccabees fought. 'The abomination that is desolation' may in Daniel refer to the cessation of worship in the Temple or the introduction of a Greek cult there. But after Jesus' prediction of the Roman destruction of the Temple, the abomination of desolation also takes on a different force: it is now not so much cultic desecration as the total annihilation of the site. The Temple will be burned down (and a temple of Venus eventually erected in its place). What they are to see, therefore, is not just the empirical evidence of destruction, but also – a deeper vision – how this destruction is a heightened fulfilment of an earlier prophecy, an earlier vision, with different implications to be played out in the later supersessionist historiography. Hence the surprising addition: 'let the reader understand' (or as Jesus enjoins his enemies in John (5.39): ἐραυᾶτε τὰς γραφάς, 'search the scriptures'). We are encouraged not just to see, but to read through scripture to reach understanding of what the time portends.[13]

After Jesus, echoing the prophets, anticipates the clash of nation upon nation (the 'beginning of the birth pangs' (Mat. 24.8)), and predicts the arrival of the Son of Man in great glory with the trumpet calls of angels (24.29–31) – a vision of the end – he also tells the disciples explicitly and through parables that 'About that day and hour, no-one knows, neither the angels in heaven, nor the Son, but the father only' (24.36), which leads to the repeated injunction: 'Therefore watch (γρηγορεῖτε); for you do not know on what day the lord will come' (24.42); 'Therefore watch (γρηγορεῖτε), because you do not know the day or the hour' (25.13). Mark adds 'Look! Stay awake! For you do not know when the time

---

[13] John Chrysostom *Adversus Iudaeos* 5 is exemplary of this politicized reading of Daniel, in his case also through Josephus' reading of Daniel.

(καιρός) will come' (Mark 13.33) – the arrival of the Messiah is the very definition of *kairos* as event or turning point. It could, he explains, be 'late or midnight or cockcrow or early' (13.35), therefore 'Watch!'. 'You must be ready', says Luke, 'because the Son of Man is coming at an hour you are not expecting' (Luke 12.40).[14] (Clement of Alexandria in his genial Hellenism bathetically parses this as the danger of oversleeping in too soft a bed *Paed.* 2.77–82.) The prediction of the future demands that our future is one in which we watch and wait. Vigilance, a vigil, is required. For Christians, the Gospels demand, anticipating, waiting, expecting is what the faithful do, all the time. That is how the time of everyday life is to be inhabited.

The Gospels were written after the destruction of the Temple by Rome in 70 CE.[15] Well before this date – the earliest sections of the text are probably third century BCE, the later parts around the turn of the first century BCE or even later – the book known as 1 Enoch stages a long apocalyptic vision of the cosmos from its beginnings, through the fall of angels, to the destruction of the wicked in a final all-embracing judgement of God. Its fascination with angelology, eschatology, demonology, and the final judgement indicates its radical difference from the Pentateuch and most of the Hebrew prophets.[16] It is indeed not included in the Jewish canon. It exists now in full only in Ge'ez, the language of the Ethiopic church, but there are extant fragments in Greek, Hebrew, Aramaic and Latin. Enoch may not be canonical for authoritative Judaism, but it is quoted in Aramaic in the Dead Sea Scrolls, and in later Greek Jewish apocalyptic literature, as well as in the New Testament book of Jude. It is also cited by Christian writers including Justin Martyr, Irenaeus, Origen – and others up to John Cassian. Jewish apocalyptic literature, and especially Enoch, is treated by the magisterial Gershom Scholem as foundational testimony of a tradition of Jewish mysticism, but it is thereby also the literature of the marginal – mysticism always insists on the elitism and separation of the elect – and, most obviously after the fall of the Temple and the destruction of Jewish political self-determination, it is the literature of the dispossessed and defeated, too. The Talmud has its moments where the humiliation of the Romans or other foreign tyrants is dramatized as a self-authorizing and deeply gratifying victory, but rabbinical writing's more standard approach to the culture of Greece and Rome is a studied ignorance and pointed ignoring.[17] The rabbis turn their back, even if it is to

---

[14] Becker (2017).
[15] A much-contested matter: see Burridge (1992); Vinzent (2014); Robinson (2001).
[16] J. Collins (1998) remains the best introduction. See also Himmelfarb (2010).
[17] Goldhill (2020) 194–235 with bibliography.

look over their shoulders, on empire culture, even and especially as Judaism continued and even increased its Hellenisms. But the apocalyptic tradition takes violent revenge on the powerful for the dispossession of the defeated. 'The stories incorporated visions of the beginning and the end of the world, and gave the faithful an idea that in the end the righteous (us) would be rewarded with eternal resurrection, when the wicked (you, and the Romans) would be damned and eternally punished'.[18] ('You' in this sentence addresses an imagined Christian author, 'Keith Hopkins', as a type for the Christian tradition, as corrected by a fictionalized 'Seth Schwartz', speaking on behalf of the multiplicity of Jewish voices, under the sobriquet, Avi – as written by Keith Hopkins.) 'This was another set of latish Jewish ideas', continues the passage, 'that the early Christians took over'. So in the Gospel of Matthew, when Jesus anticipates that 'you will be hated by all the nations for my name's sake' (Mat. 24.9) – the future persecution which will include not just the immediately addressed disciples but also the future readers who hear themselves called in the word 'you' and when he anticipates that 'the end (*telos*) will come' (24.14), he turns designedly to the promise of apocalyptic literature in its starkest form (25.46): 'And they (the cursed) will go away into eternal punishment, but the righteous into eternal life'. The end of days, the telos of time, is to be a comfort for the everyday humiliations and violence of precarity, marginalization and persecution. Just you wait . . .

The book of Revelation is the most extended expression of Christian apocalypse, with the most extraordinary account of the punishments of the wicked and the blessing of the righteous. It is framed by the declaration (1.3) ὁ γὰρ καιρὸς ἐγγύς, 'For the time is nigh' (*kairos*, of course), and the promise ναί, ἔρχομαι τάχυ, 'surely, I am coming quickly/soon'. The cry of waiting in the Psalms, 'How long, O Lord . . . ?' in Revelations becomes the promise of immediacy. For Paul of Tarsus, indeed, the present moment was urgent. 'For the form of the world is passing away' (1 Cor. 7.31). The crucifixion of Jesus and the resurrection were the transformative events that heralded the Second Coming. Paul's insistence on the need for every human, and not just the Jews, to transform themselves is a response to the promised imminence of the return of Christ. Jesus himself in Mark (1.15) announces his mission with a demanding prophecy: 'The time is fulfilled, and the kingdom of God is at hand; repent and trust in the good news'. As in Daniel, *kairos* is the time of revelation and the time of the end; it has reached fulfilment, the fullness of time (πεπλήρωται: a perfect tense,

---

[18] Hopkins (1999) 236.

fulfilled and being fulfilled); and, Jesus declares, the kingdom of God is 'nigh' (ἤγγικεν): it is this prophecy which Revelations echoes – and which makes Paul so insistent.

For Paul writing in the middle of the first century, salvation has already come through the crucifixion and resurrection, but the kingdom of God has not yet arrived. Between the already fulfilled and the not yet completed is the *parousia* of Christ and the promise of the end: this is the present: *ho nun kairos*.[19] But how is this eschatological present to be lived? Paul's struggle with the sinfulness of his fleshly existence produces an extraordinary expression of his daily existence: 'And as for us', he asks 'why are we in danger every hour? I die daily . . .'. On the one hand, Paul is rhetorically displaying the apostles' willingness to risk their lives hour by hour as a proof of the firmness of their belief. On the other hand, his personal declaration, 'I die daily . . .' takes the risk of death in a different direction. John Chrysostom understands this claim to mean 'By his readiness and preparation for the event' of death, that is, by his attempt to live with a constant eye on the Christian promise of the resurrection. The danger thus is also the danger of slipping from the path of faith that 'we too', καὶ ἡμεῖς, all risk. Life thus is to be lived hourly with an anxious eye on our participation in the coming of the kingdom of God.

For Luke, writing in the first century, the present has already been extended. The Gospel of Luke does not include Mark's recognition that the time is fulfilled, and his version of the parable of the faithful servant, unlike Mark's (though like Matthew's), imagines that the returning master is late (χρονίζει, 12.45). Luke 'continually tamps down and reshapes the vibrant apocalyptic traditions that originally shaped what he had inherited'.[20] The Kingdom of God is not impending, but already inside us (17.20): 'When he was asked by the Pharisees when the Kingdom of God was coming, He answered them, "The Kingdom of God is not coming with signs to be observed; nor will they say 'Lo, here it is!', or 'There it is!'. Behold the Kingdom of God is within (*entos*) you"'. The standard translation 'to be observed' underplays the force of the Greek *ek paratēreseōs*: the verb *paratēreō* means not just to observe carefully but to watch out for. Luke's Jesus here does not encourage waiting and watching, but transformation. This move is in itself a sign of things to come: 'eventually, as the years stretched on, evolving traditions would de-eschatologize the meaning of Jesus' resurrection . . . the resurrection itself, later tradition will assert, says nothing about what time it is

---

[19] See Nasrallah (2019) 199–223, building on Agamben (2005).   [20] Fredriksen (2018) 93.

on God's clock'.[21] Over time, as Christianity became institutionalized into the empire's dominant religion and as theology expanded into a discipline, the delay of the Second Coming developed its own continually lengthening history.

*

Augustine waited a long time for his conversion to happen: '*Quamdiu quamdiu*', 'How long?!, How long?!' he cries out even as the moment of Grace itself approaches (*Conf.* 8.12).[22] Augustine takes the notion of waiting down to the level of the moment and theorizes it fully as part of his philosophy of time. Like Aristotle, he recognizes a tripartite system of past, present and future, and that the notion of the present is both constantly turning into memory as it happens and is constantly looking forward in anticipation to the future. The present of human time becomes thus a less and less stable idea, constantly evanescent. 'So there is no "present" as such. And yet, paradoxically, we know the past only as *present* memory, and the future as *present* anticipation. There is, then, no real present and nothing *but* present.'[23] But since the past 'is not' and the future 'has not yet been', he concludes painfully, 'time tends to non-being' (11.14.17).[24] This conclusion, however, leaves Augustine with the problem of how he can tell the story of his past now in the present, how he can relate to God, here and now, and how his self over time has the consistency necessary to recognize the change of conversion. His tentative answer to these questions extends the word 'tends' *tendit* – as Catherine Conybeare has expounded with exemplary clarity and style. *Tendit*, she notes, is a richly layered term, that gives us 'tense', which is how verbs relate to past, present and future, but also 'attention' and 'intention'.[25] Augustine builds on this word to suggest that 'time is a sort of *distentio*', a 'tension', or 'distension'. Typically – and emblematically of his tentativeness – he immediately adds 'but (dis)tension of what, I do not know, and I would be very surprised if it is not (dis)tension of consciousness itself'. Time is measured and registered in the mind – which leads Augustine to demonstrate his idea of *distentio* through reading a Psalm, where each word is understood in relation to the previous words and the memory of their meaning and in anticipation of the words to come. Reading and interpretation of the act of reading in the final books of the *Confessions* become a test-case of experiencing time.

---

[21] Fredriksen (2018) 106.   [22] Discussed below, pp. 190–1.   [23] Wills (1999) 90–1.
[24] See Ross (1991).   [25] Conybeare (2016) 124–5. See also Pranger (2010) 35–54.

'Who would deny that the present has no duration, since it passes in an instant?', writes Augustine boldly. 'Yet', he continues, 'our attention (*attentio*) *does* endure, and, through our attention, what is still to be makes its way into the state where it is no more' (11.28.37). 'The mind expects, and attends and remembers', and this is not just a phenomenology of consciousness but a theology of waiting. The prefix *dis-* in *distentio*, as Conybeare writes,

> suggests distraction, dispersal, straining apart. *Distentio* attempts to capture the simultaneous presence and absence of time: the way in which we may be focused in a particular moment which yet extends out into past and future. Our awareness of time in its elusive instant is always shaded and inflected by the *dis-* of other, evanescent times.[26]

Chadwick, more simply, a 'stretching out on a rack'.[27] *Attentio*, by contrast, is the attempt to turn oneself towards the immutability of God's time, to recognize and search for something beyond such dispersal of the now. James Wetzel captures the contrast superbly: 'as the soul enfolds time into a threefold presence (attention), so does time disarticulate the soul, resolving memory and expectation back into the "no longer" and the "not yet" (distension)'.

Augustine describes his life before conversion and his embrace of the love of God as simply *distentio*: *distentio est mea vita*, 'my life is tension', or 'nervous distraction', as Conybeare translates *distentio* here. Augustine describes this feeling with passionate intensity (11.29):

> I have leapt down into the flux of time where all is confusion to me. In the most intimate depths of my soul my thoughts are torn to fragments by tempestuous changes until that time when I flow into you, purged and rendered molten by the fire of your love.

To be in time, in the human experience of time, is to not know the order of things (*ordinem nescio*) and to have thoughts that are violently ripped apart (*dilaniantur*). 'Time's imperfection carries into his mind's efforts at self-definition.'[28] Yet Augustine also explains how this fragmented and miserable experience of time can be transcended through God's intervention, his love:

> I may be gathered from my old days (*veteribus diebus*) following the One; forgetting the past (*praeterita oblitus*), and stretching undistracted (*non*

[26] Conybeare (2016) 125. See O'Daly (1977). [27] Chadwick (1991) 230 n. 19.

[28] Wetzel (1992) 37.

distentus sed extentus) not to future things doomed to pass away, but to my eternal goal.

Distentus, the distraction or distention of the human temporal experience, is ameliorated by becoming *extentus* – extended, that is, stretched out towards the future, but not a future that is the continuity of the present experience of fragmented time, but the future that is the eternal future of eschatological promise. Faced by a congregation, Augustine sermonizes this into a stunning, summary, paradoxical injunction: *ut ergo et tu sis, transcende tempus:* 'So for you too to be, transcend time'.[29]

Augustine's exposition of the fragility of the present moment, which leaves the experience of time as the fragmentation of thought into tempestuous change, finds salvation in the attention that produces a condition of being *extentus*, focused on the future of the life to come. For Augustine, to inhabit time is made bearable only by translating the *distentio* of the present, through *attentio*, to an *extentio* towards God's time: time must be extended. Waiting is transformed into the salvation of eschatology. Waiting becomes the necessary condition of the Christian living within God's love.

*

The power and subtlety of Augustine's argumentation is highlighted by a writer such as Tatian, from the second century, and not the sharpest theologian in the Christian choir.[30] In his *Address to the Greeks*, an angry apology for Christianity, he writes as follows (26):

> Why do you divide time, saying that one part is past, and another present, and another future? For how can the future be passing when the present exists? As those who are sailing imagine in their ignorance, as the ship is borne along, that the hills are in motion, so you do not know that it is you who are passing along, but that time (ὁ αἰών) remains present as long as the Creator wills it to exist.

Tatian is attacking contemporary philosophers for borrowing others' words, like the jackdaw in borrowed feathers, and it is telling about his era's expectations that the cliché he sniffs at concerns the division of time. He seems to mock the Aristotelian tradition of worrying about the separation of time into past, present and future. For Tatian, stronger on rhetorical

---

[29] *In Evan. Joh. Tract.* 38.10. Augustine's sermons have not usually made it into discussions of Augustine's views of time.

[30] Notwithstanding Hunt (2003). Nasrallah (2005) 299 notes the standard description of Tatian as 'vicious and brutal', but defends his 'passion and humour'.

outrage than coherent argument, it is not the future that is advancing, but rather humans that are travelling towards death, while the one constant is *aiōn*, the unmoving era of the present, dependent on the will of God. If Augustine takes it for granted that the present has no duration, for Tatian, the unending present alone exists, which makes any further discussion of a philosophy of time unnecessary.

How to imagine the time between the Resurrection and the Second Coming requires increasingly varied stories, with different complicities of delay and action, and different comprehensions of what the mundane thus is and how it should be experienced. (Recognizing how Paul's insistence in the immediacy of the end of days was just wrong is an ongoing negotiation of theology.) Adopting and adapting Jewish apocalyptic, Christianity reshapes the necessary temporality of waiting. For epic heroes, waiting or delay is not exactly dead time, though it is a time of threat, the threat that deferral will block the heroic action necessary for immortal glory. So, with typically Homeric precision of evocative imagery, when he refuses to fight, Achilles sits by his tent and sings the *klea andrōn*, the 'famous deeds of men', 'epics of men': his very act of performing these tales is keeping him from the story that will guarantee his own *kleos*, though in retrospect, after his return to fight, it will be the beginning of a roll-call of heroic fame that he is to head. Odysseus is beguiled into sexual relationships that keep him from the household he craves; he must fight monsters to reattain the normality of his life on Ithaca; Aeneas' love affair with the queen and builder of a city prevents him from founding the city that will sing his fame. But for the Christian theologian, the transcendence over death, promised by the salvation that the Resurrection signals, transforms life into an arena where the struggle for humility and the avoidance of sin make mundane existence a constant danger and source of anxiety. 'I die daily . . .' But above all, waiting and watching are not choices of withdrawal to be confounded by an explosive return to action. Waiting and watching have become the human condition of the faithful, in the extended present between the promise of salvation and the not yet of the end. How such waiting and watching are to be enacted becomes an ethical question of Christian practice. What, then, happens to narrative if waiting, waiting for the end but waiting with no end, becomes the story to be told? If we concluded the last chapter with the question of how the ordering, measuring, counting of time changed the experience inhabiting time, now we can see how the *distentio*, or *extensio* of time – the very sense of being in time – becomes for Christianity a theory and practice of waiting.

CHAPTER 5

Time and Time Again

The discipline of classics is unthinkable without the notion of the exemplary.[1] On the one hand, for centuries in the West, the classical past has provided models of the best in literature, style, political system, sexual freedom – and many other idealisms. The very name 'classics' exists because the great texts of the past give us the first, second and third classes by which we classify and evaluate the modern. The lure of classicism is its image of the ideal in the past – a perfection or grandeur or beauty to strive after. The repeated challenges to the privilege of the Greek and Latin past in the education system, the hierarchy of cultural value, or in the very assumption of how the past matters, have sought again and again – often with a self-defeating obsessiveness – to dethrone this position of classics in the tradition of the West. In English, the apparent connection between classics and class has become a byword for such a challenge, insisting on the complicit and corrupting link between social elites and the fantasies of entitled genealogy that ground classicism.[2] A similar recognition of the role of classics in the ideological projections of empire, of racial thinking, of religious dominance, has tried to invert the privilege of classics into a negative exemplar of the dangerous links between an education into a sense of superior value and a politics of oppression. Here, too, however, classics becomes the exemplary discipline of blame or self-laceration.[3]

On the other hand, the ancient cultures of Greece and Rome, not least thanks to their own commitment to rhetorical training within their educational systems, both utilized the *exemplum* with creative flair and theorized such use with self-conscious acumen. From Homer onwards, Greek and Latin writing mobilized the mythic or political or heroic figures of the past to make normative claims on the present. An iconic example of

---

[1] Goldhill (2017); Langlands (2011), (2015), (2018); The Postclassicisms Collective (2019); Güthenke (2020). See also, for the tradition, e.g. Hampton (1990).
[2] Goldhill (2002); Richardson (2016); Hall and Stead (2020).
[3] See from a growing bibliography e.g. Malamud (2016); Zuckerberg (2018).

examples? The chorus of Aristotle's exemplary tragedy, Sophocles' *Oedipus Tyrannus*, sings 'Now I have your example (*paradeigma*), yours, Oedipus, I count nothing mortal as blessed' (*OT* 1193–5). The chorus insist that what they have seen in Oedipus' story is exemplary of the horror of mortal existence in general; Aristotle takes this play as exemplary of tragedy as a genre; Freud takes Oedipus as the psychological map of every man. I am taking it as an example of how exemplification multiplies. Classics as a discipline is constantly mining the ancient world's use of examples to make the ancient world exemplary for the present. If waiting links the now to the future, the exemplary makes the classical past significant to the present.

How the past is made exemplary, however, has become for modernity a highly problematic question.[4] Classicism, the turn back to antiquity to find paradigms of excellence, has traditionally depended on genealogy and idealism. That is – and, evidently, there would be many nuances and variations of these generalizing claims in any full-scale history of classicism – classicism depends, on the one hand, on treating ancient Greece and Rome as the origins and foundation of Western culture; and, on the other, on not just privileging this past but idealizing it as the paradigm to which we should aspire. Modernity has increasingly challenged both vectors. The genealogical affiliation to the past of Greece and Rome has been questioned by recognizing the self-serving ideological commitment of the very search for such origins, as well as by marking the differences and ruptures between antiquity and modernity. The idealization of the past has been undermined both by a recognition of the slavery, sexism, imperialism, violence and social divisions of the ancient world and by a symmetrical and self-serving assertion of the values of modernity. Nonetheless, antiquity has remained integral to the contemporary intellectual imaginary. This has a profound effect on the discipline of classics and its institutional epistemology. The challenge to the dominance of philology as the privileged science of *Altertumswissenschaft*; the challenge mounted by literary theory's questioning of the canon, the status of the text, and the status of literary knowledge and its inherited values; the challenge mounted by rhetoric's question to history, and post-processual theory to archaeology; in short, the challenge by modernity to classicism, in questioning both the idealization and the genealogy of classicism, sets out to undermine both the evident object of knowledge and the established subject who knows. Consequently, the constantly repeated

---

[4] If it is not clear that I am referring to a Western/global north sense of modernity, Chakrabarty (2000) will make it so.

questions of modernity to the classical tradition have become: If this is an example, what is it an example of? If this is a paradigm, how does it establish a politics and an epistemology? How am I to be part of your machinery of exemplification?

The exemplarity of the past, thus, is one crucial and contested route of organizing modernity's temporal relationship with antiquity. How, then, should the temporality of the exemplar be understood? At one level, scholars of reception theory have insisted on the *present* of any interaction with the past, even and especially when taking an example from the past. So, as Charles Martindale has been particularly influential in asserting, if 'meaning is realized at the point of reception', meaning is not integral to or inherent in a text of antiquity; and, what is more, later traditions create the meaning of texts of antiquity, as the filters of the tradition of reading produce a community's understanding.[5] Thus, with a wilful reversal of the arrow-like directionality of time, it makes sense to talk of Hegel's influence on Sophocles. A recognition of the constitutive role of the history of reading in the construction of the object of knowledge will inevitably change its status as a paradigm. At what point, then, it may be asked, does studying the classical past become no more than a cultural history of a modern era? At what point is an example only of and for the present?

Yet the move away from genealogy and idealism that marks the moment of post-classicism insists on a gap between the here and now of modernity and the past of antiquity, a gap constitutive of critical self-consciousness, based on an awareness of how the present is different from the past ('familiar strangers' to use Kanaan's expressive phrase)[6]. As Aristotle writes, it is 'when the mind pronounces that the "nows" are two, one before and one after, that we say there is time'. The exemplar plays a key role in this process of seeing oneself otherwise. The exemplar is, certainly, a mode of encounter with antiquity, a mode of knowing and of organizing knowledge of the past. It is also a persuasive term designed to construct a particular and particularized form of attentiveness: it is activated as a normative template for a present moment. But the exemplar, rhetorically the figure of a past historical character, event, or locus, emphatically interrupts context.[7] An exemplar is not invoked as a representative cross-section of a historical point in time and space: it is a singular case designed to

---

[5] Martindale (1993) 3. See also Martindale and Thomas eds. (2006). Nice to be able here to acknowledge more than thirty years of discussion of such issues with Charles Martindale.

[6] Kanaan (2019).

[7] This paragraph was originally collaboratively drafted with The Postclassicisms Collective and in particular Constanze Güthenke (see also Güthenke 2020).

illuminate a single case. The exemplar insists that the past is necessary to explain and frame and comprehend the present. Whatever its claim on a specific historicity – and an exemplar so often speaks of a lost time with a hankering for an idealized past – the exemplar speaks *across time*. The exemplar, then, is first of all a persuasive genealogy that asserts the past as a model, that interrupts the context of the present with the claim of an elsewhere and other time, that reads selectively, normatively, hopefully (there is, as John Dunn writes, a 'reckless vulnerability' in exemplarity).[8] Making an exemplar for the here and now is also a way of going beyond the here and now, of seeing oneself from elsewhere, in the light of another time.

This very engagement with antiquity as an exemplary model requires, therefore, a particular dynamics of *untimeliness*. It requires an explicit hypostasization – which need not be positive – of a particular past; it requires a sense that the present is disrupted from that past, while it is also genealogically linked to it, and that the present needs this ancient past to find itself; it needs its sense of untimeliness to find self-expression. Classicism must know it is not classical, that it is out of time with the models it is adopting and adapting, it must know its own untimeliness. Untimeliness, in this sense, marks the self-consciousness – in political, aesthetic, psychological, cultural terms – of one's self in construction as a historical subject.

Yet what classicism finds in antiquity is also a highly developed rhetoric and theory of exemplification. The treatise *On the Sublime* attributed to Longinus offers a particularly incisive and insightful summary.[9] It begins with the idea that whenever we strive for a particularly high style we should imagine in our inner being (*anaplattesthai tais psuchais*) how Homer, or Plato or Demosthenes or Thucydides would have striven to express themselves (*De sub.* 14):

> Therefore it is good for us also, when we are labouring on some subject which demands a lofty and majestic style, to imagine to ourselves how Homer might have expressed this or that, or how Plato or Demosthenes would have clothed it with sublimity, or, in history, Thucydides. For by our fixing an eye of rivalry on those high figures they will become like beacons to guide us, and will perhaps lift up our souls to the fullness of the stature we conceive.

---

[8] Dunn (2011) 310. See also, less pithily, Rood, Atack and Phillips (2020) 146–68.
[9] The exemplary analysis of Longinus is Porter (2016).

The writer must fictionalize the exemplars of poetry, philosophy, rhetoric and history – the four greatest genres – in a spirit of rivalry (*zēlon*). 'Comparison is inherent in exemplarity', and here the comparison is competitive in the search for 'fullness of stature'. Yet note the 'perhaps' (*pōs*). There is always a gap between the example and the case, a snag in the fit, as Montesquieu famously put it: 'tout example cloche', 'every example snags'.[10] Especially when you aspire to the greatest literary models of all. This slight hesitation leads immediately to Longinus imagining these masters not just as his examples but as his judges (*De sub.* 14):

> And it would be still better should we try to realise this further thought, How would Homer, had he been here, or how would Demosthenes, have listened to what I have written, or how would they have been affected by it? For what higher incentive to exertion could a writer have than to imagine such judges or such an audience of his works, and to give an account of his writings with heroes like these to criticise and look on?

Now the writer is to see himself before the jury of the past, or in a theatre (*dikastērion kai theatron*) with heroes as both his judges and witnesses (*kritais te kai martusin*) for his evaluation (*euthunas*). Setting oneself against the exemplary greats makes an example of you.[11] It is not just the past, however, that sits in judgement (*De sub.* 14):

> Yet more inspiring would be the thought: with what feelings will future ages through all time read these my works? If this should awaken a fear in any writer that he will speak in a way unsuited to his own life and time it will necessarily follow that the conceptions of his mind will be crude, maimed, and abortive, and lacking that ripe perfection which alone can win the applause of ages to come.

More important still is the idea of becoming an example for the future (*pas met'eme . . . aiōn*, 'every age after me'). This hope of posterity should not produce a fear of being untimely, of saying something 'unsuited' – *huperhēmeron*, 'beyond the temporal limit' – of his own life and time (*chronou*)', because this will prevent him reaching the time (*chronon*) of the 'applause of the future'. As Isocrates expressed it much earlier, a man's actions should look forward to 'the memory that goes hand in hand with time' (*Phil.* 134) – you should anticipate a future time remembering your past excellence. The process of seeking the great examples of the past produces a vivid sense of seeing oneself judged by

---

[10] Montesquieu, *Essais* III.13; Gilby (2006), especially 109–11.
[11] Dio of Prusa, *Or.* 18.12 finds this prospect frightening and recommends against it.

both the past and the future. Exemplification is also setting oneself in time, seeing oneself in history: the *eme* in *pas aiōn*.

Thucydides asserts that *his* writing of history is a *ktēma es aiei*, 'a possession for always' (1.22). It is to be a paradigm that will always work – exemplarity without loss, without the snagging, across time. It is a claim based on the universality of human nature – *to anthrōpinon*. ('Humanity', like 'nature', is a term designed to create an unbroken continuity across time.) Pindar, by contrast, will link the victory of today to the victor's familial past in a genealogy of excellence, as well as to the inimitable transcendence of the heroes of the past, with Heracles, say, as the limit case of contrast. Pindar is fully aware, too, that he, like 'Longinus', is writing not just for the praise of the present but also for the eyes of the future.[12] Exemplarity is in this sense 'situational and dependent on timing'. After many centuries of the institutions of rhetoric – teaching and performance – however, *exempla* are also layered with the history of their use. As Rebecca Langlands writes in her study of Valerius Maximus, the first-century CE anthologist of *exempla*, where each of his cases is an example of how examples are used,

> Roman exemplarity offered a simultaneous display of a wide variety of exemplary tales that can be envisaged as a heterogeneous and evolving whole where the individual elements – the *exempla* – could be played off against and compared with one another; an *exemplum* is never interpreted on its own, but always in the context of others.[13]

The *exemplum* interrupts the present with an invocation of the past, which acts as a richly variegated, normative positioning of the subject in and against time, both the past and the uses of the past in the past.

So when Livy or Tacitus, say, in their narrative histories offer brief or longer comments on individuals who star or feature momentarily in the unfurling of events, there is a broad cultural expectation that such remarks are to be read within the developed structure of the theory of *exempla*, the long history of the role of *ēthos* within rhetorical exposition, and the equally long and intricate history of the representation of character as a mode of constructing and understanding literary narrative. Livy's Cato is a perfect example of the complexity of this process of exemplification.[14] Livy sums up Cato's character thus (39.40):

> He was undoubtedly a man of a rough temper and a bitter and unbridled tongue, absolute master of his passions, of inflexible integrity, and

---

[12] Spelman (2018); Agocs (2019).   [13] Langlands (2015) 69.   [14] See Chaplin (2000).

indifferent alike to wealth and popularity. He lived a life of frugality capable
of enduring toil and danger, with a mind and body tempered almost like
steel, which not even old age that weakens everything could break.

Cato's censoriousness and upright opposition to the direction he thought
Roman cultural life was taking under the corruption of its own imperialism
is itself based on his view of a better past. He speaks out in favour of the
Oppian Law against extravagance, in Livy's account, partly because women
have protested against it, and partly in sarcastic contrast with the good old
days of the early Republic. In those times, when true Roman values
obtained, if offered extravagant bribes no Roman man or woman accepted
them, and there was no need for the Oppian Law against extravagance,
because 'there was no extravagance to be restrained'. Cato's exemplary, old-
style virtue is predicated on his portrayal of an exemplary past.

His opponent in the debate, however, Lucius Valerius, attacks Cato
precisely for his dismissive and aggressive attitude towards the women,
and, with brilliant rhetorical flair, theatrically unrolls Cato's own history,
the *Origines*, and claims to read out from it examples of how often women
have acted thus properly for the benefit of the Republic. The *Origines* has the
status of the first piece of Latin prose, the first history of Rome. The *Origines*
told the story of the beginning of Rome (and the cities of Italy). Valerius sets
Cato's own ideological foundational story of Roman beginnings against the
history Cato himself mobilizes in his rhetoric. Livy's history depicts his
historical characters arguing through historical exemplars – and allows
Cato's own written history to be quoted against the historian. It is a scene
that demands therefore a certain self-reflexivity about a self-positioning in
history. Livy, looking back at the Republic, has Cato set himself up as
a figure of old and now lost virtue – but he is ambushed by his own history
of Roman antiquity. By dramatizing these conflicting uses of *exempla*, Livy
offers a carefully layered question to his Roman readers about their own use
of the past in their projection of virtue. The theory and complex practice of
exemplification, with its historical placement of the subject in time, change
the expectations of how stories are written and read.

When Marx writes of the French Revolution that it was enacted in
Roman dress, he epitomizes – exemplifies – the self-positioning in time
that the logic of classicism's *exempla* establishes. For Marx, the French
Revolution is itself a complex paradigm of the necessary revolution that
marks the inevitable juncture of a historical process that situates every
citizen in their moment of historical significance. To tell the history of the
French revolution is, for Marx, to locate ourselves in the unfurling of

history. Yet the republicanism that drove the revolutionary ideology, argues Marx, is a strategic appropriation of a paradigm from antiquity. The example of the past informs the action of the present. Yet this very politics of the past is already structured by the past's rhetoric of exemplarity which organizes its heroes of republican fervour in line with the heroes of a Roman past. The temporality of exemplarity, for Marx, is thus integral to understanding history. In such an argument, Marx is being faithful to his educational training within the discipline of classics.

*

In 1 Corinthians 10, Paul speaks out against underestimating the power of temptation and the danger of worshipping idols. To reach his conclusion he constructs an elaborate parallel between the Israelites in the desert and the contemporary Christian. 'Our fathers', he enjoins, 'were all under the cloud and all passed through the sea' – they all experienced the miraculous exodus from Egypt. Indeed, they all ate food from God ('spiritual food' πνευματικὸν βρῶμα) and drank from the Rock (Moses hit the rock miraculously to produce water). 'The Rock', states Paul baldly, 'was Christ'. In Jewish exegesis the water is to be allegorized as the refreshing nourishment of the Torah: Paul redrafts this easy symbolism to his own agenda. But, Paul declares, the majority of the Israelites still died in the desert and thus the Israelites have become '*tupoi* for us', examples, models, warnings. We have to learn not to sin as they sinned; and he duly lists the many sins of the Israelites to be avoided. Thus, he concludes with a sentence that comes to mark a radical shift in Christian thinking about time and narrative (10.11), ταῦτα δὲ τυπικῶς συνέβαινεν ἐκείνοις, ἐγράφη δὲ πρὸς νουθεσίαν ἡμῶν, εἰς οὓς τὰ τέλη τῶν αἰώνων κατήντηκεν, 'These things happened to them *tupikōs*, and were written in scripture for our instruction, for whom the end of times has come'. *Tupikōs* is translated sometimes with 'as a warning'; the King James version has 'for examples'. But the language of *tupoi*, 'types', will take on a special force in Christian exegesis, which goes beyond the idea of *exempla*. For Paul, *tupoi* are for moral instruction – *nouthesia* – an instruction which only now makes full sense because theirs is the generation for whom the end of times (*aiōnōn*) has come. As we have already seen, the end of time structures Paul's thinking. Here, it makes his contemporary Christians the fulfilment of a story which will turn out to have always already been prophetic. As the end of time approaches, they can close the circle. Similarly, Paul calls Adam *tupos tou mellontos*, 'the figure of him that was to come' or 'the figure of the future' (Rom. 5.14). The grammatically simple Greek is semantically strained, at least for the first century: elsewhere Paul can use *tupos* in its more familiar contemporary sense of 'model', something to be imitated (Phil. 4.17). For later theologians, Paul's phrase became clear and fundamental, however. Christ is a second Adam, Adam a first Christ. They are *tupoi* of each other, figures linked as paradigms of each other. What, we can ask with Giorgio Agamben, is the 'transformation of time implied by this typological relation'?[15]

Now, typology – this device to close the temporal gap between old and new testaments in Christian thinking by making each an analogy of the other – has a long history, going back to the beginnings of allegorical reading in one genealogy, and also, perhaps more pointedly, back to Philo's Platonizing conception of the Hebrew Bible, which forms the basis of so much of Clement's thinking, through which it becomes a central plank of Christian argumentation.[16] So, when the Israelites arrive at the Grove of Elim (Exodus 15.27), this story of a staging post in the Hebrew Bible in Prudentius' eyes finds fulfilment in the Christian Bible through his typological reading of scripture, which reveals that the twelve trees in the grove stand for the twelve apostles. The grove depicts what in chronological terms would be the far future, but what in theological understanding is always already present. The text of Exodus is already replete with its future embodiment in the language of the Gospels.[17] Such a poetics of typology, as we will see, is one frame for understanding late antiquity's repeated use of analogy, proleptic and retrospective figuration.

Christianity's reinvention of the Hebrew Bible as the Old Testament entails that theological interpretation repeatedly discovers a pattern of anticipation and fulfilment in scripture. But how such a technology of reading should be comprehended and utilized became a long-running theoretical debate among early Christian intellectuals, and consequently for modern scholars who have argued fiercely the boundaries between allegory and typology.[18] ('Typology' itself is a modern term, unlike *allēgoresis*, but the vocabulary of *tupos* and *antitupos* is prevalent in early polemics.) Origen – to take an example of a figure who is an excellent guide

---

[15] Agamben (2005) 74.
[16] See Dawson (1992), (2002); Struck (2004); Niehoff (2001), (2011). The following paragraphs reprise Goldhill (2020) 129–33. On Clement, see now Heath (2020).
[17] On Prudentius' use of allegory, typology and symbol, see most recently Hardie (2019) 188–222, and below, pp. 124–6.
[18] This is a hugely intricate and contested area: for guides to the controversy, see Daniélou (1948), (1950); Young (1997); Herzog (2002a); Dawson (2002); Hanson (2002); Edwards (2002); Martens (2008).

to where controversy is, if not always to the development of orthodoxy – contrasts those who read 'superficially' (*parergōs*) with those who read with 'care and attention' (*met'epimeleias kai epistaseōs*) (*Peri Pascha* 12).[19] The contrast depends on how the Passover of Jewish scripture is to be understood as a *tupos* of Christianity. The Passover (*pascha*) should be understood as a *tupos* of Jesus, Origen argues, but not of the Passion (*pathos*). That is, despite the imagined linguistic link between *paschein* and *pascha*, '*passion*'/'*Passover*', the dissimilarities between the paschal sacrifice and the passion are too great: the paschal lamb is sacrificed by holy men, Origen points out, while Jesus was crucified by criminals. Therefore, and more importantly, 'it is necessary *for us* to sacrifice the true lamb; if we are ordained or bring sacrifices like priests, *we* have to cook and eat its flesh' (*Peri Pascha* 13). There are two crucial strands of argument intertwined here. On the one hand, the importance of the typological reading is in its internalization for the Christian reader who wishes to aspire to the holiness of a priest (either as an ordained member of the church or as an individual whose holiness is to be like a priest's): Passover acts as a type not just for events in the New Testament but for the daily life of the Christian – the constant work of sacrificing for God. On the other hand, Passover as a *tupos* has a non-literal referent, not to another event simply, but to something itself non-physical but spiritual. Typology is also a way of seeing beyond even the miracle of an event into its spiritual significance. For Origen, then, Passover is not only replete with a future meaning, but also the meaning demands that the event in its spiritual significance is always already present in the spiritual struggle for the freedom of service that the Passover announces. Typology turns temporality away from any direct linearity.

One figure whom Origen might be engaging with is the second-century bishop Melito of Sardis.[20] His text, also called *peri pascha*, also theorizes typology but in a quite different way. He begins by announcing that the Hebrew scripture of Exodus has been plainly stated, and proceeds with the general theoretical position that the text is both 'new and old' (*kainon kai palaion*), 'everlasting and temporary' (*aïdion kai prokairon*), 'perishable and imperishable' (*phtharton kai aphtharton*). In line with his hostile supersessionism, he explains that it is ancient with regard to the law, but new with regard to the Logos; temporary with regard to the type (*tupos*), eternal

---

[19] See Hanson (2002); Ayres (2004) 20–31.

[20] For a text and translation see Hill (1979); for discussion see especially Lieu (1996) 199–240; also Lieu (2004) 81–2.

through grace; perishable through the sacrifice of the lamb, imperishable through the life of the Lord. A *tupos* is described with the strikingly rare word *prokairos*, which means not just 'temporary', but – etymologically at least – 'before the right time', 'untimely', and Melito will go on to expound exactly what he means by such an expression. The remainder of the text explores this logic through a reading of Exodus, centred on a statement of principle about typology. 'Nothing is said or happens without a parable or a prior etching ... the said through parable, the event through prior modelling/prefiguration (*protupōsis*)' (35). This leads to an extended analogy between an artist's modelling in clay or wood, as a preparation for an art work.[21] This sketch or preparatory model becomes useless (*achreston* 37) once the work comes into being: what was once valued becomes without value. With a redrafting of Ecclesiastes, he sums up 'To everything there is a proper season (*kairos*); a proper time (*chronos*) for the model (*tupos*); a proper time for the material; a proper time for the truth' (38). A model is necessary (and prior to the model the material) because 'you see in it the image of the future': however necessary it is, for the process, it is also, precisely, *prokairos*. He explicates the implications at length: the people and the law were the *tupos* which is fulfilled in the Gospel and the Church (40). 'The model (*tupos*) surrenders its image to the true nature and is voided' (43) ... the model *tupos* was dissolved when the Lord was revealed'. As he rewrites Ecclesiastes to make a new message about supersessionism – denying the revelation on Sinai as a *kairos* – so he theorizes typology not as Origen had, as a key to how the present still reverberates with the past, but as an aggressive destruction of what was once valued. 'If you stare at the *tupos*, you will see Him through the outcome', he writes (58), and so Abel, Isaac, Joseph, Moses and David are each simple types by virtue of being (in turn) murdered, bound, sold, exposed, persecuted (the sort of simple analogy Origen criticizes). But this leads finally to a bitter and lengthy rejection of the Jews as the killers of Christ – and as a misunderstanding of the scriptural typology. For Melito, typology's temporality is both linear and mutually implicative with a hatred of Israel.[22] The violence of typology, which forces past and present into co-temporaneous models of each other, comes hand in hand with a supersessionist violence against the Jews. As we will see especially in chapters 14 and 15, the assumptions of supersessionism continue in Christian readings of late antiquity with a blithe

---

[21] Kessler (1994) 74–96 argues that Cyril's use of *skiagraphia* 'foreshadowing' (xli.21) in his discussion of typology becomes crucial in Christian defences of the use of imagery in churches.

[22] See Lieu (1996) 199–240.

failure to recognize the continuing impact of the anti-Judaism of Christian self-definition. Typology's political force is politely ignored.

This sense of the buried life of the present and future coming to view in the reading of the past is enacted at the level of the word. The Prophets are the place where typology is most pressingly debated. Hosea 1.1 announces 'The word (*d'var*) of the lord that came to Hosea son of Berri'. Origen notes the straightforward historical understanding of this phrasing (*kata tēn historian*) but adds that it should be read 'according to a mystical logic/ argument', *kata mustikon logon*. For, he interprets, Hosea in Hebrew means 'saved' (*Jo.* 2.4). Thus, we can hear not just a general message of salvation in the prophets but an echo of Jesus, whose name comes from the same Hebrew root. The prophet's Hebrew name turns out to speak a Christian truth.

Such an interpretation then becomes an integral part of the Latin Bible itself, the Bible of the church for centuries, thanks to Origen's great defender, Jerome. The prophet Habakkuk (3.18) wrote (in Hebrew): 'Yet I will rejoice in the Lord, I will joy in the God of my salvation' (the King James version). The Septuagint translates the final phrase accurately enough with ἐπὶ τῷ θεῷ τῷ σωτῆρί μου, 'in God my saviour'. Jerome translates the whole sentence into Latin as *Ego autem in Domino gaudebo et exsultabo in Deo Jesu meo*, 'Yet I will rejoice in the Lord, I will joy in my God, Jesus'. *Sōtēri*, 'saviour', has been translated 'Jesus'. The Hebrew for 'my saviour' is *ye'shi*, and to Jerome's eyes this is close enough to the name *ye'shua*, 'Jesus', which is indeed formed from the root for 'salvation'. For Jerome, reading through the Greek to the Hebrew opens a new potential for seeing the truth of the text. Jerome's translation is a theologically creative rendering which not only sees in the name Jesus the root of (all) salvation, but also regards it as inevitable and right to imagine that a prophet who lived hundreds of years before the birth of Jesus could rejoice in Jesus, because in the temporality of Christianity, which through typology could see Jesus in Adam and Adam in Jesus, it was right and proper to uncover the timeless truth of Christianity in what was now an Old Testament. His Latin reads through Greek and through Hebrew to a deeper truth. Translation into Latin makes patent what the translation into Greek had buried. The timelessness of God, the co-eternity of Son and Father, finds here in typology its strategy of reading, where any sense of past and present becomes rather a revelation of the always already.

Typology, even though *tupoi* is sometimes translated in the King James Bible as 'examples', changes the temporality of exemplarity. The *exempla* of Latin historiography – for example – construct layered and complex interactions between past and present that depend on the disruptive difference as well as a genealogical link between past and present, and locate the subject thus in a self-conscious historical positionality, with the need to negotiate the snagging of fit between case and example. Typology by contrast, even allowing for the differences in method that we have traced, closes the gaps of time and signification between past and present, resisting linear time, and finding a revealed always already in mutual figuration, without loss. Typology is not merely a hermeneutic strategy, however. There is a violent normativity in this redrafting of how time and exemplars interrelate, which is integral to a politics of supersessionism – and, as we will see in chapter 15 in particular, allows for a redrafting of what historical narrative can look like. (The long established and authoritative power of such a resistance to linear time is inadequately recognized by theories of 'queer time', which seek to re-privilege temporal non-linearity as challenging to dominant cultural models.) The question which remains, then, is: how does typology redraft the horizon of expectation for how stories are told and understood in late antiquity? What is the consequent impact on the self-understanding of historical positioning? At the very least, any specificity of Jewish history is subsumed as a version of Christian narrative. What politics – politics of history and politics of disdain – are made possible by this version of temporality?

CHAPTER 6

Making Time Visible

For any discussion of the politics of history evoked thus by typology it is basic to ask how a community – a city, a nation, a sect – curates its memory.

Athens in the classical era is instructive, exemplary even. The *polis* of Athens in the fifth and fourth centuries BCE was a community which repeatedly represented itself to itself as a continuing exploration of civic identity. In a period of rapid social and political change, when the very constitution of the city as a direct democracy was actively promoted and resisted, and when the international status of the city depended on its pronounced differences from other neighbouring cities, the self-consciousness of what made Athens distinctively and essentially Athenian was a political project to be performed.[1] As ever, war – and imperial Athens was always at war – sharpens the dichotomies of self-representation. How did Athens make its memory – its sense of time past – visible? How did temporality take shape as a political project?

Memory made visible – the display of civic time – took, first of all, architectural form. Athens, with the new-won wealth of empire, like so many modern and ancient imperial forces, adorned its own civic frame-work, and most famously with the Parthenon, that rose above and amid the ruins left by the Persian invasion and sack of Athens.[2] The Parthenon frieze, the band of statuary that runs around the whole temple, is the only temple sculpture in the ancient Greek world to represent directly the human agents of the city that built the temple.[3] It depicts a long procession of Athenians, ideally and iconically pictured, proceeding towards the presentation of the sacred robe to the goddess Athene, the patron goddess of the city, whose image dominates one end of the building dedicated to

---

[1] See for the general case: Shear (2011), (2012a), (2013), (2020); Steinbock (2013); Castagnoli and Ceccarelli (2019) and the chapters in Castagnoli and Ceccarelli eds. (2019) 93–178; on relation to orality, see Thomas (1989).

[2] For the pre-Parthenon site, see Rous (2019) 84–125.   [3] Osborne (1987); Neils (2006).

her. That is, the central religious site of the city represented the city itself performing a rite of civic, mythological genealogy. Each time a citizen walked around the Parthenon, he joined the ritual procession above him, mirrored and mirroring himself in a transition between his own ritual life and the mythological world that the ritual evoked. As ritual makes time a commemorative sequence of repetitions stretching back to a mythic origin, the new temple of the Parthenon set its citizens in line with their marble, perfected icons of citizenship, still and always processing towards the annual ritual commemoration of the city and her goddess. After the Persian invasion was defeated, Greek cities agreed in the oath of Plataea that burnt temples would be left as visible monuments of Greek survival, or, as Pausanias 500 years later described them, 'memorials of hatred'.[4] The Parthenon's very newness stood – as time rolled on – as a sign of the city's memory of victory, monumentalized in a ritual of stone.

One side of the *agora*, by contrast, was bounded by the Painted Stoa (*stoa poikilē*). This long colonnade had three large murals, representing the sack of Troy by the Greeks, the defeat of the Persians by the Greeks at the battle of Marathon, and the defeat of the Spartans by the Athenians.[5] The three pictures are set up as analogies of each other. The modern soldier-citizen, ever at war, is to see himself as the avatar of the Greek heroes who fought at Troy, their own ancestors who defended the city against the Persian invaders, and their own military heroes who fought and continued to fight against Sparta. It is not hard to see this as a full-scale normative expression of the military commitment to the city which undergirded democratic expectations of civic duty. Again, a sequence, ritualized, but now of three singularities, three historical events, made into a sequence by the act of civic building and self-representation – a sequence for citizens to join. Even walls, built out of dismantled graveyards and buildings, recalled how the material of the city was reused in its own defence.[6] The topography of the city in democracy is rebuilt as a topography for the performance of citizenship, for the citizen to see himself in the timeline of the city.

Citizens also took part in regular ritual processions, which conceptually linked the various sites of civic infrastructure into a map of civic engagement, and also watched them. Social and cultural memory of the collective and for the collective is conjoined with embodied 'habit memory' as the citizen processes or watches the procession, observing oneself observing, together.[7]

---

[4] Paus. 10. 35.3: examples were still visible to his historical gaze.   [5] Castriota (1993).

[6] Rous (2019) 31–61, which is also useful (1–30) on the history of '*spolia* studies'.

[7] For 'habit memory', Connerton (1989); see also Assmann (2000a), (2000b); Erll (2011); Funkenstein (1989); Mendels (2004); Nelson and Olin eds. (2003).

This dynamic of participation and representation between the images on the temple, marching, and watching the marching, makes each procession a performance of the 'law' or 'custom of the fathers', the *patrios nomos* – something *time-honoured*, that is, a ritual which, by performing it, maintains an ideological as well as an aetiological continuity between present time and the past, even and especially the deep time of myth. In the next chapter, we will see how the distinction between mythic time and historical time is an invention of the era that sculpted the Parthenon frieze, its contradiction.

One occasion in the calendar of Athens, which brilliantly used its ritual to insist that the most singular historical event – the death of a citizen-soldier – must be seen in a self-defining, repetitive sequence back into the deepest past of the city, was the public, collective funeral of the war dead, an event which Thucydides makes a paradigm of the city's *patrios nomos*, 'ancestral custom' (although the ritual was, in crude chronological terms, a relatively recent introduction of democracy). The Funeral Speech was delivered by a distinguished politician each year over the collective dead of the city who died fighting for the state. Nicole Loraux in particular has analysed how this event contributed to the civic ideology of Athens, with its praise of unnamed warriors as the latest to 'prove themselves good men'.[8] Pericles, in Thucydides' version, makes the line of citizens back into times past his starting point of memorial (2.36): 'I shall begin with our ancestors first: it is both just and proper to give them this honour of memory (*mnēmē*) on such an occasion. They dwelt in the country without break in the succession from generation to generation, and handed it down free to the present time by their valour.' He moves on from this deep genealogy to their own fathers, founders of empire, before praising the political glories of Athens in the present. It was typical of such speeches to recall the defeat of Persia and other victories back in time. The speech did not memorialize any individual heroic success or name any Athenian, but insisted that, like the figures on the Parthenon observed by the citizens, the dead warriors were celebrated as icons of valour in a line of icons of valour. The Funeral Speech idealized a history of the city, making any individual memory of families subordinate to the ongoing history of the State. 'We have no need of Homer', declares the Thucydidean Pericles (2.41): these warriors have not died for individual glory, memorialized in song; but unnamed, they find their honour in the continuity of the city. The time of death is fully politicized as a duty to the state's necessary timeline. As Hobbes and Puffendorf in the seventeenth

---

[8] Loraux (1981a).

century argued, states in their self-conceptualization *require* 'an endless succession of time'.[9]

Athens was, in Demosthenes' words, a 'state made up of speech-making', 'a constitution of discourse', *en logois politeia* (19.184). Memory, not least in contrast to these institutions of civic promotion of an idealized procession of marmoreal statues, was also profoundly contested, in law, politics, theatre and the very invention of historiography, the arenas of *logoi*. I will return more fully to the self-announcement of the historians in subsequent chapters, but there is no more evident example of revisionism in a city's memory than Thucydides' acerbic account of the statues of Harmodius and Aristogeiton. This pair of heroic marble figures was for many years the only statue in the *agora* of Athens, and this very placement exaggerated their celebration as the tyrant slayers.[10] The archetypal hate-figure of ancient democracy is the tyrant; Harmodius and Aristogeiton were lauded as the liberators of Athens because they killed the city's last tyrant; and they were remembered in drinking songs, popular sayings and the city's self-understanding of its political history. Thucydides, however, insists that they didn't actually kill a tyrant but his brother, and it was a lovers' tiff, anyway, not an act of political idealism. Thucydides is arguing for his own place in the contests of authority, and his intellectualized corrective had little effect on popular drinking songs. But his insistence that Athenian civic memory was faulty is testimony of a significant strand of Athenian democratic culture.

In both law courts and political speeches, there is both a repeated rewriting of the past in the cause of a persuasive case for the present, and a rhetorical awareness of such manipulativeness.[11] 'Remember!', declares Andocides (1.69), 'with regard to the truth of my words, and those who do know much teach the rest that don't': truth is defended by an appeal to memory, an appeal which anticipates multiple possible engagements with the salient past. Such appeals become part of the rhetorical contest. 'If my opponent says a fact is well known to you', states Demosthenes in a magnificently disingenuous demand for a cautious self-awareness of manipulation (40.53), 'what anyone of you does not know, let him assume that his neighbour does not know it either'.[12] The contest over the past is a conflict of knowledge and knowingness – of carefully curated memory.

---

[9] Hobbes, *Leviathan*, ch. 46. Quentin Skinner opened my eyes to this. See also Gorham (2014).

[10] See Shear (2012b), with relevant bibliography.

[11] Hanink (2014); Canevaro (2019); Steinbock (2013); Thomas (1989); Nouhaud (1982); and on inscriptions see below pp. 132–7.

[12] See the excellent discussion of Hesk (2000) 202–21; Canevaro (2019).

So, the memory of political violence in particular, because it is repeatedly instrumentalized for the fervour of the present, in ancient Athens as in the modern world, prompts 'memory sanctions' and gestures of amnesty – politicized, compulsory forgetfulness. With varying degrees of success, of course. The political and emotive instrumentalization of the past go hand in hand with conflicting narratives of memory, and with competing strategies of making memory authoritative.

The past embodied in mythic and epic narratives too was subject to scrutiny, not just in the high intellectual analysis of historiographers (again deferred to another chapter), but also in the most public arena of theatre. Theatre, an institution on which Athens prided itself as its inventor, certainly represents the city to itself at multiple levels – in the rituals which surrounded the performance, in the organization of the space of the audience, and in the plays themselves, tragedy, comedy and satyr play. The comedies displayed a carnivalesque version of Athens's political and social system (women taking over the Assembly, old men behaving deplorably in the courts, women stopping war by refusing sex); the tragedies searingly uncovered the contradictions in the normative language of the city, again and again exposing fissures between the ideal of family care and the exigencies of civic duty. Tragedy in particular becomes a way of rewriting the myths of the past for the contemporary political scenario of the democratic city – and rewriting them as questions. If the Funeral Speech idealized the death of the citizen-soldier as the performance of civic duty, tragedies again and again exposed the internal tensions in such ideals, setting duties against each other in violent, self-destructive conflict. Tragedy asks – in public – how the stories and values of the past can and should speak to the present.[13] Tragedy provokes an emotionally raw and uncomfortable awareness of how fragile the aetiology of the myths we live by is: in tragedy we are also brought to face the precarity of the political present.

In classical Athens, then, we see how times past are made visible in civic architecture, and how the city's 'endless succession of time' is encoded in ritual's sequence, and articulated – constructed – in civic self-representation, both in material form and in the speeches and rituals of civic ceremony. Yet the fifth-century enlightenment is also characterized by the self-reflexive questioning of *aitiai*, the causes of things: the past becomes a site of conflict, both in how myth as aetiology of value is

---

[13] See Ceccarelli (2019); Scodel (2008); and for the best study of how tragedy enters social memory, Hanink (2014).

understood in the democratic city, and in the epistemology of producing a narrative of the past.[14] Not only are the past and the memory of the past argued over in the personal battles of the court, and the political debates of interstate conflict, but also this argument is conducted with a sophisticated rhetoric of self-consciousness, which is further embedded in treatises or theories that discuss the epistemology involved in understanding the past. In the fifth-century city, the awareness of the gap between the past and present is not simply troped as decline or progress, but explicitly theorized, self-consciously manipulated, aggressively instrumentalized: time made visible.

So can a city be imagined where the foreign country of the past is not open to the imperialism, border raids, fantasies and mapping that foreign countries experience? Plato tried such an exercise of the imagination in his *Republic*. To establish his state without internal conflict and without change, he needs a 'noble lie', a lie that shuts off his citizens from the past. Plato's Republic is designed to embody the 'endless succession of time', where the future will not be different from the present. But to get to this ideal point of changelessness, Plato also needs to make a lie of the past, a lie that cannot be challenged.

Except in Plato's utopian ideal, states inhabit a time of conflicted, selective, self-serving, mistaken, demanding and embracing memories. Curating memory is integral to the politics of the state. In ancient Rome, 'a memory culture par excellence',[15] the institutions and practices of the state strive to make time visible – the *horologium* of Augustus in the forum is a convenient icon – and work to locate the Roman citizen in the genealogy of historical time. If Athenian processions integrate the citizen in a time-honoured ritual performance that joins each person to the endless timeline of the city, Roman funerary processions require an even more intense impersonation.[16] When a Roman noble, who had achieved significant political office, died, the funeral procession included actors who wore lifelike wax masks of his ancestors. At the time of death, thus, in what became a 'lavish drama and magnificent spectacle as well as being a medium for popular history', the 'ancestors welcomed and received' the dead man 'as one of their number',[17] an enactment of a 'ritual joining of the long line of ancestors'.[18] The procession, complete with *tituli*, titles or explanations of the images, displayed the family line as a continuing

---

[14] See, most recently, Pelling (2019); Baragwanath (2008).   [15] Galinsky ed. (2014) 17.
[16] Flower (1996) is the standard analysis. See also Gowing (2005).   [17] Flower (1996) 141, 91.
[18] Hölkeskamp (2014) 69, not dented by Wiseman (2014).

exhibition not just of status but also of nobility across time – a 'continuous rehearsing and reconstructing of family traditions'.[19] These masks, called *imagines*, were kept in the atrium of the noble's house, and each branch of the family line had similar masks. When a noble's clients visited their patron for the rituals of *salutatio*, they were greeted thus by a man surrounded by the images of his ancestors – the family line made visible, the status of the family on display. Hence the shock of Marius, no noble, who taunted the senate and paraded his own bravery as testimony of his contribution to the state: 'My scars are my *imagines*', he boasted, collapsing the possibility of family history into a single person's singular military achievement.[20]

In Roman culture, 'memory [was] a space, filled with monuments, inscriptions, portraits, written accounts and other testimonies to the life of a Roman citizen', designed 'to ensure the survival and continuity of the community and the particular culture of its political families'.[21] Statues displayed 'the whole city's ancestral actors', and a string of *lieux de mémoire*, from Romulus' hut to Nero's Colossus, enacted the 'monumentalization of cultivated memory' in the civic topography.[22] Cities were 'memory theatres'.[23] Imperial authority changed the possibilities and requirements of such memorialization, and with the conflicts over dynastic succession came 'memory wars', the attempt of each ruler to 'define and control public memory', not least by destroying, defacing or changing the materiality of memorialization of their predecessors.[24] The decree of the Senate from 10 December 20 CE, against Cn. Calpurnius Piso specifies that despite the *mos maiorum*, the custom of the ancestors, which frames and embodies normative collective memory, no mourning would be allowed at his funeral; that all statues and busts of him should be removed, destroying his presence in the public view of the city; that his *imagines* could not be processed at any funeral, thus excising him from the public acknowledgment of his family line; and that such *imagines* could not be kept at home, removing him from even the domestic history of his own family. Such *damnatio memoriae* is a public display of the redrafting of memory. Some statues were physically destroyed; some were defaced; some, because marble was expensive and sculptors efficient, were re-carved into new portraits. Such re-cutting, which always leaves

---

[19] Flower (1996) 274.   [20] Sallust, *Bellum Jug.* 85.29–30, with Leigh (1997) 229–30.
[21] Flower (1996) 276. More generally, see Bettini (1991).   [22] Hölkeskamp (2014) 70.
[23] Alcock (2002) 54; Nasrallah (2010); Favro (2014); Hughes (2014).
[24] Flower (2006); Varner (2004); Vout (2008) and for the later period, Hedrick (2000); also Elsner (2003) for the Greek case.

a 'legible reminder'[25] of the previous portrait, lets the past be viewed through the active work of the present: it makes the changing times visible, embodied. Turning an image of Domitian into a portrait of Augustus seems even 'to reverse the march of time'.[26] The erasure of sentences or words in an inscription demands attention to their absence. ('Images', writes Joseph Koerner, 'are never as vociferous as when handled by an iconoclast'.)[27] *Damnatio memoriae* certainly allowed the violent expression of vengeful feelings – Pliny tells us the overwrought story that the statues of Domitian spurted blood and screamed when beaten – but the destruction or alteration of imperial images also 'constituted an active mechanism in the ongoing process of historical emendation', the paraded re-construction of memory.[28] As the Pentateuch insists, with productive paradox, about the enemy of Israel, Amalek, 'You must blot out the remembrance of Amalek ... You shall not forget.'[29] One must *remember* to block out the memory. But, most saliently, such a performance of *damnatio* in imperial Rome also highlighted the transition of dynastic rule, not just the passing of time but the announcement of a new era. Romans knew in what time they lived by such performances of destruction, and the history of *damnatio* turns each new era into the sequence of historical time. The imperial era is conceptualized by contrast with the Republican era;[30] emperors give their name to time. Through the practice of *damnatio*, the materiality of the city is made to announce the era, a violent curation of time and memory. Through the continuing history of *damnatio*, and the traces and relics that *damnatio* always leaves, the very constructedness of imperial time is made visible.

*

How, then, are such sites and rituals of memory to be responded to? How is the viewer of such memorials to engage, react, express emotion?

For Pausanias, travelling around the major cult sites of mainland Greece, his perspective is that of the educated, sophisticated observer of the internal nuances and niceties of Greek culture.[31] He wants to lead his Greek readers to view *panta ta Hellenika,* 'everything Greek' with him: he curates curated memory. He recognizes the privileged monuments of the

---

[25] Varner (2008) 134.   [26] Vout (2008) 164.   [27] Koerner (2003) 106.

[28] Pliny, *Paneg.* 52.4–5; Varner (2008) 129.   [29] Deut. 25.17–19.

[30] Gowing (2005). For the literary history, see the incisive reflections of Vessey (2015).

[31] See Elsner (1995) ch. 4; Alcock (1996); Alcock, Cherry and Elsner eds. (2001); Hutton (2005); Pretzler (2007).

past, and their rituals, but his political investment in such history is strategically subordinated to a distanced stance of intellectual evaluation and description. He will look on the Parthenon, he will fail to see the Roman buildings in front of him, which are thus not permitted to block his historicizing gaze, but he will not draw aggressive or even explicit attention to this selective viewing or its implications: he is happy to 'read the Roman elements out of the landscape'.[32] For Pausanias, seeing the history in the Greek monuments of the past is not to engage directly the Roman imperial politics of the here and now. 'The impious present' may be 'a lamentable ruin' in comparison with the glorious past, and 'the list of ruins scattering the land' may be 'long and sad', but as a 'Greek nationalist response to the fact of Roman dominance', Pausanias nonetheless emphasizes 'fortune' as the prime cause of change.[33] Pausanias' journey invites his readers to share his melancholic but educated view of 'everything Greek' as a recognition of a lost past, the visible signs of how times were once other.

Such a grand tour of the past is in sharp contrast to Lucian's roughly contemporary engagement with the temple of his homeland. In *de dea Syria*, 'On the Syrian Goddess', Lucian writes in the language and style of Herodotus – in Ionic Greek, with a fascination for the cults of other peoples, a nice recognition of his sources as a mixture of first-hand experience and priests, and with sexy inset narratives along with ethnographic description.[34] The 'mock innocence' of such a stance is belied both by its literary knowingness and by Herodotus' complex reputation as the lying father of history and as the lover of barbarians. Yet the author of the piece not only refuses to name himself but also identifies himself as a Syrian, who has worshipped at the temple and left his name inscribed there (for us to imagine). Typically of Lucian's wonderfully ironic use of self-concealing masks, this piece claims to be written by an unnamed Syrian, in Greek, a Herodotean Greek of the classical era, in which he describes his homeland as if from the perspective of a Greek outsider but who is actually a Syrian worshipper at the temple. This self-dramatization of a response to a temple and its cultic practice of self-memorialization uses thus the Herodotean past to express a sense of the twisty self-defining strategies of insider and outsider in the multilingual, multicultural empire. The historian, as a figure of the past and teller of the past, becomes an affordance of self-representation: playing Herodotus, staying Syrian. A sophisticated

---

[32] Hutton (2005) 305.   [33] Elsner (1994) 249.
[34] See Elsner (2001), with Lightfoot (2003) for philological commentary.

response to such a *lieu de mémoire*, suggests Lucian, takes shape by incorporating such foreign historical perspectives from another time.

Such complex imperial responses to visiting sites of worship form a contrastive anticipation of the pilgrim narratives of late antiquity.[35] Gregory of Nyssa captures the normative experience of the Christian traveller in one telling question: 'How will it be possible to travel through places full of passion without passion?'[36] How, that is, when faced by sites of suffering (*pathos*), replete with the memory of the martyrs dying in imitation and fulfilment of Christ's passion (*pathos*), it is possible to react without emotion (*apathōs*), without re-experiencing the suffering? How could the *pathos* of the place (*topos*) not affect or infect the feelings of the viewer? Gregory is careful to warn against pilgrimage if it turns into a party (of a Chaucerian kind), which he saw in Jerusalem – his polemic was consequently used by Protestants in the Reformation to attack Catholic practice – and he worries that ascetics may be distracted, and women come into too much contact with men, but despite his own political, theological disputes in Jerusalem, he cannot repress the joy he felt at seeing the sites where Jesus had walked.[37] Pilgrimage cannot but engage the Christian in a re-visioning of the past.

Prudentius, writing at the end of the fourth century, dramatizes the emotionally overwhelming experience of the pilgrim with vivid intensity, especially in his poem on the martyr Cassian of Imola, *Peristephanon* 9.[38] In this poem, a micro-version of the whole collection, Prudentius travels from his native Spain to Rome and back, participating in the developing map of Christian pilgrimage.[39] Egeria had already written her account of her trip to the Holy Land, and Jerome too had already sent to Marcella his powerful description of pilgrimage to sacred sites (*Ep.* 46). Jerome uses scripture to justify pilgrimage. Elsewhere (*Ep.* 58), like Gregory, he strongly advises a monk – Paulinus of Nola – to avoid leaving his ascetic life for the city, even and especially Jerusalem. God is everywhere and travel

---

[35] To move in a discussion of time to a discussion of memory and memorization, to pilgrims in the Holy Land, is to trace the intellectual history of Maurice Halbwachs, one father of memory studies, who studied with the great philosopher of time, Henri Bergson, then under Durkheim's influence wrote on social and collective memory, and published, posthumously, a book on the construction of pilgrim sites in the Holy Land as a test-case of the social frameworks of memory.

[36] *Ep.* 2.5–7. On pilgrimage see Hunt (1982); Sivan (1988), (1990); Ousterhout ed. (1990); Leyerle (1996); Frank (2000b); Maraval (2002); Elsner and Rutherford eds. (2005); Dietz (2005); the extremely useful Bitton-Ashkelony (2005); Bader (2018), and more broadly Coleman and Elsner (1995). On Christian responses to the space of empire, see Nasrallah (2010).

[37] Bitton-Ashkelony (2005) 30–64 has a full discussion of Gregory and Basil. See in general Kirk ed. (2005).

[38] Goldhill (1999); Clarke (2020).   [39] For a full background, see Hershkowitz (2017).

a misprision of religious duty and understanding. 'The true temple of Christ is the believer's soul' (*Ep.* 58.7), he writes, echoing Origen's 'the holy place is the true soul'.[40] But to Marcella, he outlines a string of injunctions to travel to find God, starting with the instruction to Abraham to 'Get you out of your country' (Gen. 12.1); he lists all the amazing sites to be seen, from the very spot at Golgotha where tradition has it that Adam lived and died (immediately glossed with a reference to the 'Second Adam'). His apologetic also analyses and dismisses the counter-arguments against pilgrimage to Jerusalem, including the instructions to the Apostles to leave the Holy Land and spread the word to the gentiles. Like Augustine, Jerome can see pilgrimage as an image of the soul's rising from its alienation in the physical world to the spiritual City of God, an *itinerarium animae*, 'a journey of the soul', where the struggle of the journey itself signifies: 'Our life is here, but our citizenship is in heaven',[41] declares John Chrysostom, and Augustine tellingly calls his readers 'my citizens and pilgrims with me' (*Conf.* 10.4.6). Above all, however, with increasing intensity Jerome repeats that they will see not just the sites but the events themselves: 'We will see Lazarus come forth bound in grave clothes; we shall look upon the waters of the Jordan purified for the washing of the Lord ... we shall see the prophet Amos, upon his crag blowing his shepherd's horn'. The vividness (*enargeia*) of ecphrasis becomes here a religious vision. So Paula (*Ep.* 108.10) – and it might be worth noticing that both Jerome's paradigmatic pilgrims are women, while it is male monks he advises to study – when in Bethlehem, 'could behold with the eyes of faith the infant Lord wrapped in swaddling clothes and crying in the manger'; indeed 'she declared she could see the slaughtered innocents, the raging Herod, Joseph and Mary fleeing into Egypt'. Pilgrimage is to change one's experience of time, to inhabit another time's present: 'To be in the place of revelation is to be in the time of revelation.'[42] Finally, he imagines how Marcella and he will return to the cave where he lives as a hermit: 'then we shall sing heartily, we shall weep copiously, and pray unceasingly'. Jerome provides a template for Christian pilgrimage: justified by scripture, not just to experience the sites but to re-experience the *pathos* of the sites, and to recall them in song, tears and prayer, as an image of the Christian life, necessarily an alienated journey. As with the hermeneutics of typology, Jerome closes the gap between the past and the present in his

---

[40] *Homily on Leviticus* 13.5 [41] John Chrysostom, *Homily 15 on 2 Cor. 5.*
[42] Leyerle (1996) 131. On Jerome see Frank (2000b) 96–101; Newman (1998).

passionate vision: this is to make place not so much a *lieu de mémoire* as a *topos* of *pathos*, a permanent present of shared tears.

Prudentius is not a monk (yet). He is on his way to Rome with a petition. He stops at Imola to pray at the tomb of Saint Cassian and there sees a picture of the saint's martyrdom. The sacristan explains the picture to him in the long central section of the poem. Prudentius prays to the martyr for help, and the poem ends with Prudentius declaring that his mission in Rome was successful, and he returned to Spain to praise Cassian, praise instantiated in this poem. The recollection of the trip to Rome thus frames his encounter with the sacristan and the sacristan's inset description of the picture. The sacristan explains that the picture 'recounts a history (*historiam*) that is handed down (*tradita*) in books and demonstrates the true faith of olden times (*vetusti temporis*)' (19–20). So, Prudentius recalls in the poem hearing a history, now with the status of scripture (*tradita libris*) that shows the faith of 'olden times'. The temporality of the poem is carefully layered: a memory of being told a history of a former age. But his response is not so distanced, so framed: *Stratus humi tumulo advolvebar*, he begins, 'I was stretched out on the ground, prostrate before the tomb' (7–8):

> dum lacrimans mecum reputo vulnera et omnes
> vitae labores ac dolorum acumina

> While in tears I reflected on my wounds and all the labours
> Of my life, and the pricks of grief

Prudentius weeps, as he recalls (*reputo*) – another layer of time – his 'wounds' and the 'pricks' of grief. Cassian was martyred by being stabbed to death with pens: the 'wounds' and 'pricks' do not merely anticipate Cassian's story programmatically or for aesthetic effect, as we might say, but structure Prudentius' experience as a form of *imitatio*, just as every martyr's suffering is an *imitatio Christi*. After he has been told Cassian's story, he again weeps: 'I embraced the tomb, I poured tears too ... then I reflected on all the hidden parts of my distress, then on what I sought, what I feared, muttering'. Unlike a Roman orator or exegete, strutting and expounding before a work of art, Prudentius depicts himself in emotional turmoil, weeping, humbled, prostrate – muttering. As Jerome advised: here is embodied song, weeping, prayer. As Prudentius himself writes (*Perist.* 2.529–36), 'blessed is he ... who prostrates himself by your bones, who sprinkles the place with tears, who presses his breast to the ground, who pours out his prayer, murmuring'. Prudentius before Cassian's tomb is representing himself in the authoritative, humbled, physical and mental position for a good Christian.

The poem also asks a question, however. The poem offers the prostrated and crying Christian imitator of the saint as a model of response: but how is a reader of the poem to respond to this response? How tearfully is Prudentius to be read? Is Prudentius' poem, as an exercise in devotion, to become itself a prompt to devotion? How catching are the tears of the faith of olden times?

Prudentius' poetry, like Jerome's prose, is the product of a traditional training in Roman literary culture, for all that it struggles precisely with the tension between its Christianity and the values encoded in such training. Behind the weeping Prudentius stands the most famous response to temple art in Roman culture, namely, pious Aeneas before Juno's temple in Carthage in Book 1 of the *Aeneid*. When Aeneas sees the pictures of the Trojan war imaged on the temple, he too cries (*lacrimans*, 1.459), as he observes *sunt lacrimae rerum*, here too 'the tears of the world exist'. He 'groans greatly, and moistens his face with abundant flow'. Aeneas sees his own story, and like Odysseus in the *Odyssey* hearing a song of his own exploits, is moved to weeping. Aeneas stands transfixed by the picture. Prudentius' bodily performance is quite different from that of Aeneas – who will meet Dido at the very moment he stands thus in manly reflection before the temple – just as Prudentius' murmuring of prayer contrasts with Aeneas' address to his faithful Achates, with whom he stands. Christian *tapeinōsis*, 'abasement', redefines the heroism of the body. Yet at another level it may seem harder to distinguish Aeneas and Prudentius. Aeneas weeps to see an image of his own experience, and understands that the Trojan sufferings speak to the Carthaginians too (*et mentem mortalia tangunt*, 'the human condition touches their minds'); Prudentius sees the suffering of another as the shared sign of his own human woes. But Augustine suggests that there is also a crucially different sense of identification at work in this Christian response to a martyr story. How, asks Augustine, is the spectacle of a martyr story different from theatrical spectacles – the tragedies where we cry at the woes of others? 'We', he answers, 'in as much as we have a sane mind in us, want to imitate the martyrs we watch'.[43] For the Christian there is no Aristotelian pity and fear at the spectacles of suffering; rather the Christian's sanity, his good sense, is to be seen in a desire for imitation of suffering: to turn the layerings of

---

[43] Augustine, *Sermo Denis* 14.3.

history, the gaps between the present and the past, into as close as it is possible to get to a form of typology, where each life is an *imitatio Christi*.

The cult of the martyrs and the draw of the Holy Land created a new map of pilgrimage and new places of memory. What distinguishes this changing topography, however, is a normative expectation of a different type of enacted response from Greek and Roman culture's traditions of travel writing, festival attendance, ritualized embassies, and strategies of memorialization. The weeping, praying, humbled, prostrate Christian embodies this contrast. But when Jerome writes, 'We will see Lazarus come forth bound in grave clothes', what his intense religious vision also demands is that time is made visible, otherwise. This *lieu de mémoire* works to efface memory and replace it with the present vision of the eyes of faith; to suffer alongside; to inhabit Christian time.

*

One tetchy commonplace response within memory studies to questions such as 'How does a city curate its memory?' is that only individuals have memory and, what's more, everyone's memory is different.[44] Such a response is not necessarily shoddy empiricism masked as sniffy common sense, but also makes a claim to privilege the phenomenology of memory. It is a claim that has a long history, too. Yet, here too, in late antiquity Augustine marks a watershed.

Plato, a necessary starting point, does not evidence a systematic theory of memory, but, as ever, his scattered remarks are hugely productive for later writers, in at least three areas. His image of the 'wax-tablets of the mind', conceiving of memory as the stylus scratches in the mind's wax, founds a long tradition of attempts to explain how impressions enter the mind, leading up to Freud's 'Mystic Writing Tablet' – the twentieth-century device with a celluloid sheet set over waxed paper, over a waxed tablet.[45] You can write a note or draw a picture on the celluloid and it will appear as etched letters or an image. Raise the celluloid and it will erase the note or picture. The inscription of the note or image, however, is still left in the lowest level of the wax. For Freud – and later Derrida – this was a marvellous, technological image of the subconscious – a series of overlapping, confused, readable, traces, beneath any explicit message of the conscious mind, the mere celluloid of the psyche. Although the language

---

[44] Klein (2000).
[45] On Plato see Lang (1980); Sassi ed. (2007) especially Cambiano (2007); King (2019); on Freud's mystic writing pad Derrida (1967) 293–340.

of 'impressions' is still pervasive in everyday speech, the study of the physical process of memory is currently dominated by the different language of cognitive neuroscience; it remains to be seen, however, whether cognitive science and cultural history will find its long-promised but still undeveloped integration, in which the phenomenology of mind and social process can adequately converse.

Secondly, Plato links memory to epistemology. Plato's depiction of *anamnēsis*, 'recollection', insists that knowledge must exist within the mind if a person can be led to realize something he could not immediately state. Knowledge exists as an incorporeal Form in a time before it is expressed or accessed: it is then 'recalled'. For later Neo-Platonists, whose influence on Christianity is formative, *anamnēsis* is connected to the descent of the soul into the physical world – and the desire of the soul to rise again, a theme Porphyry's allegorical analysis can find in Homer's description of the cave of the Nymphs in the *Odyssey*.[46] The epistemology of memory, however, has remained a standard topic of philosophy and psychology, in particular with regard to the errors and distortions of memory: the epistemology of memory is bounded by forgetfulness, ignorance and the processes of learning.

Thirdly, Plato establishes an argument, which Aristotle takes forward critically, about the respective roles of the senses and the intellect in memory, and thus distinctions between memory, recollection, reminiscence.[47] Aristotle's distinction between memory (*mnēmē*) and recollection (*anamnēsis*) redefines Plato's analysis to understand recollection as an active search through the storehouse of memories, a sort of reasoning that also involves the body because of the sensory basis of the images in the memory. Proust's wonder at the flood of memories released by the taste of a madeleine – a highly intellectualized wonder at his senses – is a distant echo of such a tradition of philosophical discussion.

Augustine picks up all three of these argumentative traditions in his *Confessions*, which redefines how memory is shaped in the history of the West.[48] Augustine begins by spatializing the faculty of memory in an argumentative journey that begins with an Aristotelian idea of a storehouse but ends in a quite different topography. His mind, he argues first, contains images of prior sensations, which can be re-accessed in his

---

[46] See Lamberton (1986), (1992); and for its polemics Edwards (1996).

[47] Lang (1980); Annas (1992), Castagnoli (2019a); and the editions and commentaries of Sorabji (2006) and Bloch (2007).

[48] See Hochschild (2012) on Augustine's view of memory as 'a right ordering of the soul in relation to creation' (2) and 'the embodied soul's mode of approaching God' (139).

mind, without the experience of the sensation itself, so that he can sing
a memorized song internally without using his physical voice, or remember
joy when sad. He expresses this in a stunning bodily image: the memory is
'the stomach (*venter*) of the mind', which does not taste but stores. But
Augustine continues to strive to capture a fuller and more nuanced topography of inwardness. He talks of the 'fields and vast palaces of memory',
campos et lata praetoria memoriae (10.8.); the 'huge cavern of memory'
(*grandis memoriae recessus*, 10.13) with its 'mysterious secret and indescrib-
able nooks and crannies'; a 'vast hall (*aula*) of memory' – a hidden place,
a *crypt* within.[49] But he also recognizes that his knowledge of the liberal
arts, mathematics, say, is different from the *imagines* of prior sensations.
With mathematical principles, he knows the thing itself: it does not matter
if it is expressed in Greek or in Latin, the principle does not depend on its
expression, its *imago*. This memory, writes Augustine, is *quasi remota
interiore loco, non loco*, 'as if moved back into a place that is deeper, a non-
place' (10.9.16). Topography here is strained to the limit, as place becomes
a no place.

There are good reasons for this impasse of topography, this journey to
a nowhere. Discussing memory is prompted by the startling question – ten
books into the *Confessions* – 'I turned myself to myself and I asked, "Who are
you?" and I replied "a man" (*homo*). Look, there are a body and a soul in me,
one exterior, the other interior' (10.6.9). The 'interior is superior', and it is in
the interior that Augustine must seek God. The search to understand
memory is a search to understand the where of his knowledge of God. But
that search necessarily starts with the self. For it is in memory – the cavern,
the hall, the palace of memory – where 'I meet myself, and recall myself
what, when and where I acted, and in what way I was affected when I was
acting' (10.8.14). On the basis of what has been experienced, however, 'I
reason about future actions and events and hopes, and again think of all these
things in the present. "I shall do this and that", I say to myself within that
vast recess of my mind.' Memory is integral and fundamental to the process
of change – of desiring to act and making decisions to act. Aristotle saw
anamnēsis as a dynamic activity; Augustine sees it as the core of moral
decision-making. 'Recollection' for Augustine is a 'matter of will' and thus
is 'pre-eminently a moral activity'.[50] The 'dynamics of recollection' here
challenge the 'static containment' model of the cavern of the mind.[51]

[49] On the crypt, see Dillon (2007) building on the psychoanalytic theories of Abraham and Torok
      (1994).
[50] Carruthers (1998) 68; see also the fundamental Carruthers (1990) and behind that Yates (1966).
[51] Stock (1998) 216.

Memory is not a stroll around a building (as the *Mnemotechnik* of rhetoric suggests, with its *loci* of memorization) but a precarious, insistent journey of self-discovery.

'I myself, O Lord, labour on this, and I labour on myself' (10.16.5). For Augustine, the search for morality, which is also the search for how he knows God, is to work on oneself – on the interiority of one's being. 'This effort of self-discovery is part of his ongoing journey towards integration', writes Andrea Nightingale.[52] But to do this, Augustine recognizes, is to seek also to transcend his memory, mired as it is in the continuing physical impressions from his body and recollections of sensory pleasure and pain. Memory has immense power (*vis*), huge and terrifying in its complexity: 'this', he declares 'is my mind, this is what I myself am': memory is self-definitional (10.17). It is impossible to plumb the depths of all its complexities, and, listing the multiplicities of the interior world of the mind, Augustine, in increasingly intense self-expression, concludes: *per haec omnia discurro et volito hac illac, penetro etiam, quantum possum, et finis nusquam*, 'I run through all these things, I flit here and there, and penetrate their workings as far as I can; but the end, nowhere!'. The search for the self in the memory reveals the fragmentation and dispersal of the self, running and flying through its dark caves, unendingly. Consequently the desire to transcend memory: 'I will transcend even my power of memory, I will transcend. I will rise beyond it to move towards You, sweet light. I am ascending through my mind up to you who are constant above me ...' Augustine seeks the constant of God to set against the dispersal that his discourse reveals (with what Foucault would insist is etymological fate: *discours* is derived from *discurro*).[53] Augustine knows that God is not in his memory and therefore he has to go beyond memory to find him. Three times he repeats *transibo* 'I will transcend'. Yet he is thrown back into questioning as (three times) transcendence escapes through his grasp (10.17.26): 'I will transcend my memory so that I might find you – but where? ... If I find you outside my memory, I am not mindful (*immemor*) of you. And how shall I find you if I am not mindful (*memor*) of you.' Augustine ends in his self-defeating paradox. 'He cannot locate God in his memory either temporally or spatially',[54] but he cannot escape the need to be *memor* – to care, to have in mind, to find in the moral activity of recollection – if he is to locate God. Augustine cannot escape his endlessly fallible, endlessly regressive human state of mind.

---

[52] Nightingale (2011) 67.   [53] Foucault (1969).   [54] Nightingale (2011) 69.

This condition is resolutely bodily, for all the turn to inwardness. Hence the remainder of Book 10 – less rarely included in discussions of Augustine's writing on memory – outlines at length the temptations of the body and his attempts to control sensuous and sensory pleasures. Augustine turns to the practical necessity of living morally as a human: for him, the phenomenology of memory is absolutely integrated, through his theological vision, with the fundamental social question of how life should be lived. The failed struggle to transcend the condition of memory is a sign and symptom of the inevitably failing struggle to transcend the condition of being human.

Memory thus for Augustine is fully part of the structure of *confessio*. To say what one has been and what one now has become is a fundamental narrative commitment of a Christian. For Augustine, the before and after of confession structures his narrative and prompts his reflections on memory, and the deep complexity of his representation of memory, its power and fallibility, informs and is informed by this narrative, full of incipient change, false starts and ignorance, along with the chance interventions that betoken grace: the theory and exercise of memory are set in a mutually challenging, interactive dynamic in the *Confessions*.[55] The time of confession is necessarily retrospective and apologetic: a re-telling of the passing of the past from the perspective of the here and now. Augustine's discussion of memory both theorizes this double temporality and produces an intense self-consciousness about it: he makes anxiously visible his struggle with time. The insistence that the Fall is definitional of humanity, and that Christ's redemptive power is available to humanity – twin pillars of Christian theology – makes the telling of sin and redemption the necessary grounding of confession; this narrative requires that living in Christian time is constantly searching backwards in the hope of a transformed future. Unlike the proliferation of many later triumphal conversion narratives, however, Augustine's *Confessions* shows a remarkably intricate self-consciousness about how such a search for the self is necessarily fractured and dispersed by our living in human time, constantly experiencing a body we must wish to transcend. It is not by chance that Augustine's discussion of memory precedes his extended discussion of time. Both discussions explore the painfully incomplete and fragmented present of Christian time. We have moved in this chapter from memorial and memorialization to

---

[55] See Hochschild (2012).

memory: if the 'memory wars' of civic time and *damnatio* show humans attempting to transcend or politically instrumentalize time's ravages, Augustine's lasting legacy – an irony of sorts – is his determination to recognize that the fragility and force of memory, the very grounding of memorialization, is inevitably a changing and unstable dynamic of fallible human time.

CHAPTER 7

At the Same Time

To stand in Rome before the Pantheon and stare up at the huge bronze letters of the domineering inscription M. AGRIPPA L F COS TER FECIT is to recognize the memorializing power of the ancient epigraphic habit – and the degree to which the ancient city was full of sites that made the past physically, visibly present. The Parthenon parades its anonymity, proud in its generalization of the democratic ideal:[1] the Pantheon broadcasts its maker's name. Even or especially here, however, the act of making memory visible turns out to be more complicated than it might at first seem.[2] The temple itself, originally built by Agrippa, had been burnt down twice, and the second rebuilding was completed by Domitian, an emperor so despised by those who remembered him that his statues were said to bleed and scream (it will be remembered) when beaten in violent *damnatio*. There is no record of his name on the Pantheon. This facade was rebuilt by Hadrian, who nonetheless has also not left his name anywhere on the building. This may seem an act of surprising restraint, but later history (*HA Hadrian* 20.3) records that Hadrian did not care to have his name proclaimed on buildings that he established or restored (though he was happy for cities to be named after him). He imitated what the propagandists cared to call the pious humility of Augustus – no less a performance of self-assertion, for sure, but a different display of power, both by Augustus, the first emperor for whom restoration inevitably had a different ring, and by Hadrian aligning himself with the first emperor's memory. It was probably Hadrian who memorialized Agrippa with these huge bronze letters. The fascinating suggestion has been made that Hadrian's display of Agrippa's name should be read as a politicized recalibration of Augustus' own famous boast to have transformed Rome's cityscape, adding another agent of

---

[1] Suggestions that Phidias had carved recognisable faces was therefore a scandal: Rood, Atack and Phillips (2020) 59–60.

[2] Thomas (1997) provides the facts for the following paragraphs.

power to the history.[3] Underneath the proclamation of Agrippa's name, however, there is another long inscription, now barely visible, which attests that Septimius Severus and his son, Caracalla – they are named with full honours, in contrast to Agrippa's 'consul three times' – 'restored with full refinement the Pantheon, which had been damaged by age (*vetustas*)', memorializing a further rebuilding more than fifty years after Hadrian. (The words 'damaged by age' are ... damaged by age, a melancholy to warm the love of ruins in any Romantic poet.) For a Roman emperor, the act of restoration is charged with political assertiveness. It announces that there has been the need for such work – which implies the failings of previous regimes; it denotes a return to what in such political rhetoric in Rome is always the 'good old days' of the *mos maiorum* (though not in more cynical minds); it associates the emperor with such values of the honoured past. The building is 'entangled in a web of diachronic cross-referencing'.[4] Thus the phrase *corruptum vetustate*, 'damaged by old age' is not merely a surveyor's report: as *vetustas* 'manifested time's continuing power'[5], so *corruptum* 'damaged' (as opposed to *delapsum*, say, 'fallen down') suggests a moral framework for the emperor's restorative intervention. The facade of the Pantheon, in its inscriptions and in its silences, displays at the same time a multiform history of memory-making and imperial attempts to record their interventions in the public life and memory of the city. It makes visible a politics of time.

Such imperial endeavours to create and display a topography of sacred architecture for the city and the empire, and to use it to monumentalize also a *cursus honorum*, express the position of the emperor with regard to the divine and with regard to the institutions of human power. Other types of epigraphical display in both Greece and Rome construct quite different frameworks.[6] The Lindos Chronicle (as it is known) is a monumental inscription, set up in 99 BCE, from the stunningly beautiful site of the temple of Athene at Lindos on the island of Rhodes.[7] It describes first the decision to research and record the dedications made at the temple, and then, at far greater length, offers not only a list of dedications but also brief historical comments including in particular the textual sources which attest

---

[3] Boatwright (2013).

[4] Howard (2018) 64 – of Renaissance Venice, on which Trachtenberg (2010) contrasts 'building-in-time' with the idealism of 'aesthetic immutability'.

[5] Thomas and Witschel (1992) 143.

[6] Ma (2002) and particularly (2015) are especially good on Hellenistic materials; on classical see most recently Low (2020); on Roman material see the exemplary Woolf (1994), (1996), (1998).

[7] Higbie (2013) is excellent on this.

to the dedications. It does not record all the dedications that were in the temple, however. Pliny the Elder (*NH* 35.71) notes austerely that the temple had an amber cup dedicated by Helen of Troy, and that 'history adds it was the size of her breast', *adicit historia, mammae suae mensura*, but this surprisingly does not get a place in the inscription as we have it. How, then, does the selectivity of the Chronicle and its specific expressivity construct its own historical framework? It certainly takes the history of Lindos and Rhodes back to the Homeric poems. In Homer, Rhodes appears only briefly and for the most part inconsequentially, with a handful of ships under Tlepolemus, who had founded the three cities of the island, of which Lindos is one. (Tlepolemus dies in the first battle of the *Iliad*, though his wife gets a murderous and lurid afterlife in much later mythography.) These three communities in the classical period founded a new city called Rhodos, to which they willingly ceded political authority. By recording gifts from Homeric heroes, Rhodes, which by 99 BCE, was central to the grain trade, demonstrated a more central place for itself in the deep past, a genealogy of significance. The Chronicle does not mention Homer by name, but does create an unbroken time-frame from the time of the Trojan war forwards.

It is particularly striking for such a public document that the provenance of many of the gifts is supported by a specific citation from a string of ancient authors. These writers are largely what we would call historians or local historians; but this scholarly apparatus has the effect of locating Lindos as an integral node in the network of the intellectual history of Greek culture. The difference between 'local historians' and 'historians' is not just a nicety of historiography in this case: *how local* a history Lindos should have is precisely the question this inscription poses. Rhodes became well known as an intellectual centre.[8] The orator Aeschines, after his defeat by Demosthenes, retired to Rhodes, still a celebrity from a world much larger than the purview of the island. Later, Mucius Scaevola studied in Rhodes; Julius Caesar visited, and Cicero himself chose to spend time there to study with the rhetoric teacher, Apollonius. The very form of inscription, with its network of citations, speaks to this self-representation of the community. This is a site that has been *written about*, and this is a community that prides itself on its ability to research and record such writings. That the list records gifts also from great political figures – Alexander, say – is to be expected, but such a roster also establishes a significant continuity between the Homeric heroes and the political

---

[8] On local history see Clarke (2008) which is not superseded by Thomas (2019).

grandeur of a more contemporary history. In a way that the later Greek writers of the Roman empire would recognize, there is also what appears to be a studied silence about any gifts from Roman notables (though such dedications certainly existed). The temple of Athene is to have therefore a Greek history, placed within the history of Greek historians, and thus Lindos is to find its place on a Greek map of significance, its place in time.

It is not possible to trace in Lindos as detailed a political intervention as is evidenced in the restoration of the Pantheon by Severus; and in the drive towards memorialization as well as in the inevitable expression of the power of divine articulated in the record of the epiphanies of Athene, there are continuities with other such inscriptions in this era across the Mediterranean. Nonetheless, the Lindos Chronicle demonstrates a revealingly specific framework for understanding the timeline of a temple and the cultural significance of such a timeline – in a way which is quite different from the Pantheon.

The Parian Marble (*Marmor Parium*), another large-scale Greek inscription that, like the Lindos Chronicle, is exceptional and exceptionally surprising in its style and content, indicates a further way of proclaiming how a local history relates to a broader political world.[9] The Parian Marble lists events of 'world history' down to the 260s BCE; it dates these events according to the list of archons in Athens, which is one of the authoritative chronological schemata (used in Athens in particular, of course).[10] Because it includes stories from deep history, such as the flood and Deucalion, and because of its record of the archons, the Parian Marble has most often been discussed in terms either of the distinction between myth and history – not a concern of the Marble but one to which I will return – or of chronography – how the ancient Greek communities calibrated and recorded time. Yet what is also crucial is how this chronicle places Paros in the world. By linking its history to Athens, Paros is located side by side with the dominant power of the region; by placing this history on the grandest of scales – from the beginning of time forwards to the contemporary moment – it gives Paros a place in such a history. 'Whoever governs simultaneity controls the temporal dependencies derivable from it.'[11] It is a place Paros never held: it is an island of quite undistinguished political impact. But it was the island where Archilochus was born and about which Archilochus wrote. Alongside the development of his tomb as a cult site, we

---

[9] For description and discussion see Rotstein (2016), with full bibliography.
[10] For the later history of 'world history' as synchronization see Jordheim (2017), (2019).
[11] Nowotny (1994) 10.

can see strong hints in the Parian Marble, as in the Lindos Chronicle, that the culture of Greece – its literary tradition – is also a salient framework of history. For the battle of Marathon, for example, it is recorded that the playwright Aeschylus fought in it, but not the names of the generals. Just as Pausanias finds in poets' tombs a 'root system . . . that in [his] imagination nourishes the sacred Greek landscape',[12] the Parian marble links its literary life to the outside world, synchronically. It demands a historical perspective that includes how Athens is a *paideusis* to the world. The Parian Marble is not exactly 'a continuous time map'.[13] Its vision of time, of what counts in the passing of history, and of Paros' place in it, is a form of self-placement in time, a placement where the types of event recorded construct a map of displayed significance – where Aeschylus is privileged over a military leader – a map on which Paros is to find its place.

The epigraphic culture of antiquity can be seen thus to create a framework of self-representation, a framework that is revelatory about the broader concerns of social change, within its acts of memorialization. It is not by chance that the democracy of Athens recorded in stone so many decisions by its institutional bodies of civic authority, or listed its war dead generally without marks of social status, nor when 'anxiety about social mobility occupied a significant place in the collective imagination of early imperial Italy',[14] is it arbitrary that so many inscriptions from this time of rapidly expanding empire wealth record in detail the social status and changes of social status of the country's citizens – or, for example, that Pliny can write with such twisted political ambivalence about the honours accorded to and denied by a eunuch freedman, Pallas. (Much writing from within the British empire's expansion shows a similar obsession with social distinction, hidden truths of identity, and the possibilities of social mobility.) The chronology of an inscription such as the Parian Marble, as much as the grandest texts of ancient historiography, functions, as Reinhart Koselleck would argue, to locate the self in time.

*

For Koselleck, the late eighteenth century marks a turning point in this self-defining function of historiography, and it is difficult to avoid the impact of this modern sense of history.[15] It is hard now, it is often

---

[12] Hanink (2018) 249. See also Clay (2004); Höschele (2018); Platt (2018).   [13] Feeney (2007) 81.
[14] Woolf (1996) 35.
[15] Kosselleck (2004): for him too, historiography is the problem of *die Gleichzeitigkeit des Ungleichzeitigen*, 'the simultaneity of the non-simultaneous'.

asserted – with some justification – to imagine the experience of living without a single, unified calendar or clock-time, even when it is the case that our current scheme of dating, which takes Jesus' (imagined) birth year as a starting point, became prevalent only in the seventeenth century, and, as I have already mentioned, a national clock becomes a requirement only after the railway system is up and running in the nineteenth century. But it *is* a failure of the imagination. Simply taking Jesus' birth as 'an arbitrary point' from which 'Newtonian absolute time' can be 'indefinitely extended forwards and backwards' on a single scale,[16] not only ignores the imperialism of this necessarily Christian perspective on how time is to be organized, but also represses the Enlightenment values from which Newtonian absolute time emerges. There is no doubt about the usefulness of such a shared scale. But we also organize and express time in many other ways. We Jews and Muslims (and many other societies) have a different calendar, where festivals shift against the Christian calendar, though are regular enough to our own communities. Ramadan moves through the seasons predictably and on a fixed time-scale, part of time's order; Easter moves because of the regularity of Passover in the lunar calendar. To tell the story of our time to ourselves and others, we can overlay university or school routines; a sense of a career or places of work; before and after marriage; children; partners; ageing. Self-placement in time is far more complex than the familiar single scale of counting years allows. How counting time and experiencing time interrelate is not a mathematical expression.

The usefulness of the single clock and single time-scale is predicated on the problem of simultaneity. How to know when trains start at different places in the rail system? How to know what day to agree to hold an international meeting? How to understand the connection between events in different places? What, in short, is 'the same time'? (This problem will also motivate the development of post-Einsteinian quantum theory, but that is another story.[17]) The politics of 'the same time' is a grounding problem of historiography. *How local* is history to be? The Lindos Chronicle and the Parian Marble show how smaller polities could strive to associate themselves to larger-scale narratives, to inhabit the same time; and there was, it seems, a flourishing genre of such parochial histories – stories of a local community, from foundation to contemporary recognition. Yet the connection of events in different places across 'the same time'

---

[16] Wilcox (1987) 7–8. Herder (1998) 360 imagines it: 'There are (one can say it earnestly and courageously) in the universe at any time innumerable different times.' See Jordheim (2014).

[17] Well-told by Galison (2003).

is the founding claim of so-called universal history. Polybius provides the *locus classicus*. His opening methodological statement needs quoting at length (*Hist.* 1.3):

> The date (*chronoi*) from which I propose to begin my history is the 140th Olympiad, and the events are the following: in Greece the so-called Social War, the first waged against the Aetolians by the Achaeans in league with and under the leadership of Philip of Macedon, the son of Demetrius and father of Perseus; in Asia the war for Coele-Syria between Antiochus and Ptolemy Philopator; in Italy, Libya, and the adjacent regions, the war between Rome and Carthage, usually known as the Hannibalic War. These events immediately succeed those related at the end of the work of Aratus of Sicyon. In the times (*chronoi*) before these events, the actions of the world had been, so to say, dispersed, as they were held together by no unity of initiative, results, or locality; but ever since this crucial juncture (*kairoi*) history has been like an organic whole (*hoionei sōmatocidē*), and the affairs of Italy and Libya have been woven together (*sumplekesthai*)[18] with those of Greece and Asia, all leading up to one end (*telos*). And this is my reason for beginning their systematic history from this crucial juncture (*kairoi*).

Polybius offers a starting point for his history and justifies it. He determines that at this one juncture (*kairos*), in Greece, in Asia, and in Italy and North Africa, three different battles for imperial supremacy were being fought out. But the importance of this juncture is that from this time (*chronoi*) the three theatres of political power were integrally woven into a single, organic whole – 'like a body', a natural, and in Aristotelian terms, perfected whole. Previously history was necessarily local, because it was dispersed, without any unifying narrative to link different aims, consequences, or places. What is now possible is systematic history – a single time-scale, to capture the significance of events in different places 'at the same time', to tell the single story of the rise of Rome – which is the remarkable transformative world-making event that Polybius, like other Greek writers, wishes to explain.[19] Local history from now on can only be insufficient in its partiality (1.4). To make sense of history, argues Polybius, is to recognize the significance of what happens 'at the same time'.

Polybius here provides a systematic view of his own commitment to systematicity. He includes further methodological comments scattered throughout the unfurling of his history that reflect back on this opening.

---

[18] On this term see Maier (2018).

[19] On universal history see the somewhat dyspeptic Sacks (1981) 96–121; on Polybius, Hartog (2010); for a brilliantly stimulating reflection on what 'universal history' entails, see Garcia (2014) 287–302.

He recognizes that this interest in simultaneity has implications for his narrative. 'I am not unaware that some will find fault with this work on the grounds that my narrative of events is imperfect and discontinuous', he writes (38.6), noting that he began the story of the siege of Carthage but broke it off to go to Greece and then Macedon before returning to what he had left in suspense. But, he disarmingly adds, he 'obviously grants liberty to students (*tois philomathousi*) to carry back their minds to the continuous narrative and the repeated interruptions of the action so that nothing is imperfect or deficient to the careful listeners of what has been said before'. Simultaneity may come at the price of immediate continuity, but the intelligent reader – the demand for complicity is disarming – will put together the necessary continuity.[20] The breaks in the narrative are not an issue of digression, but a vision of causality and interconnectedness, in which the reader plays a necessary role. The aesthetics of discontinuity, which Callimachus in the Alexandrian Library made a watchword of contemporary poetics, here is further intellectualized as an engagement with how events may be understood in and through time.

Polybius is himself a juncture from which any telling of the history of historiography could go back towards Herodotus and Thucydides and forward to Livy and Tacitus (and far beyond) to look at how different understandings of the rise and fall of powerful states incorporate and express their theories of time, both at the grandest levels of the *translatio imperii* and at the more local level of the interconnectedness of events and the structuring of narrative form. There will indeed be more to say about the time of historiography, if not an account to match the hundred-book scale of ancient histories; but there are immediately two further issues opened by Polybius' challenging prose.

First, he begins by determining boldly a fixed starting point by reference to the 140th Olympiad. The scheme of Olympiad dates appears to offer a single time-scale, like the Christian dating system of BC and AD, to match Polybius' claim of a unified, organic history of things. The Hellenistic scholar Eratosthenes seems to have proposed the usefulness of precisely such a way of systematizing events to a single temporal scale. Scholars of chronography have been assiduous in collecting and comparing such ancient techniques for recording events against a time-scale, an intellectual project that finds its paradigmatic opening salvo in Thucydides' rejection of the Athenian archon list as inadequately discriminating for his desired

---

[20] On the failures of Polybius' narrative teleology and its 'horizon of uncertainty' (251), see Wiater (2016), with bibliography to earlier discussions.

level of accuracy.[21] Yet even Polybius, like many other writers, refers only occasionally to the Olympiads for dating specific events in his historical narrative, and it is the confluence the three streams of history that dominates the logic of his narration. Donald Wilcox puts the point with sharp simplicity: 'The time did not give meaning to the events, they gave meaning to the time-line.'[22] The recognition of a single story – Rome's rise to power – produces a single time-zone out of what had previously been local histories with their own temporalities. Rome sucks the temporality of others into its orbit, as, once again, power defines time. Yet even when the value of a single chronological scheme is acknowledged; even when empire provides a condition of possibility for the adoption of such a temporal scheme; even when annalistic historians, Livy or Tacitus, say, continue to mark each year by the elected consuls; it remains an open question to what degree any one absolute timeline functions for any individual historian let alone historians of different times and places in the empire. It is not a question of a failed teleology – why did ancient historiography not develop a single timeline like us? – but rather a recognition of how these multiple and relative time-scales continue and create differing possibilities for narrative form in historiography. As Plutarch concludes (*Life of Solon* 27.1):

> I do not think it is right to give up [the chronologically unlikely meeting of Solon and Croesus] to any so-called fixed chronological tables (*chronikois kanosi*). Although countless people have been revising them right up to this day, they have been unable to bring their contradictory arguments to any point which is agreed amongst themselves.

Time-scales are recognized as sources of contention and interest, rather than simply useful measures.

Second, Polybius (12.11) cites the historian Timaeus as a predecessor, who, he says, compares different time-scales, the dates of the kings of Sparta, the archons at Athens and priestesses of Hera at Argos, three regularly used lists, against the Olympiads – and finds a damaging inaccuracy or inconsistency between them. It is hard to tell how systematic Timaeus' work was from this citation. But another fragment of Timaeus shows that he established that Rome and Carthage were founded in the same year. The date of the foundation of Rome was intensely contested by ancient authorities, however, and Timaeus' case is most likely to be making

---

[21] E.g. Bickerman (1968); Samuel (1972); Wilcox (1987); Salzman (1990); Grafton (1995); Möller (2001); Grafton and Williams (2006); Mosshammer (2008) – all with bibliographies.
[22] Wilcox (1987) 89.

a symbolic point. As rabbinical writings insist that the first and second Temple of Jerusalem were destroyed on exactly the same date in the calendar, or as Syncellus' chronology states that the creation and the crucifixion took place similarly on the self-same date, so the parallel foundation of the two great imperial enemies suggests providence at work. 'At the same time' here implies: fated to go hand in hand.[23] (The 'axial age', beloved of some modern historians, shows a similar moral and political imperative of simultaneity.) The imperial success of Rome prompts a sort of reverse teleology. Once it is determined that the local histories are 'all leading up to the one end' of Rome,[24] it is hard to resist seeing local histories *from elsewhere than* such a teleology, and thus narratives are reshaped towards such an end. Universal history retrospectively makes Rome central from the beginning, as an attempt to explain Rome to Greece.

Diodorus Siculus, however, utilizes such a strategy to develop a counter-genealogy of political and cultural affiliation. Like Polybius, Diodorus announces that his is a universal history, which starts with the antiquities of Greeks and barbarians and moves up through to the Roman empire. He states that he has deep knowledge of the empire, that he has learnt Latin well (a rare admission by a Greek writer) and visited the countries about which he writes over thirty years. Nonetheless, when he comes to write about Sicily, a local history of his homeland within his universal modelling, he links its story to events at the same time in mainland Greece – and significantly not to Rome or Italy. He makes his 'choice of chronological frameworks to create a Sicilian historiography with aspirations to Greekness', writes Katherine Clarke:[25] simultaneity is designed to form the Greek political world which Diodorus Siculus wishes to inhabit, to assert a cultural genealogy against a political reality. 'At the same time', thus, functions as a trope of the political rhetoric of history. Its organization of temporality is in service of an ideologically laden narrative.

*

'Thucydides invented historical time.' So, trenchantly, Bernard Williams, in his pursuit of truth and truthfulness.[26] By this he means that it is with Thucydides first that there is 'a rigid and determinate structure for the past', which in narrative form is open to critical analysis; and thus categories of truth and falsehood can be applied rigorously to such narratives. This

---

[23] Feeney (2007) 47–51. For Orosius, a Christian version, see van Nuffelen (2012) 49–53.
[24] Polybius 1.4.1.   [25] Clarke (2008) 236.   [26] Williams (2002) 162.

'historical time' is to be contrasted with myth or legend: 'of many events in myth or legend, there is nothing to be said about when they are meant to have happened'.[27] Indeed, it makes no sense, we know, to ask if Thumbelina lived at the same time as Jack and the Beanstalk. It may be trivially true that Thucydides and Thumbelina do not inhabit the same time, but, for Williams, it is a significant category error to see myth and history as referring to the same temporality.

The distinction between myth and history is an invention of the fifth century BCE, a persuasive if self-serving claim of historians, philosophers and natural scientists within the battle over authoritative discourse in the conflicts of the rapidly changing culture of the intellectual hot-house of the classical city.[28] So, natural scientists and historians, and Thucydides prime amongst them, contrast the research and accuracy of their methodology with what is denigrated as mere myth, stories that circulate without the groundings of truth and authority. Similarly, Plato inaugurates a philosophical method which sets the procedures of dialectic, always aimed at truth, against the vivid stories, but no more than stories, of myth. Yet, as Bernard Williams knew as well as anyone, Plato also utilizes myth profoundly within his philosophical system.[29] When Socrates cannot fully persuade his audience of the immortality of the soul by logic, he turns self-consciously to myth with a depiction of the underworld and its punishments, a story that has its desired effect of luring his listeners to his case. More tellingly, however, there is no more seductive – or epistemologically influential – passage in the *Republic* than Plato's description of the cave with its figures staring fixated at the shadows on the wall – a myth, if ever there were one. The boundary between myth and method is less well fortified than Williams would have it.

Demosthenes – not a favourite source for philosophers such as Williams – shows the porosity of the boundary between what is explicitly called myth and the events of contemporary military and political history with vivid self-awareness. Demosthenes' Funeral Oration, his version of the institutionalized speech over the war dead, is inevitably a text where self-conscious acknowledgement of tradition is high, both because it is a heavily conventional genre, and because the conventions demand a tour of the historical traditions of the city. (As Loraux writes: 'the funeral oration reveals an ever more imaginary installation of the city in a time

---

[27] Williams (2002) 162.
[28] Detienne (1981); Lloyd (1987); Marincola (1991); Morgan (2000); Grethlein (2010); Baragwanath and de Bakker eds. (2012).
[29] E.g. Collobert, Destrée, Gonzalez eds. (2012); Morgan (2000); Brisson (1998).

that is ever more timeless').)[30] After Demosthenes, with suitable generic deference, has recalled the Athenians' fight against the Amazons (familiar from the statues on the Parthenon), the Athenian support of the Heracleidai, and of the Seven Against Thebes, he transitions to events closer to the present time (60.9):

> Now, I have left out many deeds that are classed as myths (*muthoi*). Of the events that I have recalled to mind, each has provided many charming stories (*logoi*), so that our writers of poetry, whether recited or sung, and many historians, have made the deeds of those men the themes of their art. But now I shall be describing deeds, which, though in merit they are not inferior to these former events, because they are closer in time, have not yet become myth (*memuthologeitai*) or even been raised to heroic rank (*tēn hērōikēn taxin*).

Demosthenes does not merely refer back to what he has discussed as myth, but specifically as events 'classed as myth', τῶν … εἰς μύθους ἀνενηνεγμένων ἔργων: he acknowledges the work of classification between myth and other categories. These events have provided *logoi* for the multiple public voices of the city, different genres of writing. Myth is not opposed to *logos* here, as Plato would often have it. But the more recent endeavours of the city, Demosthenes continues, because they are closer in time, have not yet become myth – μεμυθολόγηται – 'been told as myth', nor even have they entered the rank of the heroic. The implication is that as time passes so contemporary events in history will become myth. 'Heroic rank' may imply that in the future events of today will be the subject of epic poetry, or that its heroes will become cult figures as with the men who fought at Marathon. As time passes, the status of stories changes. Myth is reserved for the stories of the deep past. Where Thucydides demands that his history is a 'possession for all time', Demosthenes suggests that history passes into myth over time.

Thucydides' austere definition of what counts as history denigrates the mere pleasure of *to muthōdes*, 'the mythical', 'the fabulous', which has all too often been taken to refer to Herodotus' *Histories*. Yet Herodotus' own stance on what constitutes historical authority has often been underestimated in its sophistication (not least, it seems, by Thucydides). Bernard Williams' contribution is, again, provocative. He turns his attention to what has become a key passage for many critics, where Herodotus states (3.122): 'for Polycrates was the first of the Greeks whom we know to aim at the mastery of the sea, leaving aside Minos of Cnossus and any others who

---

[30] Loraux (1986b) 131. On Demosthenes, see now Westwood (2020) 45–7.

before him may have ruled the sea; of what may be called the human race, Polycrates was the first.' What Williams nicely calls 'the problem with Minos'[31] turns on the phrase *tēs anthrōpiēs legomenēs geneēs*, which is given one of its traditional translations above as 'what may be termed the human race'. For Williams, this shows that Herodotus is only just beginning to become anxious about the status of the deep past and the stories of Greek tradition – in contrast to Thucydides who invented historical time. The participle *legomenēs*, 'so-called', 'what may be termed', draws attention to the phrasing and demands that we read with especial care. *Geneē*, we have seen, is usually translated 'generation'; or in Herodotus it appears also to have the sense of 'nationality' ('group linked by common generation'?), when applied to a specific community. He also uses elsewhere the phrase *anthrōpion genos*, 'human race', which, as we noted with Plato's use of the idea, is not in general a common expression. The expression '*anthrōpiē geneē*' specifies, therefore, that Minos is of a different *generation* from Polycrates, which may indicate both that there is a significant temporal difference between the two figures and that this difference is related to their generation, that is, that Minos has a divine father, whereas Polycrates has two human parents. *Legomenēs* is also a characteristic Herodotean self-distancing: it is another's wording. This may imply contemporary intellectual discussions about how the past is to be divided, Hecataeus' interest in genealogies, for example,[32] or it may indicate a more general sense of tradition, which must then include the most famous use of the 'generations of humans', namely, Homer's, the archetypal representation of the past where men with a divine parent strode the battlefields alongside the divinities, and where the boundaries of the human are a repeated theme. History also separates itself from epic, and the sorts of stories 'that Homer or one of the poets from past times invented' (2.23), that sort of myth. Herodotus' gestures of self-distancing are always also self-authorizing.

The dismissiveness towards Minos (and his ilk, 'any others') is connected not just to his status in time, however, but also to epistemology. Herodotus uses the word *muthos* only twice, on both occasions to dismiss stories connected to Egypt that he thinks have no authority. In the first case, he rejects a theory because 'it cannot be tested' (*ouk ekhei elenchon*); in the second case, because it reveals the ignorance of Greeks about Egyptian culture. The use of the word *muthos*, as in Thucydides, is a persuasive definition of where authority lies, part of the historian's self-authorization. The 'problem with Minos' thus is also that, unlike Polycrates, we do not

---

[31] Williams (2002), the title to ch. 3.   [32] Dunn (2007) 29.

know about him, do not know, that is, in the way that the historian constructs acceptable forms of knowing.

Herodotus' 'presentation sensitizes readers to the difficult question of whether the difference' between Minos and Polycrates 'is purely temporal, or whether it runs deeper than that; or indeed, whether we simply cannot know'. The problem with Minos thus becomes 'the problem of whether our ignorance or his status is at issue'.[33] This ambivalence about what can be called mythic narratives runs deeply in Herodotus' history from his opening redrafting of the first conflicts between Greece and Persia, revealed but studiedly distanced under the name of Persian *logioi* and narrated under the aegis of rationalizing de-theologization. The recognition of the role that such stories play in the self-understanding of the actors of history is balanced against a critical and sceptical awareness of the unreliability of these narratives. Herodotus does not excise the doubted past but mobilizes it. Herodotus circulates stories he does not credit, talks about the instrumentality of circulating stories, criticizes the foolishness of circulated stories, attributes such stories to different sources and audiences, and thereby explores and creates the imaginary of cultural difference that grounds the narrative of war he tells. Griffiths captures something of the effect of this, jauntily: 'Herodotus not only rides the two Phaedrian horses, *muthos* and *logos*, with ease, but he knows it, delights in it, and consciously exploits it. And the listeners collude in the enterprise.'[34] But this should not efface the historian's careful self-positioning as an authority, and certainly does not conceal the seriousness of the endeavour. His desire to stop the amazing deeds of the present becoming *exitēla*, 'wiped away', like the fading inscription on a temple or the lost features of a statue, is to stop events turning, as Demosthenes has it, from history into myth. No less than Thucydides, Herodotus seeks to stop the effacement of events by time, events that, in his account, are constantly becoming embroiled in mythologization.

Herodotus' understanding of the past and how divisions in it may be conceptualized is not, then, a step on a journey towards Thucydides' invention of historical time. Rather, he is acutely aware that the divisions between *temps des dieux, temps des hommes*, or *spatium mythicum, spatium historicum*, are fractious and fragile; negotiating his discourse of history becomes a complex cultural map of self-discovery for the reader. And nowhere more clearly than in his description of Egypt and its time. Herodotus' engagement with Egypt demands a recalibration of his

[33] Baragwanath and de Bakker eds. (2012) 25; see also Munson (2012). [34] Griffiths (1999) 180.

audience's expectations of time.[35] Herodotus dramatizes this shock of the old most tellingly when he tells us the story of Hecataeus, his rival and predecessor in writing history.[36] The book opens programmatically with Psammetichus experimenting to find out the oldest language to challenge the Egyptian belief that they are the oldest people on earth, and with the announcement that the Egyptians were the first to understand how the solar system gives the markers of time. To visit Egypt is always for a Greek to stand at the edge of the abyss of time. Hecataeus boasts to Egyptian temple priests that he is descended from a god sixteen generations previously. He is taken into the temple and shown a line of wooden statues, each dedicated as a representation of the temple's high priest in a line of patriarchal genealogy. Herodotus tells us that he has had the same demonstration, though he had not searched out his own genealogy. The priest pointed to each statue in turn and counted (the prose is as sticky as the process), listing 345 priests, father and son, all humans. It was to them unbelievable that a man could be descended from a god: and they had the record to prove it. Before these priests there was a time when the gods ruled the land. But there is an uncrossable divide between the time of the gods and the time of human priests. Hecataeus' response to this crushing demonstration is not recorded. But Herodotus here allows for a quite different time-frame for Greeks and Egyptians, and faces Greek readers with the insufficiency of their self-understanding of time ('and I myself . . .', writes Herodotus, putting himself in the same line). Plato has Solon have a similar experience: 'he discovered that neither he nor any other Greek knew anything at all' about the depth of time: the Greeks were, in the eyes of Egyptians, 'all children' (*Tim.* 22a–b). The claim of the anthropologist Arjun Appadurai that 'cultural consensus as to the relative values of different time-depths' is necessary for regulating the resources of the past only functions in a society that has no contact with another culture, no access to different sciences of the past.[37] The simultaneity of lining up Hecataeus' genealogy and the priest's genealogy becomes a demonstration not just of ignorance about the self in time, but of cultural difference; of not inhabiting the same time (scale).

Egypt offers a topographical and chronological othering to give Greece a different perspective on their war with Persia, and its polarizations. Above all, its manifestation of deep time means that history, for the Greeks,

---

[35] Vasunia (2001) 133–5.
[36] The nature of his disagreement with Hecataeus is much discussed, most recently by Dillery (2018) with references to earlier positions: see in particular West (1991); Fowler (2006); Munson (2012).
[37] Appadurai (1981) 203.

always starts somewhere else and before its own narrative can begin. Diodorus Siculus' universal history faces the problem head on. He starts, he says, with Egypt because this is where mythological accounts insist the gods first appeared and because the very oldest observations of the stars were made there – two points that Herodotus emphasized. But he starts with barbarian stories, he also claims, not because they are older than the Greeks, 'as Ephoros said', but to avoid bringing in such foreign elements when he does turn to the earliest history of Greece. Even if it is important to be seen to correct the claim of Ephoros, one of the founders of universal history, that barbarians are simply older, the precedence of Egyptian civilization, as the following chapters demonstrate repeatedly, is taken for granted – and his narrative puts Egypt first. Much as wealth is re-estimated and moralized when Greeks encounter the Persian empire, and space changes its ideological coordinates when Scythian nomads enter Greek historiography,[38] so time – a sense of scale and hence simultaneity – is dangerously recalibrated when Greeks visit Egypt. To look down over so great an abyss of time is dizzying, as Victorian geologists discovered. As Hecataeus is set up by Herodotus' narrative to find out, this perspective from Egypt challenges how the self in time is to be comprehended.

*

Eusebius' *Chronicon* begins with a robust declaration of inevitable failure. 'I warn and advise everyone from the start, that no-one should ever pretend that he can be completely certain about matters of chronology.' He has biblical authority for this view: 'It is not for you to know the hours and seasons which the Father has set under his own authority.' This quotation is from the opening of Luke-Acts, the first historical account of the earliest days of the community of followers of Jesus, and constitutes Jesus' final warning that humans will not know when the end of days will come; Eusebius takes it to refer to knowledge about dates in general. 'It is not possible to gain an accurate knowledge of the whole chronology of the world.' In contrast to the naive and certain chronological predictions of Julius Africanus, whose dated prediction of the end of days meant 'time itself would fairly soon be no more',[39] and in contrast also to the deep time of Berossos or Manetho, whose mythological past of tens of thousands of years no Christian could accept, Eusebius parades caution. Nonetheless,

---

[38] See Hartog (1988) for the Scythians. In general, Purves (2010).
[39] Grafton and Williams (2006) 152. See also Feeney (2007): 7–43; Kelly (2010); Corke-Webster (2019) 40–2; de Vore (2020).

his project – with the provisos that no-one should be deceived about accuracy and that the whole is 'for discussion' – is first to provide in his *chronographia* chronologies of the major kingdoms (Assyria, Egypt, Israel, Persia, Greece and Rome) through extensive quotation from earlier historians, and second in his *chronikos kanōn* to represent these chronologies comparatively in tabular form. The form echoes his master Origen's use of tables in his *Hexapla*, but in his alignment of national histories he seeks to demonstrate key moments of synchronism, in order to prove the superiority of Judeao-Christian antiquity. 'By comparing individual histories to one another . . . the reader could see the hand of providence at work':[40] or as Arnaldo Momigliano, with a characteristic widening of the question, puts it, 'unlike pagan chronology, Christian chronology was also a philosophy of history'.[41] The technical expertise of tracing what happened 'at the same time' in world history is to demonstrate the new order of world history in Christianity. 'Chronology should serve eschatology.'[42]

So, Eusebius duly notes that the Septuagint and the Hebrew Bible differ greatly in calculating the years between Adam and the flood. The Hebrew Bible has the shorter span of 1,656 years, the Septuagint 2,242. The difference is accounted for by shorter times of life before marriage in the genealogies of the Hebrew Bible. Eusebius suggests that the Septuagint, which he calls 'ours', that is, the privileged Christian version, is right, and that the Hebrews deliberately shortened the years to encourage early marriage, presumably in unstated opposition to the ideal of Christian celibacy. Unlike the rabbinical writings, Eusebius can imagine that the Hebrew Bible, the original, has been altered in response to the Greek translation, despite the apparent temporal order of their production. This academic polemic demonstrates its underlying supersessionism. Tabulating time, again, serves an ideological purpose: how does Christianity, with its claim on eternity, fit into history's time? What happens – we will find out in chapter 15 – when history, starting from 'In the beginning' no longer allows for the contrast between *spatium mythicum* and *spatium historicum*, when (no doubt to Bernard Williams' dismay) the invention of historical time is disinvented? For Censorinus, summarizing a long tradition in 234 CE, there were still three 'distinctions of time' (*discrimina temporum*), the 'unclear', the 'mythical' (from the first flood to the first Olympiad), and the 'historical', from the first Olympiad

[40] Rosenberg and Grafton (2010) 15.   [41] Momigliano (1977) 110.

[42] Grafton and Williams (2006) 151. For Jerome's Latin redrafting of the *Chronikos Kanōn* see Vessey (2015). For the Byzantine reception and Syncellus' importance for understanding Eusebius, Sevcenko (1992). For later versions of such tables, Jordheim (2017).

'to us' (tellingly, this technical vocabulary is in Greek).[43] For Christian historiography, all was a clear and single story, from the beginning.

The fractured, horrified, sardonic vision of Tacitus' *Annals*, far from Eusebius' tabulations, can allow no such ordered godly system. Despite the repeated annalistic structure of consular years, his common temporal markers are *sub idem tempus*, 'about the same time', or '*interim*', 'meanwhile', or the ablative absolute, which does not specify its temporal relationship with any precision. And 'about the same time' can refer to a single year in Rome and about six years outside it, even when the single year is the 800th anniversary of the foundation of Rome. Such disjointedness is part of the disordered and disordering world that is Tacitus' image of imperial corruption. By contrast, the apologetics of simultaneity are immediately on display in the opening of Eusebius' *Ecclesiastical History* (1.5.2):

> It was then in the forty-second year of the rule of Augustus – and the twenty-eighth after the subjection and death of Anthony and Cleopatra with whom the dynasty of the Ptolemies over Egypt met its end – when our Saviour and Lord, Jesus Christ, was born in Bethlehem in Judaea, following the prophecies about him, at the time of the first census, when Quirinus was governor of Syria.

Although this may sound like similar moments of significant dating in the tradition of Greco-Roman historiography, Eusebius is also picking up on earlier Christian recognitions of the parallel between the *pax Romana* and the dawning of the age of Christianity, and, as James Corke-Webster has argued so well, taking such claims further. 'By Eusebius' logic', Corke-Webster writes, 'the *pax Romana* did not spark the spread of Christianity, but the reverse. The nine books of the *History* that follow merely prove the point: Rome had always owed its greatest successes to Christianity.'[44]

This goes to the heart of Eusebius' project. He announces in his prologue with great flourish that his is the first history of the church, that no-one has ever taken this lonely and unworn way before, and that he has 'anthologized' – 'plucked', *apanthisamenoi* – from 'intellectual meadows' (*logikōn leimonōn* – 'textual fields') and historians (*sungrapheōn*) of antiquity, and he will 'try with a historical guiding line to make of them an organic whole (*sōmatopoiein*)

---

[43] Censorinus, *De die natali* 21. Scaliger, whose discussion of Eusebius was so important (Grafton [1983–93]; Grafton and Williams [2006]) was fascinated by this passage: Rood, Atack and Phillips (2020) 107. *Primum tempus*, he writes, *sive habuit initium si semper fuit, certe quot annorum sit, non potest comprehendi*, a blitheness unavailable to Christians: 'It is not possible to understand if the first time had a beginning or was always existent; certainly how many years it lasted.'

[44] Corke-Webster (2019) 83.

(1.1.4). The language here is deeply layered with the history of critical discourse. The 'path' – beaten or empty, narrow or broad – is an image for literary production that runs from Homer through Parmenides up to Callimachus' search for exclusivity. 'Plucking flowers' and 'anthologizing' joins the prose of Plato and the lyric verse of Pindar with the characteristic poetics of late antiquity, where anthologizing, collecting, selecting – the poetics of *spolia* – becomes a dominant literary strategy not just for the high tradition of Greek and Roman literature but also for Christianity, especially in the shape of Clement of Alexandria, a constant influence on Eusebius.[45] The historical 'guiding line' (*huphēgēsis*) is the term used by Plato for the lines on the page to help a budding writer (*Prot.* 326d), and, as we have already seen in Polybius, *sōmatopoiein* does not mean simply to 'embody' but to structure into an organic whole, the universal historian's task *par excellence*. In his very claim of near complete novelty, Eusebius' rhetoric embeds him intimately in a long tradition of methodological reflection.

This double assertion of newness and tradition is both an aesthetics *and* a politics. Eusebius is writing at a turning point in imperial history. The third century experienced a military, political and economic crisis, with multiple emperors vying for powers, and multiple experiments in power sharing collapsing into violence. The 'great persecution', inaugurated by the imperial edicts of 303 forbidding Christian worship and demanding conformity with traditional Roman religious practices, had been over-turned by the Edict of Milan in 313, and Licinius, the prime mover of the persecution, was removed in 324. Constantine's regime brought some stability and allowed Christianity to practice, but Christianity's institu-tional position remained not fully established. The complexities of Eusebius' own role in church and state politics, especially with Constantine himself, do not need rehearsing here, fascinating though the reception of the *History* through such a lens is, especially with regard to the more recent and necessary sea-change away from the sneers of a Gibbon ('less tinctured with credibility, and more practised in the arts of the courts, than almost any of his contemporaries') or Joseph Burckhardt's equally dismissive judgement ('the first historian of antiquity dishonest to the bone'), or even Overbeck's nasty 'stylist to the emperor's theological wig'.[46] Nor is it straightforward to determine how the bare outline of elements of third- and fourth-century history listed above constitute

[45] For a discussion of the significance of this language see Goldhill (2020) 71–113; Goldhill and Greensmith (2020). See also Elsner and Hernández Lobato eds. (2017), for the Latin side of things.
[46] Quotations from the very stimulating Corke-Webster (2019) 2.

a historical contextualization specifically for Eusebius, studying away in Caesarea. Indeed, even the dating and form of publication for the *History* is highly contested. Rather, this turning point is the condition of possibility for Eusebius' mix of new intellectual perspective and self-conscious awareness of a long intellectual tradition grounding it. As Corke-Webster concludes, Eusebius was 'not just writing history. He was world-making.'[47] A new history, then, for a new time: 'People learnt a new history because they acquired a new religion. Conversion meant literally the discovery of a new history from Adam and Eve to contemporary events.'[48]

Eusebius' history is driven by many small character sketches of Christian leaders and clerics, with the longest reserved for Origen. These figures are offered as exemplary, and Eusebius is explicit about the values by which they are to be judged. The Christian man 'through the knowledge and instruction of Christ is outstanding in self-control and justice; and in endurance of life and in manliness of virtue; and in the confession of piety to the one and only God above all' (1.4.7). These qualities are evident in the precursors from the Old Testament too, Abraham and the other founders of God's covenant. These values have a specific Christian commitment, of course – in the 'knowledge and instruction of Christ', the *paideusis* that Gregory emphasized too, and in the language of confession and the one God. But the qualities are also the most traditional of Greco-Roman values: *sophrosunē, dikaisosunē, karteria, andreia, aretē,* 'self-control', 'justice', 'endurance', 'manliness', 'virtue'. In Eusebius' vision, the Christian combines Roman and Christian values and excels in them all: Eusebius 'set out to argue that it was the Christians who best represented traditional Roman values and who were thus best placed to inherit Rome'.[49]

Eusebius was faced by many Christianities, and many possible routes through the past to get to where the church now was, in his view. Although Origen, his leading example of a Christian intellectual, was at the centre of arguments about zealotry and heresy – whether the story that he castrated himself because he read scripture's instructions literally is true or not, the fact that it circulated as an exemplary warning is telling – nonetheless Eusebius constructed a picture of the church that brought it closer to the emperor's court and further from the extremism of the increasingly evident desert ascetics. Throughout the *History*, Eusebius praised not just education but specifically Greek learning – a *paideia tēs tōn Hellēnōn philosophias*, 'an education in the philosophy of the Greeks' (7.32.6), especially for

[47] Corke-Webster (2019) 85.   [48] Momigliano (1977) 110.   [49] Corke-Webster (2019) 57.

bishops, which, as with Gregory and Basil, was both contributing to an internal argument among different Christian groups, and speaking to the wider Roman imperial culture. For Corke-Webster indeed 'responding to elite Greco-Roman prejudices about Christian status and education' and thus positively 'comparing the intellectual abilities of Christian and non-Christians' was 'the first – and most fundamental – aspect of his new vision of the Empire and its place in the Empire',[50] a place negotiated actively and dynamically in the cities of the empire.[51] It made his history a 'roll-call of effective academics'[52], a demonstration of the growing linkage between *paideia* and power.[53] As Grafton and Williams lay out with such elegance, Eusebius set himself in a line through Pamphilus to Origen, and constructed in Caesarea a library and a way of living in and through a scholarly book collection that changed the image and practice of Christian scholarship.[54] Eusebius combined this view of the Christian intellectual with a carefully nuanced picture of asceticism and martyrdom. Both commitments had deep roots in the practices of Greco-Roman society, and its philosophical schools; and while Eusebius praises the spiritual heights of asceticism, and praises the bravery of martyrs, he also resists more extreme forms of self-isolating behaviour. He celebrates the Christian family as a model for Roman society. In this way, Eusebius' *History* provides 'a carefully constructed moderate picture of the Christian past designed for its new fourth-century world'.[55]

Genealogy is crucial to how Eusebius shapes this moderate and well-ordered picture of the Christian Church – a picture which, it has often been noticed, designedly smooths over the rough intellectual and physical violence that continued to be endemic to the development of the church. Eusebius took part in the church councils that were the sign and symptom of such dissention; his *History* takes a clear line, however, which emphasizes the clerical succession lists and the transfer both of authority and of doctrine from generation and generation back to the unchallenged knowledge of the disciples. Heresies were minor splinter groups in this story, who relied on what Irenaeus, the master of heresies, programmatically terms 'useless genealogies', *genealogias mataias* (*Adv. her.* I pr. I). A heresy is a failure in the pedagogical transfer of authoritative knowledge from generation to generation. The connection between the *Chronicon*, with

---

[50] Corke-Webster (2019) 91–2.   [51] Eshleman (2012); Sandwell (2011).
[52] Corke-Webster (2019) 120.   [53] Brown (1992).
[54] Grafton and Williams (2006); on Caesarea, see also Levine (1975). The archaeology has advanced: see e.g. Patrich (2011).
[55] Corke-Webster (2019) 72.

its lists of events across time, and the narrative of clerical lists in the *History* is telling. In the *History*, 'the basic building block of time ... is the bishop holding office',[56] and the passing of time becomes meaningful as the passing of authority and authorized knowledge.

Even so, Eusebius cannot finally and absolutely conceal 'the fault line between the seemingly permanent edifice of the succession of the Church and the practical, inevitable plasticity and imperfection of the bishop lists'.[57] Nor can he repress the differences that constantly threatened the Church's harmony. Yet the way in which he tries to do so, itself signifies. In Book 5, Eusebius records an argument about the dating of Passover and its implications for when Christians should end their ritual fast of Easter (arguing about the date of Easter would not stop for centuries).[58] A flurry of letters is taken as the source evidence for the disagreements ('A new chapter of historiography begins with Eusebius not only because he invented ecclesiastical history, but because he wrote it with a documentation which is utterly different from that of the pagan historians').[59] Epistolary exchanges exhibit the network that links the Christian community as a community, just as synods, for Eusebius, are not public displays of difference but rather, in their decisions, demonstrate how (5.23.2) 'all, of one mind, through letters between every place, regulated by agreements an ecclesiastical decision (*dogma*)', and all 'declared the same opinion and judgement and cast the same vote' (a view that scarcely captures the bitter infighting and long-lasting conflict that the synods failed to resolve). The word translated 'regulated by agreement' is *diatupoun*, which by basic etymology means 'to put a thorough stamp on', 'to mould perfectly', and is used for a range of significant or authoritative utterance, often hierarchically (*anōthen* 'from on high'), including, as in John Chrysostom, God's commands to man. Eusebius finally quotes the end of the letter of the Palestinian synod, which in his eyes closes the matter. The bishops speak, he writes, with the authority of the apostolic succession (*diadochē*), giving the genealogy of authoritative knowledge that is required (5.25.1):

> We are making it clear to you that on this day in Alexandria also they keep Passover on the very same day (τῇ αὐτῇ ἡμέρᾳ) as us. For letters are brought

---

[56] Johnson (2019) 207. [57] Johnson (2019) 198.

[58] Mosshammer (2008); Beckwith (1996) 51–70. John Chrysostom *Adversus Iudaeos* 4 is a good example of such arguments entering wider polemics.

[59] Momigliano (1977) 118. A view not contradicted by Becker (2016), (2017).

from us to them and them to us so that we keep the holy day harmoniously and at the same time (συμφώνως καὶ ὁμοῦ).

To celebrate 'on the same day' is made possible by the exchange of letters – enacting the community as a community – and the aim is harmony, to speak in one voice – *sumphōnōs* – which means to celebrate 'at the same time' (*homou*). Synchronicity is the sign of a single, harmonious community.

This wish for the one self-same day is at the heart of Jewish liturgical hope too. Eusebius is concerned with ordering a succession (bishops, knowledge, authority) to transcend contentiousness. The *aleinu* prayer, already attested in late antiquity, and still part of daily services, shows the universalizing desire that lies behind Eusebius' pragmatics.[60] The *aleinu* imagines the future day when all the world will be perfected under the rule of God, 'when all humanity will call on Your name'. This will be a time of eternity – 'You will reign over them soon and for ever.' For, 'on that day, the Lord will be One, His name One'. For both Jews and Christians, the hope is for a single regime of a single God, a single day when the differences between peoples will be erased, and time will be an undifferentiated eternity: the idealism that makes synchronicity a non-issue.

Eusebius' *History* is world-making in that it sets out to tell a history that synchronizes the Roman imperial state and the Christian Church through its exemplary clerics and their work. This is not a history of triumph or takeover so much as a narrative of integrating different orders of time, place and power. It imagines a world where Christian time makes sense to the Roman empire: the history of the church and the history of empire can rhyme.

Here, then, we see how the historiographical question of what is at stake in 'at the same time' has fully taken on a new theological framing. Synchronicity becomes an issue – a problem or a potential – only when there are different regimes, different perspectives, different possibilities of power, different histories in play. Christianity building on the Jewish prophets, makes an ideal out of a single regime of God's kingdom, and a single theological present, where the son and the father are co-eternal, typology collapses past and present in a single model, and the threat of change or conflict is absent, as the lion and the lamb lie down together. Now and for evermore. Christianity, however, takes shape and has a history of itself to write only when there are different regimes, different

---

[60] Historical evidence in Reif (1993) 208–9.

possibilities of power, different perspectives visibly and dynamically present. Eusebius' intellectual work between the *Chronicon* and the *History* is formed in this tension between hope and pragmatics.

Eusebius gives a crucial nexus of questions, thus, for this book, which will become most pressing when we turn from Nonnus' flamboyant stories of the beginning of things to the Christian histories of the world discussed in chapter 15. How is the self to be placed in historical time? How, that is, is the self to be comprehended and represented in this new Christian history, when Christianity is now becoming no longer an illicit religion of alienated resistance, and when Paul's promise of the imminent end of things is continuously delayed? The hermeneutic thrust of typology – to make present and past and future models of each other in a single theological normativity –finds a historiographical equivalent in the hope of a single narrative, an integrated story of the world, from the beginning to the end of days. As typology encodes a violence, so too this is an authoritarian projection. But subtending it, demanding it, is the deep anxiety of a lasting question: can the Christian live a Christian and a Roman or Greek life *at the same time*?

CHAPTER 8

Timelessness and the Now

The messianic religions that came to dominate this lived life of late antiquity made *waiting* central to their sense of temporality, as we have seen. As the poets of erotics have always known, there is certain headiness in the combination of fervour and deferral. Waiting, however, structures the sense of the present – the now – with a question of its value, its temporariness. 'Who would deny that the present has no duration?', asked Augustine. In the nineteenth century, William James tried to answer this anxiety about the duration and thus evaluation of the 'nowness' of the now with an empirical, experimentally tested answer: 'the practically cognized present is no knife-edge', he concluded, 'but a saddle-back, with a certain breadth of its own on which we sit perched and from which we look in two directions into time'.[1] It was possible to count in seconds, and then in fractions of seconds, a human experience of now, a breadth measured 'from one hundreth of a second to twelve seconds'.[2] James was engaging with a host of neurological scientists who were exploring the space between sensation and its mental recognition; but this was also part of a moral argument, with deep roots back to Plato and Augustine in particular, about 'momentary pleasure' in contrast with a timeless ideal.[3] Must waiting inevitably and always turn the now into empty time? Can 'the now' be reinvested with value?

Modernity – or, to be precise, the self-proclamation of modernity – has indeed repeatedly tried to rediscover the now, the 'tyranny of the moment'.[4] The self-help gurus who currently encourage their readers to 'live in the moment', are a pale echo of a passionate movement, for which

---

[1] James (1890) I: 609.   [2] Zemka (2011) 209.
[3] See Zemka (2011) building on Dames (2007). Daston and Galison (2007) is a crucial overview here.
[4] Hartog (2011) 217.

William James is an iconic scientific authority. D. H. Lawrence provides a characteristically (over-)heated statement of principle:

> Give me nothing fixed, set, static. Don't give me the infinite or the eternal: nothing of infinity, nothing of eternity. Give me the still, white seething, the incandescence and the coldness of the incarnate moment: the moment, the quick of all change and haste and opposition: the moment, the immediate present, the Now.[5]

For Lawrence, the moment, the Now – the capital letter marks the intensity of his striving for a general point here – is full of potential: it is still and seething, burning and cold, and it is what gives burgeoning life ('the quick') to speed, conflict, alteration, and this is in contrast to eternity and infinity which are unmoving, in all senses, without emotion, without transition: 'the moment that remains and blows up the continuum'.[6] Lawrence rejects what is 'fixed, set, static'. The religiosity of this incantatory rhetoric is evident, not just in the 'incarnate moment' – the use of incarnate rather than embodied, say, is a pointed rejection of the Christian dismay at the fleshliness of life – but also in the very opposition of eternity and infinity to the immediate present, again redrawn to dismiss the standard Christian tenet of focusing on eternal life as the goal. Lawrence establishes the Now as the rejoinder to such idealism:

> The ideal – what is the ideal? A figment. An abstraction. A static abstraction, abstracted from life. It is a figment of the before or the after. It is a crystalized aspiration, or crystalized remembrance: crystalized, set, finished. It is a thing set apart, in the great storehouse of eternity, the storehouse of finished things.[7]

The ideal is something invented for a time before or a time after – rather than the Now. In language that recalls Augustine's topography of memory, the ideal exists only as hope or as a 'crystallized remembrance ... in the great storehouse of eternity'. Unlike the presentness of the now, the ideal is always lost in a past or future. So much for Plato.

It may be something of a surprise that in this paean to the Now, despite his familiar turn to raw and bloody nature (and, in his final poetry, even to the gods of Olympus), Lawrence echoes most intently Walter Pater, an 'illusive, inscrutable, mistakable self' of a writer,[8] and, in particular, Pater's

---

[5] Lawrence (1994) 616. This passage is discussed in Dillon (2007).   [6] Nowotny (1994) 152.
[7] Lawrence (1994) 618.
[8] Hext (2013) 184. On Lawrence and Ruskin, see Landow (1985). On Pater and classics, see Martindale, Evangelista and Prettejohn eds. (2017), and especially Porter's chapter in it.

infamous final section of *The Renaissance: Studies in Art and Poetry*. (Lawrence railed at the 'deadly Victorians', but read Pater and Ruskin avidly.) In these paragraphs, Pater lauds the 'exquisite intervals' of physical life, 'the moment' of joy. He revels in the sensuality of his body's sensations and the flow of thought and feeling. Each of these sensations, he recognizes, is 'limited by time', 'a single moment, gone while we try to apprehend it'. These vanishing moments are what make the self a shifting, changing, sensible being: 'It is with this movement, with the passage and dissolution of impressions, images, sensations, that analysis leaves off – that continual vanishing away, that strange, perpetual, weaving and unweaving of ourselves.' Pater offers a profoundly challenging image of the unstable subject, like Penelope at the loom, weaving and unweaving a self, a 'continual vanishing away', without even the closure of the *Odyssey's* narrative. How, then, to live?

> How shall we pass most swiftly from point to point, and be present always at the focus where the greatest number of vital forces unite in their purest energy?
>   To burn always with this hard, gem-like flame, to maintain this ecstasy, is success in life.[9]

To many of his readers, especially those from a normative Christian commitment, Pater seemed here to be scandalously proposing an amoral life that privileged sensation over duty, ethics, or even a career: the moment without consequence. Pater, deeply upset by the reaction, censored the paragraphs from the second edition, thus ensuring their continuing celebrity and influence, especially in modernism's search for epiphanic 'moments of being'.[10] He still has Marius, his hero in *Marius the Epicurean*, declare that 'the little point of the moment alone really is'.[11] No stranger bedfellows than Lawrence and Pater, but both reveal not merely how time is moralized, especially in a capitalist system where time is money, but also, and more specifically, how the conceptualization of the here and now becomes an ethics of existence – how to live one's life with an eye on the passing of time.[12] Not merely: can the now be fully enjoyed without a sense of consequence, as Pater was accused of promoting? But also: can the now be comprehended *without* the idealism of eternity?

---

[9] Pater (1910) 235–6.
[10] The phrase is Virginia Woolf's (Woolf (1939/1986) 70). See Zemka (2011) for its context, and a fine history of the momentary.
[11] Pater (1900) 143.
[12] On capitalism's moralization of time, see from a large bibliography Thompson (1967); Schivelbusch (1986); Chakrabarty (2000) 47–113; Jameson (2003); Zemka (2011).

Pater, 'the patriarch of aestheticism's epiphanic figures',[13] is important to Lawrence, but we could add Kierkegaard, whose modernity George Pattison defines as 'the culture of those whose horizons are completely filled by "the-time-that-now-is", the momentary, the shock of the new ',[14] or Nietzsche (in some moods) – 'He who cannot sink down on the threshold of the moment and forget all the past, who cannot stand balanced like a goddess of victory without growing dizzy and afraid, will never know what happiness is',[15] and others, back to Schleiermacher, a font as so often: 'it is an infinity of past and future that we wish to see in the moment (*Augenblick*) of utterance'.[16] We shall recall this visual language of epiphanic ecstasy when we turn in chapter 10 to the sudden, the momentary, the rupture in time, and its religious instantiation in the leap of faith.

Pater's *joy* in the now, however, was anticipated with a symmetrical *anguish* in Thomas De Quincey. For De Quincey, who read Augustine's *Confessions*, of course, before writing his *Confessions of an Opium Eater*, the intensity of 'the now' led down into the pits of despair. To ask how long the now lasts, De Quincey reflects, leads to an inescapable paradox, a paradox outlined most vividly in Augustine's *Confessions*. De Quincey uses the image of the *klepsydra*, the ancient water-clock, to try to capture the passing of time, and imagines a drop of water squeezing through the funnel: 'You see, therefore, how narrow, how incalculably narrow, is the true and actual present.'[17] It is barely there, but that it passes away. 'Yet,' he adds, 'even this approximation to the truth is *infinitely* false.' The solitary drop which represents the present can itself be broken into tinier and tinier droplets. 'Therefore the present, which only man possesses, offers less capacity for his footing than the slenderest film that ever spider twisted from her womb.' The idea of the present, all we have, is no basis to ground the self: our footing is slipping above the abyss of time. Thus, concludes De Quincey, turning, as ever, his misery into stunning phrases, 'All is finite in the present; and even that finite is infinite in its velocity of flight towards death.' For De Quincey, to seek the finite present is to end up in the infinity of its divisibility, and to view such instability as a sign of the rapid human journey towards the end of death. Infinity and eternity are not as easily escaped as Lawrence's exclamations insist. The paradox of 'the now' for De Quincey, following Augustine, is that it must and cannot be inhabited. The contrast with Pascal here is particularly striking: for Pascal too 'the present is the only time that is truly our own'; the past

---

[13] Zemka (2011) 220.   [14] Pattison (2002) 19.   [15] Nietzsche (1997) 62.
[16] Schleiermacher (1998) 23.   [17] De Quincey (1998) 159.

and future are beyond our control. But Pascal concludes from this that we ought to live 'according to the will of God' and seek 'repose'.[18] For De Quincey in this *Suspiria de profundis*, 'Sighs from the Depths', the trauma of reflecting on time's Now leads not to the wild joy of Lawrence or the ecstasy of Pater, nor to 'repose', but to melancholia and the horrors of his deepest, most lasting memories – and a recognition of how slender the films that support the self are.[19]

For Martin Heidegger, the positivity of Lawrence or Pater and the negativity of De Quincey about the experience of the now are equally the regrettable heirs of Aristotle, whether any of them read Aristotle or not.

> Ever since Aristotle all discussions of the concept of time have clung *in principle* to the Aristotelian definitions ... Time is what is 'counted', that is to say, it is what is expressed and what we have in view, even if unthematically, when the *travelling* pointer (or the shadow) is made present. When one makes present that which is moved in its movement, one says 'now here, now here and so on'. The 'nows' are what get counted. And these show themselves in every 'now' as 'nows' which will 'forthwith be no-longer-now' and nows which have 'just been not-yet-now'. The world-time which is 'sighted' in this manner in the use of clocks, we call the '*now-time*' [*Jetzt-Zeit*].[20]

Now is a moment, and time – 'now-time' – is made up of a series of nows that can be quantified. Heidegger wants to find another way to understand time and being, another way of being in the world, without such service to clock-time, its travelling hands or sundial shadows.[21] For him – and you can hear the cry of Lawrence behind this – 'Temporality ensnares itself in the Present, which, in making present says pre-eminently, "Now!", "Now!".'

For psychologists, philosophers and poets what it means to cry out 'Now! Now!' – can the now last? Can the now be invested with meaning? Is the now the only place of meaning? What is 'being in the world'? – has become, then, a defining question of modernity, the now of our time. To be modern is to be self-conscious about the nowness of the now – and thus needs such explication, brief though it is, as the place from where this discussion takes shape. The discipline of anthropology, however, has given another, critical, painful perspective on what we might thus call the ethics of the present tense – which demands of us a further level of scholarly self-consciousness.

[18] Pascal (1963) 270, given a context by Prendergast (2019) 141–2.   [19] See de Maniquis (1985).
[20] Heidegger (1962) 473–4; a much-discussed passage; for our purposes, see Kennedy (2013) 139–52.
[21] For a stimulating attempt to escape the now, based on a wonderfully lucid explanation of its difficulty, see Garcia (2014) 177–88.

Johannes Fabian's agenda-setting book *Time and the Other: How Anthropology Makes its Object* offers a pointedly different, politicized analysis of the use and abuse of 'the now', which will lead us back slowly into the texts of antiquity. Written in the 1980s at the height of a crisis of self-criticism that characterized the anthropology of these years, Fabian's attack on ethnography trenchantly exposes its 'moral complicity', 'ideological and even epistemological',[22] with a distanced objectification and thus exploitation of other communities. Specifically, and with bravura rhetoric, Fabian anatomizes the tradition of the 'ethnographic present', the depiction in the present tense of the culture of Others. He calls this language 'the denial of coevalness'.[23] The Other does not inhabit the same 'now' as the observers. The ethnographic present tense is used 'for the purposes of distancing those who are observed from the Time of the observers'.[24] Such discourse establishes the temporality of the West in a privileged position with regard to the 'primitive', 'savage', 'ritualized', 'ancient', 'unchanging' time of Others. In this way, 'the anthropology of Time becomes the politics of Time', a strategy of colonizing self-definition in a historical period of political decolonization.[25] In the period of such modern political dominance, which is the era of anthropology too, imperialism requires 'Time to accommodate the schemes of one-way history: progress, development, modernity ... In short, *geopolitics* has its ideological foundations in *chronopolitics*.'[26]

Like other critiques from this era, Fabian's analysis is set in a history of approaches to time in anthropology.[27] Two main lines of this history can be articulated (with many interstitial positions and differences of detail between the dozens of significant theoretical expositions). The beginnings of anthropology, first of all, are closely intertwined with stadial and then evolutionary historical theories that argued that all human societies develop according to the same schema, beginning in savagery and gradually achieving the heights of civilization, represented by the Western imperial powers. The discovery of communities of 'undeveloped' peoples allowed a vision of the childhood of all human culture, and also justified attempts to raise their levels of civilization through missionary work and imperial improvement. A lack of coevalness is integral to such theorizing, and to the politics with which it is enmeshed: the tribes of anthropological enquiry simply did not inhabit the same moment in the time-scheme of human

---

[22] Fabian (1983) 96.   [23] Fabian (1983) 31.   [24] Fabian (1983) 26.
[25] Fabian (1983) 51. For critiques of Fabian's periodization of secular modernity, see the introduction to Davis (2008), Harstrup (1990) and, generally, the stimulating Chakrabarty (2000).
[26] Fabian (1983) 144.   [27] Gell (1992).

development as the anthropologists. It is also fascinating that many of the founders of the discipline of anthropology turn out to have stridently evangelical parents, as Tim Larsen has detailed, and the consequent interface between eschatological religious commitments, secularized history, and personal intergenerational strife creates an extraordinarily complex environment for research.[28] This history of the discipline inscribes the different time of its objects of research at its core.

Secondly, Durkheim's sociology of time – the analysis of 'the time common to the group, a social time'[29] – is foundational for a long history of anthropological projects that record how different societies have their own sense of time, from Gurvitch, Zerubavel, Schutz, Geertz to Lévi-Strauss, Leach and so on.[30] (It has often been noted that the English translation of Durkheim appeared the same year as Einstein's seminal papers (1915), though without noting the irony that synchronicity was the very problem.) Modernity's time-discipline, where the factory's 'machinery requires the kind of mentality that concentrates on the present and can dispense with memory and straying imagination' makes industrial society an alienated example of its own theorizing.[31] For Fabian, both functionalism and structuralism 'put on ice the problem of Time', because both involve 'a freezing of the time frame'.[32] As Lévi-Strauss states, his aim is to reveal 'a system that is synchronically intelligible',[33] a system, for example, that wilfully represses the different times in which different versions of mythic narratives take shape, as classicists have often complained about his analysis of the Oedipus myth. Lévi-Strauss generalized about 'cold' and 'hot' societies, where 'cold' societies did not possess the complex sense of historical change and self-consciousness about history that distinguished the societies from which anthropologists came. Even Clifford Geertz, in his celebrated discussion of time in Bali, described an intersubjective sense of time that allowed social interactions to take place 'in a motionless present'.[34] For Fabian, the theoretical turn in anthropology towards synchronicity, with its concomitant dismissal of history, distorts the necessary analysis of 'social time'. In fieldwork, anthropologists engage in dialogic interaction with other communities and participate in their now; but in writing ethnography, the 'you' becomes a 'they', and this

---

[28] Larsen (2014).   [29] Durkheim (1915) 11.
[30] Gell (1992) has been particularly influential in writing this history; see also Adam (1990); Nowotny (1992). The references are to Gurvitch (1961); Schutz (1962); Geertz (1973); Lévi-Strauss (1963); Leach (1961); Zerubavel (1981).
[31] Quotation from Horkheimer (1994) 22; 'Time-discipline' is taken from Thompson (1967).
[32] Fabian (1983) 20.   [33] Lévi-Strauss (1963) 216.   [34] Geertz (1973) 404.

'specific cognitive stance towards its object' creates for these other societies another time.[35] The ethnographic present, for Fabian, is a grammatical process of objectification.

Now, Herodotus' history has been regularly enlisted as a precursor, even a founder of anthropology since the first reflections of the discipline on its own development. About the same time as Fabian also rehearses this unreflective genealogical cliché, François Hartog and James Redfield, directly in engagement with the discipline of anthropology, and in Redfield's case, his father's contribution to the field, were demonstrating the complex, self-aware and world-building constructions of the other in Herodotus, the sophistication of which discredits the more naive and patronizing appropriations of Herodotus as the father of anthropology (as well as of history and lies).[36] Herodotus' present tense is fundamental to his rhetoric. On the one hand, his description of Egypt, say, takes place in a timeless present, for all that it recognizes the profound difference between the time-scales of Greece and Egypt.[37] Greece and Persia are countries with histories of change: change of constitutions and the shifting fortunes of grandeur and humiliation are structuring principles of the history for these protagonists. But the opening four books, with their descriptions of other communities, are primarily constructed through the observer's eye as a synchronic picture, as if Herodotus was looking as a scientist at the unchanging objects of natural history.[38] On the other hand, Herodotus repeatedly uses *legetai*, 'it is said' (and other such phrases) to mobilize the various stories that make up the swirl of events or the comprehension of a phenomenon. Rumours, lies, theories from different sources or eras are all viewed as making up the event, or the framing of a phenomenon. The present is where such stories come together; the present is formed in and by the circulation of such historically layered but now contemporary narratives. Herodotus' grammar performs the effects of the past on the present.

The plupast – the time before the past of the war, which remains the explicit subject of the history – is manipulated in multiple ways both by the characters who speak in Herodotus, and by the narrator's account of things, to construct persuasive normative paradigms;[39] the future, too, is a brooding element in the history's rhetoric of exemplarity, as imperial Athens emerges as the potential subject of a warning tale of excess and

---

[35] Fabian (1983) 86.   [36] Hartog (1988); Redfield (1985).   [37] Vasunia (2001) 133–5.
[38] Thomas (2000); more generally, Daston and Galison (2007).
[39] Grethlein and Krebs eds. (2012); Baragwanath and de Bakker eds. (2012); Rood, Atack and Phillips (2020) 119–43.

collapse.[40] But Herodotus' utilization of the present tense shows how what has often been criticized as his indiscriminate repetition of unsubstantiated or simply false anecdotes or observations is a strategic element in his mapping of the cultural forces he sees as necessary to understand the war, now.

If the present tense marks Herodotus' performance as a scientific observer, the present tense also marks the scene of performance in Homer – to go right back to the beginnings, again – through the moment of apostrophe or invocation. That Homer's heroes live in a past lost to now is explicit: 'ten men of today could not lift the rock that Hector lifted with ease'. We should recall how the failure of memorialization is troped by the washing away of walls by the waves, much as Achilles' singing of the *klea andrōn* foreshadows the epic's own promise of eternal glory. When Homer invokes the Muse at the beginning of each epic, 'sing, Muse', and even more strikingly when he addresses the divinity in a hymn, 'Come here, appear …' – and all the Homeric Hymns have such ritual apostrophes – he inhabits his own scene of performance, of course, but more importantly he invokes the god into presence. The hymn welcomes the god into the sacred space of performance and bids farewell at its close. This performed 'control over absence and presence'[41] images the scene of epiphany. Sappho's prayer to Aphrodite most vividly dramatizes such epiphany in the stunning counter-address of the poet-lover by the goddess – 'Who, Sappho, is wronging you?' – an exchange of names to match the dangerous exchange of roles promised in the poem's narrative of desire ('if she flees, she soon will chase'). The hymnic invocation, however, also projects the circle of an audience, participants in the ritual moment. This projection of an audience – a community sharing in the ritual – is fundamental also to Pindar's poetics of praise, which, ever aware of the dangers of the victor's success, is careful with the boundaries of a community's welcome of the hero's return.[42] This sense of bringing the addressed hero into the community of the performance may also help us appreciate the strange moments when in the epic narratives of the *Iliad* and the *Odyssey* Homer apostrophizes his heroes in the second person. Menelaus and Patroclus in the *Iliad* and Eumaeus in the *Odyssey* are addressed by name. For Patroclus in particular this invocation occurs when he appears to be at mortal risk and the poet, like Achilles, reaches out in sympathy and care towards him in the second person.[43] Such structures of sympathy seem harder

[40] Redfield (1985); Raaflaub (1987); Moles (1996); Fowler (2003); Buxton (2012).
[41] Bergren (1982) 90.   [42] Kurke (1991).
[43] Bergren (1982); Block (1982); Allen-Hornblower (n.d.); Klooster (2013).

to maintain for Eumaeus or Menelaus even. But in all cases, the second-person address pulls the narration into the scene of the present performance, and projects, as it were, a ritual circle of shared song in which the hero is present, like the epiphanic god of cult. The 'as it were' is crucial here. The desire for presence in both lyric and epic poetry is mediated by the poet's projection. Future performances, like the past of the *klea andrōn*, already haunt the here and now. It is hard simply to live in the present.

Or to appreciate fully what the apparently self-evident claim to live in the now means. From Homer to Heidegger, as we might say, to insist on 'the now' turns out to be an ideologically fraught moment of self-positioning. Against the now of performance, the now of the scientific gaze, the now of the moment of battle glory, the now of immediate sensation, and, worst, the now of momentary pleasure, Christian thinkers, looking back to Plato, strive to create a different sense of the present, turned towards a timeless ideal (as Lawrence knew well and so resented). Perhaps now we can begin to see more clearly what the 'now' of waiting demands of us, in its requirement to turn ourselves towards a future; to make the performance of the now such a turning, such a conversion.

*

The present time, 'which only we have', may nonetheless not be the time we wish to live in. In Hesiod's hard world, the five ages of man move from the golden age to the iron age, which is ours. 'Would that I had been born earlier or later', laments Hesiod (*W&D* 175), since in today's world 'men never rest from work and sorrow during the day, and being destroyed at night' (176–8). And it will get worse before it is over. Humans will be born grey, intergenerational strife and disrespect will be rampant, legal and moral order will collapse, justice will be no more than force, and the wicked will harm the good (179–200). 'There will be no defence against evil' (201). The horror of the present is not mitigated by the thought of the future or the lost glory of another, better time. 'Better', indeed, 'not to have been born', as Greek wisdom depressingly repeats.

Plato's response to what he saw as this instability of the mundane world – physical, moral, epistemological – is to turn away from it towards the timeless eternity of the Forms. In the *Timaeus*, a dialogue which had a deep and long-lasting influence on the development of Neo-Platonism and the reception of Platonic idealism by Jewish and Christian writers[44] –

---

[44] Reydams-Schils ed. (2003); Runia (1986). Proclus' commentary on *Timaeus* is one of the most significant contributions to fifth-century thinking on time.

indeed it is 'the most seminal philosophical or scientific text to emerge from the whole of antiquity'[45] – Timaeus begins his account of the genesis of the world with a founding distinction between, on the one hand, what is 'ever being', 'continually existent', (*to on aei*, 27c–d), 'has no coming into being' (*genesis*), and, on the other, what is 'coming into being' (*gignomenon*) but is 'never existent' (*oudepote on*). That which is existent is always the same and is apprehended by the mind; but that which comes to be and is destroyed (*gignomenon kai apollumenon*) is perceived by the unreasoning senses: it is not 'really existent', *ontōs on*, its material existence is not reality. It is a hierarchical opposition. When a creator wishes to make something, continues Timaeus, if he looks at a model (*paradeigma*) which is in a state of sameness (*to kata tauta echon*), the outcome will be fine (*kalon*); if he looks at something which has come into being and thus uses a created model, the created object will not be fine. So, is the universe, which, as it is made, must be an image (*eikōn*) of something, created according to a generated or a self-same, ungenerated model? The arguments they use to broach such a question, he adds with gentle irony, must be like the objects themselves, 'stable, firm' – unchanging in time.

So, how did God – the Demiurge – create the cosmos? The universe is conceived by the mind of God as a living creature with a soul; a single, unified, perfected animal (*zōon*). Timaeus spends many paragraphs showing that the universe must be single, and that it must be perfected (*teleon*), to which he adds 'unageing' (*agērōn*, 33a) and 'without sickness' (*anoson*), adjectives traditionally applied to divinities. Timaeus takes even more paragraphs of dense mathematical exegesis explaining how God created a third form of being, a mixture of indivisible being and divisible being in order to allow for the ordered movement of the cosmos.[46] It is his far from easy conclusion that is crucial here, however (37c–d):

> When the Father that engendered it, recognized it in motion and alive, an object of joy to the eternal gods, he rejoiced, and in his happiness decided to complete it all the more like its model. Now, the nature of the ideal being was everlasting (*aiōnios*), but to bestow this quality in its fullness upon a living thing was impossible. But he resolved to have a moving image of eternity (*aiōnos*), and when he set in order the heavens, he made this image eternal (*aiōnion*) but moving according to number, while eternity (*aiōnos*) itself is stable in unity; and this image we have named time (*chronon*).

It is, first of all, evident how such language attracted later Jewish and Christian thinkers. This is an account of genesis with a Father who

[45] Sedley (2007) 96.   [46] Lloyd (1968).

engendered (*gennēsas*) the universe, who rejoiced at his creation (as in Genesis, 'he saw it was good', *kalon*), and who brings order to chaos (the starting point of Genesis). Plato, alone of Greek philosophers, it seems, does not take for granted the impossibility of *creatio ex nihilo*, the requirement of Jewish and Christian narratives of creation. Philo, and later Christian writers too, take up the idea of the creator using a model, an archetype, or an image in mind, to create the world. Justin Martyr is programmatic in claiming the *Timaeus* as a proof that Plato was following a Mosaic cosmology.[47] The problem Plato faces, however, is how to have a physical, moving universe, if it is to be an image of a stable, unmoving ideal. He mediates this tension by invoking order and number, an eternal, regular movement. This he gives the name Time. Time, for Plato, requires such motion; there is no time before the heavens with their regular moving parts come into being. 'It is only when the regular motion of the heavenly bodies comes into being that time exists'.[48] Alongside the creation of the heavens, were created the divisions of time: days, nights, months, years, which the heavenly bodies demarcate. Aristotle, in turn and in response, as we saw, goes on to define time as 'the number of motion with regard to the before and after', but not before, as we have seen, he wonders whether time exists, which Plato's birth story takes for granted (29a–b);[49] and, against Plato, Aristotle will go on to argue that time has no beginning: it is eternal, infinite. In the *Timaeus*, however – and it is a profoundly disconcerting claim – Plato's story of the genesis of the universe is the story of the birth of Time (*chronou genesis*, 39e).

It is disconcerting not least because if there was the motion of disorder before the creation of the heavenly bodies, as Timaeus allows, how can there *not* have been time, in this before? How can motion be conceived without time? David Sedley, who here follows Gregory Vlastos – both deeply uncomfortable with the very idea of the 'birth of time' – argues that for Plato before the creation of heaven there is actually a sort of unmeasured, chaotic time.[50] This is a 'compromise' (Vlastos' term) that cannot quite remove its own inconsistencies. Such a 'before' is 'disorderly but not altogether': 'it is not utterly disordered change. Wholly devoid of form it would be, on Platonic standards, wholly devoid of Being; i.e. nothing at all. But obviously it is not that. It is something.'[51] Thus, concludes Vlastos, the birth of measured time 'is not the contrary of timeless eternity, but an approximation to it': it has come to be but is now regular and eternal. The

---

[47] *First Apology* 60.   [48] Vlastos (1939) 75.   [49] Annas (1975) 97.   [50] Sedley (2007) 95–132.
[51] Vlastos (1939) 76.

work philosophers have to undertake to find a precarious coherence to Plato's thinking here is telling. Plato, indeed, immediately indicates the continuing difficulty of finding expression for his understanding of this timeless eternity. 'Was' and 'will be', he underlines, are unconsciously but wrongly applied to eternal being (*aidios ousia*). Only *esti*, 'it is' or 'it exists' is properly applicable to being. Only the present tense . . . The language we use of time cannot capture the nature of timeless being, Plato writes, but he quickly and ironically cuts short the discussion: now might not be the *kairos*, the right time, for clarifying such matters (38b). As Augustine too will find, the everyday language of time is a barrier to the comprehension of time.

Nonetheless, this is the very language Timaeus immediately uses. The model (*paradeigma*) is existent for all time (*panta aiōna*); the universe, its copy, has come into being (*gegonos*) and is (*on*) and will be (*esomenon*) for all time (*ton panta chronon*). (The asymmetry of the world coming into being but being eternal – no passing away, as the Stoics demanded – seemed impossible to the point of craziness to many later philosophical writers.) Plato concludes firmly, however: 'such was God's thinking with regard to the birth of time (*genesis chronou*), so that time should be born (*chronos gennēthēi*), with the result that the sun, moon, stars and planets were set in the sky in regular circulation.' That was what was required for the collaborative construction of Time (*sunapoergazesthai chronon*). The creation of the universe and the creation of time are coterminous; the existence of the so-called heavenly bodies in the heavens gives us regulated time, for all time. Plato's commitment to the stable, permanent, timeless model cannot do without time, and thus the possibility of change, which is nonetheless controlled by the regularity and order of number as embodied in the motion of the sun and moon and stars, now eternally in place.

Timaeus offers the grandest scale of Plato's idealism and its turn to timelessness as a constant and stable present. The theory of Forms supposes that any example of a table is recognizable as a table because it partakes of an abstract, unchanging, timeless model of a table, its Form (*eidos*). The *Timaeus* itself continues into a supremely complex argument about the nature of materiality and space, and the connection between Mind and Necessity; but its difficult, strained discussion of the 'birth of Time', with its repeated striving for eternity, underlines the degree to which the timeless present is integral to Plato's idealism. The universe itself is created according to a timeless model; so too are the mundane realities of the everyday. To contemplate the Good itself – one ideal that Socrates, at a party, imagines, before he is interrupted by the very physical presence of the drunken Alcibiades – is to turn away from this material world with its instabilities and to focus on what is truly stable and truly real: to find what is the timeless.

Cicero, central to the mediation of Greek philosophy to Roman culture, seems to respond directly to these passages of the *Timaeus*. In his *On the Nature of the Gods*, his Epicurean speaker, Velleius, engages with the problem of the time before the creation of the heavenly bodies, and the intervention of divinity, which, for Plato and for the Stoics, is the necessary creative change for measured time to come into being. He writes in direct criticism of Plato (1.21):

> For if there was no world, it does not follow that there were no centuries (*saecla*). By 'centuries' here I don't mean the ones made up by the number of days and nights (*dierum noctiumque numero*) as a result of the annual orbits (*annuis cursibus*). These, I concede, could not have been produced without the world's rotation (*conversione*). But there has existed a certain eternity from an infinite time past (*ab infinito tempore*), which no bounding of times (*circumscriptio temporum*) measured, but in its extent (*spatio*) it can be understood what sort of thing it has been, because it is not even thinkable that there existed a time (*tempus*) at which time (*tempus*) was not.

Velleius, the speaker, admits that without the heavenly bodies there would be no measured time, no regulation, no order. But, he argues, there must still have been unmeasured time before this creation (as Sedley and Vlastos argued was the implication of Plato's argument for the birth of time in the *Timaeus*) – though the language of *saecla*, however redefined, makes it hard not to hear a trace of measurement. Eternity is a time without bounds, without delimitation (*conscriptio*) which has infinite extension into the past. We can understand what sort of thing this eternity is, its extent, because, Velleius' argument runs, it is impossible even to conceive of a time when time does not exist. Velleius does not simply assert the necessity of the infinity of time as an Epicurean principle, but insists – tautologically? – that the attempt to imagine unmeasured, formless time is not possible because it is not possible for human consciousness (*cogitatio*) to imagine a lack of time. Humans cannot get out of time, even in the imagination. We can try, haltingly or intently, to imagine infinity or eternity or God's being beyond and outside time, but we cannot in our minds inhabit no time.

Lucretius has no truck with Plato's longing to escape materiality. His is a resolutely material universe, grounded in his reading and promoting of

Epicurus' atomistic understanding of nature. Time in this vision, Lucretius asserts directly, does not exist in itself (1.459):

> tempus item per se non est, sed rebus ab ipsis
> consequitur sensus, transactum quid sit in aevo,
> tum quae res instat, quid porro deinde sequatur.
> nec per se quemquam tempus sentire fatendumst
> semotum ab rerum motu placidaque quiete.

> Even Time in and of itself does not exist, but from events themselves
> Sense follows what happened in the past;
> Then what event is insistent, what then after follows.
> No person, it must be agreed, senses time in and of itself,
> Separate from the movement and the calm resting of things.

Time, although it is not a quality like colour or size, does not exist except when human sense perception recognizes it through events that have happened, are happening or are anticipated; it is an accident produced by the impression of movement or lack of movement of things. It would be hard to construct a view more trenchantly opposed to the notion of Providence.

Lucretius' language here strains against itself, as with so many discussions of time. *Aevum* 'age', 'era', might be thought to imply that time does stretch backwards in a significant way. Lucretius continues, however, immediately to worry away at this idea of the past, as if a counterargument might attempt to use the self-evidence that some things did happen before now as a proof of the necessity of time itself (1.464–71):

> denique Tyndariden raptam belloque subactas
> Triugenas gentis cum dicunt esse, videndumst
> ne forte haec per se cogant nos esse fateri,
> quando ea saecla hominum, quorum haec eventa fuerunt,
> irrevocabilis abstulerit iam praeterita aetas.
> namque aliud terris, aliud regionibus ipsis
> eventum dici poterit quodcumque erit actum.

> Next, when people say that Tyndareus' daughter was kidnapped,
> Or the Trojan peoples were beaten, we must take care
> They do not force us to say that these things exist in and of themselves,
> When those generations of men, whose accidental properties these were,
> Have now been stolen away by the irrevocable passage of time.
> For whatever will have taken place will be able to be called an accident,
> In one case, of the earth, in another case, of particular regions.

These are difficult lines to follow.[52] It does not seem that Lucretius is arguing that only the present exists ('presentism'), as some philosophers have essayed.[53] Rather, he is suggesting that when people retort, 'Helen was taken', 'the Trojan war took place', it does not require him to say that such events existed in and of themselves (*per se*) – and thus agree that this is an argument that time exists in and of itself. He can say – and simply does – that 'time' – now *aetas*, 'age' – has passed, and, in the most traditional of language, it cannot be summoned back (*irrevocabilis*), and the generations (*saecla*, another term for time) have been lost. There were 'accidents', *eventa* in the past, as there will be in the future. For 'whatever will have taken place, will be able to be called an "accident"'. That is, it is a category error to think that events in the past, just because we say they 'exist', have a different status from the 'accidents' in the present or future. The infinitive *esse* also plays with the grammatical necessity in Latin that to say 'it has happened', *actum est*, will use *est* 'it exists': it is hard in Latin to use the past and not assert its existence. All events are *eventa*, 'accidents', the coming together of atoms in movement and in rest.[54] So, although Lucretius uses *aetas*, *aevum*, *saecla*, *quando*, the past tense, none of these expressions imply for him the existence of time *per se*. Discussing events from the past – stating 'Helen was taken' – is no evidence that time in and of itself exists.

Time, especially *vetustas*, 'old age', and other words which 'manifest[] time's continuing power',[55] indeed recur throughout Lucretius' depiction of the materiality of things precisely as an agent of change. So, in Book 5 'time' (*aetas*) is said to 'change the nature of the whole world, and one condition after another must overtake everything, nor does any thing remain the same as itself: everything changes place, nature changes every-thing and compels it to turn' (528–30). Where Plato had demanded that the only real truth must be firm and remain the same and be unmoving, Lucretius describes reality itself, the nature of things, as necessarily in a constant state of transition and alteration. Even time itself seems open to such transformative swerving. So Lucretius writes of how precious metals shift in value (5.1276), *sic volvenda aetas commutat tempora rerum*. It can be translated 'Thus rolling age changes the seasons of things', which makes the verse a truism at best (the dictionary gives only Lucretius as a source for the meaning 'seasons' for *tempora*); it has a more unsettling and

---

[52] See Warren (2006); Zinn (2016) for discussion and bibliography.
[53] Warren (2006) and Zinn (2016); also Berns (1976).   [54] On *eventa*, see Wardy (1988) 117–21.
[55] Thomas and Witschel (1992) 143.

transformative force, however: what is changed is 'the circumstances of things', even 'the crises of things' – how things inhabit time is changed by the very passage of time. And time's passage is infinite, unceasing: 'Nothing which is in a mortal body, because of the infinity of time (*ex infinito tempore*), could still now have had the ability to deny the powerful strength of immense age (*aevum*).' The infinity of time and immensity of age are set here against even the sea and the heavens and the sun, no less open to violent destruction and the reordering of matter (*Civ. Dei* 20.25). (An end that is no surprise to Jews or Christians: 'The heavens are the work of Thy hands: they shall perish' (Ps. 102), though the causality and anticipation of destruction are quite different.) Where Plato saw an image of eternity and order in the natural world, an ensouled creation of a beneficent Demiurge, Lucretius splinters this vision, this language, the expressivity of language, into clashing and crashing and swerving *stoicheia*, elemental particles, on the move.

The atoms themselves cannot be destroyed, however: matter remains. The eternity of matter is at the deepest level of difficulty for later orthodox Christian readers of Lucretius. If matter itself cannot come to be or pass away, what then of God's creation or the end of the world? Synesius, bishop of Cyrene, and a fully-fledged philosopher, especially of the Platonic school, before and, to a degree still debated, also after his conversion and immediate appointment as bishop, was quite clear that philosophy was not easily refuted here.[56] Nothing could convince him that creation could be made out of nothing (*nihil ex nihilo*) or that matter would pass away, even though such a position stood against the founding principles of Christian theology. Synesius' unorthodox stance reveals a precise tension between normative Christianity and its appropriation of Greek and Latin wisdom, and especially Epicurean physics with all its implications for theology. In Lucretius, then, it is not immaterial Forms but matter itself that is timeless.

Neither Plato nor Lucretius would have you fear death, for the very different reasons of the immortality of the soul for Plato, and the dissolution of the body back into insensate matter for Lucretius. Yet both seek to find the timeless, the everlasting, against which to view the instability of the here and now. For both writers, how to live, and how to face death as an ethical question of living, are issues shaped by what is determined to transcend time's vagaries. So, to be in the now, does it require the timeless?

*

---

[56] Bregman (1982).

The geographical and political ambition of empires is enacted at the level of temporality, too. The topographical and the chronological are easily over-lapped from the beginning. When Xerxes in Herodotus declares (7.3) 'we shall extend the Persian territory as far as God's heaven reaches; the sun will then shine on no land beyond our borders', he inaugurates the rhetorical tradition of lauding the 'empire on which the sun never sets' – the boast, first made popular in the Holy Roman Empire of Charles V, that one's territory is so extensive that it is always daytime somewhere in it. The imposition of a single time of empire redefines the lives of its subjects. Imperial time reconstructs not just the daily lives of the empire's inhabit-ants, but also their sense of history, the past and the future.

The Seleucid empire of the second century BCE, according to Paul Kosmin's superb analysis, has a claim to have invented imperial time, and thus to be 'the first truly historical state'.[57] It achieves this status for Kosmin because it constructed the first numerical system of ordering years, rather than using regnal titles to mark eras, as is usual in the region. It structured 'the kingdom's visibility and institutional practices around a continuous, irreversible, and predictable accumulation of years'[58] (and, Reinhart Koselleck reminds us, a sequence of years, human order, always indicates an 'historically and philosophically impregnated experience of time').[59] Although the centralized power of an empire such as the Seleucids was always mediated through continuing local authorities, and was flexibly attuned to regional variation, its fiscal control was more standardized. 'The arrangement of fiscal life into an annually serialized grid established an orderly geometry of imperial time at the heart of the city's public life, a locus of maximal temporal conformity and rationalization that idealized and euphemized the messiness of the greater societal pattern'.[60] The imperial temporal regime reorganizes the timing of administrative pro-cesses and thus reshapes the social life of its citizens under administration.

Empires also need to narrativize their own rule across time. Unlike the birth of time in Plato, or the creation as told in Genesis – or, for that matter, unlike the claims of nationalism to a timeless, continuous and unbroken national character – empires come into being in a year one, and there is a pre-story to be told (305 BCE, in our calendar, is year one for the Seleucid kingdom). There is also always a projection of the future. As J. M. Coetzee, in *Waiting for the Barbarians*, evocatively writes: 'One thought alone preoccupies the submerged mind of Empire: how not to end, how not to die, how

---

[57] Kosmin (2018) 76.   [58] Kosmin (2018) 76.   [59] Koselleck (2002) 149.   [60] Kosmin (2018) 59.

to prolong its era'.[61] The *Aeneid* may have promised *imperium sine fine*, 'empire without end', and Horace may have imaged the everlasting life of his poetry through 'for as long as the priest ascends the Capitoline Hill', but Polybius and other historians encouraged the darker picture that Coetzee draws: 'by night [empire] feeds on images of disaster: the sack of cities, the rape of populations, pyramids of bones, acres of desolation'. Polybius himself, an eye-witness to the event, was there when Scipio, the Roman general, looked down over the city of Carthage as it was being destroyed by his troops.[62] Scipio, with tears in his eyes, quoted Hector from Homer's *Iliad*, 'A day will come when holy Troy will fall, and Priam and his people'.[63] Polybius who had been tutor to Scipio, asked him what he meant by the quotation, and Scipio, who had been musing on the fall of the empires of Assyria, Media, Persia and Alexander's Macedon – 'whose brilliance had been so recent' – freely admitted he was thinking of Rome itself, and its inevitable fall. The scene has a deeply layered sense of the transience of empires. 'He sees the fate of Rome in Carthage and Troy is the model for both'[64] – with the added *exempla* of the other empires of the eastern Mediterranean. Polybius' *History* is prompted by his wonder at the rapid rise of Rome; he offers as one cause the flexible stability of the Roman constitution. But he not only anticipates a general reversal of fortune as a pattern of existence (Herodotus, too, of course, sees such a turn in the lives of kingdoms), but also sees the time of empire as necessarily 'the time before the end'.[65] Rome's triumph must of a necessity prefigure its fall. The duration of empire is no more than the deferral of an inevitable end.

Julia Hell has wonderfully mapped how the fall of Rome, including its anticipation at the acme of its triumph by Scipio, became a repeated model for empires to discuss their own impending history. 'Romans invented the very concept of *imperial ruins*' and became the privileged example of such necessity.[66] In her richly evidenced and sophisticatedly argued book, Hell takes us from Charles V (on whose empire the sun never set) marching in celebration along the roads of Italy flanked by statues proclaiming him to be Scipio, through to the proclamations of the Third Reich (along with its authoritative philosophers, Heidegger and Schmidt). 'Obsessed with the

---

[61] Coetzee (1980) 133. His title is taken from Cavafy.   [62] Dio. Sic. 32.24.
[63] *Il.* 4.164–5 (Agamemnon); *Il.* 6.448–9 (Hector) – once by each side.
[64] Feeney (2007) 55. See also Rood (2007) 181; Wiater (2016), especially 257–8 for other scenes of tears over victory.
[65] Hell (2019) 44.   [66] Hell (2019) 87.

problematic of duration and ruination',[67] Hitler's team used the model of Rome to fortify their empire, militarily, ideologically, symbolically. 'Hitler alone', wrote Simone Weil prophetically in the midst of the war and his military success, 'has understood correctly how to imitate Rome'.[68] The imagination of modern empires is haunted by Roman nightmares – not just its fall but also the inevitably failing attempts to maintain its security, made physical in the ruins across the landscape of the now. If the anthropologist's gaze is articulated in the present tense of objectification, empire is always self-aware of its own future perfect – of what its ruins will declare it to have been. Thus Macaulay in 1840, great historian of the British imperial project, famously imagines a Maori sketching the ruins of St Paul's from a ruined arch of London Bridge – a colonized barbarian now calmly, *civilizedly*, observing the ruins of the former imperial power.[69]

Yet the subjects of empire do not need to wait for their oppressors' ruination to write back against imperial control of time and history. Much of the evidence Paul Kosmin utilizes to demonstrate the Seleucid kingdom's regulation of chronology comes from resistance to it, and a good deal of that is from Jewish groups. The festival of Chanukah, now part of the Jewish calendar, is inaugurated as a memorial of the armed resistance against the Seleucid regulation of Jewish ritual life. The book of Daniel is paradigmatic in that by prophetically framing the here and now as an age that is one of a series of empires past and to come, it constructs a symbolic historiography that stands aggressively against the chronological reach of the Seleucid rule (as Kosmin analyses in great detail). Daniel is 'an interrogation of political change, Hellenistic imperial rule and the nature of meaning and justice in history', with 'an overriding concern with the periodization of history and the temporal location within it of the Seleucid kingdom'.[70] Daniel's polemic is to redraft the place of the Seleucid empire in time. Indeed, the Babylonian Talmud's intense focus on the regulation of daily time, especially the Sabbath, the dating of festivals, and even the broadest ideas of chronology, aims to construct a counter-world of self-enclosed time, counter to the dominant cultures of Persia, Greece and Rome in which the rabbis were living. As Sacha Stern has argued,[71] Jewish groups in Judaea/Palestine and in the diaspora, as an act of cultural resistance, continued to maintain their lunar calendar even and especially when the Roman empire made the Julian calendar the standard of imperial

---

[67] Hell (2019) 309.   [68] Weil (1962) [1940] 101.
[69] See Dingley (2000); Skilton (2007), with further bibliography – and for a contrast with Thucydides, see Rood (2016).
[70] Kosmin (2018) 139, 140.   [71] Stern (2001) 157–75.

rule for their subjects; and, what is more, Stern shows that through late antiquity there was a conscious, concerted effort by the rabbis to regularize the previously flexible and empirical calendrical systems, in which different communities celebrated festivals on different days:[72] 'The gradual fixation of the Palestinian rabbinic calendar was thus the result of an attempt to unify and standardize the calendar of the rabbinic communities of Palestine and Babylonia',[73] as a function of the disintegration of society after the destruction of the Temple and the gradual development of rabbinical authority over the subsequent centuries.[74] The calendar becomes a way of regulating the community and setting it apart from the surrounding society. In a remarkable act of writing from below, the rabbis even superimposed Jewish history on Roman time. Thus, the origin of the Roman festival of Kalends, the first day, is found in Adam, the first man. Adam was terrified when the days first shortened; he quotes the Psalms to express his fear of encroaching darkness; but when he realizes that light returns, he exclaims, '*kalon dies*', 'Good is the day', a mixture of Greek (*kalon* = good) and Latin (*dies* = day). In this rabbinical time Adam can use the Psalms, Greek and Latin – and the result is an appropriation of the Roman calendar to a Jewish narrative through aetiology and etymology.[75] 'The Jews wrote more about their calendars . . . than just about anyone else in the ancient world until the late Roman period',[76] when Easter started to become a major issue for Christian chronographers.[77] This is not an intellectual idiosyncrasy of a marginal group but a continuing politics of community through telling time.

What, then, happens if the resistant group becomes the power in place? Will Macaulay's Maori also have sketched an anticipation of his own people's future ruin? Augustine's *City of God* is written after Rome has been sacked by the barbarians in 410 and also after Rome has become a Christian city, and broaches, therefore, precisely these issues.[78] The *City of God* begins as an explanation of how it is possible that Rome could have

---

[72] Including the 'marginal and dissident' 364-day calendar in the Dead Sea Scrolls of Qumran (Stern (2012) 359–79, quotation 362; Ben-Dov (2008)); the use of time to mark 'inter- and intracommunal difference' (228) by Jews is the central thesis of the fine work of Gribetz (2020).

[73] Stern (2012) 335. For the continuation of pagan festivals in the Christian empire, and attempts to stop them, see the pungent account of MacMullen (1997) 32–72, and from the Jewish perspective Gribetz (2020) 55–91.

[74] Heszer (1997); Schwartz (2001); Goodman (2000).

[75] *Y. Avodah Zarah* 39c, excellently discussed by Gribetz (2020) 55–91; see also Schäfer (1996).

[76] Stern (2012) 331.　　[77] Mosshammer (2008).

[78] Wetzel ed. (2012) is an excellent introduction with bibliography to a huge field of study. General background in Kahlos (2007).

been sacked by the Visigoths – nuns raped, churches desecrated. With a battery of arguments, Augustine rejects the pagan response that the sack has been a demonstration that the Christian God could not protect his own pious ones. This immediate defence opens into a 22-book contrast between the city of heaven and the city on earth, which constitutes no less than a systematic theological engagement with the prospect of world history through a reading of scripture, with a vision of the end of days as the justification and culmination of the whole narrative. What, then, is the historical now for Augustine?

Augustine responds to Paul's insistent, intense waiting for the Second Coming by quoting the repeated paradox of St John's Gospel, *erchetai hōra kai nun estin* (4.23, 5.25), 'the hour is coming and now is here', *venit hora et nunc est*. The existence of the now is always and necessarily marked by what is to come. This sense that the now is to be experienced through the future is explained through a full-scale chronological scheme across the history of the world. As the world was created in seven days (and that 'as' is a sign of the typological strategy of reading), so the world's time is organized into seven ages. The first runs from Adam to the Flood; the second from the Flood to Abraham (these both have ten generations). The span from Abraham to Jesus has three periods each of fourteen generations, as the genealogy from the beginning of the Gospel of Matthew calculates – Abraham to David, David to the Babylonian captivity, the captivity to Jesus. We are now thus living in the sixth day. How long is this day? Augustine answers that 'It cannot be measured by any number of generations, as it has been said, "It is not for you to know the times, which the Father hath put in His own power".' But he also says, following the book of Revelation, that it will last a thousand years. The kingdom of the Saints will last a thousand years, he asserts, and this is *now* (20.9). (It is debatable whether the final three and a half years of persecution by the Devil which herald the end are included or not (20.14).) After this sixth day closes, then we shall be in the seventh day, the Sabbath, when 'God will cause us to rest in Him', *in se ipso Deo faciet requiescere*. This seventh day will not end with an evening but with an eighth day, an eternal day when the body as well as the soul will find eternal rest, a day when 'we will rest and see, and see and love, and love and praise' (a beautifully simple promise after so much detailed theological analysis). Augustine summarizes: *Ecce quod erit in fine sine fine*, 'Behold, this is what will be in the end without end.' The end will be endless. Time will become timeless. That is what it means to inhabit 'the kingdom without end'. The fall of Rome becomes finally the story of the everlasting kingdom.

To get to this conclusion Augustine has had to work hard. In Book 11, he has reprised his argument from the *Confessions* that time, which requires motion, has a beginning, in contrast to eternity which is motionless. In Book 12, he argues with pagan historians about chronology, denying that the thousands of years of human history in pagan accounts can be a true figure. Similarly, and at even greater length in Book 12, he earnestly strives to refute claims of any circularity for time. He has already demonstrated that time has a beginning, of course, but now – with a nice reference to the poetry of the Psalms that 'the sinner walks in circles' – he argues against any notion that times repeat or return. His final salvo that 'the eternal life of the saints completely refutes any idea of the circularity of time' will convince only the faithful.

Seneca, the Roman philosopher whose Stoicism was felt to be so close to the austerity of Christian commitment that letters between himself and St Paul were creatively imagined, sums up perfectly the pervasiveness of the language Augustine is striving to redraft. He expresses the temporality of human life precisely in terms of a set of concentric circles (*Ep.* 12.6–7):

> Tota aetas partibus constat et orbes habet circumductos maiores minoribus: est aliquis qui omnis complectatur et cingat – hic pertinet a natali ad diem extremum; est alter qui annos adulescentiae excludit; est qui totam puer-itiam ambitu suo adstringit; est deinde per se annus in se omnia continens tempora, quorum multiplicatione uita componitur; mensis artiore praecin-gitur circulo; angustissimum habet dies gyrum, sed et hic ab initio ad exitum uenit, ab ortu ad occasum.

> Our span of life is divided into parts; it consists of large circles enclosing smaller. One circle embraces and bounds the rest; it reaches from birth to the last day of existence. The next circle limits the period of our young manhood. The third confines all of childhood in its circumference. Again, there is, in a class by itself, the year; it contains within itself all the divisions of time by the multiplication of which we get the total of life. The month is bounded by a narrower ring. The smallest circle of all is the day; but even a day has its beginning and its ending, its sunrise and its sunset.

This system of analogous circles will be reprised throughout later literature in Greek and Latin, as the 'circling years' of Homeric diction are rearticulated into a broader discourse of temporality. In these books, then, Augustine sets himself explicitly and most intently against the philosophers of the Greco-Roman tradition, as he has against the historians and other learned traditions. The comprehensive vision of the *City of God* explains the universe against the authoritative discourses of antiquity. Including Christian theologians: not only does he refute Origen on the circularity of time, but he discusses in Book 15 the different lengths of time that the Hebrew manuscripts and Greek translations have for the ages of the figures who lived before the flood. These detailed arguments about aspects of time throughout the *City of God* are all subordinate, however, to the overarching problem of how to conceive of the now of the earthly city in relation to the final timelessness of the heavenly city. Can and should Christian Rome be different from other empires that come and go? An individual's life and ethics are judged by accession to or rejection from eternal life among the *sancti* – what, then, of a kingdom?

In Book 20, Augustine analyses scripture's depictions and prophecies of the 'end of days', including the prophecies of Daniel with their promise of the collapse of kingdoms. In this discussion, Augustine frankly admits that he is uncertain whether Paul refers to the Roman empire in his warnings of the destruction of evil, but, he says, it is 'not absurd' to assume some of his prophecies allude to it (a very cautious approach). In Book 21 he describes the end for the wicked; in Book 22, he outlines the end for the saved. The eternal punishment of the damned is contrasted to the eternal life of the *sancti*. In these final three books, the end, Augustine is concerned with the eternal city of God at the end of days; in the previous books, Augustine has traced the presence of the city of God in earthly guises. In contrast with his own earlier views of *tempora Christiana*, or, say, Prudentius' typically triumphant *vicimus: exultare libet*, 'We have won: we can rejoice', or even of Ambrose's or Origen's views that the unification of the empire under Augustus was providentially designed to enable the Gospel to be spread more easily, Rome, for Augustine, is not the earthly representative of the city of God, an *imperium Christianum*. Rather, the heavenly city can only be at best a *peregrina* on earth (19.17) and even in this earthly *peregrina* city of Rome its citizens are *peregrini*, 'foreigners', 'sojourners' or 'pilgrims' – not just incapable of being full citizens, belonging to this city, but also both inhabiting the now (sojourners) and on a journey (pilgrims) to a greater sanctification.[79] The church at its best is a community of these *peregrini* on earth, living together in peace or striving to do so. For Augustine, however, a Christian empire can only ever be a precarious regime, in a time before the end. Yet unlike other empires' nightmares of dissolution, it is this end of destruction that Augustine positively seeks, to

---

[79] Vocabulary back to Tertullian at least: *De cor.* 13: '*peregrinus mundi huius et civis civitatis supernae*', you are 'a foreigner of this world, a citizen of a higher state' – where it is also linked to the idea of the *saeculum*. See Vessey, Pollmann and Fitzgerald eds. (1999); McLynn (1999), (2009).

discover the endless peace that will follow it. This anticipated, perfected, desired end, for him, is history's story, time's narrative.

Aristotle declared that tragedy was more profound than history because tragedy expressed *to eikos*, what is likely, probable, natural, whereas history only told what had actually happened. Historians, ever since and even before, have looked to find the general laws of *to anthrōpinon* in contradiction of Aristotle's dictum. For Augustine, however, there is only one history, which is both what has happened and what must happen: that is the logic of providence. Or as Augustine writes in *De vera religione* (7.13): 'the essence of Christianity is the history and prophecy of the temporal dispensation of divine providence for the salvation of the human race which must be reformed and healed into eternal life'. Augustine's is a narrative that seeks its end in timelessness, a contradiction of the condition of narration. Simply to live in the now is not a possibility: to say *nun esti*, 'it is now', is already to have said *erchetai hora*, 'the hour is coming'. The promise of timelessness haunts the now: the time which the Christian inhabits is between an already and a not yet.

We began with God's time and Augustine's project of imagining divine timelessness, and moved through Christian attempts to rewrite the flow of time both as a constant truth of the present through typology, and as experiencing the past in the present in the vision of pilgrimage, through the exploration of what 'at the same time' means, to (now) Augustine's all-embracing history, a single passionately assertive story that seeks to end in the ideal timelessness of the eternal kingdom. Christian temporality. Yet timelessness can only remain an impossible projection for human striving and failing. The unimaginable imagined, barely, hopefully, as an end. The question Augustine places on the agenda of this book, then, is to wonder about the consequences of such a self-placement in historical time. What does it mean to live thus in this regime of a still unending time between the already and the not yet? How is the now to be experienced if your eyes are turned always towards the promise of timelessness?

CHAPTER 9

Life-times

If we ask, then, what it means to live in this regime of a still unending time between the already and the not yet, one answer to the question must come from how a life is represented, how life-writing is shaped by the times of its production and how it shapes time for the readers and writers of life-times. So, how does Augustine – to continue where we left off – begin such a biographical story? With a contorted statement of ignorance (*Conf.* 1.6.7):

> What is it that I want to say, lord, except that I don't know from where I came to this place (*nescio unde venerim huc*), I mean, into this life that dies (*vitam mortalem*) or death that lives (*mortalem vitalem*)? I just don't know (*nescio*).

This is the first question that Augustine asks about himself in the *Confessions*, and it begins with a stumbling into speech. He does not know where he comes from.[1] This is the question which stalls Sophocles' Oedipus in his domineering argument with Teiresias, starts his search for his parentage, and thus begins his downfall into knowledge and self-destruction. Oedipus does not know where he comes from, an ignorance displayed even and especially when, with multiply-layered ironies, he calls himself 'the know-nothing Oedipus'. It is also the foundational question for Freud, reader of Oedipus, who insists that for all the productive work of analysis of the self we can never fully and properly know our own self, and certainly not the answer to where the self comes from. Augustine specifies *huc* 'to here', which he immediately glosses as 'this life that dies or death that lives'. The horizon of expectation is defined – in a way that is alien to Sophocles or Freud – by this definition of a life-time as a hesitation between a journey towards death, or an already living death: a theologically defined time shaped between the already and the not yet.

---

[1] Great thanks to Catherine Conybeare here, who discusses this passage in a forthcoming article (and discussed it with me at length). I use her translations here.

Augustine continues (1.6.9): 'Look, my one-time infancy is dead, and I'm still alive', *Et ecce infantia mea olim mortua est et ego vivo*. The *ego*, the *I*, lives on although its past is declared dead. As Augustine goes on to explore, 'infancy' has its full etymological significance: the loss of infancy is a coming into speech, the ability to ask the question of where he comes from, to ask what it means to say *vivo*. Although he has not yet in the *Confessions* laid out the theology that lies behind such statements (and that I discussed in the first chapter), he addresses God as the creator who exists before it is possible even to say 'before', and transcends all instability of time and change, and who can therefore answer Augustine's increasingly insistent and recessive questioning:

> tell me whether my infancy succeeded some previous dead age of mine. Is that the one that I lived in my mother's womb? For something of that has been made evident to me, and I have seen pregnant women. And then what was before that, my sweetness, my god? Was I anywhere or anyone (*fuine alicubi aut aliquis*)?

Augustine does not know where to start. Should his life-story begin in the womb, which, he suggests, is a 'previous dead age' (*mortuae aetati*) (before his now dead infancy). Each successive age, it seems, for Augustine, *dies*, as he progresses towards the death that opens its way to eternal life. The time in the womb is a potential beginning – and his apparently naive, if strongly expressed empiricism, 'I have seen pregnant women', seems to be because he can have no memory of this time, no words for this experience – but what, then, about the time before conception? He does not so much cue the long philosophical tradition, with its Platonic source, of debating the pre-existence of souls, as allude to it in the designedly childlike question, 'Was I anywhere or anyone?'.

The paragraphs that immediately follow show how different from Plato's Augustine's response to this question is. First, he acknowledges not just that God is the only possible source for such knowledge, but also that God's time, the fact that God has no today, no here and now, is the limiting frame for Augustine's all too human attempt to understand the mysteries of his own timeliness. How Augustine became flesh is not comprehensible without God's incarnational power of creation. Second, even more arrestingly, he turns to man's inevitable sinfulness: *Exaudi, deus. Vae peccatis hominum*: 'Listen, God! Alas for the sins of man'. Original sin takes Augustine's search for where he comes from, back to the beginning of humanity itself, because he is still marked by that moment of the fall. How the self comes into being in time and through time – the narrative of the

Confessions – is shaped by the continuing theological compulsions of incarnation and original sin. Telling his own life-story, for Augustine, is structured through a relation to the divine time of God.

*

Augustine's *Confessions* is a unique text, atypical in structure and purpose in the life-writing of late antiquity (and barely with direct impact, it seems, through the Middle Ages). Yet in one respect at least it is paradigmatic: its focus on conversion as a central, life-changing event. The turn towards God, which is the turn towards the future from the here and now, is the condition of *conversio*. Conversion changes the temporality of the narrative of a life-time.

From the cry of John the Baptist, *metanoeite, metanoeite*, 'Repent! Repent!', as it is usually translated, Christian narratives have maintained the process of conversion as a fundamental vector of how a life can be narrated. Augustine's autobiographical story of conversion is especially long and complex, and performs his later belief in the agency of Grace in the repeated encounters which seem to be by chance (*forte*); and, although the moment in the garden, when he hears the voice of what may be a child's game crying '*tolle, lege*', becomes the critical turning point, the hesitations, questions and slow recognition of change stand against many conversion narratives, ancient and modern, which imagine a sharp rupture of 'before' and 'after' (Augustine's theory of time will inevitably make any notion of 'the moment' problematic).[2] Conversion, however, becomes a retrospective, teleological narrative that organizes a life-time around a pivot of the fundamental alteration of self-understanding and perspective of how to live this life, here and now. It becomes the paradigmatic story of a Christian life-time.

Metanoia, often translated into Latin as *paenitentia*, 'repentance', or even 'penance', in its earlier usages, means a 'change of mind', and in its *locus classicus* is tightly tied to timing.[3] Thucydides relates how the Athenians voted to execute all the men of Mytilene after its revolt from the empire; they then had a change of mind (*metanoia*) the next day, reconvened the Assembly, changed their decision, and sent a boat in a race against time to overtake the boat carrying the original order (3.36). *Metanoia* is a change of judgement. The word does not occur in the Septuagint's translation of the Pentateuch (however familiar the language of repentance is in the King James Bible), and in the Prophets it usually

---

[2] See Wetzel (1992) 112–60 for a wonderful discussion of grace and will.
[3] Goldhill (2020) ch. 5 for a fuller discussion of *metanoia*.

means a change of mind either of God towards punishing humans, or of humans towards following the law: for humans, it marks a return to what is the already agreed system of God's laws, a coming back to the true path of observance. Yet, through the texts of Christian late antiquity, building on the Septuagint as well as the paradigm of Paul on the road to Damascus, its range of significance stretches to mean something close to what Arthur Darby Nock's celebrated book on conversion defines as a 'renunciation and a new start ... not merely acceptance of a rite, but the adhesion of the will to a theology, in a word faith, a new life in a new people', a will to 'belong body and soul' to a godly community.[4] Nock's definition certainly privileges his deeply held Protestant understanding of inward commitment – what Lambert nicely calls 'the evangelism of interiority'[5] – and there are certainly many more complexities to the philology of this language of change, as well as to the anthropology of conversion over the centuries. I have written about this transition of *metanoia* specifically between Jewish and Christian texts in late antiquity, and Christopher Prendergast's sharply funny and erudite book on counterfactuals takes forward the analysis of penitence, regret, remorse, reproach and moral luck beyond the Enlightenment into modernity – and demonstrates how unconvincing claims to any essential stability in this vocabulary and its underlying presuppositions are.[6] But here and now I want to focus on the temporality of the narrative of a life with conversion at its heart. How does the injunction to repentance and conversion invite us to live in time?

A contrast with the standard biographical models of Greco-Roman elite culture is emphatic and revelatory for this question.[7] There are, of course, texts from the non-Christian literary tradition that centre on transformation: Apuleius' novel, *Metamorphosis*, is structured between the episodes of the hero being turned into an ass, and finally becoming a priest of Isis, with the remarkable erotic tale of Cupid and Psyche ('Desire' and the 'Soul') in between, with its precarious allegoresis. The status of the final religious transformation – a conversion of sorts – cannot escape its destabilizing relationship with the picaresque and filthy adventures of the human soul inside an ass's body.[8] Augustine marks the problem for the reader when he calls Apuleius' story of the transformed self 'either

---

[4] Nock (1933) 14.   [5] Lambert (2015) 6.   [6] Goldhill (2020) ch. 5; Prendergast (2019).
[7] Discussed in Goldhill (2020) ch. 6. For the background of what follows see Cox Miller (1983); Momigliano (1983); Edwards and Swain eds. (1997); Hägg and Rousseau eds. (2000); McGing and Mossman (2006); de Temmerman and Demoen eds. (2016); Hägg (2012); Fletcher and Hanink eds. (2016); Burridge (1992).
[8] Winkler (1985); Harrison, S. (2000).

demonstrated or made up', 'either real or fictional' (*aut indicavit aut finxit*, *Civ. Dei* 18.18). Apuleius' novel is not so much a biography, as, at best, a calque or parody of the genre's scope for teleology. Similarly, Ovid's *Metamorphoses* revels in the materiality of bodily change, but has no interest in a biographical model of inward repentance or affiliation to a new community. A rhetorician like Dio of Prusa may praise his own discovery of philosophy as a sort of personal transformation, but such stories are rare, and the process of transformation is articulated less forcibly than the self-congratulation.[9] In any case, our main evidence for such a 'conversion' comes from a Christian, Synesius, who may have reframed Dio's self-dramatization in a more Christian manner. Augustine, too, has his transformative discovery of philosophy, more fully described as part of his intellectual and emotional journey. (Unfortunately, there is no evidence that Synesius in Cyrene and Augustine in Carthage ever knew each other, although Charles Kingsley in his novel *Hypatia* wonderfully dramatizes them presenting their contradictory views on marriage.)[10] From the *Odyssey* onwards, the disguise of a hero, which can be a full bodily transformation, demands revelation, but here too 'conversion' is scarcely at stake. None of these texts from the Greco-Roman tradition is straightforwardly biographical; many tell a story of change, but none shows a strong exemplary model of what will become the paradigm of conversion.

Within the self-defined genre of biography, there is a recognition that people do fundamentally change. Tiberius, for example, was a successful and respectable young politician who became a reprobate tyrant in old age. But Tacitus merely notes in summary *morum diversa tempora*, there were 'different times/periods of his behaviour'.[11] Suetonius has an extended arsenal of explanatory factors for Tiberius' character from his opening account of his family background to the emperor's early predilection for heavy drinking. He notes how at the high-point of his career as a young man, he suddenly entered retirement – and suggests several possible causes that circulated to explain this change, without offering his own privileged explanation. The shifts into his foul behaviour on Capri are listed with horrified attentiveness but scarcely related to any cause beyond the cruelty of corruption and opportunity of power. In these biographical accounts of Tiberius there is no narrative of a moment or process of change to organize his *diversa tempora*, no rupture between his noble youth and corrupt old age.

---

[9] Moles (1978); Whitmarsh (2001) 156–66.
[10] Kingsley's novel finds academic heritage in Lane Fox (2015) who compares Synesius and Augustine.
[11] *Ann.* 6.51.3. Hoffmann (1968).

Nor would we expect such a 'conversion' narrative from Plutarch's *Bioi*, or from a historiographer's inset character descriptions and diffused stories of the lives of the actors in their histories.[12] Plutarch is paradigmatic both in locating moral, political, intellectual character as instructive, and in explaining such character as a mix of education, natural talent, and circumstance. The writer's training in rhetoric's mobilization of the stereotypes of personality and the expectations of probable narrative make character a determining causal factor in the narrative of a life, but it is a character that is formed to act and be judged – to be, as we have discussed in chapter 5, exemplary. Plutarch's standard practice of *synkrisis*, 'comparison', between a Greek and Roman figure is designed to set two formulated models of cultural and political virtue against each other. As in tragedy, a *peripeteia*, a sudden reversal, happens to a figure and the figure responds, but the reversal of fortune is not a conversion.

The depiction of Socrates accepting his death as a fulfilment of his choice of life as a philosopher provides one model for making the *prohairesis* 'the choice of a way of life', a key decision in such biographical narrative (and this is the obvious model behind Dio's paraded, new commitment to philosophy). Josephus describes how he travelled through Palestine, visiting different groups of philosophers, as he calls them, to decide which offers the best regimen for life (*Vit.* 9–12). The schools of philosophy competed for adherents. Both Origen, from the Christian side, and Galen from the pagan, complain that all too few people make such a rational choice between rival sects (Orig. *Cels.* 1.10; Galen, *Libr. Ord.* 1). Stories of instant conversion to a philosophical teacher appear in Diogenes Laertius' *Lives of the Philosophers*, to boast how a philosophy can become a total transformation of a life: the drunk, domestic abuser, Polemo, becoming the sober pupil and lover of Xenocrates is iconic – afterwards, he could be bitten by a dog or even listen to Homer without being emotionally disturbed (DL 4.16.19).[13] Lucian, with his usual leery eye, mocks the pretensions of dress, behaviour and language that such converts or leaders adopt, skewering the hypocrisy of the multiple sects of empire society, though some modern critics have strangely preferred to take him as a convert himself, or even to use him to postulate a 'rich tradition of protreptic conversion literature in the Hellenistic and Imperial eras'[14] of which there is very scant indication.[15] The educated man reveals his education by the choice he makes of a life-style, and its underlying principles. This

---

[12] See Duff (1999); Goldhill (2002) 246–93.   [13] See Eshleman (2007/8).

[14] Grethlein (2016) 257–8.

[15] See Schäublin (1985); Cancik (1998); Grethlein (2016), and, more generally, Branham (1989); Goldhill (2002) 60–107.

discourse of 'choice', *hairesis*, will become in later Christianity the accusation of 'heresy', the wrong choice of a way of life, or of a dogma or belief – formalized and institutionalized not so much in texts such as Irenaeus' *Against Heresies*, or Hippolytus' compendious *Refutation of All Heresies*, an obsessive, projected, anxious map of the other, as in the Councils, which, as we saw in chapter 1, attempted to formulate and regulate God's time for the faithful. *Prohairesis* provides a cultural link between Christian lives and the practice of the educated elite of Greco-Roman society, but the strong emphasis on *metanoia* or *conversio* remains a Christian narrative trope of life-writing. Despite the occasional story of instant change towards worshipping the one God in rabbinical writing, and the rabbinical commentaries on the story of Ruth, an intense focus on conversion also distinguishes Christian life-writing and Christian practice from Jewish normative expectations. The *catechumen*, the person institutionally undergoing conversion, is a specifically Christian development of religious life in late antiquity.

It would be easy to continue to trace examples of the multiple forms of life-writing in Greco-Roman antiquity, both within self-determined genres of biography – the brief sketches of Philostratus' *Lives of the Sophists* or the full-scale *Life of Apollonius of Tyana* – and within other genres – Horace's poetic explorations of *Heu ... fabula quanta fui*, 'alas, how big a story I became', for example, or Cicero's self-aggrandizing epic of his own career. We could add Aelius Aristides' diary of his painful medical history; Pliny's or Cicero's letters are edited and partial accounts of an embedded political life. The foundational models for the culturally dominant forms of life-writing, however, are to be found in the classical city. Xenophon's *Cyropaidia, The Education of Cyrus*, is repeatedly echoed in the Greek prose of the Roman empire as an example not just of the didacticism integral to representing moral character, but also of erotics as a key scene of self-control. Likewise, Xenophon's *Memorabilia*, his recollections of Socrates, in its structure of brief paragraphs of biographical anecdote not only stands behind the form of a miscellany or anthology such as Philostratus' or Eunapius' *Lives*, but also is at the start of the tradition of *Socraticoi*, so-called disciples of Socrates who preserved his memory in prose, often dialogues. The variety of forms and long history of life-writing is complex enough, then. Yet it is possible to summarize without too much distortion that biography is a major and recognized genre of Greek and Roman writing, albeit with very porous boundaries; it is a genre with a powerful agenda of how to conceptualize the narrative of a life, what values are at stake in telling a life, and with a recognizable pattern of education leading to character formation leading to the revelation of virtue and vice in a competitive national and cultural context.

Yet what one strikingly does not find in this tradition are stories of conversion such as Paul on the road to Damascus. Paul's own account has no vocabulary of *metanoia*, though Acts already rewrites the story with a new emphasis on conversion. Paul falls from his horse and experiences a silent, three-day unconscious trauma (nothing could be more different from Augustine's highly intellectualized and reflective, halting passage towards conversion). Paul is transformed by his trauma, and like the patriarchs of Genesis, his transformation is marked by a change of name as well as by a change of life. In the 'Acts of Paul and Thecla', a long-lastingly influential narrative, Thecla hears Paul preach through an open window and goes into a three-day silent trance, from which she emerges as a committed Christian, to reject her marriage and family, and to leave home to follow Paul and spread the word of the Lord.[16] In both cases, the three days of silent trauma echo the three days between the death and resurrection of Jesus as they are reborn into their Christian identity. In neither case is there any discussion of any internal process, or any access to the subjectivity of the convert (as is relentlessly explored in Augustine). For Paul and for Thecla, there are threats from social authority and eventually martyrdom as a consequence of their new status. Likewise, Perpetua, as represented in the 'Passion of Perpetua and Felicity' must reject her father and her own role as a mother to bear witness in death to her Christianity, and alongside her, Felicity must give up her generative possibilities as a pregnant woman.[17] In these three texts, Acts, the 'Acts of Paul and Thecla' and the 'Passion of Perpetua and Felicity', conversion comes from outside, from God or God's word; what then appears to the world as a choice is challenged and threatened from the forces of a dominant culture, and the converts bear witness to their Christianity to the point of death. Life is divided between a before and after of conversion, and after conversion, life is structured as a process of witnessing with the constant anticipation of death as its climax. In contrast to the life-writing of Greco-Roman culture, education and character are not formative in the key moments of life. Because the key choice of a life is compelled or at best accepted, the models of agency are different: although, as we have seen, suffering, fearlessness, and calm have their philosophical and cultural roots in Greek and Roman virtues, the logic of suffering has been redrafted to

[16] Johnson, S. (2006); Cooper (1999).
[17] Heffernan (2012) is now the standard edition; Shaw (1993).

a new theological agenda. The timeline of a life-time is now structured around a moment of rupture. As human history is structured between the Fall of Adam and the Resurrection of Jesus, the Second Adam, so an individual life needs its crisis of change and rebirth.

Such conversion narratives are necessarily retrospective. Although these paradigmatic accounts of conversion present a transformative moment of change, it is a moment that exists only in retrospect. As Paula Fredriksen has written, 'To see a content-filled moment of conversion is to have constructed a narrative whereby the moment emerges as the origin of (and justification for) one's present'. The here and now is shaped by this past transformation. But, continues Fredriksen, 'the convert thus sees the subsequent events in his life in light of his conversion, but *à l'inverse*, his description of his conversion should be read in light of those subsequent events'. In short, 'the conversion account is both anachronistic and apologetic'.[18] Conversion stories rewrite the past, and such rewriting is always open to re-reading (and further rewriting) by its audiences. In particular, a first-person account of conversion has therefore a certain precarity, and a consequent heightened investment in integrity. The third-person narratives of conversion, destined as models for future first-person enactment, engage readers in an anticipated retrospection of their own experience – as beautifully captured by Augustine's first-person recollection of hearing the story of Victorinus' conversion and then the story of Ponticianus reading the *Life of Anthony*, as part of Augustine's retrospective narrative of his not yet happened experience of conversion – a chain of conversions, which Augustine and then his readers can join through this 'internalization of *exempla*'.[19] Ponticianus, Augustine tells us, reads the *Life of Anthony* and, he says, immediately rejects his secular, civic life to be a friend of God: *ecce nunc fio*, he concludes, 'behold, I become so now': the moment of change (*fio*) displayed (*ecce*) precisely as a moment (*nunc*), in the present tense. Augustine is fired up, humiliated, ashamed by this story of transformation – but spends the remainder of the book recalling how God brought Ponticianus' story to prompt Augustine to turn back to himself (*retorquebas me ad me ipsum*, 'you were turning me round to myself'), to face himself, to look at himself intently, to struggle with his own self-loathing story. He is tortured with doubt and indecision, *volvens et versans me in vinculo meo*, 'rolling and turning myself in my chain', as he

---

[18] Fredriksen (1986) 33; see also Kennedy (2013) 1–42.
[19] Ayres (2009) 263; see also Johnson (1991); Keevak (1995); Kotzé (2004). On Petrarch as one in the chain see Beecher (2004); Robbins (1985).

'hesitates to die to death and live in life', *haesitans mori morti et vitae vivere*. His opening biographical question about the death of his infancy now has its full moral impact: he must kill not just his past but the deathliness it is (in his retrospect) to live in the present of a true life. He is still desperately broken by the lure of the pleasures he must deny, however, and in his misery he cries out *quamdiu, quamdiu cras et cras? quare non modo? quare non hac hora finis turpitudinis meae?*, 'How long?! How Long?! Tomorrow and tomorrow? Why not now? Why is this not the hour of the end of my sinfulness?' Conversion has become a question of time. The Psalmist's cry to God ('How long, how long, O Lord' (Ps. 13)) is Augustine's tortured recognition of his own delay, his inability to act, which is also the deferral of Grace. 'Tomorrow and tomorrow' in its very utterance performs the delay it laments. What Augustine desperately demands is 'the now', the *nunc*, that Ponticianus celebrated: 'Why not now?'. He needs the immediacy of the moment: 'Why is this not the hour?' (Is 'the hour' in fact coming? Is it now? *Venit hora et nunc est*). Augustine's extraordinary narrative is a desperate plea to experience the immediacy of a present that because of his divided self, his human interior life, he cannot reach by himself without God's grace.[20] In the garden, his turnings finally make the turn of *conversio*. He hears the children repeating *tolle lege*, 'Pick up and read', and 'immediately' (*statim*) his faces changes; as Anthony had read one verse and become immediately converted (*confestim . . . conversum*), so Augustine picks up a text of Paul and reads one verse at random, and 'immediately' (*statim*) it is as if his heart is flooded with light. After all the delays (*cras, cras*), the moment of immediacy (*statim . . . statim*) is fully experienced. 'You converted (*convertisti*) me to you', concludes Augustine, and 'you converted (*convertisti*) my mother's grief to joy'. Now, now Augustine will no longer seek a wife, nor any 'hope of this age (*saeculi*)', nothing the age can offer. He and his mother will no longer seek the immortality of family, 'the grandchildren of my flesh' (*nepotibus carnis meae*) – which his father too had hoped for, to adolescent Augustine's dismayed embarrassment. Immediately on conversion, Augustine and Monica, son and mother, recognize that he has rejected the possibility of the future of their family, a rejection of the most insistent injunction of both Greco-Roman and Jewish communities, namely, the continuity of the family – a fulfilment of the most shocking radicalism of Christianity's call to change. Conversion, that immediate moment, is to transcend the *saeculum*, the time that is the mundane world, in the name of *vitae vivere*,

---

[20] For the relation between grace and will, see the crucial discussion of Wetzel (1992).

to live for a life beyond; and conversion is a rejection of the immortality of the generations which drives so much of Greco-Roman and Jewish moral and social expectation. Augustine makes visible what has to die for his future to take shape. Augustine's theoretical exposition of the unstable, disappearing moment of the present, his 'philosophy of time', which we have already discussed, comes after this extraordinary narrative of his search for the transformative immediacy of the now of conversion. The tension between the theory and the narrative is profoundly eloquent.

Augustine's longing for the now of grace is, then, also a longing for an instant of rupture that stands out against his long-drawn out process of gradual change over time. For later Latin monastic writing, however, conversion – the turning towards God – becomes a daily process, an ongoing and continuous dynamic of religious growth, through a striving repetition (a process that goes beyond Fredriksen's model of retrospective rupture). So, Cassiodorus writes that 'the soul [is] converted by an unstable and shifting will, nor does it abide in one firm purpose of the will, but even against its own disposition is changed in its orientation (*se conversione mutari*)' (*De an.* 4.214–18). The temporality of the monk's *regula* is a constant effort of conversion: the precarious construction of the self over time in its dynamic relation to God. Here, the moment of conversion is constantly deferred, or rather the moment of now is always ongoing.

Augustine boldly manipulates the time of his retrospective narration. Ponticianus' intervention is followed by several chapters of tortured reflection on his process of tortured reflection, his conflicting desires, his pain, before returning to the scene of his tearful conversation with Alypius. By contrast, when Alypius is given the same passage of Paul to read, he reads a little further, but also converts 'without any turbulent delay', a phrase that marks the narrative style as much as the psychology of Alypius. They go to see Augustine's mother together; they tell her what has happened; and she celebrates – but this part of the story is told in just three verbs (*ingredimur, indicamus: gaudet*). It is as if after the moment of conversion everything now has a narrative immediacy (the verbs are all in the present tense). It is also an immediacy that emphasizes how atypical the *Confessions* is as a narrative of conversion. Paul and Thecla, like Perpetua in her martyrdom, are more typical of saints' lives: they have no doubts, and no hesitations, and their conversions are not framed by any conflicting desires for a former life. (As we will see, later lives of saints certainly emphasize the temptations of former pleasures of the body, in the case of Mary of Egypt for fully seventeen years of solitary torture.) For Paul and Thecla it is what happens after their conversions that is significant and the narrative is shaped to this agenda. The 'before' has little importance for Perpetua's story, either, except in the form of her baby or her father, which are also rhetorical strategies to emphasize the total commitment of the martyr to her witnessing: already a Christian but not yet having won the crown of martyrdom. She too, like Augustine, denies the continuity of her family in the name of a Christian future for herself. Augustine is profoundly out of line with these other ways to tell a life in that he goes back to ask where he comes from and continues his story in the *Confessions* not just to his mother's death but also into the theoretical discussions of memory and time and, further, into the analysis of the book of Genesis, the narrative of where we all come from. All these texts are designed to change lives through their normativity – as Augustine dramatizes in the reception of the *Life of Anthony* and his *sortes sacrae* in Paul – but they differ not just in theoretical sophistication, intellectual scope or literary power, but also in their conceptualization of how conversion takes shape in time and how a narrative of conversion embodies that temporality. Augustine not only looks back in time in a ceaseless search for the causes he knows will escape his human comprehension, but also theorizes such a retrospective narration through his discussions of memory, time, interpretation, a theorization which is part of his experience of his human temporality and its attempt to reach towards God and the eternity of God's promise. If Paul's vision leaves humans between the already and the not yet, the hour that is coming and the now, Augustine frames this as a tension between the overwhelming now of God's grace at the transformative moment of conversion and the constantly unstable and slipping now of mundane time in which human life is necessarily lived. The very first question asked after the climactic moment of conversion is still *Quis ego et qualis ego?*, 'Who am I and what am I?'. Whatever conversion did for Augustine, it did not stop the insistent pertinence of these questions, or their uncertain answer.

*

Augustine's theologically laden description of the death of his past times is strikingly intense, and provides our first answer to how the 'now' is inhabited with a gaze on timelessness, but it will also have recalled to some readers, ancient or modern, a particular argument of Plato. In his *Euthydemus*, Socrates recalls a discussion he had with two sophists, the brothers Euthydemus and Dionysodorus. The dialogue is dramatic, funny, parodic and, although it starts with a testing of the claim of the brothers to teach virtue, which was a central concern of contemporary didactics, its extended focus is on a series of problems of self-refutation and contradiction, a set of issues that run through the philosophical literature to Augustine.[21] In *Euthydemus*, these arguments, conducted with both playful hooliganism and serious intent – and both tones are discussed explicitly in the dialogue – focus in particular on education and change: what is the dynamic of alteration between not-knowing and knowing, between being wise and being ignorant? Augustine's *Confessions* obsessively reconstructs the process by which his self became what it is, which includes reading Platonic or Platonizing philosophy; Plato establishes the philosophical agenda of exploring how the self alters or stays the same over time. Am I still the same person when I change? *Quis ego et qualis ego?*

What, then, does the *becoming* mean in 'becoming wise'? That is Plato's question.[22] After the long opening skirmishes, Dionysodorus gets the ironic Socrates to agree to be serious and asks him if he wishes to make their young friend, Cleinias, wise (283c–d). Socrates agrees to this aim. Dionysodorus gets Socrates to assent that, as Cleinias is not wise now, but ignorant, Socrates wants him to become what he is not, and not to be what he is now. Socrates therefore wants to destroy (*apolôlenai*) the Cleinias of the now. 'What friends and lovers they must be who would give anything utterly to destroy their beloved!'. To want to change a person from one state to another is to aim to destroy the person of now. Cleinias' lover, unlike Augustine when he recognizes the necessary 'death' of his past, is outraged by such a suggestion that he wants to destroy his lover, and the dialogue nearly collapses into chaotic dissolution. The aim of the sophists' display of technique is to '*turn (protrepein)*' the young Cleinias 'to philosophy' (274e), to convert him. Augustine, too, had his life-changing turn to philosophy; but Augustine's willing recognition that he lives (*vivo*), while his infancy and other former conditions are 'dead', in Plato's dialogue is seen as a challenging paradox about the continuity of the self through change and time. For Cleinias to be destroyed as he now is, is seen as a threat both to his being and to his lover's care.

The sophists' case seems to depend on a determination that qualities or conditions undergo change by replacement rather than by process. Either one is wise or one is ignorant; one knows or one does not know (or, as Augustine might say, either one is Christian or one is not). One cannot be both at the same time. For Dionysodorus and Euthydemus, as M. M. McCabe argues, 'truths are episodic'.[23] Thus, as Socrates himself recognizes, there is no place in their argument for learning as a process,

---

[21] Castagnoli (2010). [22] I follow here McCabe (2013/14). [23] McCabe (2013/14) 493.

only for knowing or not knowing. The dialogue opens with Socrates talking about how he is learning to play the lyre, old though he is, and this apparently passing mention of musicianship carefully cues one area where the possession of knowledge is least convincingly explained without a process of gradual acquisition. For Socrates, McCabe continues, wisdom itself is a quality which alters the areas of life in which it is embodied: it is a 'transformative good'. By contrast, the sophists argue that even contradiction is impossible because it assumes a continuity across truth's episodic conditions. The sophists themselves have moved from city to city (Socrates, of course, almost never left Athens), and have changed professions, reinventing themselves as they progress. Their characters and their arguments are of a piece. To understand what it is to become wise, then, to have a particular character, depends on negotiating a tension between a continuous process of change over time, and a recognition of the ruptures of difference. In a life-story, is there a point of change or only a process? If there is only a process, how can you articulate the difference between contrasting states of being?

The beginning of *Euthydemus* is marked by a surprising focus on ageing. Yet ageing – as Augustine with his insistence on the death of infancy underlined – is the very basis of the question of continuity and change in the self. Cleinias is introduced by Crito, who is talking to Socrates, as the 'youth (*meirakion*) whose father is Axiochus' (217b). *Meirakion* is a specific age-class in Athenian culture: not a child, and not yet reached the military age of the ephebe.[24] Cleinias has grown a good deal, notes Crito, and, although he is the same age (*hēlikian*) as his own son, is bigger and better looking. One of the other young men is described as 'arrogantly aggressive' (*hubristēs*) 'because of his young age'. Crito asks Socrates if he is not worried to fence with the new younger sophists, as 'he is rather old'. Socrates retorts that the brothers themselves, Euthydemus and Dionysodorus, are 'old men', because they became sophists only 'a year or two ago'. Socrates tells Crito that he has already been mocked by boys because he goes to lyre-playing class with them, and they call the teacher *gerontodidaskalon*, 'tutor to the old'. And Socrates adds that he brought some old men with him for support, and asks Crito to help make up the team, and remarks that his own boys will be a lure to encourage them to come along. Banter, for sure, but designed. The opening paragraphs remind the reader of the transitions of life, growing up, passing through the prime of life into old age, the distance between the young and the old.

---

[24] On the importance of age-class see Davidson (2006).

Biological continuity grounds the continuity of the self. When Dionysodorus actually argues that he knew how to dance with swords as a new-born, even Socrates sniffs his dismissal of such a claim without the need for an argument. There is a recognized pattern of *to eikos*, what is natural, likely, in the biological continuity of the body. It is a sign of horrific chaos if humans are born grey-haired, as in Hesiod's vision of the future of the Age of Iron. The dialogue has already ambushed the claims of the two sophists, by establishing a continuous process of ageing as the norm for social and biological life.[25] Isn't ageing a necessity of any life-story?

Cleinias is the object of erotic attention by the men who accompany him (erotics and education are so often overlapped in the elite pederastic environment that Socrates and his chums inhabit), and erotics also negotiates this tension between continuity and rupture through its familiar patterning of a life of loving. Plato's dialogue *Protagoras* begins with an unnamed friend asking Socrates if he has just come from 'the hunt to catch Alcibiades' moment (*hōran*, 309a)'. I have translated *hōra* as 'moment' because the word implies the right or perfect time of desirability ('the freshness and vigour of youth', as the standard dictionary has it). A boy is attractive only for a short season. Sappho lyrically captures this connection between *hōra* and erotics, but from the perspective of ageing: 'The moon has sunk, and the Pleiades; it is the middle of the night. The *hōra* has passed by (*para d' erchet' hōra*). And I sleep alone'.[26] The time of this night, like others, the season of the year, and the moment for love in her life as Sappho ages, have all passed (*hōra* has these senses), and the juxtaposition of the fading time and sleeping alone is eloquently poignant.

In the *Protagoras*, the friend goes on, indeed, to suggest 'between us, as it were', that Alcibiades is actually a man now, with a beard – and thus according to normative expectations no longer fit for such a hunt. Socrates, performing the elite banter of such erotic discourse, quotes Homer in response, retorting that actually the most attractive time is when the beard *just* appears, as Alcibiades now is: the margin, the moment of transition, the scene of impending loss is the most erotic ... The gossip

---

[25] Augustine uses the ages of man further to express the ages of the world in *In Gen. contra Manichaeos* and *De ver. relig.*: see Ladner (1959) 232–8. This parallelization continues up into medieval Irish texts: Clarke and ní Mhanoaigh (2020).

[26] Fr 168b Voigt. It has been much debated whether this fragment is by Sappho, as our ancient source attributes it. I take *mesai nuktes*, translated usually as 'it is the middle of the night', to suggest by its plural that this is not the only night for such a reflection. Which *de* answers *men* structures the reading(s) of the poem.

continues: *ti oun ta nun*, asks the friend, 'so, what's the case now?': how is the moment for the moment (*hōra*) developing? The 'now' is the *hōra* which desire seeks: a *kairos* to be grasped or missed, in the moment, the lover's now.

The timing of desire is a cliché that organizes not just erotic discourse but also the lived life of the citizen (and the women of the city too). In classical Athens, the social organization of the biological requires that the young male who is the object of desire, as he becomes a man, will become the one who desires; manhood expects marriage and children; a girl when she becomes a *parthenos* is expected to marry as soon as possible. So much of the literature of desire in antiquity is concerned with this timing. New Comedy's plot (in all senses) is to get the right girl married at exactly the right time. Or in epigram the constant repetition of the *kairos* of the epigram's wit creates the *kairos* of the erotic moment. Or in the novel or other extended love stories there is the constant delay of erotic fulfilment: the waiting for the moment that becomes its own erotics of fervour and deferral. Desire may be episodic – *amores* plural – but there is a continuity of ageing that structures each self's expected position. Aristophanes' joke scenario, where a young man is forced to sleep with increasingly older women, before he can sleep with his own young girlfriend, has the shape it does partly because of these expectations of timing, partly because of its amused glance at the power of the old in enforcing social norms, the *senum severiorum*, 'the all too severe old men'; Horace's satiric aggression towards old women – or the epigrammatists who write without sympathy about the decrepid fate of old prostitutes – have a similar mix of not just misogyny, but also a fierce conservatism about the proper sense of time, the necessary timeline of erotics.[27] To have sex with someone whom society judges too young or too old is scandalous, then and now. The *chronos* of a life of loving depends on its recognized moments of *kairos*.

Imagining time is organized through the body: a biological clock, time of the month, onset of puberty, arrival of bodily hair, loss of hair, menopause and so on. 'Any ethics we might wish to derive from a consideration of temporality must contend with the irreducible force of time's movement on our bodies', writes Valerie Traub.[28] Both the language of ageing and the

---

[27] For how temporality gets queered, nowadays, by sexuality and its corporeality, see e.g. Edelman (2004); Halberstam (2005) and, with a medieval spin, the excellent Dinshaw (2012), and, for a slightly later era, Traub (2016); Freccero (2006).

[28] Traub (2016) 75. On women's time, see the influential Kristeva (1981); also for summaries of more recent discussions, Radstone (2008) 56–111, on the body, 71–111. For the Jewish story of gendered time and the body see Gribetz (2020) 135–87.

language of erotics in *Euthydemus* project a culturally embedded timeline of the lived experience of the body – a continuity of the bodily self – to set against the episodic truths of the sophists. How can a self-consciousness of bodily continuity be denied, without denying something basic about a human sense of the self in time? The Gospel may proclaim that 'the hour is coming', *erchetai d' hōra*, but Sappho knows that *para d' erchet' hōra*, 'the hour is passing by'.

*

The rupture of Christian conversion, however, is not so much a bodily change as a turning of the soul towards God. A turning of the soul which demands a denial or repression or suppression of the bodily; which makes life in a body – the biological journey from birth to death, and its erotics – a deathly experience, a *vita mortalis*. Conversion disrupts how time is organized through the body. The life-writing of Christianity sets out to alter the culturally embedded narrative of the body, its organization of time.

Augustine in the *Confessions* describes his erotic desires with a focused fleshliness, from his recollection of how his father happily recognized in the bathhouse that he is now capable of siring children, to his own admission that even after conversion he has longing sexual dreams that result in fully physical experiences. Augustine testifies that his turn from the body is a painful, ongoing struggle. In later, ascetic writing, especially in the lives of saints, there is a more violent intimacy with the bodily.[29] The Life of Mary of Egypt, as written in the seventh century, tells of one Zosimas who thinks he has nothing left to learn about ascetic practice, but travels to the desert beyond the Jordan in the hope of finding a holy father who can inspire him further.[30] He finds instead an old naked woman, her skin blackened by the sun and her flesh emaciated. She tells him her story: how she left home in Egypt when young because of her desire for sex, and how she lived a life of complete debauchery, offering men sex for the pleasure of it – she never took money – until she arrived in Jerusalem, where at the Church of the Holy Sepulchre, thanks to the virgin Mary, she has a religious experience of conversion that leads her to immediate chastity and her ascetic life in the desert. She took only two and a half loaves of bread with her, after which she has existed solely on what she could find in the desert. For seventeen years, she recounts, she struggled with the demons of temptation, as she recalled her experiences of wine, song and sex. There is a lurid narrative of remembered sin; but after conversion, the only story is temptation.

[29] On the role of saints lives, see especially Perkins (1995); Clark (1999); Castelli (2004); Kelley (2006).
[30] Burrus (2004) 147–54; Burrus (2019) 93–117.

Temptation is the poetics of asceticism. When temptation is conquered, there is no story for the remaining thirty years. She describes the thirty years of her life after temptation simply as 'Lasting calm after a violent storm', a calm defined by negativity: she has seen no one, read no book, heard no voice. The life ends with Mary's miraculous death and burial (with the help of a passing lion). Zosimas has learned of a greater ascetic than himself (asceticism is always competitive). Mary has crushed her bodily desires by her suffering and repentance in the desert, and this leads to a timeless present, ended only by the transformation of death. In the life of another Mary, the niece of Abraham, this Mary, after being seduced, goes in shame to a brothel, from where she is rescued by her uncle Abraham, and returns to an uneventful solitary life of penance. In the 'Life of Pelagia', Pelagia, an actress lavishly adorned with jewels, after her conversion – a complex, erotically charged story – cuts her hair, takes on male dress and becomes a solitary ascetic monk, unrecognizable as herself, and lives without event until her death, when her gender is revealed to a shocked world. In all three cases, the women separate themselves from the expected timelines of the female body; their bodies are transformed; and they finally enter a life without differentiated events, without the measurement of time, except for the memory of the before and after that is repentance.

Simeon Stylites is even more aggressive towards his bodily condition.[31] He lives on a series of high pillars: the height of each pillar and the years he sits on it are carefully recorded, as if to compensate for the indiscriminate repetition of daily suffering. His ascetic violence ties ropes so tightly to himself that they fuse with his flesh; he stands on one foot for two years because of an ulcer on his thigh; his flesh crawls with worms and stinks foully. His body oozes from open wounds. Here the mortification of the flesh takes on a grim literalness: like a corpse, his stinking body crawls with animals and is decaying. When worms fall from the pillar, he has them gathered back and welcomes them again onto his flesh. For Simeon, the holy man, the arrival of other humans for advice or blessings or reproach is no more than an interruption of his desired continuity of physical suffering, his turning of his body into a display of mortification, a *vita mortalis*; an interruption of his unmeasured time. Fasting, the primary practice of asceticism, as Peter Brown eloquently demonstrated,[32] is designed to control physical desire and, for a woman, to stop menstruation, the physical symptom of the Fall, in order to get back to the state of the

---

[31] Burrus (2019) 123–34.   [32] Brown (1988).

Garden of Eden. Asceticism aims to break the social and biological time-lines of bodies, hoping to reverse time to a time before, while waiting for the time to come.

'The *Life* of a saint', writes Virginia Burrus, 'breaks with the chronologies inscribed by both the biographical and novelistic conventions'.[33] 'Birth' she continues, 'does not anchor a social identity', as it does in Greco-Roman literature. The social conditions of the beginning of a life are 'unknown, ignored, transformed or abandoned'. The standard time-lines of ageing, with the social expectations of marriage and children, are rejected. The careers of civic duty or family property are resisted. The body and its desires become an enemy of the self. It is not, however, quite that 'time, for the saint, is unhurried, open, meandering'.[34] Rather, against threat or temptation, there is a narrative hope for the timelessness of waiting, which is also the continuity of unmeasured suffering ('to be Christian is to suffer'[35]); a hope that until death the *vita mortalis* should be uninterrupted by any irruption of the *saeculum*. In such a challenge to biological and social norms, such Lives of the Saints offer their faithful readers a new narrative of a life, a precarious, impossible idealism of how the self can inhabit time.

*

In retrospect – as all narratives of conversion are (re-)written and (re-)read – these later lives take to a further, more extreme level trends that are already evident in earlier Christian life-writing. Gregory of Nyssa, for example, wrote a Life of his sister Macrina.[36] Her birth is marked by her mother having a vision of a guardian angel who instructs her three times to call the child 'Thecla' – after the saint who converted at the sound of Paul's words. This secret name, reveals Gregory, was a sign of Macrina's future life as a holy virgin. So, indeed, when Macrina is engaged to be married to a young man of talent, who promptly dies, she takes this engagement as binding and refuses all other possibilities of marriage – sexual threats which, unlike the case of Mary, niece of Abraham, or, indeed, Thecla, disappear easily and without enactment. The story of her life is framed as a commitment to 'philosophy', the term often appropriated by Greek Christians for their faith, and which we have seen in quite different guise in Augustine's self-transformative reading of

---

[33] Burrus (2019) 140.   [34] Burrus (2019) 140.
[35] Perkins (1995) 32 – the conclusion, she suggests, to be drawn from such hagiographies. Boyarin (1999) 93–126 focuses rather on identity formation.
[36] Burrus (2004). On Gregory of Nazianzus on *his* sister see Burrus (2006); Elm (2006), and, more generally, Momigliano (1985).

Neo-Platonic texts. In his dialogue *On the Soul and Resurrection*, also set at Macrina's deathbed, Gregory depicts himself as her pupil, a Socrates to her Diotima, or, since the dialogue is a deathbed discussion on the immortality of the soul, a Cebes or Simias to her Socrates.[37] He calls her 'The Teacher', *hē didaskalos,* throughout, a displayed commitment, as ever, to *paideia*, typical of the Cappadocian fathers' 'project of cultural reclamation',[38] though it is particularly worth noticing that here the teacher of philosophy is female.[39] Macrina stays by her mother (as Augustine does with Monica), and encourages her too to greater levels of 'philosophy', as well as comforting her at the death of her son, Macrina's brother.[40] In a strikingly self-reflexive phrase, Macrina, unable to rise from her bed, tells Gregory her life-story 'as if a written history' (*sungraphē*, 20.3) – the story he then writes for us (*sungraphē*, 18.13). Gregory's biography of Macrina, with its inset account of Macrina's autobiographical story, 'as if a written history', presents 'a dying woman willing herself into prose, easing her own transition from life to memory',[41] which turns hagiography's 'narration itself into a pious act',[42] a liturgy where the text replaces the body, and becomes the stimulus to spiritual change, and a transformation of the reader's lived life: memorial *as* a continuing praxis – the *askēsis* of a life becoming the *askēsis* of writing and reading a life to encourage the *askēsis* of a life . . .

Gregory himself has returned home at the point of Macrina's death, and the scene of death and burial takes up a good proportion of the whole life. This life is full of stories of death (the destruction of this holy family's future as a family). Surprisingly – to him and to us – Gregory at the end is shown his sister's partially uncovered body where there is a small mark on her breast. It is explained to him that this is a sign of a miraculous cure. Macrina had a tumour, but refused to allow any doctor to view her body out of her heightened sense of modesty; she treated herself with mud moistened from her own tears, shed in an all-night vigil, at the foot of the altar, and when her mother made the sign of the cross on the breast, she was cured. The mark remained as a sign of the miracle and a proof of her sanctity. As with Odysseus, as Georgina Frank has outlined, Macrina's heroic identity is

---

[37] See Ludlow (2015); Muehlberger (2012); both with extensive bibliographies to earlier discussions.
[38] Muehlberger (2012).
[39] Ludlow (2007) 202–19 has a full discussion of responses to the issue of gender.
[40] For the family/private frame of these scenes see Bowes (2008) 202–16.
[41] Krueger (2000) 492, an excellent article more broadly contextualized in Krueger (2004).
[42] Krueger (2000) 497.

proven by a scar.[43] Her body is transformed by being cured rather than by rotting (though Simeon would no doubt say his mortified flesh is a cure of the sins of life). As with Mary and Pelagia, the truth of her body can be seen only at death. A saint's body does not corrupt, it outlasts its own temporality, and, Gregory suggests, here is the first sign of this sanctity to be witnessed on his sister's breast. The slow, withdrawn life of Macrina, committed to a home-life without marriage, and marked for sanctity on her body at death, anticipates the more extreme isolation and bodily transformations of a Mary or Pelagia.

In a similar way, Jerome describes an idealized Paula at her death. Jerome knew a thing or two about the temptations of the ascetic. He admitted to seeing visions of dancing girls during his own time of retreat in the desert ('How often I longed for the delights of Rome!', *Ep.* 22.7), and, in his *Life of Paul the Hermit*, conjures a more detailed, self-exposing scene in which a young man is tied up naked in a beautiful grove, and then visited by a prostitute. Unable to move, he bites off his own tongue and spits it at her to avoid kissing her, but not before she has stimulated his genitals and mounted him. For the ascetic, the necessity of internal life is a temptation to acute erotic fantasy. Jerome had, of course, lived at the heart of Roman society, and, unlike Mary, did not stay in the desert. Paula's temptations are not like Jerome's, however. He describes Paula leaving her children as she sailed to the Holy Land, without a tear in her eye (*Ep.* 108).[44] No temptation for her to stay home and play out the role of mother, even though 'no mother ever loved her children so dearly' (*Ep.* 108.6). Jerome's letter emphasizes that Paula was so holy that she had no longing memories for her wealthy past in Rome, troped through the biblical narrative of the Israelites' inability to forget Egypt in the travails of freedom: 'To the day of her death, she never returned to Chaldaea, or regretted the fleshpots of Egypt or its strong-smelling meats' (32). 'Strong-smelling' is Jerome's addition to the biblical language, increasing the sensuality of longing. Her settled life in Bethlehem – I have already discussed his description of her pilgrimage in chapter 6 – is also elevated, in a way that the later saints will dramatically embody, into a form of martyrdom: 'It is not only the shedding of blood that is accounted a confession: the spotless service of a devout mind is itself a daily martyrdom' (32). The mundane becomes the exceptional, as the passing of time is a continuous martyrdom. The transvaluation of waiting ... The only temptation he does allow Paula is

---

[43] Frank (2000a); Burrus (2004) 69–76, differently; and Macrina's ring, Burrus (2019) 151–6.
[44] Standard edition is Cain (2013).

the argument of an Origenist Christian, whose views on the resurrection of the body are called heretical by Jerome. This leads him to speak passionately about the continuity of biological time and the body as a grounding of the self (24):

> No difference of age (*aetatum*) can affect the reality (*veritatem*) of the body. Although our bodies (*corpora*) are in a perpetual flux, and lose or gain daily (*cotidie*), will we be therefore as many persons as there are daily changes? I was not one person at ten years old, another at thirty and another at fifty; nor am I another now when all my head is grey.

His *epitaphium*, 'funeral oration', for Paula is a life-story, and for him the continuity of the life of the body and the daily work of self-sanctification constitutes a narrative of martyrdom, crowned now by her death. Her journey from Rome culminates in the long, still years of prayer and worship, a time without the change or eventfulness of narrative: 'day and night alike a time of almost unbroken prayer' (15). Her very obscurity is her glory, declares Jerome. Jerome rewrites the celebrated phrase of Pericles' Funeral Oration in Thucydides, that 'a woman's greatest glory is to be least talked about by men'. Paula is to be talked about and praised volubly, as he here performs, precisely because she turned her glorious birth and wealth into this humility of withdrawal from social process, her turn to daily martyrdom.[45] Similarly, in the company of virgins, she was 'least remarkable', *minima omnium*, which pointedly rewrites the classical praise of Nausicaa in the *Odyssey* or Electra in Aeschylus' *Choephoroi* or Artemis and her devotees, each of whom stands out in stature and beauty from the band of women they are surrounded by. Her life-story turns on her conversion. She prays that she must disfigure her face that had worn makeup; she must afflict her body that has had pleasures; she must cry because she has laughed; because she has pleased a husband, now she must please Christ. She makes her journey from Rome and her noble life, through pilgrimage, to the nunnery that she builds at Bethlehem and whose order is described at length by Jerome. Jerome even goes so far as to suggest that Paula herself in her daily piety is an object of pilgrimage for others.

Paula's life-story in Jerome's letter tells of her repeated illnesses, and her patience and survival of such difficulties, and of her care for the other virgins in the nunnery: her daily round. The structure of the narrative is clear, however: her married life in Rome is broken by her conversion to an

---

[45] Gregory of Nazianzus, *Or.* 43.5 uses the phrase 'living martyrs' on which Rousseau (1994) 5 comments: 'In those simple phrases, fearful refugees were brilliantly transformed into pioneers of a new age'.

ascetic ideal; her pilgrimage – described at length – leads her physically and symbolically to Bethlehem, where she stops, and, in ceaseless prayer about her conversion and commitment to an 'afflicted body', lives out her liturgical life over many years of daily martyrdom. Again, these are the terms that later Lives of the saints will take to heightened levels of absolute withdrawal, continuous supplication, and violent mortification of the flesh.

Perhaps the fullest expression of autobiographical writing before Augustine is found in Gregory of Nazianzus. He writes three long poems on his own life; he edited and published a collection of his own letters (the first Greek writer to do so); he writes funeral orations for family members, sermons for his community, and as discussed in chapter 2, his funeral epitaphs include dozens of poems about himself, his mother and his extended family. Gregory, as we saw, for all his own commitment to virginity, is constantly negotiating an ameliorative and assimilating position between his vision of learned and civilized urban Christianity and the tradition of Greek culture in which he has been educated.[46] In their different ways, Gregory of Nyssa, Jerome, Gregory of Nazianzus, and Augustine show how in both Greek and Latin communities these earlier life-stories are consciously and intricately engaging with the traditions and practices of the life-stories of the dominant Hellenic and Roman culture of empire, finding a space for a Christianity that recognizes the institutions and lures of contemporary society, and strives to articulate a new place for itself that will engage Christians and non-Christians in the 'philosophy' of an educated life. The later Lives of saints, written in an established Christian empire, offer more extreme acts of social rejection and bodily punishment, narratives which are both lived up to in ascetic practices and good to think with for their reading communities – maps of a life to find one's self on. *How* such life-writing becomes normative, *how* such life-times are lived up to, forms one crucial history of the formation of Christian sociality over time.

The two strands of this chapter, then, are intimately connected. On the one hand, how to narrate the turning point of conversion? How much is it a question of process, how much a blinding flash? On the other, how is the life after conversion to be narrated? Is it to be a mission of converting others? Or a constant repetition of repentance, a performance of self-punishment? In both strands, time plays a fundamental role, not least in

---

[46] Hofer (2013) argues well that Gregory's theology can only be understood through this autobiographical mode, 'an appropriation of Christ's life to his life' (6).

how narrative is shaped. Is conversion to be expressed as a 'before' and 'after' where the 'now' of process is minimized? Is there a continuity across time in the self or the body which remains beyond conversion? How is the time after conversion to find a narrative form? Fairy stories accustom Western children to an ideal or a fantasy of the changeless time of marriage with the closural formula 'they lived happily ever after'. Saints' Lives, the lives of the converted *sancti*, work to find their own formulae, with death as their closure. 'She lived with an awful but necessary recognition of sin ever after', 'He lived in constant and increasing self-inflicted physical anguish to find increasing spiritual transcendence ever after.' The stretch of time I characterized in chapter 4 as 'waiting' becomes thus the problem and resource of the narration of a life-story. To discover a certain timelessness in the now.

*

The history of Jewish life-writing in late antiquity contrasts with both the Greco-Roman and the Christian traditions. I have written at length elsewhere about the complexities of this story.[47] Although there are early cases of biographies and even autobiographies by Jewish writers – Philo's *Life of Moses*, say, or Josephus' *Life* which is a turncoat's *apologia pro vita sua* – these examples come from the most Hellenized strata of Jewish culture, written in Greek in Hellenistic Alexandria in Philo's case, and in Rome in Josephus', and they follow the conventions and expectations of Greco-Roman genres of life-writing: Greek texts for a Greek-speaking community. But in the rabbinical writings of late antiquity, written in Hebrew and Aramaic, there is a refusal to write continuous life-stories, even when the material is easily available. Unlike the scriptural accounts of David or Moses, the celebrated rabbis of the Talmud have no birth to death narratives, no explanations of behaviour according to character or education; little motivation by family circumstance or even social setting, beyond the recognition, say, that the upper classes communicate better with the rulers of Greece or Rome. Nor are the lives of scriptural figures retold as continuous stories, though midrash produces anecdote after anecdote of discrete biographical events that do not appear in the Bible. Nor is there much interest in conversion either as a narrative of transformation or as a ritual possibility, until Christianity makes such an issue unavoidable. With the exception of the strange novella *Joseph and Aseneth*,[48] the representation of such internal or social change is at best

47 Goldhill (2020) ch. 6.   48 See Goldhill (2020) ch. 5.

scattered and without the anxious obsession that Christian texts demonstrate. The ideological and narrative focus is on the precarious and porous boundaries of community and how they are to be policed. Life-writing focuses on episodic incidents with an implication for *halacha*, Jewish law. The narrative style of the Talmud – quite unlike anything in Greek, Roman or even Christian traditions – places the self differently in history, as we have already discussed, and insists on a constant rabbinical perspective that life is to be measured by *halacha*. The narrative style of the Talmud engages the Jewish reader in a constant set of discrete questions of what it is to be a good Jew, which thus also performs the demand that such questioning is itself part of the answer. The self in time here is represented in and through such repeated questions of ritually and legally correct behaviour. This is how belonging and commitment are regulated. Christianity in late antiquity, in its aggressive and self-defining gestures of separation from Judaism, repeatedly argues against what it thus defines negatively as the legalism of Judaism. Polemics between Judaism and Christianity are enacted in and through these different narratives of what a life-time looks like.

Biography is a form of writing where narrative most evidently depends on an idea of time, where the self in time finds narrative expression. Rabbinical texts show with remarkable sharpness how a particular sense of self is constructed by a particular style of temporal narrative. But both Christianity and Judaism reveal how rewriting the form of life-writing – a contesting rewriting that contrasts and engages with the traditions of Greco-Roman culture – is integral to the development of their sense of community, belonging and commitment. Life-writing projects models to live by: how life-writing is lived up to, however, is always a question. In what ways these changing models of the self in time inform the literature of late antiquity will remain a structuring concern of this book.

CHAPTER 10

The Rape of Time

One of the most surprising expressions that captures a self-conscious positioning in time comes in a remarkable poem in Book 1 of the *Palatine Anthology*. The poem is not an epigram at all, despite its inclusion in the anthology, but rather a 76-line poem of dedication celebrating the construction of the church of St Polyeuctos in Constantinople.[1] The church was rebuilt in the early sixth century by Anicia Juliana, who came from one of the most distinguished family lines in eastern Greek Christian nobility. Indeed, the original church had been built in the fifth century by the empress Eudocia, who was Juliana's great-grandmother. Juliana, in her act of rebuilding, was certainly engaging in the familiar competitiveness, aimed at both contemporary and historical rivals, that continued Hellenistic euergetism – the public display of wealth and authority through the sponsoring of public buildings – in order to contribute to the splendid redesign of Christian urban space. Juliana was also inevitably engaged in the still rumbling arguments between the supporters and opponents of the settlement at the Council of Chalcedon, and it has been plausibly suggested that such building projects were likely to be embroiled in such public statements of religious affiliation as well as political clout. The first book of the *Palatine Anthology* is entitled 'Christian Epigrams', which justifies the inclusion of this poem in itself, but there are also several epigrams collected in the book, which are written as if they were intended to be inscribed on the walls of a church, and some of which may have been so.[2] Thanks to a stunning archaeological discovery back in the 1960s, it has been demonstrated that this long poem, or at least the first 41 lines for certain, were actually inscribed in the church of St Polyeuctos, making it one of the longest verse inscriptions to have survived from antiquity – and testimony of the lavish scale of the church and its decoration, which other

[1] See Whitby (2006); Bardill (2006).
[2] See Agosti (2018) and Agosti (forthcoming) for other examples of church inscriptions.

evidence confirms and extends.[3] This was a major building project in the capital of the eastern empire, and it made a statement, both in itself and in the words with which it was decorated.

The moment which is so surprising, however, comes in the second section of the poem for which there has been no archaeological evidence for its presence in the church yet discovered, although the manuscript indicates that it was also inscribed in the building (although exactly where is less clear). Let me quote the surrounding lines, too, as the whole passage is salient (*AP* 1.10.43–51):

> ποῖος Ἰουλιανῆς χορὸς ἄρκιός ἐστιν ἀέθλοις,
> ἢ μετὰ Κωνσταντῖνον ἑῆς κοσμήτορα Ῥώμης,
> καὶ μετὰ Θευδοσίου παγχρύσεον ἱερὸν ὄμμα,
> καὶ μετὰ τοσσατίων προγόνων βασιληίδα ῥίζαν,
> ἄξιον ἧς γενεῆς καὶ ὑπέρτερον ἤνυσεν ἔργου
> εἰν ὀλίγοις ἔτεσιν; χρόνον ἥδ' ἐβιήσατο μούνη,
> καὶ σοφίην παρέλασσεν ἀειδομένου Σολομῶνος,
> νηὸν ἀναστήσασα θεηδόχον, οὗ μέγας αἰὼν
> οὐ δύναται μέλψαι χαρίτων πολυδαίδαλον αἴγλην

> What chorus is sufficient to the efforts of Juliana?
> After Constantine, who ornamented his Rome,
> After the all-golden holy eye of Theodosius,
> After the royal tree of so many ancestors,
> She completed a work worthy of her descent and even grander
> In a few years. She alone raped time
> And surpassed the celebrated wisdom of Solomon,
> When she rebuilt a temple to receive God, for which a great span of
>   years
> Could not hymn the multiform splendour of its beauties.

I have translated the Greek verb *ebiēsato* with the unpleasantly violent word 'raped' for its full shock value. It has been translated more decorously as 'overpowered',[4] or, even more blandly, 'conquered'.[5] It is a very strong term, however, most often used of sexual or physical violence, or aggressive compulsion ('force'), and is a very strange and unparalleled verb to take 'time' (*chronos*) as its object.[6] What does it connote here? It indicates that Juliana's building project must be seen as an intervention in history, an intervention that not only turns history on its head – violently – but that also she alone (*mounē*) could achieve. This intervention has different instrumental trajectories. First, Juliana is placed within a particular genealogy (*geneē* – the

---

[3] Harrison (1983), (1986), (1989).   [4] Whitby (2006).   [5] Bardill (2006); Whitby (2006).
[6] Clement of Alexandria (*Stro.* III.8.61.1) accuses those who select texts self-interestedly to support their pleasures, of 'forcing', 'raping' scripture: *biazomenoi graphas*.

line of generations celebrated by Homer and rejected by Augustine's Christianity, as we saw). This genealogy includes noble ancestors by name – Constantine, Theodosius – and the many unnamed nobility who make up the chorus evoked in the first line; it also specifies the great builders of previous generations, Constantine who decorated 'his Rome', Constantinople; and Theodosius, probably Theodosius II, who built the walls of Constantinople. Juliana's project is said not just to be worthy of such ancestry but also to surpass it – celebrating her competitiveness with the heroes of the past. And her project has been achieved in only a few years. The speed of her transformation of the cityscape is part of her attack on normal time.

But beyond this, her temple is praised as a triumph over the lauded wisdom of Solomon. *Aeidomenou*, 'lauded', 'sung', 'celebrated in epic verse', recalls both scripture's praise of Solomon especially in the book of Kings, and also the long pagan tradition of recording fame in song, which we discussed in chapter 2. *Sophiēn* is not only the iconic quality of Solomon, his 'wisdom', but also comes trailing clouds of the long tradition of calling not just Christian theology, 'wisdom', but also specifically the transcendent mind of God, as incarnated in Jesus. This claim to surpass Solomon's wisdom also and most saliently indicates that this *naos*, this temple built by Juliana, rivals the Temple of Solomon in Jerusalem, and is the Christian spiritual triumph (*sophiē*) over the Jewish Temple of ritual. Solomon's Temple of Jerusalem, destroyed by Nebuchadnezzar in the sixth century BCE, was reimagined by the prophet Ezekiel in a heavenly form of perfection, and the second Temple, rebuilt by Zerubbabel after Cyrus' return of the Israelites to Jerusalem, and reconstructed by Herod, was razed to the ground by the Romans in 70 CE. The book of Revelation prophesied that the Temple, as imagined by Ezekiel, would reappear in the eschatological era. It would descend from heaven. The church of St Polyeuctos is being heralded thus as the earthly copy of the new and better Temple foreseen in the scriptures (Hebrews 8.5). Juliana is not the first to make such a claim. Eusebius in the *Life of Constantine* declares that Constantine's basilica over the site of the crucifixion – what we now call the site of the Church of the Holy Sepulchre – was a 'New Jerusalem' (3.33), and he praised Paulinus' temple at Tyre as the New Temple to surpass Solomon's or Zerubbabel's (*Hist. Eccl.* 10.4.2–72).[7] The praise of Juliana is also competitive with the praise of Constantine and Paulinus. Her Temple, and it is not without point that this is a woman as builder, is overwhelming (*ebiēsato*) previous times of praise. An age – *aiōn* – would not be enough to

---

[7] On Paulinus' Temple see Wilkinson (1982).

praise it adequately: this project not only surpasses the past but also requires the future for its adequate celebration, conquering competition to come. Ironically, the most famous claim to have surpassed Solomon is meant to have taken place only a few years later in 537, when the emperor Justinian is said to have declared when he entered his new church of Hagia Sophia in Constantinople, 'Solomon, I have defeated you (*nenikēka*)' – at least according to the ninth-century *Diegesis* in which the story is first attested, as part of the legend of the construction of the church.[8] Hagia Sophia, too, is imaged as the fulfilment of the prophecy of Ezekiel.[9] St Polyeuctos and Juliana are overtaken by Justinian and Hagia Sophia in the competition to defeat time.

Juliana thus 'forcibly overwhelms time' at multiple levels: she embodies and surpasses her own genealogical excellence; she has bettered the great city-builders of the past; above all, she has triumphed over the wisdom of Solomon, in embodying a Christian spiritual vision over the ritualism of the Jewish deep past, thereby going beyond the praise of former generations, and defeating in anticipation future rivalry. The startlingly strong vocabulary of *ebiēsato* captures the scope of this vision of supersedence. Juliana does not merely inhabit time, but bestrides it like a colossus.

*

Ebiēsato is a surprising word not least because its subject is emphatically female, and, in Greek discourse of all periods, *bia*, 'force', 'violence' – especially with the sexual sense of 'rape' – is strongly associated with men, and rarely in a straightforwardly positive sense. Especially in a public context of praise for an authoritative female figure, the term is arresting, and the very surprise has a functional effect in drawing heightened attention to exactly how Juliana's achievement is temporal. The function of surprise in language is integral both to Aristotle's discussion of metaphor and to the practice of comedy (to take but two specially well theorized examples).[10] For Aristotle, in contrast to the *eikos* of semantics, the normality and normativity of expression, the surprising conjunction of terms in metaphor is productive of new meaning. In comedy, the twist of the end of a phrase to an unexpected sense is a staple of jokes, captured simply in the standard recognition by commentators of the *aprosdokēton*, 'the unexpected'. Freud, of course, discusses the pleasure of such a transgression of

---

[8] Dagron (1984) 191–314; Mango (1992); Brubaker (2011).   [9] Gerhold (2018).

[10] For Aristotle see Kirby (1997) with huge bibliography 518 n. 3; Lloyd (1987); Laks (1994). On laughter, see for classical material Beard (2014); Halliwell (2008). In modern theory, Bergson (1911) has been especially influential. No surprise to see theorist of time also writing on comedy – timing.

the normal pattern of language and the desires it embodies. The hooligan-ism of wit and the sublimity of lyric poetry share their desire to disrupt the authoritative ordinariness of language.

The unexpected, the sudden, the unanticipated goes to the heart of how narrative and time are interconnected. Viktor Shlovsky, we noted in chapter 4, made the manipulation of time the defining characteristic of literariness. Suddenness – which Karl Heinz Böhrer has analysed with regard to the moment or the flash of a glance (*Augenblick*) as a key term for modern literature under the rubric of *Plötzlichkeit* – depends on an intervention in the normal or expected timing of events or their telling.[11] For historians in antiquity, and for rhetoricians alike, responding to the unexpected is a sign of extemporizing brilliance which is a product of character and training. History itself, said Polybius, was a 'spectacle of surprise', *paradoxon theorēma*, which requires us to recalibrate the lesson that its narrative can provide.[12] Communication between gods and men is a key site of the unexpected – either in the epiphanic disruptions of a god's sudden presence or absence in Homer, say, or in the demand for the discovery of the hidden meaning of an oracle.

Plato, however, makes 'the sudden', *to exaiphnes*, a technical term in his philosophy of time and a crucial device in his narrative of change. In general, Plato insists that philosophy, unlike the time-bound and competi-tive performance of the law court, requires the slow, careful labour of dialectic towards the truth. Yet at crucial moments, dialectic gives way to the sudden. In the *Republic*, the crucial transition of the prisoners of the cave is a 'sudden' (*exaiphnes*, 515c) and unexplained freedom. In the *Symposium*, Diotima describes the celebrated ladder of desire as a slow process of growth, but the vision of the Beautiful itself is a 'sudden' (*exaiphnes*, 210e) flash of recognition. This sudden, transformative inter-ruption, however, is immediately performed in quite a different way by the 'sudden' (*exaiphnes*, 212c) interruption of the *Symposium*'s discussion by the arrival of the drunken Alcibiades; and his rambling, charismatic greeting is itself interrupted by his 'sudden' (*exaiphnes*, 213c) spotting of Socrates, a sign of his own philosophically longing desire for the master. Just as Aristophanes' hiccoughs disrupt not just Aristophanes' ability to

---

[11] Bohrer (1981). For the nineteenth-century history of 'the moment' with regard to psychology, philosophy and the novel see Zemka (2011); for the background in neurology, see Dames (2007); for the seminal importance for aesthetics of Lessing's idea of the 'pregnant moment', see Mitchell (1984); Lifschitz and Squire eds. (2017).

[12] Polybius 1.1.2: see Maier (2018), with Miltzios (2016), especially 147–51.

give a speech but also the planned order of speakers in the dialogue, the pointed repetition of the word 'sudden' draws attention to suddenness in transition as a question of the relation between rational argumentation and insight as a key tension in understanding the philosophical process.

Plato's dialogue *Parmenides*, however, goes on to make 'the sudden' into a technical term with regard to time and change. The argument (155e4–157b5) – extremely complex and much debated by modern philosophers[13] – is struggling to understand the relation between movement and rest for the One, and introduces 'the sudden' (*to exaiphnes*, 156d – the definite article is used to make it a technical term or concept) as a 'strange' concept (*atopos* – out of place, like Socrates). 'The sudden', it is argued, is the moment when movement stops and when rest starts, which is itself 'not in time' (*ouk en chronōi*, 156e). 'The sudden', it seems, is an instant without duration, which divides states or processes: the very moment of conversion. This remarkable idea of a moment of being that is not in time was apparently hailed by Heidegger as representing 'the deepest point to which Western metaphysics ever penetrated. It is the most radical advance into the problem of being and time, an advance that was afterwards not taken up (by Aristotle), but rather closed off.'[14] (Only Herbert Marcuse's notes on Heidegger's course exist for the detail of this argument, hence 'apparently'. Heidegger was due to teach this course with Wolfgang Schadewaldt, whom classicists will recognize as a celebrated classical scholar, like Heidegger deeply tainted by a Nazi affiliation.) Heidegger's dismissive treatment of Aristotle's version of 'the now' which I quoted earlier, takes on a further colouring from this sense of the whole history of metaphysics taking a wrong turn. Surprisingly, Heidegger appears not to have referred to Kierkegaard here, who in his *Concept of Anxiety* used Plato's notion of the sudden (*to exaiphnes*) – and precisely this passage – not just as a route towards thinking through his critical idea of the 'leap of faith', but also as an anticipation of a specific Christian theological understanding of time. From Kierkegaard, that is, we can see a forged link between Plato and Augustine's immediate moment of conversion as well as the Christian debates about the immanence of temporality and timelessness in the Incarnation: 'only with this category [of the moment] is it possible to give eternity its proper

---

[13] Strang and Mills (1974); Bostock (1978) indicate the classical philosophical interest; more telling here are Backman (2007); Rangos (2014); Gonzalez (2019) on the impact of Plato's formulation. Cimakasky (2017) is pedestrian.

[14] Marcuse's notes translated and cited by Gonzalez (2019) 329; discussed also by Backman (2007) and Rangos (2014).

significance ... It is only with Christianity that ... temporality, and the moment can be properly understood, because only with Christianity does eternity become essential'.[15] Both of Plato's necessarily interrelated questions – What is the process of transition? What is an instant? – echo back through the chapters of this introduction. For Plato, 'the sudden' becomes a necessary but necessarily disruptive – violent – element in philosophical process: is the flash of insight the culmination or the contradiction of methodical argumentation?

Examples of how the unexpected or the sudden is mobilized in ancient and modern narratives could easily be multiplied, but Frederic Jameson brilliantly generalizes this pervasive relationship between temporality and narrative. Suddenness, he argues against Bohrer's depoliticized aesthetics, constructs 'a privileged relationship between violence as content and the closure or provisional autonomy of a temporal form'.[16] This is a dense and expressive comment. Suddenness is a way of doing violence to timing: it disrupts the expectation of the flow of events. For Jameson, this can embed also a political violence. 'Meaning, violence and the moment categorically infect each other'.[17] (Heidegger engages in a profoundly melancholic narrative of buried violence when he wrote in 1950, still in Germany, in a letter to Hannah Arendt, anatomist of Nazism, his former lover, in New York: 'the sudden ... requires a long time to be delivered'. What grim complicities would need to be unpacked in the infected politics of such a remark about time!)[18] But, in Jameson's account, what such disruption of 'the sudden' makes visible is how the expected embodies the closure or provisional autonomy of a temporal form: that is, the expected is aimed at closure, closure of a sequence of events or a generic form. The unexpected may be provisional and autonomous – it was not expected that the dead body in the cave was the servant and not the mistress, to take an example from the narrative of Achilles Tatius' novel, *Callirhoe and Cleitophon* – or it may involve closure in a stronger generic sense – marriage at the end of a novel, say,

[15] Kierkegaard (1980) 84.
[16] Jameson (2003) 174; see also Jameson (2002) 195: 'intimate relationship between violence as content and the "moment" as form'. See also Friese ed. (2001).
[17] Zemka (2011) 216. She overstates her case when she writes 'Modern violence pollutes the purity of suddenness' (221).
[18] Letter 47, 8 February 1950, in Arendt and Heidegger (1998) 74–5. See Villa (1996) for the philosophy, with Sluga (1993) for the contextualization, and Maier-Katkin and Maier-Katkin (2007) for the fascination with the salaciously personal.

disrupted or reinstated ('Reader, I married him'), with its full ideological bearing. But what is crucial is that narrative *as a temporal form* is made visible in a privileged way by the disruption that is *Plötzlichkeit*. Literary narrative, for Jameson, does not just manipulate time, as Shlovsky has it, but *forces* it into the ideology of form, revealed by the willed and complicit disruption of the unexpected or the sudden.

Aristotle made movement and change essential to the recognition and definition of time, as we saw in chapter 3. In part, his argument engages, implicitly at least, with Parmenides and Zeno, the philosophers who challenged in the most extreme and vexing way common-sense notions not just of movement but also of becoming, and even of time itself, therefore, as a measurement of movement and change. Zeno's well-known paradoxes, which prove that an arrow never reaches its destination, or that swift-footed Achilles will never overtake the slow tortoise, depend on infinite division, which we have already seen repeatedly recurring as a crushing problem for those who try to define the present through duration. But Parmenides' insistence on the primacy and unity of *esti* 'it is' overlaps the ontological and the temporal in such a way as to deny the possibility of change in and across time: what is, is. Arguments among scholars from Plato onwards about exactly how to engage with the logic of Parmenides are motivated initially by the recognition that his conclusions do a certain violence not just to 'common-sense notions of time', but to the most fundamental phenomenological sense of change; violence, that is, to the personal, human experience of being in time.

I have already discussed how sacred calendars organize the pattern of the year, and how the theology of Christian eschatology changes the experience of time, and how the present of everyday life can become a ritualized, self-absorbed waiting. Since the work of Arnold van Gennep in the early years of the twentieth century, expanded by the anthropology of Victor Turner, it has also become a commonplace that rituals, and especially rituals of initiation, divide time into a sacred period with passages of transition into and out of such a period, and the otherwise usual passage of days and weeks, where within sacred time the usual rules of exchange and hierarchy may be inverted or distorted.[19] Dividing the flow of time into different types of times, *diversa tempora*, is a constant of social process. Such acts of separation in the sequence of events find full intellectual expression in the practice – and debates about – periodization in historiography, which have become in turn part of the politics of self-recognition. Achille Mbembe, for example, to take

---

[19] van Gennep (1960); Turner (1967), (1969).

a particularly influential and provocative contemporary theoretical intervention, calls for a resistance to the inheritance of colonial time systems, to find
an African time that is a precursor to another not just post-colonial but
uncolonized or decolonized history: 'The time has finally come', he writes,
'to begin-from-ourselves', as if dechronolization (to apply Freccero's coinage) could take place by fiat.[20] Mbembe wants to force time into a new story.

Classicism itself, whether it is conservative or revolutionary, is an
aesthetics and politics of untimeliness. In chapter 5, I indicated how the
long tradition of classicism in Western culture mobilized two main strategies of affiliation, namely, genealogy and idealism: finding an authorizing
origin for contemporary society, identity, ideas or art, in the cultures of
Greece and Rome, and finding in the antiquity of Greece and Rome an
idealized art, society, identity or body to aspire to. Both of these strategies
of self-expression recognize a rupture between the past and the present
(even in the assertion of genealogical descent). The now looks back to the
idealized and lost past of classical antiquity. Revolutionary classicism insists
that the ideal of the past is exemplary and drives the need to make new for
the future: a new society in Shelley's or Marx's political use of Greek
freedom ('We are all Greeks') or Roman Republicanism ('The French
Revolution was enacted in Roman dress'); a new art-form in Wagner's
opera-festival or Gluck's new emotional opera; or a new sexuality for the
nineteenth-century sexologists and *littérateurs* who avidly read Plato. But
both for each of these revolutionary claims – which could, of course, be
extended – and also for the more conservative branches of classicism that
use the past to resist the claims of the modern and of change, there is an
insistence not just on the insufficiency of the present but on a sense of
untimeliness, of not inhabiting the right time, of not being at home in the
now.[21] Nietzsche most fully theorized this sense of what he called
Unzeitgemässheit (untimeliness) in his search for disruptive wisdom.
Classicism is a rupture in the self-sufficiency of the present; it asks to
make the now otherwise; it forces a new self-awareness of time; it forces
time.

To see thus revolution as a politics that demands a rupture in time – the
new calendars of the French or Russian Revolution mark this conceptual
insistence – remind us how the Christian revolutionary invention of time,
embodied in the violence of typology or the church's authoritarian,

---

[20] Mbembe (2017) 7; Banerjee (2006); Freccero (2006), on which see Holmes (2020).
[21] The Postclassicisms Collective (2019) 161–81; Grosz (2004).

eschatological history, has also been integral with a violent politics of exclusion or missionary compulsion, both between different Christians as in the Reformation and Counter-Reformation, and between Christians and others in, say, the imperial expansion of Europe; just as the history of classics, entangled as a subject with Christian institutions, in its claim to genealogical value has also been complicit as an educational and justificatory force in the darker forces of Western nation-states and their imperial engagement with the other. The history of temporality – its revolutions in time – is never without self-implication – or self-justification.

I have been looking so far in this chapter at the elite poetry of memorialization, at literary theory, at philosophy, anthropology, religion and modern post-colonizing agendas, but the necessity to mark out time in divisions is pervasive in human culture. 'I have measured out my life in coffee spoons', writes T. S. Eliot, and the repeated gestures of smoking a cigarette to mark the passing of time, or glancing at a watch or pouring a cocktail to emphasize a point in time or a process of transition is vividly captured in all its repetitive insistence, as we discussed, in Christian Marclay's installation, 'The Clock'.[22] As Bakhtin wrote, 'Time, as it were, thickens, takes on flesh, becomes artistically visible'.[23] When Hobbes describes the life of uncivilized man as 'solitary, poore, nasty, brutish and short', amid the signs of no civilization – lack of agriculture, architecture, navigation, arts, letters and so forth – he specifies 'no account of time'.[24] It is a basic principle of human society to break up time into eras, periods, divisions. Parmenides' extremism is unsettling because it denies not just the evident shared experience of time passing, but also the essential act of civilization, to give an account of time. Without an account – as ever, both counting and telling – it is hard for humans to recognize themselves. Time's expected order is what makes doing violence to time the threat and promise of a potentate like Juliana, a philosopher like Parmenides, or a writer like Augustine. Modern Western culture has inherited thus a triple compulsion: the desire to see time as a natural and inevitable linear flow from birth to death; the desire to impose the order of measurement and regulation on this flow; the desire to break free from this linearity and transcend the bounds of time.

*

The praise of Juliana that she 'forced time', 'violently overwhelmed time', may be a semantically rich expression with implications about different

---

[22] Smith (2011).   [23] Bakhtin (1981) 84.   [24] *Leviathan* 1.13.

understandings of a placement in time, from her family background to the history of Solomon's Temple, but it scarcely constitutes or even embodies a systematic theory of time, even if in its theological assertion of supersedence it cannot but allude to such systematization. Where, then, do we look for a theory of time? Or, rather, what levels of abstract thinking about time can be traced in and through ancient texts and practices? How systematic, how coherent can any society's expressions of being in time be?

One answer from within the disciplinary silo of philosophy would have no difficulty in tracing the technical, theoretical expositions of time as a philosophical problem, with high points in antiquity that start with Aristotle, with a nod to the influence of Plato's *Timaeus*, or even earlier notions of the creation of the universe as a question for a theory of time, and move through a host of contributions, including the masterpiece of Augustine – with the potential to link contemporary philosophy back to this tradition, as Heidegger does with his recognition of Aristotle's importance, or Wittgenstein, who starts, famously and obsessively, with Augustine. The value of such disciplinary thinking is not undone by Ricoeur's melancholic conclusion to his three-volume study of time and narrative that thinking about time must fail, that each theory produces a new aporia. But the role of Neo-Platonism and Stoicism in the development of Christianity also allows for an intricately shared platform of exposition, which can move such a history away from merely a roll-call of theoretical responses to a theoretical question into an appreciation of how theological arguments about time come to ground the experience of time in daily practice and the engagement of individuals in a temporal world (which is where the discipline of philosophy all too often starts to part company, and leave discussion to theology or church history). Simeon Stylites, to take an extreme example, may have had little theoretical understanding of time and the written Lives which describe his painful experience never focus on such theorizing; it seems evident that he has no place in the history of philosophy, but his actions make sense only with a Christian comprehension of temporality. So, if we do trace a theory of time through philosophy, the more such theorizations have purchase in a community and have a formative impact on the experience of time, the less philosophy as a discipline seems likely to provide the right tools for analysis.

The buildings that I looked at in chapter 6 are some of the loudest expressions of time, in that they dominate the cityscape and fill the imaginations of citizens, and express forcibly their patrons' will to define how the history of the city or empire is to be comprehended. Yet they are rarely explicit in this aim, and certainly not systematic or theoretical in their expressivity. The boast of Augustus to have changed the cityscape of Rome may help define the era as Augustan, but it makes no contribution to a theory of time. Yet for Roman historians and other writers, the divide between the Republic and the Principate, which Augustus embodies and his policies enact, becomes the defining precipice of the political history of the state, the key 'before' and 'after', a necessary way of organizing time. Much as the Christian theology of time becomes embedded and embodied in Christian practices of living, with varying degrees of tacit or explicit engagement, so the silent physical markers of a claim over time can enter the more theorized written reflections of those striving to explain the unfurling of history to themselves and their readers.

It is because of this porous dynamic between theoretical expositions and social practices and behaviour; between explicitation and tacit knowledge; between intellectual debate and physical embodiment; that I have concentrated in these essays on what might be called 'the discourse of time', and not dedicated chapters to such discrete areas as the anthropology of time, the philosophy of time, the time of historiography, the history of timepieces or calendars, the representation of time in epigrams, and so forth. Rather, what has interested me have been the cross-cutting, interconnected questions that explore the porous dynamic in which a discourse is shaped. To see how time changes in late antiquity for a Christian community in comparison both with a Jewish and with a Greek or Roman community, with their different histories, zones of contact and antagonism, forms of normative writing and social regulation, it is necessary to cast the net of analysis much more broadly than such discrete disciplinary projects would allow. Every attempt to discuss so broad a field commits a certain violence to its integrity by dividing it into delimited subjects – and a further violence to its mess by categorizing, collecting, organizing its disparate elements. Any discussion, that is, including, of course, this one, forces time into a form. Can formless time, time without the normativity of form, without regulation or measurement, be imagined? Perhaps not, not fully.

The day, the month and the year are categories shared across the world because the circulation of the earth around the sun gives every community the appearance of sunrise and sunset, the repeated pattern of day and night; the moon provides the lunar months; and the year as a period is also tied to the movement of what we still call the heavenly bodies. Yet the week, used throughout the modern, industrialized world, is a determinedly artificial construction with no referent in the natural world.[25] The seven-day unit does not divide months easily (except February, three times out of four), nor the year. The week exists because – in the beginning – in six days God created the world and on the seventh he rested. The invention of the Sabbath is the invention of the week. For the Greeks and Romans, who did not have weeks until they became Christians, the oddity of the Sabbath was one of the paradigmatic ways in which they saw the Jews as incomprehensibly different from themselves. It is the Sabbath that stops seven being just an arbitrary figure, however useful or familiar, like the number of innings in a baseball match or the number of balls in a cricket over. We can thus end these essays where we began, with the translation of the opening of Genesis, in this case into the practice of communities, an inherited theology: our time, still.

But, before we conclude, this ring composition itself also signifies. Yukai Li writes, 'Ring composition is the structure of untimeliness', by which I understand him to be pointing towards something that reading Homer tells us: ring composition expects us to read from the beginning with an anticipation of retrospective recognition, and to end not just with a sense of fulfilled closure but with a turn back to the beginning to find significance in the juxtaposition of two scenes, and in the transition between them.[26] Ring composition, as Daniel Mendelsohn has it, embodies the elusive melancholia of recognizing that things can, temporarily, cohere.[27] The first word of Western literature is *mēnin*, 'wrath', which in itself and in the founding, divisive, opening scene of argument poses a question to which the final scenes of the *Iliad* offer a variety of closures: Achilles' reconciliation with Agamemnon; the funeral games, where the earlier combative rage is turned to the reconciliatory order of prize-giving; the return of the body of Hector to Priam; Achilles' tears of regret and mourning; Achilles' recognition that he could still be angry with Priam if pushed too far; the funeral of Hector as the end of Achilles' wrathful revenge. The narrative itself dramatizes what is at stake in the rage that is to be assuaged by this narrative of multiple ends. How much is Achilles' fame the result of his violent willingness in his rage to transgress the norms of others – by refusing to fight, by refusing to listen to his companions' entreaties, by refusing supplication, by refusing in his grief to sleep, eat, or rest? Ring composition sends us back to the beginning to retrace the coming of the end.

---

[25] Zerubavel (1981) 11, expanded fully in Zerubavel (1985); Salzman (2004); Henkin (2018).
[26] Li (unpublished).   [27] Mendelsohn (2020).

It is, then, not just with the first word of Western literature but with the problem of how narrative makes sense of time, its time, that I end these essays, as an introduction to the readings to come, readings that aim both to extend what is currently understood as the institutional canon of Western literature, and to explore how time is reinvented within the writing of late antiquity. This discourse of temporality is crucial not just for classicists or theologians, but for the self-understanding of Western historical self-placement. In this first section, we have seen how the formative categories of temporality are reshaped under the jurisdiction of Christianity in late antiquity. It is time now to turn to see how the imaginary of time is formulated in detail across the texts and genres of the era.

PART II

CHAPTER II

Beginning, Again
Nonnus' Paraphrase of the Gospel of John

We ended with first words. Let us start there again. And with a startling beginning.

There is no more influential opening to a piece of Greek prose – no shock in this judgement – than the first five words of St John's Gospel: ἐν ἀρχῇ ἦν ὁ λόγος, 'In the beginning was the Word'. The Gospel begins with the word 'beginning', which, as we have already discussed, is an assertive reappropriation of the opening of the Septuagint's translation of Genesis: ἐν ἀρχῇ ἐποίησεν ὁ Θεὸς τὸν οὐρανὸν καὶ τὴν γῆν, 'In the beginning God made the heaven and the earth' – the beginning of everything. The opening of John marks a new start, in all senses.

But it is truly startling, therefore, when we turn to Nonnus' *Paraphrase of the Gospel of John*, to read its first word: *achronos*, 'timeless', 'non-time'. It would seem that Nonnus, from the start, has aggressively replaced the proclamation of a beginning with an assertion of *no beginning* – timelessness: an infinite scope. How can a retelling of the word of God bear so drastic a reformulation? In an era when an assertion of theological principle could risk both one's career in this world and one's soul in the next, why does the poet hazard such a different framing of temporality for his story of Jesus? This chapter is an attempt to explain the power and logic of Nonnus' extraordinary gesture, which could seem to challenge the very narrative moorings of the Gospel it paraphrases.

We will need to circle back to this originary moment, if we are to understand it. Let us start with the act of rewriting the Gospel. What is the status or authority of such acts of recomposition of holy writ? One strand of response to the fixity – the unalterable authority – of scripture stems from within text of the Bible itself. In Deuteronomy 4.2, the people of Israel are instructed 'you shall not add to the word which I am commanding you, nor take away from it, that you may keep the commandments of the Lord your God which I command you', an instruction which is reprised in Deuteronomy 12.32 with 'What thing soever I command you, observe to do it: thou shalt not add thereto, nor diminish from it'. The Gospel of Matthew seems to echo this injunction (5.18): 'For truly I say to you, till heaven and earth pass away, not an iota, not a dot, will pass from the law until all is accomplished'. Jesus is explaining that he has not come to abolish the laws but to fulfil them. He continues with the warning that a person who relaxes a single law is the least (*elachistos*) to be called to the kingdom of heaven (5.19). Both Deuteronomy and Matthew insist that it is essential that the materiality of the text is not changed: no words or letters or even marks are to be added or taken away, because such material consistency is integral to observing the laws inscribed in the Bible.[1]

These injunctions demand that the text of scripture is seen to be unalterable and determined – and, by one trajectory at least, such demands culminate eventually in the Evangelical insistence that the literal and certain message of the word of God is evidenced in the Bible, a text 'written in stone', like the tablets that Moses brought down from Mount Sinai. In the bitter and violent theological arguments which prompted and followed the Council of Nicaea in the fourth century, and especially in the subsequent disputes between Cyril and Nestorius, the rhetoric of 'not adding or diminishing' became a compelling commitment – that orthodox theology should not add a word to or remove a word from the Nicene creed. This was upheld as a principle even when it was demonstrably not the case: Cyril, whose twelve anathemas redirected understanding of the Nicene creed in the most provocative fashion, declared that he followed a 'simple and undefiled tradition', and 'we follow the opinions of the holy Fathers in all things ... We refuse to differ from them in any respect'.[2] 'Change was most effectively achieved', summarizes Mark Smith, 'by denying that there had been any change at all'.[3] Theological conflict, itself deeply tied up with the power structures not just of the church but of the empire too, made language – the precise words used and their meaning – a battleground of politics. So in turn, earlier opponents of the Nicene creed attacked its wording for including terms such as *homoousios* 'consubstantial', that were not in the Bible, terms which could thus be thought to have been 'added' to scripture. They also claimed to follow 'the undefiled apostolic tradition'.[4] More generally, Tertullian tellingly calls the Greek philosopher an *interpolator*, the figure who adds words into the text and thus falsifies the word of God.[5] The Talmud argues that even the finials on the tops of letters, the

---

[1] Both Judaism and Islam add an 'oral law' to temper – extend, modify – the written law.
[2] Cyril, *Letter to John of Antioch*.   [3] Smith (2018) 209. See also Wessel (2004).
[4] Thedoret, *Eranistes* 63.3–4.
[5] *Veritatis interpolator*, Tert. *Apol.* 46.18 – not the *interpreter*, philosophy's usual role.

ornaments without apparent semantic content, must be reproduced accurately because they too, like any letter of the text, are open to interpretation.[6] Rewriting the Bible seems banned from within the Bible itself, and this fixity is integral to the proclamation and observance of its laws. The material certainty of the text provides the grounding of the moral certainty of its guardians.

The Codex Sinaiticus from the middle of the fourth century, and the earliest surviving full-scale manuscript of Christian scriptures when it was discovered in the nineteenth century, tells a slightly different story. The manuscript contains some 23,000 corrections by six different hands. Of course, the scribes routinely overwrite faded letters or correct spelling errors; but they also insert words or lines that have been omitted, change wording and even delete material. In some cases, phrases are deleted by one corrector, and then re-added by another. The Codex Sinaiticus proved that the ending of Mark in the received text was an interpolation. At very least, the Codex Sinaiticus shows the belligerent polemics involved in how 'early Christian groups or individuals', around the time of the Council of Nicaea, 'read and altered the text'.[7] That is, it was already recognized that even a luxury manuscript of the Bible such as the Codex Sinaiticus was corrupt, alterable, altered and open to dissent. The material text of scripture, like the canon, was not fixed and certain, but needed to be made so.

There is, however, a further strand of response to fixity that also finds its source in the Bible itself. The Greek name 'Deuteronomy' ('Second (recitation of the) Law') indicates how the fifth book of the Pentateuch retells the history of the Israelites and the regulations outlined in the first four books. Since the nineteenth century at least, Deuteronomy has been read as an ideological reconstruction of the Mosaic law along new and more committed priestly lines, a rewriting. Yet from the earliest commentators also, who did not follow the agenda of critical history, it was noticed and debated that even the Ten Commandments have a different wording in Exodus and in Deuteronomy. The books of Chronicles and Jubilees also retell the same Israelite history from their own agendas. At a more granular level, specific laws are told more than once, in different contexts and thus with different emphases: the law not to seethe a kid in its mother's milk is listed among sacrificial regulations, as if it were a matter of cultic propriety, and among dietary laws, as if it were a question of the rules of food taboos.[8]

---

[6] b*Menachot* 29b. Seth Schwartz has pointed out to me that there are actually no examples of such analysis. It is a limit case of future interpretative possibilities.

[7] Parker (2010) 3. The figures in this paragraph are taken from his discussion.

[8] Ex. 23.19; Ex. 34.26; Deut. 14.21.

The synoptic Gospels, since their canonization was established and since the Marcionite solution of producing one harmonized Gospel was rejected as a heresy, provide a fertile space for interpretation precisely because they have different narratives, different wordings, and different agendas in their announcement of the good news.[9] The history of canonization is an integral trajectory in the politics of Christian language, the authorized *forms* of Christian narrative. John, in turn, provides a further telling of the story, a different engagement and perspective. The Gospels quote lines from the Septuagint – turning the Hebrew Bible into the Old Testament by appropriative recomposition. The texts of the Hebrew Bible and the Gospels perform their own rewriting.

Indeed, numerous examples of rewriting or expansion of biblical texts, complete with those scriptural claims that nothing should be added or removed, remain fully part of the authoritative tradition of both Judaism and Christianity. Philo of Alexandria in the second century BCE, as we discussed earlier, rewrites the texts of the Hebrew Bible into Greek prose through his Platonizing, allegorical reading, his assimilationist intellectualism.[10] Allegory – 'reading otherwise' – enables Philo to straddle Greek and Jewish cultures, and both his methodology of reading and his ready adoption of Plato into scripture allow him later to become an authority for Christian Neo-Platonist thinking from Clement to Origen and beyond – though not for the rabbis, who largely ignore Philo, even when they themselves approach the subject of allegory or the wisdom of the Greeks. Josephus – again, as we saw earlier[11] – also redrafts the Hebrew Bible into Greek prose in his *Antiquitates*, as he translates himself from leader of the Jewish revolt into an imperial historian. We began this book, too, with the Septuagint's designed mistranslation (according to the rabbis) of the creation story of Genesis: the Septuagint itself is a sign and symptom of a society which was unfamiliar with the Hebrew of the Pentateuch and needed its own rewritten version. Similarly, the Aramaic *Targumim*, glosses or translations of the Hebrew words for an Aramaic-speaking congregation, are markers of the transitional cultural and linguistic vectors of a society in change, when Hebrew is no longer the vernacular of the community committed to the Hebrew Bible.[12] The Bible is rewritten not just into different languages but into different forms.

---

[9] It was much argued from the early modern period onwards whether the Gospels themselves were Greek rewritings of earlier Aramaic or Hebrew texts: see Levitin (forthcoming) with the general background of Shuger (1994) and Sheehan (2005).

[10] See pp. 21–2.   [11] See pp. 21–2.

[12] See Alexander (1992b); McNamara (2010); Alexander, Lange and Pillinger eds. (2010); Najman (2010); Hayward (2013).

Reading the Bible also prompts the further rewriting of *midrash* and commentary.[13] Much *midrash*, specifically *aggadic* or narrative *midrash*, starts as a form of interlinear re-reading, finding the gaps in the texts and filling them with stories. Many stories which contemporary Jewish communities now take as Bible truth are in fact midrashic invention. Indeed, the very status of *midrash* – just *how* authoritative such stories are – is a deeply contested issue. By the thirteenth century, rabbinical writing, self-reflexively, discusses four levels of reading, known by the acronym formed by their first letters, as *Pardes* or 'Garden'/'Paradise'. The four levels, each valorized and not hierarchically opposed to the others, are *Peshat*, 'literal' or 'surface' meaning; *Remez*, 'allegorical' or 'symbolic' meaning (the space Philo inhabits); *Derash*, 'comparative', 'expository', 'homiletic', the space of *midrash*; and *Sod*, 'secret' or 'mystical', where eventually kabbalah will take root. Rabbinical reading flourishes and sprouts in this garden of meanings through variety and hybrid vigour. Commentaries – Cyril's 600-page commentary on John will be especially salient for Nonnus' *Paraphrase* – often follow a line-by-line exposition, expanding, like *midrash*; glossing linguistically like *targumim*; exploring allegorical senses, like Philo's treatises; arguing theological import, like a sermon of Gregory. From Origen onwards, commentary becomes a standard form of internal Christian polemic.[14] The idea that the Bible 'speaks for itself' in an unmediated clarity is starkly contradicted by the religious tradition it underpins.

If we put together the translations, the retellings, the paraphrases, the commentaries, it is evident that the Bible is something other than the authoritative, original, fixed text that some fundamentalist fantasies demand. Rather, it is a textual world in a constant state of retelling, recapitulation, redrafting. When Augustine turns in the final books of the *Confessions* to his theoretical exposition of reading and interpretation, focused predictably on Genesis, granted his fascination with creation and beginning throughout the *Confessions*, his profound sense of the insufficiency of human language grounds the need for this constant labour of reading and re-reading.[15] Scripture's very nature, according to Augustine, requires that we recognize that it is unfinished and unfinishable. Rewriting is what the Bible does and what it produces, in theory and in practice.

Christian paraphrases – and there are significant examples in both Greek and especially Latin before and after Nonnus[16] – must, therefore, be

[13] Boyarin (1994), (2009), (2015); Zornberg (1995); Kugel (1997); Rubenstein (2003).
[14] See Goldhill (2021).   [15] See Ando (1990); Conybeare (2016); Stock (1998); Wetzel (1992).
[16] Latin examples of paraphrase and cento have been particularly extensively discussed: see e.g. Herzog (1976); Roberts (1985); Springer (1988); McGill (2005); Green (2006); also Kartschoke (1975); Kirsch

located between the desire to regulate meaning, control interpretation, assert theological authority, and the recognition of the slippages and insecurities of language that both demand and hinder such regulation. What, then, is the status of a paraphrase? Within this arena of rewriting scripture, paraphrase strives to distinguish itself generically and polemically from centos and translations (as well as commentaries), an act of definition that is itself, as we will see, part of the politics of language. In centos, Homer's and Virgil's verses, the very materiality of the foundational language of the dominant culture, are cited and deracinated to make them speak otherwise. Centos denaturalize the ideological apparatus of classical epic, by rebuilding its material parts into a new construction for a new community. Translation, by contrast, makes the texts of one culture available to another, and by such very activity makes evident the linguistic and cultural differences, even and especially when, as we saw with Philo, the gap between different languages and cultures is earnestly denied. The boundaries between these forms of cento, translation and paraphrase, particularly between paraphrase and translation, are fragile, and, as we will see, become increasingly contested when the politics of language – the authority of authoritative texts – becomes an insistently and publicly fraught cultural issue, as is the case with the long and bitterly fought arguments over the Nicene creed in the Christian East. In contrast with the transformative strategies of both cento and translation, however, paraphrase rewrites an authoritative text *within* and *for* its own community. It performs its new authority while insisting on the continuing authority of the paraphrased text. This doubleness is its specific ideological affordance, and its danger. Fear and rejection of 'novelty' in the spreading of the revolutionary, Good News are dominant impulses in the fourth and fifth centuries – *kainotomia*, 'novelty', is a thoroughly negative accusation, along with a corollary insistence on the positive value and authenticity of tradition.[17] What better genre, then, to make an intervention than paraphrase, a form which disavows its necessary newness and proclaims its dependence on the authority of tradition?

(1989); Bažil (2009); and, more generally, Pollmann (2017); Vessey (2004); Schottenius Cullhed (2015); Fontaine and Pietri eds. (1985); on Greek examples, see Usher (1998); Jarick (1990); Golega (1960); Bevegni (2006); Smolak (2001) – and the works cited below.

[17] As Pope Stephen I said (Cypr. *Ep.* 74), *nihil innovetur nisi quod traditum est*. This theological rejection of novelty is not contradicted by political claims, especially in the West, of Christianity as a *nova Roma*: see Hardie (2019) 135–62 (coins minted by Constantius II and Constans had the slogan *felicium temporum reparatio*); nor by the constant Christian interest in reform – *ecce facta sunt nova* 2 Cor. 5.17 – discussed by Ladner (1959).

Paraphrase redirects – remediates – the reading of the Gospel; it calls for the reader's complicity with a new framing of the Gospel's sense; it triangulates the reader between the Gospel and the *Paraphrase*. The redirection – the triangulation – is predicated on the Gospel's openness to re-reading and on the need to make reading the Gospel better, though *how* it is to become better remains contested. Erasmus, for example, – and it is an example which vividly demonstrates the *politics* of how the boundary between translation and paraphrase is determined – saw his Paraphrases of the New Testament placed in every church in England by fiat of the king, at a moment of bitter and violent contention over the text of the Bible, when all too recently it had been a capital offence to translate the Bible into the vernacular.[18] The striving of authority to distinguish between translation and paraphrase could not be better displayed than this case: translation is banned on pain of death; paraphrase is a compulsory mediation. In self-justification, Erasmus wrote, 'it is not I who speaks in the *Paraphrase* anywhere', which did not stop one of his critics sniffing at one of his interpretations, with good reason, 'Luke never said this; it is Erasmus speaking with his usual audacity'.[19] Paraphrase can only and hopelessly disavow its act of rewriting the word of God. To read a paraphrase of the Gospel requires, that is, an uncomfortable shuttling between the Gospel and its new form. The Paraphrase is read with an eye on the Gospel; the Gospel through the Paraphrase. Scheindler's standard modern edition of Nonnus' *Paraphrase* prints the Gospel beneath the Paraphrase so that the shuttling between the two texts is performed bodily by the eye.[20] Without the physical presence of the Gospel, as in the ancient manuscripts we now possess of Nonnus' *Paraphrase*, the shuttle is between memory and enactment. When Augustine in *Confessions* analyses the reading of a Psalm, you will remember, he talks about how the presence of meaning is constructed in a shuttle of memory between the recollection of the words that have passed and the anticipation of the words to come in the Psalm. Reading the paraphrase of a Psalm overlays a further temporal structuring, as we shuttle also between the words of the paraphrase and the words of the psalm, as

[18] On Erasmus, see in particular Pabel and Vessey eds. (2002); also Boyle (1977); Chomarat (1981); Ferrier and Mantero eds. (2006); Henderson ed. (2013); Bloemendal ed. (2016); on Erasmus in parish churches see Craig (2002); on vernacular translation of the Bible see Mandelbrote (2016), (2018); François (2008), (2009) and especially (2016).
[19] On Béda's critique of Erasmus see Rummel (2002); Bedouelle (2002); Phillips (2002); Leushuis (2016).
[20] Scheindler (1881).

well as the sequence of words in the paraphrase itself. Paraphrase makes for re-reading.

One particular translation of the psalms from late antiquity demonstrates vividly how the recalibration demanded by Christian rewriting is also and explicitly an agonistic engagement with the inheritance of Greek and Roman culture. Ps-Apollinaris' *Metaphrasis of the Psalms* is difficult to date with certainty but is probably from the 450s, in the decade after the Council of Chalcedon, perhaps as late as the 460s.[21] The Psalms are a foundational text for Christians, the basis of a Christian education and an engagement with the past of inheritance and the future of prophecy, as well as of Christian liturgy.[22] To turn the familiar *koinē* Greek of the Septuagint, embedded in memory and ritual, into the high verse of hexameter is thus an especially combative cultural act: it transforms the furniture of the Christian mind. This hexameter translation has a long introduction, known as the *Protheoria*, which explains the work's genesis and purpose. The work is titled a *metaphrasis*, 'translation', but its transformation not just of Hebrew into Greek, but the prose of the Septuagint into verse – unlike the versions of Philo or Josephus, which retell the Septuagint in prose – bring it close to the form of a paraphrase, an immediate indication of how precarious the boundaries between different forms of rewriting can be.[23] It does not merely render Hebrew verse into Greek verse, but triangulates the Hebrew text between the long-inhabited Greek of the Septuagint and the long tradition of elite Greek culture. The *Protheoria* indeed sets out to indicate the theory of language and language's transformative nature, the thinking which grounds the work's cultural practice. The very fact that the translation opens with such an explicit and enacted agenda to explain itself as a form indicates the politics of language in action – and the text's intervention in the contest of cultural authority between Hellenic and Christian cultures.

Strikingly – and I know of no parallel for this – its first word is *elpomai*, 'I hope'.[24] In the Homeric hymns, one inevitable touchstone for any

---

[21] See Golega (1960); Agosti (2001) 87–91; Faulkner (2014). Faulkner (2020) most recently has argued cautiously to attribute the piece to Apollinaris and to date it thus in the fourth century. I remain unpersuaded that the Nicaean theology expressed in the *Protheoria* is reconcilable with the accusations of Apollinaris made by Gregory in *Ep.* 101 and 102. Thanks to Andrew Faulkner for sending me proofs of his book ahead of publication.

[22] See McKinnon (1991); McKinnon (1994) which talks of a late fourth-century 'psalmodic movement'.

[23] On the distinction between *metaphrasis* and *paraphrasis*, see Philo, *Vit. Mos.* 2.38 – though we have already seen Philo's commitment to a perfect rendition between Hebrew and Greek which colours his discussion. On translation and paraphrase see Faulkner (2014) with further bibliography.

[24] The last line of Nonnus' *Paraphrase*, as discussed below, starts *elpomai*. For a suggestion of a link between these two verses see Agosti (2001) 95–6.

introduction in hexameter verse, it is standard to announce the subject of the poem with *archom' aeidein*, 'I begin to sing', and/or the future verb *mnēsomai*, 'I will proclaim', sometimes with an invocation also of the Muses. The explicit vocabulary of beginning, including any invocation of the Muses, is occluded in Ps-Apollinaris, now in the language of hope or expectation. This poem's beginning is in the commission of patronage, and its introduction dramatizes in direct speech the request of one Marcian, a request that the author should produce a poetic version of the Psalms in Greek.[25] The inherited language of patronage is clear enough in the opening lines (1–4):

ἔλπομαι ἀθανάτοιο θεοῦ κεκορυθμένος οἴμῃ
σοὶ χάριν ἀντὶ πόνων φορέειν καὶ κέρδος ἐπ' ἔργῳ
καὶ τυφλὸς γεγαὼς δοκέειν φάος ἄλλο κομίζειν,
Μαρκιανὲ κλυτόμητι

I hope, armed with a song of immortal God,
To bear you thanks in recompense for your toils, and profit for the work;
And although I am blind, to seem to convey another light,
Marcian, famous for wisdom.

The poet indicates that his poem is an act of *charis*, reciprocal exchange, a term that Theocritus, for example, makes central to his appeals for support.[26] So, too Ps-Apollinaris mentions *kerdos*, profit. Hope here is the mode of anticipating the success of an appeal for the rewards of patronage. The poet describes himself as blind, as if he were Homer; but, with a turn to the familiar language of Christian scripture, he is to bring 'another light', the revelation of scripture's promise.[27] The signature of Homer is in order to signal difference – an embodiment of the poetics of paraphrase. The *koinē* prose of the Septuagint's Psalms is overwritten by the verse lexis of Homer, itself redrafted by the poet's intertextual creativity: the poetics of paraphrase transforms everything it brings into its purview.

Both the reference to Homer and the language of *charis* prove important as the introduction unfurls. For Marcian requests a verse translation

---

[25] This may be the emperor Marcian, who convened the Council of Chalcedon and who died in 457. That he is called *pater* (5), assimilating the title of Augustus (*pater patriae*) to an ecclesiastical model of religious subordination, is no bar to the identification. That Marcian is depicted surpassing Ptolemy may encourage it. From Golega (1960) 5–24 onwards, however, the dedicatee is usually taken to be the *oeconomus* of St Sophia, Marcian.

[26] See Theocritus 16, with Griffiths (1979) 9–50; Gutzwiller (1983); Goldhill (1991) 273–83; Hunter (1996) 77–109.

[27] *Komizein*, 'to convey', may recall this verb's use in Pindar's poetics of exchange, a further element of the language of patronage.

precisely as an act of competitive rewriting with the great works of the Greek tradition in order that everyone can perceive what he calls the *charis* of the Hebrew songs. First he lays out the status of the Psalms (15–18):

οἶσθ', ὅτι Δαυίδου μὲν ἀγακλέος ἤθεα μέτροις
Ἑβραίοις ἐκέκαστο καὶ ἐκ μελέων ἐτέτυκτο
θεσπεσίων τὸ πρόσθεν, ὅθεν φόρμιγγι λιγείῃ
μέλπετο καὶ μελέεσσιν

You know that the ways of famous David
Are superlative in Hebrew verse, and were fashioned
Previously from divine song, when he performed with his high-pitched lyre
And songs

The Psalms are ancient (*to prosthen*), divinely inspired songs (*thespesion meleōn*) in verse (*metrois*) and Hebrew (*Hebraiois*); but David's singing to a high-pitched lyre also directly echoes Homer's Achilles (another famous hero) who sang the deeds of famous men outside his tent in *Iliad* 9 to a high-pitched lyre (*Il.* 9.186). Achilles and Homer are salient predecessors also because this paraphrase is in hexameters, when you might expect Psalms to fit more obviously into lyric metres: Achilles is a heroic paradigm of epic verse, hexameters, performed to a lyre, David's instrument.[28] Josephus, with greater interest in persuasive assimilation than metrics, reports that Moses' song at the sea and his final address to the Israelites in Deuteronomy are in hexameters;[29] and the Psalms in diverse metres including trimeters and pentameters.[30] Philo, also discovering the Greek roots of Hebrew, tells us that the Therapeutae use trimeters and also sing in antiphonal strophic choruses.[31] Most saliently for Ps-Apollinaris, Origen cites as an authority 'a certain man', probably Josephus, to claim that Hebrew verse uses trimeters and tetrameters, and the discussion is picked up by Eusebius (and, most fully, by Jerome).[32] It is unclear that any particular knowledge of Hebrew metrics grounds these remarks, but the cultural politics of bringing Hebrew verse into the culturally privileged narrative of specifically Greek literary history is evident.[33]

---

[28] For other potential echoes of Homer in these lines see Faulkner (2014) 202–4.
[29] *Ant.* 2.16.4; 4.18.44.   [30] *Ant.* 7.12.3.   [31] *De vit. contemp.* 2.6.68–9.
[32] Origen, Scholion on Ps 108; Eus. *Praep. Evang.* 2.5; Jer. *Ep.* 155; *Praef. Chron. Eus.* 2.
[33] The problem is taken up insistently in the Renaissance and beyond: see Baroway (1935), who notes (71) that 'The hexameter in some form was imputed in whole or in part to nine books and poems of the Old Testament'. Reuchlin, Scaliger and Vossius, for example, were all instrumental, until the seminal intervention of Robert Lowth (1787/1753): see Haugen (2012). Thanks to Konstantinos Lygouris who initially drew my attention to the passages in Josephus.

Marcian continues to make explicit the competition between Greek and Hebrew, prose and verse, their literature and ours (18–23):

> ἀτὰρ μετ' Ἀχαιίδα γῆρυν
> αὖτις ἀμειβομένων κατὰ μὲν χάρις ἔφθιτο μέτρων,
> μῦθοι δ' ὧδε μένουσιν ἐτήτυμοι· οὐ γὰρ ἀοιδῆς,
> ἀλλ' ἐπέων Πτολεμαῖος ἐέλδετο. ἔνθεν Ἀχαιοὶ
> μείζονα μὲν φρονέεσκον ἐπὶ σφετέρῃσιν ἀοιδαῖς,
> ἡμετέρας δ' οὐ πάμπαν ἐθάμβεον·

> but when turned again into a Greek voice
> the grace of the verse perished,
> while the truth of the narratives stayed. For Ptolemy did not expect
> song but words. For that reason the Greeks
> thought more highly of their own songs,
> and did not wonder at ours at all.

When the Septuagint was written and the Psalms were translated into Greek, their grace – *charis* – as poetry was destroyed, because their patron, Ptolemy, was interested in getting the basic sense of the stories (*muthoi*) rather than the experience of bardic song (*aoidē*). There is no hint of the stories of the divinely inspired translation of the Septuagint we discussed earlier: this is a human, pragmatic transaction. The beauty of the verse 'perishes' (*ephthito*) with the archetypal Homeric word for the loss of fame and life. Prose has its place but the result is that the Greeks – named in the Homeric manner as 'Achaeans' – continued to prefer their own literature, and, when they encountered this Christian writing, did not experience the wonder or amazement which is the response great poetry demands. Consequently, continues Marcian, now that the empire is Christian (23–8), it is fitting to provide verse to astonish the world (29–33):

> ἡμεῖς δ', ὥς κ' ἐπέοικε, τά περ πρότεροι λίπον ἄνδρες
> ἐκ μελέων, μέτροισιν ἐνήσομεν, εἰς δὲ μελιχρὴν
> Δαυίδου βασιλῆος ἐγείρομεν αὖτις ἀοιδὴν
> ἐξατόνοις ἐπέεσσιν, ἵνα γνώωσι καὶ ἄλλοι,
> γλῶσσ' ὅτι παντοίη Χριστὸν βασιλῆα βοήσει

> We, however, as is suitable, will set into metre the songs
> That earlier men have left behind, and we will raise again
> The honeyed song of king David
> In hexameter verse, so that other people too may learn
> That every type of language will shout out Christ the King.

Ptolemy only wanted 'words' (*epeōn*), but now we will have heroic metre (*hexatonois epeesin*). The task is to set the heritage of the church into poetry – the appropriation of the Hebrew to the Christian passes without notice, of course, in his supersessionist assumption – so that 'every (type of) tongue' – both every language and every sort of person – will acclaim the name of Christ. The language of 'raising again' (*egeirein autis*) is the religious promise of the Gospels, now applied to literary history, investing the composition with a fully religious command.[34] This verse version will be a religiously demanded transformation, designed to transform its readers into amazed celebrants of the Christian message.

The proclamation of 'every tongue' leads to a remarkable fulfilment in the subsequent lines of the *protheōria*.[35] For the triumph of the resurrected Christ is described first as the destruction of 'the sound of different tongues and different voices' (*alloglōsson allothroon audēn*, 59), which is the sign and symptom of the confusion of human sin (60), and secondly (and consequently), as the return of a single language for the world (62). This is not literally one language, but the all-embracing 'language' of Christianity. As with Augustine's history, this universalism is a desire for one single story for the world, the removal of difference. In contrast to the babble of the sinning opponents of Christianity, the 'speaking in tongues' of Pentecost is seen as the single message of Christianity echoing through each and every language – which is evidenced in an epic catalogue of the nations who have become Christian (67–79), and 'heard the divine virtues in their own tongue' (78), as the Christian message, even from ignorant cries, finds voice around the whole world (79–80). In contrast to the sinful hubbub of strange voices, with Christianity comes the sound of different languages all singing the same tune. The conclusion of this history of a Christian voice is Marcian's demand for these songs of praise to be translated into the Ionic language (*glōssan Iēona* 105), because it too comes from a divine beginning, because (107) every sort of language (*pantoiē glōssa* again, picking up the promise of the verse translation (33)) and every sort of song (*aoidē*) comes *from a single birthing*. The *protheoria* may not have opened with the language of beginning, but it closes by taking us back to the Garden of Eden to explain how Greek, the Ionic Greek of Homer, is a descendent of Adam's language, and therefore fit to sing the praise of God: Homeric *Kunstsprache* is tied to the *Ursprache*.[36] A single beginning to ground the single story. Remarkably, the theoretical reflection on writing a Greek verse version of the Psalms leads to a poetic account of the spreading of the word of God, which finds its beginning in the origin of language itself.

---

[34] Useful note on *egeirein* in Spanoudakis (2014a) 189 *ad* 41c.   [35] See Agosti (2001) 88–91.
[36] The Adamic language returns as a major issue in the eighteenth century: Aarslef (1982).

Ps-Apollinaris' introduction to his version of the Psalms speaks to a nexus of polemical issues, and the very act of dramatizing this self-reflection draws emphatic attention to the process of mediating scripture to its new audience. It recognizes that the educated, Greek-speaking world finds the language of scripture unimpressive – they do not share the amazement that the Gospels expect as the reaction to the story of Jesus. In particular, he admits that the prose translation of the authorized Septuagint falls short: it lacks the grace of the Hebrew poetry, and the complaint allows the religious sense of *charis* to colour the literary critical term. This version is to better the Septuagint: it promises to be transformative, for text and readers alike. Yet, not least because the Psalms in the prose of the Septuagint are so formative a text for the Christian reader, the poet also feels the need to defend the decision to write in hexameters and in Ionic Greek – epic form. He does so first by insisting that any and every language can celebrate Jesus – in contrast, say, to Jerome's return to the determinative specifics of Hebrew, or the Jewish insistence on the absolute privilege of the holy language of Hebrew – and, as his proof, he mobilizes both the image of 'speaking in tongues' and the successful spread of Christianity across the language communities of empire. Secondly, he declares that all languages come from their father in Eden, and all are thus fit for the purpose of the celebration of God. This introduction may have the narrative form of a patron's commission but it sets this commission in a theological framework of some sophistication, which performs its engagement with classical and scriptural traditions as it discusses them. Yet the poet also explicitly notes that there are actually only a few people ('one or two', 27) in the empire who are not Christian. This poem thus is primarily a work for the Christian community, redirecting their engagement with the Psalms. His *Metaphrasis* is designed to re-embed the Psalms within a classical tradition of heroic celebration, as it assimilates the heroic tradition to the celebration of the Christian God. It projects an audience that is educated, and comfortable in its mix of Greek and Christian education, and wants its Christianity in its literary form as well as in its theology to match the models of philosophy or poetry inherited through their education. It is a designedly instrumental contribution to the formation of an elite Christian culture.

Around seventy-five years earlier, Gregory of Nazianzus, whom we have already witnessed defending an elite Greek education as a fundamental strut of Christian *paideusis*, against the Christians who would spit at learning, also defended his practice of writing poetry precisely as a competition with classical tradition. With a wry elegance that is rather different from Ps-Apollinaris, in a poem that is also an apology for his practice of writing poetry, he writes:

> I know this is how I have felt – it's rather a petty thing (*pragma mikroprepes ti*), but it is how I feel – I won't grant that pagans (*tous xenous*) are stronger than us even in literature. I am speaking about their rhetorically elegant (*kekhrōsmenois*) literary composition, though beauty is for us in contemplation. Anyway, we have made these amusements for you, the men of wisdom; let there be for us too a certain leonine grace/gratitude.[37]

With a nice self-deprecation, he confesses that his feelings of competition are not quite dignified – a statement of intense confidence, of course – and his demurrals reveal a carefully sophisticated vocabulary. Pagan *logoi* are *kekhrōsmenoi*, which I take to mean 'tinged by *chrōmata*', that is, constructed out of rhetorical figures, a product of education in rhetorical schools, a formation which he sets against Christians who, with an evident Neo-Platonism, find beauty (*to kallos*) in contemplation (*theōria*), as Diotima advises in the *Symposium*. Yet his description of his poetry as *paignia* (*epaixamen*), 'amusements', is thoroughly classical in its assumption and language, as is his complicity-demanding description of his audience as 'the wise'. He requests for himself *charis leontios*, an unparalleled phrase which is not straightforward to translate. In the world of fable, the lion learns to show gratitude to the mouse or to Androcles. Is he describing himself as the apparently trivial figure who turns out to be crucial? And the *sophoi* as apparently kingly beasts who do not yet understand all they should?[38] Later in the poem (80), Gregory compares poets who are 'pretentious apes' or 'lions', again drawing on the world of fable. His defence of Christian poetry, as his opponents may have feared, is fully embedded in the language and culture of Hellenism.

Yet Gregory still feels the need to continue to defend the very act of writing poetry.[39] There is, he reminds his imagined opponents, 'plenty of verse in scripture, as the authorities (*sophoi*) of the Hebrew race declare' (80–1). But he is fully aware that Hebrew verse forms are difficult for Greek readers to access or appreciate (not for him the comfortable assimilation to

---

[37] II.1.39 *In suos versus* 47–53.

[38] My thanks for discussion of this odd phrase to Michael Reeve, Nicholas Richardson and Christos Simelidis. De Blasi (2017/18) *ad loc.* lists other, more unsatisfactory suggestions.

[39] Christians wrote hymns in non-classical metres, but in the Greek East (in contrast to Latin) only rarely in classical metres before Gregory, it seems: see den Boeft and Hilhorst eds. (1993); Schwab (2012); on his self-defence in terms of being *metrios* ('reasonable'/'metrical') see Hawkins (2014) 142–80.

hexameter or trimeter): 'If the pluckings of those strings too are not metre to you, recognize that men of old sang lyrical compositions; they composed pleasing vehicles for the good, and made models of behaviour out of song'. Again, Gregory's language is a beautifully poised mixture of Greek literary theory and Christian purpose. He assimilates the lyrical voice of the Psalms to the lyrical tradition of ancient classical tradition. The Psalms, like those early Greeks, used *to terpnon*, 'what is pleasing', to express something morally useful – Gregory utilizes here one of the most familiar oppositions of Greek literary theory, the pleasant and the useful – and, he continues, they modelled (*tupountes* – 'made types', a central Christian theoretical term, too) character (*tropous* – also a term for 'tropes') out of lyric song (*ek meleōn*). This is an elegantly doubled-edged expression. On the one hand, it recognizes that different metres in the classical tradition were said to have different characteristics – one metre was warlike, another for mourning, and so on. On the other hand, it also suggests that individual characters are made better by listening to good poetry. So, his argument's conclusive example is Saul, who was cured by listening to David, the singer of the Psalms. Gregory's own writing of poetry thus is defended by a double and interlocked tradition of Hellenism and scripture, an assimilation of what many Christians would set in opposition. He calls this *mixis eugenestera* (93), 'a nobler mingling', or what we might call a hybrid vigour. His literary theorizing is a cultural polemic.

There is one final twist to his argument, however, which takes us back to our earlier discussions of time, the moment, and conversion. Poetry can lead the young, he has argued, through enjoyment towards God. But, he states, this is not a question of 'an immediate conversion', 'a change all at once', *athroa metastasis* (92). Rather, the good needs time to achieve some fixity (94–6): 'When the good attains stability (*pēxin*) in time (*en chronōi*), we will draw away the delight (*kompson*), like struts from an arch, and will protect the good itself'. Nothing could be 'more useful' (97). Pleasure, then, is in service of the useful, defined with a Neo-Platonic gesture as 'the good itself' (that is how the rhetorical opposition of 'pleasure' and 'usefulness' is to be resolved), and the refinement of Greek learning is precisely a strut that can be removed, a swagger that leads to true goodness. That is why desired stability must come not in a moment of conversion but in the passage of time (*en chronōi*), the time of reading. As with Augustine, reading is conceptualized through a theology of time and change.

Mixis, 'mingling', is a crucial term for Gregory. In an extraordinary outpouring, that could inspire a Walt Whitman, he declares 'I sing my mixing', *melpō mixin emēn*.[40] This mixing is his intermingling of human and divine, his mortality and immortality as a Christian. This proclamation comes at the end of a *praeteritio* where he sets aside a list of classical subjects, songs he will not sing, starting, of course, with Troy. What he *will* sing culminates in this couplet (83–4):

καὶ Χριστοῦ παθέων κλέος ἄφθιτον, οἷς μ' ἐθέωσεν
    ἀνδρομέην μορφὴν οὐρανίῃ κεράσας

[I sing] also the immortal glory of the sufferings of Christ, by which he mixed
    My human form with the heavenly and made me immortal.

This is his 'mixing'.[41] Yet after his dismissal of the subjects of classicizing poetry, it is striking that he adopts the distinctive language of Homer – *kleos aphthiton*, 'immortal glory', Achilles' promise – to celebrate Christ, his own subject. He is triumphantly appropriating the privileged aim of Homer's hero and Homer's poetry for his own Christian project, as the echo of Homer is ironically turned against Homer. What immortality is, is now defined by the Christian promise, not Homer's. But it *is* the language of Homer, and the poetry's theological import depends on the recognition of it. Theology and poetics together are working to create an elite and knowing Greek Christian culture in the image of Gregory, the author. In the declaration of the *mixis* of divine and mortal in the human, the *mixis* of cultural authority is being performed.

Defending poetry itself was necessary for Gregory because there were certainly voices raised against it. Apollinaris – the fourth-century figure who gives his name to the author of the *Metaphrasis of the Psalms* – wrote poetic versions of Christian scripture and turned the Gospels into Platonic dialogues, in response, we are told by John Zonaras in the twelfth century, to the emperor Julian's edict against Christians teaching pagan literature.[42] But his work was condemned as heresy by the Council of Constantinople in 381. Gregory himself also viciously dismissed Apollinaris' works for their Arian views,[43] as his sixth-century biographer, Gregory the Presbyter, also

---

[40] II.1.34, 'On Silence at a Time of Fasting' 85. See also below, pp. 325–8, and e.g. *Or.* 38.13 'O new mixing' of the incarnation, with Beeley (2008) 131, who notes the language of *mixing* was condemned at Chalcedon; or *Ep.* 101.21.

[41] Kuhn-Treichel (forthcoming) refers it rather to Gregory's mixing of genres.

[42] The accounts of Sozomen and Socrates are discussed with bibliography in Agosti (2001) 85–7; Spanoudakis (2016) 603–4, who notes that Sozomen (*HE* 5.18), some decades later, called Apollinaris' books 'equal . . . to the works most celebrated among the Hellenes', a remark which again indicates the competition between Christian and pagan cultural achievement.

[43] See Gregory, *Ep.* 101; *Ep.* 102. Faulkner's otherwise excellent discussion of the authorship of the *Metaphrasis* (Faulkner 2020) underplays the significance of these attacks.

noted – at least in the latter parts of Gregory's career.[44] The threatening 'novelty' of Apollinaris was as much theological as in the form of his rewriting. But St Nilus of Ancyra – who may stand for the voice of the ascetics opposed to the education Gregory championed, for all that he was himself a supporter of John Chrysostom – simply dismisses poetry as 'without use' and a 'denigration of the Cross of Christ' – morally corrupt for what it embodied.[45] The *Letter of Aristeas*, many centuries earlier, insisted that Theodectes went blind because he attempted to turn stories from the Bible into dramatic verse.[46] It was not simply poetry as a pleasing form that made Christians anxious, but the Hellenic culture it encapsulated.

This anxiety is especially salient with the Gospel of John. The opening of John in particular marks a difference from the other Gospels. Matthew begins with the genealogy of Jesus and announces itself as the *biblos geneseōs*, the 'book of the genealogy', a new Genesis, and proceeds from Abraham, through David, to Jesus. Mark announces 'the beginning (*archē*) of the Good News', and takes its start from a prophecy of the Messiah in Isaiah – that is, it starts in the authority of the Hebrew scriptures. Luke, as befits the writer of Acts also, adopts the pose of a historian, and authorizes his account as an eye-witness statement, addressed to one Theophilus (and his genealogy for Jesus in chapter 3 goes all the way back to Adam: no *spatium mythicum* in this history, no mists of time in the Christian gaze, however mysterious the theology proclaims itself to be).[47] It is only John who begins not in human history, but with the beginning of everything, the first principle, and who declares this first principle to be the Logos – a term which is inevitably redolent with the tradition of Greek philosophy. If Christianity sets itself against Greek culture, as the dominant intellectual force in the Mediterranean, a threat and promise also to both Romans and Jews alike, what is the place of John in such a dynamic?

There is a very long and complex history of answers to this question, all based on the recognition that the early church gradually and increasingly assimilated itself to the dominant Greek culture from which it continued to distinguish itself, especially through the influence of Neo-Platonism.

---

[44] Beeley (2008) 285–92 offers necessary nuance to this standard view.

[45] *Ep.* 2.49. 'Do not devote yourself to verse', he continues, 'lest you neglect your own salvation through enthusiasm for poetry.'

[46] *Letter of Aristeas* 315–16. His sight was restored after days of prayer, a model obviously based on Stesichorus' palinode experiences.

[47] On the generic signs in Luke's preface, Moles (2011) is right to critique Alexander (1993), though Dawson (2019) – with full bibliography on the issue – offers a tempering of Moles.

Augustine already noted with sardonic annoyance that 'a Platonic philosopher', with equally sardonic wit, proposed that the first lines of John should be inscribed in gold on the walls of every church, presumably because they would lead the congregation towards philosophy rather than Christianity.[48] Indeed, Adolf von Harnack – to take one extreme, authoritative, modern example – declared that the Gospel of John itself already indicated the decline of the early church from its pure origins because of its engagement with Greek philosophy. The foundational Gospel as a sign of decline . . . In contrast to this doyen of Protestant theology, Pope Benedict XVI – another extreme, modern, authoritative example – in his Regensburg Address, delivered a quite different version of this history. Although the Address became notorious for its perceived dismissiveness towards Islam, it is far more radical and challenging in its attitude to Hellenism, which passed without notice in the press.[49] For the former Cardinal Ratzinger, Hellenism was integral to and necessary for Christianity. The Septuagint was 'a distinct and important step in the history of revelation, one which brought about the encounter between faith and reason in a way that was decisive for the birth and spread of Christianity', and the New Testament itself 'bears the imprint of the Greek spirit'.[50] Against centuries of anxiety and even aggression, the Pope celebrates Hellenism in Christianity as essential to Christianity's engagement with reason and faith. And his test case for the 'intrinsic necessity of a rapprochement between Biblical faith and Greek inquiry' is, of course, the opening verses of the Gospel of John. John's first words are, he writes, 'the final word on the biblical concept of God'. The many stages that lead from Augustine to this fascinating and continuing modern engagement need not be detailed here. What is crucial is that the tension between the commitments of traditional elite Greek culture and Christian faith, which we saw enacted in the defence of writing Christian poetry, finds its most compelling test-case in the opening words of the Gospel of John.

Nonnus himself remains a figure who acts as a lightning rod for debates about such cultural conflicts or assimilation. Both the fact that Nonnus wrote a huge, sexy epic about a pagan god, Dionysus, alongside his paraphrase of John's Gospel, and the fact that he composed a verse rewriting of the Gospel in itself, continue to provoke questions about his religious and cultural identity – or his political or theological agendas or

[48] Aug. *Civ. Dei* 10.29. For Neo-Platonic readings of John, see Dillon (2002); Dörrie (1976).
[49] von Harnack (1892). See Gagné (2020) for a discussion of this history; also, with a different focus, see Buch-Hansen (2018).
[50] Benedict XVI (2006) 20, 28. Benedict cites von Harnack in particular as influential.

lack of them.[51] The questions are made more emphatic by the recognition that the *Dionysiaca*, despite much searching by critics, appears to show very few explicit signs of Christian commitment in its vocabulary or moralization, although, as we will see in the next chapter, this is not the only way to calibrate the import of its narrative: in different ways, its symbolism, its narrative style, its echoes of the *Paraphrase*, have each been utilized to locate the *Dionysiaca* within a Christian context. The *Paraphrase*, conversely, is free with Dionysiac imagery, which has often both been trivialized as merely decorative and anxiously overemphasized as a sign of deep-seated discomfort with Christianity.[52] It has become a commonplace of contemporary criticism to note that previous scholarship attempted to harmonize the image of Nonnus by postulating a biographical narrative of conversion, either that Nonnus the Christian, who wrote the *Paraphrase*, lost his faith and wrote the *Dionysiaca*, or, more commonly, that the uncommitted poet of the *Dionysiaca* repented and wrote the *Paraphrase* as a sign of his new orthodoxy.[53] Such stories are predicated on an assumption, easily evidenced from one extreme brand of Christian apologetics from late antiquity, and equally easily paralleled by later evangelical fervour, that either you are for Jesus, heart and soul, or an enemy of Jesus. Conversion, thus, and loss of faith become the privileged explanatory models of difference, as we saw in chapter 10's discussion of temporal rupture. Commitment becomes the yardstick of belonging.[54] Critics in the last three decades, authors of this commonplace, insist by contrast that in late antiquity, although there are clearly ideologues who promote such extreme boundaries between the church and its enemies, there are many citizens – and many writers – who were far more comfortable with multiple cultural and ideological frameworks.[55] The parting of the ways between Christians and Jews has been dated later and later; the interaction between the church and the institutions of empire have been recognized as more and more intertwined; even the role of martyrdom, the 'blood of the church', has been argued to be evident more in rhetoric than in actual cases of physical suffering. Hybridity becomes the privileged explanatory model of difference. In this picture,

[51] Somewhat world-weary summary in Chuvin (2014); better by far is Shorrock (2011).
[52] Not by more recent critics, see Chuvin (2014); Doroszewski (2014), (2016); Shorrock (2011) 58–91, (2016); Spanoudakis (2014a) 41–51, (2016); Dijkstra (2016). This issue is further discussed in the next chapter.
[53] Bogner (1934); Cameron (1965), (2016).   [54] See Goldhill (2020) 149–93.
[55] Shorrock (2008), and especially (2011); Spanoudakis ed. (2014); Accorinti ed. (2016) and the Italian editions of Nonnus have been especially instrumental here.

Nonnus becomes a 'poster child'[56] of the vigour of Greek literary tradition within a Christianizing world.

The pursuit of the empire's 'last pagans' is always in danger of reproducing the teleological and ideological commitments of the polemical texts of the period. The sheer complexity of the map of affiliation and the network of power in the empire of late antiquity requires especial sensitivity to a writer's self-positioning. Aelia Eudocia, the wife of emperor Theodosius II, who converted to Christianity as a young woman before her marriage, in maturity supported groups in the East opposed to the Council of Chalcedon's conclusions, but then, with pressure from the imperial centre, reconverted to the orthodox church, is only one of the most high-profile cases of shifting intellectual and political roles in turbulent times – and she is also one of the best-known composers of both verse paraphrases and Homeric centos, bringing the politics and aesthetics of religious interpretation, articulated between Hellenic and Christian cultural authority, to the heart of the power structures of empire. Nonnus inhabits a culture where religious commitment is major force, and where differences of religious understanding are being aggressively enacted not just in theology but also in the state's exercise of power. On the one hand, conversion is a privileged narrative trope and a strategy of enacted authority; on the other hand, the boundaries of belonging to the orthodox church were articulated and contested in strident arguments about specific words and their use: the politics of language. Nonnus, as we will see, utilizes the poetics of paraphrase to engage both with the long literary tradition of pagan Greek culture and with the dynamics of contemporary Christian theology: to transform both. As his Dionysiac poetics puts the narrative of the *Dionysiaca* under the sway of the god of transformative othering, so the *Paraphrase* by virtue of its poetics of paraphrase becomes a strategic intervention in the development of Christian culture. When religious transformation becomes an overriding anxiety for society, and the politics of language the crisis point of such anxiety, which is the case in late antiquity as in the early modern era of the Reformation and Counter-Reformation, then paraphrase becomes a genre that speaks to its time.

*

We are now in a position finally to return to the opening of Nonnus' *Paraphrase of John*. Rewriting scripture as an interpretative intervention, within the prescription of fixity; reinvesting prose with the historical,

---

[56] Johnson (2016) 269.

heroic grandeur of hexameter verse; conceptualizing scripture within a cultural competition between elite Greek literature and the demands of Christian affiliations; the competing demands of different strands of Christian theology with regard to the value of poetry – all provide histories and frameworks to understand why this text is a compelling performance of cultural transvaluation. It is remarkable how often modern critics use the denigratory word 'exercise' to describe and trivialize Christian poetry, especially when it follows ancient Greek forms, as if the schoolroom or, at best, the literary salon, is the right frame of comprehension.[57] Better to understand 'exercise' as *askēsis*, the work of self-formation through what Gregory would call *metastasis en chronōi*, transformation in time. Nonnus' opening as a reading of John's beginning is not just stunningly bold, but also an extraordinarily intricate, theologically charged response, that requires close reading to tease out its impact.

Here, then, are the first five lines with an attempt at translation, a far from straightforward task:[58]

ἄχρονος ἦν ἀκίχητος ἐν ἀρρήτῳ λόγος ἀρχῇ·
ἰσοφυὴς γενετῆρος ὁμήλικος υἱὸς ἀμήτωρ,
καὶ λόγος αὐτοφύτοιο θεοῦ φάος, ἐκ φάεος φῶς·
πατρὸς ἔην ἀμέριστος, ἀτέρμονι σύνθρονος ἕδρῃ·
καὶ θεὸς ὑψιγένεθλος ἦν λόγος

Timeless, unattainable was the Word in the ineffable beginning;
of equal nature to His coeval begetter, a Son without a mother;
and the Word was the light of self-begotten God, light from light;
from the father he was indivisible, on the same throne on the boundless seat,
and the Word born on high was God.

John's syntactically simple opening five words, with its powerfully simple vocabulary, has become three hexameters, rebarbative in their syntax and terminology. *Achronos*, 'timeless', 'non-time', redefines a beginning within a Christian invention of time. As we have discussed already, the timelessness of God, or in this case the *Logos*, is the grounding of theology's temporality: Gregory of Nazianzus, prodigiously prolific as ever, uses this term, *achronos*, for the Son, the Father and the Holy Spirit.[59] Cyril, like Augustine, is explicit that timelessness is a denial of beginning: 'No beginning that is the least bit

---

[57] Agosti (2001) discusses the slight evidence for the actual use of such poetry in schools. Educational papyri give us paraphrases only in prose, not in hexameters.

[58] Hadjittofi (forthcoming) is the best available English version; Agnosini (2020) came out after this chapter was finished and is a useful Italian translation with comments.

[59] Discussed below, pp. 327–8. See *Carm.* I.1.27; I.1.5.55; II.1.11.649; II.1.14.41; I.1.2.21, as noted in de Stefani (2002) 104; on Nonnus' use of Gregory, see Simelidis (2016) 298–307.

temporal can be applied to the Only Begotten because he is before all time . . . he will elude any notion that he came to be in time. Through all time he was in his Father'.[60] By this beginning, Nonnus reframes the narrative, away from the human history of the other Gospels and even from the foundational moment of John's beginning, and places what is to come under the permanent sway of theology. *Achronos* makes the verb that follows ('was') not just an echo of John but also a specifically anti-Arian and pro-Nicean statement. As we outlined earlier, 'There was a time (*pote*) when he was not', ἦν ποτε ὅτε οὐκ ἦν, along with the parallel idea that 'Before he was begotten, he was not', were singled out by the Nicene creed as the most dangerous and false ideas of the Arians: different concepts of time are at the base of the different comprehensions of Christology.[61] The birth of the Logos, taken, as the second line makes clear, as the Son, is here immediately defined as timeless, no before, no after. 'What mind, tell me, can over-leap the force of the *was?*', asks Cyril with his characteristic rhetoric in his commentary on John.[62] Beyond any general 'spiritualization',[63] Nonnus is marking out a doctrinal affiliation.

I translated *akikhētos* as 'unattainable' in the sense of 'cannot be grasped' (*akataleptos* is the commoner prose term in theology) – that is, beyond human comprehension. For the pro-Niceans, an insistence on the incomprehensibility of God – an apophatic theology – was an essential and proclaimed difference between themselves and the Arians.[64] But *akikhētos* is a word that occurs only once in Homer (*Il.* 17.75), and barely anywhere else in what survives of Greek literature: it is a perfect example of how Nonnus marks, appropriates and rewrites Homeric lexis, and thus overwrites *koinē* with the language of elite verse. Apollo warns Hector not to chase what is ungraspable, that is, what cannot be touched – specifically, perhaps significantly for Nonnus' usage, the divine horses of Achilles, which Hector, mere mortal, could not hold, attain, drive; and the immateriality of the *Logos* as divine reason is fundamental. As the *Paraphrase* continues, these two senses – physical and spiritual – of this programmatic

---

[60] Cyril, *In Joh.* 1.18.2–16. Sieber (2016) is unconvincingly hesitant about Nonnus' exegetical purchase here.

[61] See Sieber (2016) for Nonnus' Christology.   [62] *In Joh.* 1.1.17.

[63] Spiritualization is the characterization of Nonnus' religiosity in Spanoudakis (2014a) 34, although he also notes well and frequently the influence of Cyril on Nonnus, concluding even that the *Paraphrase* was written with Cyril's commentary '*ante oculos*' (18); see also Franchi (2016) 244–6.

[64] See below, p. 319, where we will return to this idea with Gregory of Nazianzus and Gregory of Nyssa in particular. Hartog (2020) makes 'ungraspable' – 'insaisissible' – a key term in his opening analysis of time, but without reference to this ancient discourse.

adjective are explored through the narrative of the *theos anēr*, 'god man'. It is used a further seven times of Jesus. In the early stages of the story, each time it is used for the failure of his enemies to get hold of him, either retrospectively or proleptically.[65] Its immediate sense is physical, as if activating the Homeric sense. Jesus slips through his enemies' grasp. But as the narrative progresses, it is used only of Jesus' ascension to the house of his father – once in the words of the baffled disciples – where he is both unattainable and beyond human comprehension.[66] The movement from the bodily escape to spiritual transformation is significant, and encourages a re-reading of the earlier occurrences as more than physical safety, that is, as signs of the power of Christ to transcend his enemies, and the human condition.

These examples also culminate in a remarkable, specific human *enactment* of unattainability, where the physical is also the symbolic instantiation of the spiritual. In chapter 20 of John, Mary Magdalene goes to the tomb, sees the stone is rolled away and runs immediately to tell the disciples. In Nonnus' version of John, however, Mary has brought myrrh to anoint the corpse, and she feels around inside the tomb for the absent body: ἀλλά μιν οὐκ ἐκίχησεν, 'But she did not touch/attain him' (20.12).[67] There is a mention of spices for anointing in Mark and Luke, and in Luke a group of women enter the tomb,[68] but Nonnus has constructed this scene from a silence in John to create a scene of the pathos of Mary on her own, her feeling around the tomb, and her failure to 'attain' Jesus: the physical act here stands for the true unattainability of the risen Christ, his transcendence of the material world.[69] So, in contrast with the Gospel of Luke, where Jesus tells the amazed and terrified disciples to 'touch and see my hands and feet', precisely to demonstrate his continuing physicality (Luke 24.39), Jesus tells Mary (20.74) 'Do not touch my robes', and Thomas, who wanted to put his finger in the wounds of Jesus – he does not do so in John or Nonnus – is rebuked for believing only after he has seen with his eyes (20.134–5): 'more blessed are those who have greater faith if they do not see and lack sight'.[70] The transcendence of the physical is made integral to

---

[65] 7.123; 8.191; 10.139; 12.2.   [66] 13.137; 14.18; 14.53.

[67] Accorinti (1996) 129 *ad* 12 notes without comment Homer, *Od.* 23.524, *alla min aipsa kikhanen*, which, if it is echoed here, indicates a contrast.

[68] For Nonnus' use of the synoptic gospels see Golega (1930) 131–8; Spanoudakis (2014a) 17.

[69] Thus Gregory of Nyssa *Cant.* 89.19 writes: ἀπρόσιτόν τε καὶ ἀναφές ἐστιν καὶ ἄληπτον (unapproachable, untouchable, ungraspable).

[70] On Thomas here see Whitby (2007). Spanoudakis (2014a) 78 analyses well Mary's and Martha's different responses to the approachability of the person of Jesus: 'Martha approaches Christ as man,

Christian belief and expressed in Nonnus through this language of unattainability.

This pointed and systematic use of *akikhētos*[71] may help clarify a particular textual issue too, and reveal a second highly charged use of the word at the close of the crucifixion narrative. At 19.178, Scheindler prints the manuscript reading, ἀλλὰ θορὼν ἀκίχητος ἀνὴρ ἀνεμώδεϊ λόγχῃ 'But a man leapt up ungraspable and with a spear, swift as the wind ... '. In the apparatus Scheindler duly notes that *akikhētos* is not completed in V and corrected, and that Marcellus proposes θορὼν στράτιος λόγχης ἀνεμώδιος ἀνήρ, 'a man of the army leapt up with a spear swift as the wind', a version designed both to reintroduce the man as a soldier, and to remove the surprising use of *akikhētos* – which Hadjittofi translates 'swift', marking also her discomfort with its use here. At one level, the adjective – which has no equivalent in John – implies simply that someone might have stopped this soldier, but did not. More significantly, this is the first time the adjective is used of someone other than Jesus/Logos. Now that Jesus has died, his body is not only 'graspable' but attacked by a human, a human who 'cannot be grasped'. The act of the soldier raises the question of the status of Jesus' body and his pain during the crucifixion: does Jesus suffer as a human, is this a different body (as Gnostics claimed), does his willingness to die include a willingness to experience anguish? What does it mean to 'grasp' Jesus' body, at this juncture in the narrative? The focus on the materiality and spirituality of the flesh of Jesus is emphasized by a further linguistic detail, and by the continuation of the narrative. The spear (*longkhē*) is also described as a *makhairē*, 'dagger' in the same sentence (19.179). This is not a confusion. *Makhairē* is a familiar enough word, of course, but it is used uniquely and strangely in the Septuagint as a translation of the Hebrew מאכלת, which itself occurs only once in the Pentateuch, to describe the knife Abraham picks up to sacrifice Isaac (Gen. 22.10). Isaac's sacrifice is taken as a typological understanding of Jesus' death on the cross. The strange use of *makhairē* by Nonnus here is designed to evoke this typology. This is followed by the mysterious flow of blood and water from Jesus' wound – both symbolic liquids, the water of life and the blood of the Eucharist – an event which is described as a fulfilment of prophecy and as a 'harbinger (*proanggelos*) of his blameless flesh' (187) – that is, the body pierced and flowing with liquid is to be understood

---

Mary Christ as God' (a reading also in Cyril, less sharply expressed). On Doubting Thomas as a figure see Most (2007).

[71] *Akikhētos* is used regularly in the *Dionysiaca* but without either the spiritual sense or the systematicity of the *Paraphrase*.

as a sign of the 'body of Christ' in a fully theological sense. The description of the soldier as an 'ungraspable man', after the repeated use of the adjective for Jesus, cues this transformation of the body of Jesus. Jesus is only in one physical sense 'graspable', even on the cross, and the human act of stabbing him is 'ungraspable' only in its incomprehension of the body in front of him, a misunderstanding of its own lack of restraint. *Akikhētos* again indicates the narrative's struggle to articulate what is '(un)graspable' in the figure of Jesus, in the human world and transcending it.

The final use of the adjective is also richly evocative. Joseph of Arimathea carries the body of Jesus, 'an ever-living corpse' (21.223), to the tomb, with the help of Nicodemus, and gets home *akikhētos*. Again, the simple sense is clear enough: Joseph escapes his enemies' grasp – Jesus' enemies' grasp. Both Joseph and Nicodemus, however, were secret disciples, who visited Jesus at night and did not announce their beliefs: their religious commitment is 'ungrasped'. Their task of carrying Jesus to the tomb is described as *amarturon* (19.224), 'unwitnessed', although here it is described precisely as part of the 'witnessing' of the miracle of Jesus' resurrection. Their journey continues their narrative of paradoxically concealed witnessing of the revelation. As they visited Jesus at night, so here at night, they have carried the body of the 'ungraspable' Jesus to the tomb where its 'ungraspable' nature will be evidenced, in the light, to others (222–3). Joseph's physical escape is stressed (there is no such comment in John) as he transfers the body to the tomb, to contrast with the transcendent transformation that is about to happen. The human 'ungraspable' is not the 'ungraspable' of Jesus. The soldier, Joseph, and Mary in the tomb – three scenes within 60 lines – in their different ways each pose a question about the physical nature of Jesus, after his death, a question about how the materiality of his body is to be comprehended, grasped: the central question, that is, of Christology, the boundary between human and divine in the form of Jesus Christ.

Akikhētos in Nonnus' opening line is truly programmatic: it opens a theological perspective on both the physical nature of the body of Jesus and the comprehensibility of the mystery of Christ, a perspective which is articulated through the narrative by the salient and systematic usage of the term. Nonnus takes a term from Homer and makes it resound through his text, not just reframing and negotiating its significance through its repetitions, but also redrafting John's narrative towards a more theologically demanding understanding. This is his paraphrastic poetics at work.

Let us continue reading this first line ... The 'beginning', *archē*, is displaced from the first word in John to the last word of this first line, and qualified by *arrhētos*, 'ineffable', 'unspeakable'. The beginning which was foundational for John is an inexpressible mystery for Nonnus. Of course, 'ineffable', like 'mystery', is the language mobilized when the argument of reason fails in theology, but here it is more pointed: what can the term 'beginning' mean in the context of timelessness? If timeless is taken in its full theological sense, then, as Cyril made plain, there is no beginning, no end: beginning is unsayable. Hermes Trismegistus described the *logos* as *aporrhētos*, because the name of the Son is inexpressible and unknown to humans and will not be known till the end of days.[72] *Logos* is, in a manner which is both literal and theologically expressive, *unsayable*. Here, the ineffable is a challenge to the very idea of beginning such a narrative, because of the always already of timelessness. God's time transforms the narrative of beginning, the beginning of narrative.

Nonnus uses *aporrhētos* only once more in the *Paraphrase*, again in the context of the Incarnation, where it expresses 'a certain ineffable principle' (*thesmos*, 1.40). John writes simply and powerfully 'the word became flesh' (1.14). Nonnus expands this to (39–41):

> And the self-created Word became flesh, God, man,
> he who was born late, earlier born, in an ineffable manner
> bringing together in a common yoke the divine and the human-like form.

Nonnus again strives to express the complexity of the concept of the Incarnation. 'Self-created', *autotelestos* – another obscure word filled now with Christological import[73] – is followed by the simplest terms in bare juxtaposition, a juxtaposition that captures the paradox of this theology: *theos anēr*, 'god, man' – a semantic tension signalled as 'ineffable', echoing the 'ineffable beginning' of the opening – and recalling also Cyril's theology, who called the *genesis* of Jesus, 'the ungraspable (*akatalēptos*) and ineffable (*aporrhētos*) generation that is outside time (*exō chronōn*)'.[74] For Nonnus here, too, the issue of temporality is foregrounded. The *logos* is *opsigonos progenethlos*, 'born late, earlier born' – another juxtaposition that prompts the declaration of its ineffability. *Opsigonos* is a common enough word from Homer onwards, often in the phrase *opsigonōn anthrōpōn*, 'the men of future generations', 'men who are born later'. But *progenethlos* appears to be a word Nonnus has

---

[72] Lactantius, *Inst.* 4.7.

[73] On the text here, see de Stefani (2002) 136 *ad loc.* Nonnus is particularly fond of *auto-* compounds, a self-reflexivity.

[74] Cyril, *Comm. ad Ioh.* 1.1.22. On the use of paradox on Christian rhetoric see Cameron (1991) 154–88. Sieber (2017) 162–3 is unconvincing on this phrase.

coined for the sake of the emphatic juxtaposition.[75] Timelessness is here articulated as the apparently paradoxical – ineffable – combination of late and early. For Nonnus, John's expression of the Incarnation needs this temporal exposition and reframing: the verb *egeneto*, 'became', in order to be fully expressive of Christian theology, must be glossed with *opsi**gonos** progenethlos* so that generation extends from a moment into a timeless continuum.

John's first sentence repeats the word *logos* three times, and so does Nonnus, each time still firmly in the nominative: the *logos* stays fixed and true in its form. But each usage is glossed in Nonnus with a string of theologically charged expressions, which extend and refine the bare prose of the Gospel (the very richness of expressivity is a continuing enactment of the tension between the struggle to articulate the ineffable and the exuberance of demonstrative praise) – and which speak directly to the arguments over time to which the Nicene creed speaks authoritatively. So the *Logos* is the Son, the Son who is 'of like nature' to the father, that is, who fulfils the claim of consubstantiality, which the bitter rows over the adjective *homoousios* contest. The Son is *homēlikos*, 'of the same age' as the Father – again, the reinforcement of the theology of the Nicene creed's temporality is aggressively stated: the Father and Son must be of the same age, and thus not in hierarchical order. Timelessness is also a promise of temporal equality between Son and Father. The Son is also 'without a mother'. The specific cause of Cyril's vicious dismissal of Nestorius was the unwillingness of the church of Antioch to use the term *theotokos*, 'bearer of God'.[76] The status of Mary's motherhood was a simmering issue for decades in Christological debate.[77] The phrase 'son without a mother' significantly separates the Logos from the messy business of a human birth: 'no one hearing the expression "son of God" should conceive such evil in his mind to think that God procreated as a result of marriage and intercourse with some woman!'.[78] Hence Nonnus terms the Logos the 'light of the self-begotten God'. 'Self-begotten', like 'self-created', again is a rare, reflexive word that displays the effort of expressing a relation of son and father that can escape the standard temporal relation of creator and created. The full expression 'light, light from light' has no immediate equivalent in John, though it anticipates the use of light towards the end of John's introduction (1.5; 1.8–9). The imagery of light and darkness will flow

---

[75] See *Dion.* 47.29 for similarly pointed language of first and later birth, of Zagreus and Dionysus.
[76] Smith (2018); Wessel (2004).   [77] Cameron (1991) 165–70.
[78] Lactantius, *Inst.* 4.8. For the Jewish background of the expression 'Son of God' see García Martínez (2013) 83–100, with further bibliography.

throughout the *Paraphrase*.[79] What is telling, however, is that Nonnus glosses 'light' with the further phrase 'light from light'. The added expression is not scriptural, but it is taken directly from the wording of the Nicene creed – as if the creed is the expression of the Gospel, the Gospel contains the creed, an unbroken tradition. Cyril uses the same phrase stridently in his commentary on John. The theological intensity of Nonnus' verse is evident.[80]

The second phrase of John's opening sentence is transformed into (only) one line, but it still constitutes another extended theological rewriting. 'The Word was with God' becomes 'from the father he was indivisible, on the same throne on the boundless seat'. The openness of 'with' becomes the explicitness of 'indivisible', a theological argument also central to the Nicene creed and in particular to Cyril's defence of it. 'I and the father are one' (John 10.30) is repeatedly cited as proof that there is no distinction of time or authority – or substance – between the Son and the Father, against the Arian and Neo-Arian claims of a hierarchical relationship. Thus Nonnus specifies 'the same throne', the same seat and exercise of power, that is, the 'throne of heaven' – the site of 'sovereignty', as, again, Cyril's commentary makes abundantly clear – which is a seat that is – as an expression of power – without limit. This adjective *atermōn* occurs nine further times in the *Paraphrase*, to qualify God the Father, the universe (*kosmos*), the honour of Jesus, and time itself: it links the power of divinity to time in their shared boundlessness.[81] So, as the poem opens with the 'boundless seat', its final line declares the poet's expectation that if anyone were to try to write the miracles of Jesus into books, they could not be contained by 'the boundless universe'. Nonnus' *Paraphrase* is itself one of the future boundless library of books predicted by John – a wonderful moment in which paraphrase turns into self-fulfilled prophecy (*autotelestos*). Here in the final line, for the first time, the author enters the text in the first person (*elpomai*, 'I expect'), and John's voice and Nonnus' are coterminous, a moment of perfected disavowal, complete impersonation.[82] The narrative's end comes in the proclaimed endlessness of narrative's potential: the story of Jesus goes beyond even the 'boundless universe'. At the beginning and end of the *Paraphrase*, the limitless is thematized.

---

[79] Ypsilanti (2014); Johnson (2016) 273–80.

[80] Agosti (2009) 329 sums up the *Paraphrase* as 'an ideologically committed work into which an intense exegetical and doctrinaire effort has been poured'.

[81] *Aiōn*, 3.31; 6.147; God the Father 10.134; 19.81; *kosmos*, 17.53; 18.33; 21.143; *timē* 6.146; *pistis* as 'boundless mother of *kosmos*', 1.19.

[82] On impersonation in late epic, see Greensmith (2020).

The third phrase of John, 'and the Word was God' becomes 'and the Word born on high was God', apparently a close equivalence, with only the addition of a single adjective. But this adjective also makes a pointed difference. 'Born not made' is a strident phrase of the Nicene creed, designed precisely to attack the Arian idea that God *made* the Son, who is thus subordinate and later in time. The adjective could go with either noun, of course, and God is traditionally praised in Hebrew as 'on high', and in Greek as *hupsistos*, 'the highest': but the echo of the Nicene creed is compelling, and again shows Nonnus designedly redirecting the language of John towards a fifth-century theological significance.[83]

Nonnus' first five lines need at least this much care to appreciate the full scope of their extraordinary redrafting of the Gospel's already challenging first words. John's Greek, which already strains the agenda of the simple language of faith into an extreme of semantic expressivity, is turned by Nonnus into a glossolalia of theological normativity. On the one hand, this is intense intellectual poetry at the very highest level of elite understanding, which is closer to commentary or exegesis. It projects an audience that is educated both in Greek tradition and in Christian thinking: it creates a reader whose sophistication is demanded, a reader very different from the figures who populate Jesus' audience in the Gospels, and who match the performatively simple *koinē* Greek of the Gospel texts. The *Paraphrase* is an act of cultural transformation, a rewriting designed to effect a continuing change in the culture of Christianity. On the other hand, Nonnus' language is deeply imbued with the theological controversies of his time, and seems to embed into his verse the language of the Nicene creed and Cyril's commentary on John, already anticipated in the Cappadocian Fathers' attacks on Eunomius and his new Arianism (paraphrase here draws on and comes close to commentary as a form). John is made to reveal a pro-Nicene Christology and a theologically engaged view of temporality. This is not just polemics, or a change of perspective, but also an attempt by Nonnus to make the Bible – in his terms – *more Christian*. It is an act of theological transformation. Nonnus' *Paraphrase* thus is a cultural, political and religious intervention in the discourse of Christianity.

*

How, then, does this new agenda of theological time unfurl in the *Paraphrase*? As in the opening, programmatic lines, time is transformed by Nonnus into his version of Christian time by glossing and expansion of

[83] In Goldhill (2020) 84–5 I discussed translations which take *hupsigenethlos* with 'God', but I underestimated the force of *-gen-* in the context of arguments over Nicaea.

John's language into a new direction. Critics have often dismissed with distaste the profligate excess of adjectives and subordinate phrases in Nonnus' poetry – or occasionally, as we will see in the next chapter, celebrated such fecundity in the name of his Dionysiac poetics, where words spread and flourish like vines under the sway of the transformative god of change, Dionysus. Such expansiveness is central to the strategy of transformation in the *Paraphrase*. Consider this passage, where a single adjective works like a theological depth charge on the surrounding phrases to create a new theological perspective of time. In John 8.55, Jesus says: 'Thus I said that you will die in your sins; you will die in your sins, unless you believe that I am he'. Here is Nonnus' version (8.55–9):

> ἀλλ' ὑμῖν ἀγορεύων, ὅτι φθαμένῳ τινὶ πότμῳ
> εἰσέτι μαργαίνοντες ὁμιλήσητε βερέθρῳ
> ἀμπλακίην μεθέποντες ὁμόχρονον· ἀτρεκέως δὲ
> εἰ μὴ ἐμὲ γνώσεσθε, τίς ἢ τίνος εἰμὶ τοκῆος,
> θνήσκετε δυσσεβίης ἐγκύμονες

> But I keep saying to you that in a fate that outstrips you
> Still madly raging you will come to the pit
> Still pursuing sin cotemporaneous. Truly,
> If you do not recognize who I am and who my parent is,
> You are dying pregnant with impiety.

The term I have translated as 'cotemporaneous' is a very rare compound adjective, *homochronos*, 'of the same time'. It is an arresting expression. It seems to imply that this sin is something that goes from birth to death, that is, in Christian terms, original sin, the sin into which all humans are born. Indeed, a few lines earlier, when Jesus first declares in John 'you will die in your sin' (8.21), Nonnus (8.39–41) paraphrases this into a striking image of a life-time of sin and an old age of horror: 'you will see an end to chill you after old age, pursuing (*methepontes*) white hair the same age (*homēlika*) as your sin'. Death here is a chilling terror (this language is common from Homer onwards), but again the temporal adjective *homēlix*, 'of the same age', reframes our comprehension. *Homēlix* is used repeatedly in theological debate to insist on the coevalness of the Son and Father, as it is in the opening lines of the *Paraphrase*, but here it again implies that far from wisdom coming with old age, for the disbelievers, old age – something pursued – becomes a sign of a life-time of sin, a life lived in original sin, unchanged by the promise of the coeval Father and Son. So, as Jesus recalls this first declaration, he again states, with a repetition of the participle *methepontes*, that sin is also something that these disbelievers 'pursue'.

Humans are born into sin, that is, but can still choose sin – or choose to escape by faith. This choice of sin is their 'mad raging', a greed or lust that corrupts. This leads to the extraordinary final phrase: 'you are dying pregnant with sin'. In John the same word 'you will die' is repeated three times, each time in the future tense. Nonnus, who also uses the future tense up to this point, now puts death in the present. Death is a continuous process in the here and now for the disbelievers, who do not choose the eternal life Jesus promises. 'Pregnant' becomes therefore a richly evocative term. On the one hand, pregnancy and the travails of childbirth are the sign of the fall, the mark of original sin for women (along with menstruation).[84] Living in human time, with its materiality of the flesh, finds its female embodiment in pregnancy. On the other hand, there is a long philosophical tradition, back to Plato, of men claiming the power of reproduction through spiritual or philosophical creativity.[85] Sin is also wrong thinking, to be pregnant with heretical ideas. Dying in sin is now dying in childbirth. The adjective 'pregnant' also asks a question, absent in John, about where sin comes from.

The arresting adjective *homochronos* opens a vista onto the overlapping temporal frames mobilized in Nonnus' paraphrase. John's repeated 'you will die in your sin' is now translated into a life-time of sin till the terror of death in old age; a life-time of sin that is also a life lived under the shadow of original sin; which is also the material distress of the fleshliness of human existence; which is a constant experience of death in the present – and all of these grim experiences of human time are to be transcended by the belief that Jesus demands in his promise of eternal life, a promise evoked by the language of coevalness, which takes us back to the programmatic opening of the *Paraphrase*. Nonnus' paraphrastic technique translates John's 'dying in sin' into a narrative layered with a fully theological temporality, where 'synchronicity' takes on a fully normative force. Time made Christian.

Nonnus is particularly expansive when it comes to eternity. In John, Jesus explains that a slave to sin 'does not remain in the house for ever; the son remains for ever' (8.35). The repeated phrase *eis ton aiōna*, 'for ever', prompts this four-line paraphrase (8.90–4):

ἐν ἀθανάτῳ δὲ μελάθρῳ
δοῦλος ἀλιτροσύνης αἰώνιος οὔποτε μίμνει
ναιετάων· μίμνει φερέσβιος υἱὸς ἀμύμων

---

[84] Explicated in Brown (1988).
[85] From a large bibliography see Burnyeat (1977); Pender (1992); Sheffield (2001), each with further bibliography.

ναίων πάτριον οἶκον, ἕως χρονίῃ παρὰ νύσσῃ
ἱππεύων ἀκίχητος ἐλίσσεται ἔμπεδος αἰών.

In the immortal hall
The slave of sin never does remain dwelling
Eternal; the life-bringing son, blameless, does remain
Dwelling in the house of the father, as long as unattainable, fixed Eternity
Twists around the long-lasting post, driving his horses.

The language of time expands here dizzyingly. The house is first an
'immortal' hall (specified then as 'of the father', the everlasting); the slave
who lives in it is not an 'eternal' (*aiōnios*) dweller, he 'never' (*oupote*)
remains. *Aiōnios* recalls John's *eis ton aiōna*, but also anticipates the full
personification *Aiōn* becomes as the passage continues. It may also recall
the biblical law of the slave who, if he wishes to stay with his master, has his
ear pierced and remains a slave in his master's house *eis ton aiōna*, 'for ever',
that is, for his life-time.[86] The son is now glossed as *pheresbios*, 'life-
bringing', a word used repeatedly in traditional epic verse for the fecundity
of the earth and its goddesses, now transferred into the Christian promise
of eternal life through Jesus. The final long clause, however, depicts *Aiōn*,
which I have translated here as 'Eternity', as a divinity, driving his chariot,
like the sun, around a course.

Aiōn is an especially complex word, to which we will return in the next
chapter. Although in Homer, as we have seen, it indicates a life-time, from
Plato onwards, it is also used in contrast to *chronos*, to indicate
a continuum of everlasting extension rather than marked and measured
time.[87] In later Greek, it comes to mean 'eternity', and the transition from
'life-time' to 'eternity', and its use in both philosophical texts and in other
genres, gives it a semantic history that is intricate and far from linear in its
development.[88] In the plural, what is more, it comes to signify 'ages' and
the phrase *aiōn aiōnōn*, like *saecula saeculorum*, is used for the unfurling of
unending time.[89] *Aiōn* seems to hover between an abstract idea and
a personification (Euripides already called *Aiōn* 'the child of Time'),[90]
but from at least the Hellenistic period onwards, *Aiōn* also appears as

---

[86] Ex. 21.6; Deut, 15.17 – it is atypical in the Septuagint for *aiōn* to refer to a life-time: see Keizer (1999).

[87] See e.g. Plutarch, *De E* 392F discussed with other examples by Levi (1944) 279–80. See Degani (1961) for its polyvalence from the earliest examples on.

[88] Keizer (1999) is fullest; see also Zuntz (1992) for the imperial period, and Degani (1961) for the earliest era. Foucher (1996) discusses the material representations with bibliography. Ladner (1959) 443–8 is brief and to the point. Cullmann (1946) influentially argued that *aiōn* in the New Testament never means supra-temporal eternity.

[89] Aug. *Ennar. In Ps. IX* 6 discusses the sense of this phrase.   [90] *Heracl.* 900.

a divinity, a figure who mediates between the highest god and humans.[91] *Aiōn* is addressed in inscriptions and in magic papyri – and represented on coins and in art, including the extraordinary third-century mosaic from Antioch-on-the-Orontes, which represents, as four named figures at a symposium, 'Past', 'Present', 'Future' and 'Eternity' (*Aiōn*), under the general allegorical heading *Chronoi*, 'Times'.[92] According to Epiphanius, in Alexandria there was a cult festival where *Aiōn* was celebrated as the child of *Korē*, 'the Virgin'.[93] *Aiōn* becomes a figure that provides a way of thinking about how humans inhabit time, in relation to divinity, both through cultic activity and in philosophical thinking. This reflective mode is vividly captured by Epictetus: 'I am not *Aiōn*', he writes with beautiful simplicity, 'but a man, part of everything, as an hour is of the day'.[94] Epictetus asserts his necessary and ineluctable mortality by denying he is (the god) 'Eternity', that is, by deprecating the disabling fear of death as no more than the arrogance that would make a human think himself a god; he is rather 'a human' (*anthrōpos*), a humility given extra depth by his own history as a slave (*anthrōpe* is a slave's summons). He is part of everything – a Stoic sense of the integrated universe – expressed pointedly in the language of time, 'as an hour is of the day': the ephemerality of a human is as embedded in time as the hour is in a day. He is not *Aiōn* precisely because he is only of a day. 'I must be present like an hour and pass like an hour', he concludes: to be in time is to pass like time.

The Septuagint uses *aiōn* and *aiōnios* very frequently for the Hebrew terms *le 'olam* and *ad 'olam* (traditionally translated 'for ever', 'evermore', 'everlasting', and occasionally 'age-old', or, in the case of a human, 'for the whole of a life'). The Septuagint coins the phrase *eis ton aiōna kai eis ton aiōna tou aiōnos*, for the Hebrew *le 'olam va'ed* 'for ever and ever', Greek which would, I expect, make little sense to the Greek-speakers of the classical city.[95] Taking off from the Septuagint, *Aiōn* becomes a term with continuing, rich associations in Christian discourse ('Complication is introduced . . . in Christian literature', comments Nock, ruefully).[96] In Paul's Letter to the Ephesians (2.2), for example, as Paul recalls his audience's sins in the time before Jesus brought them from death to life, he describes their past as walking *kata ton aiōna tou kosmou toutou*, which is

---

[91] According to Foucher (1996) 30 Aion is 'virtually unknown in Rome and the West' as a divinity, and even when personified as a god in the East is not truly 'an object of cult'. He does not discuss the magic papyri.

[92] Levi (1944) is especially good; see also Nock (1935); Zuntz (1992) 12–25; and Foucher (1996) 12.

[93] *Panarion* 51.22.   [94] Epict. 2.5.13. The text is printed in full in Zuntz (1992) 12–13.

[95] E.g. Psalm 145.1.   [96] Nock (1935) 89.

normally translated as 'according to the course of this world'. But *aiōn* here too implies an 'age', even an 'eternity', which is the time of living under the power of Satan rather than in the faith of Jesus. It implies a time of dispensation that must always still be rejected, a false sense of time and eternity, like the slave of sin who cannot remain in the house, for ever. *Aiōn* becomes a term of Christian reflection, as eternity and the sense of personal time become powerful vectors of religious thinking for a Christian.

Nonetheless it is still something of a surprise to see the figure of *Aiōn* appear personified as a god in Nonnus' narrative of Jesus. The imagery of the race track – not something prevalent in Cyril's commentary – is repeated throughout the *Paraphrase* for the passing of time, and it is a very familiar nexus of ideas from classical poetry, especially, of course, for the movement of the sun and the moon, the markers of time. It is an image especially associated with poetic narrative: the racing journey of verse towards a winning conclusion; and Triphiodorus, a poet Nonnus read, plays brilliant games with its resources, as he starts his miniature account of the fall of Troy with the word *terma*, both 'end' and 'turning post', as he drives his narrative faster and faster towards its pre-announced conclusion.[97] Yet it is the three adjectives here in Nonnus' description that stand out. The turning-post of the track is termed *chronios*. *Chronios* often means 'after a time' – in tragedy it marks the missing of the *kairos*, the disaster of the too late[98] – or 'long-lasting'. It implies here that the turning-post, the archetypal site of contingency and danger in a race, is part of the very lastingness of eternity (Hadjittofi translates it simply with 'eternal'). This course lasts. *Aiōn* itself is described as *akikhētos* and *empedos*. *Akikhētos*, 'unattainable', from the opening lines has been used, as we know, to describe not just the inability of humans to capture Jesus (or other figures in his story), but also and significantly to indicate Jesus' incomprehensible transcendence of time and space. Here, for the only time in the epic, *akikhetos* is applied to another type of figure. But Eternity is an integral element of human incomprehension of Jesus, and the eternity of a faithful Christian's life is the promise which Jesus is here explaining to his uncomprehending audience of believers. *Akikhētos* here links Logos and eternity, as it narrates the promise of everlasting life.

Empedos, 'fixed', is a paradigmatic Nonnian paradox: Eternity, for all the twisting rush of its horses round the course of time, is 'fixed', 'established' – 'fixed' like Odysseus' bed is fixed to the tree, fixed in the ground, fixed in

---

[97] See Goldhill (2020) 75–7; and especially Maciver (2020).
[98] See Soph. *Phil.* 1446; *OC* 441; Eur. *Or.* 475.

place at the centre of the house. *Empedos*, and compounds with *empedos*, occur fully thirty times in the *Paraphrase*: it is a sound that knells through the narrative. It is used most regularly for the words of John the Baptist, Jesus himself, and the witnessing of Jesus: the fixed truth, that makes the promise of Jesus transcend the fluid uncertainties of language. Unlike the winged words of Homer, these are the fixed words of the Lord. It is associated with the rule of heaven. But it is also used for the flowing 'water' of truth (4.68) and the shimmering of the 'light' of truth (8.6). Immediately before this representation of *Aiōn,* Jesus declares that everything he speaks 'to the senseless world' is *empeda*, 'fixed', 'established' (8.65), and that his deeds please his father 'for the fixed circle of time', *chronon empedokuklon* (8.74). Words, time, water, light ... are all connected in this theological declaration of established certainty. *Empedokuklon*, 'of fixed circle', does not occur elsewhere in surviving Greek, and seems to suggest that time is imagined as a certain and established structure that surrounds us, like the circle of Ocean in Homer's geography – or like the circles of Neo-Platonic cosmology:[99] circles are usually imagined as ever-flowing, but here the circle of time is fixed. Nonnus indeed in the *Paraphrase* uses four *empedo-* compound adjectives, three of which he seems to have coined: *empedomuthos*, 'fixed in utterance' (eight times),[100] *empedometis*, 'fixed in plotting', *empedomochthos*, 'fixed in labour', and *empedokuklos*: he revels in the *kainotomia* of his vocabulary, even and especially when it proclaims the fixity of speech. It is fascinating that the one author to use the phrase *empedos aiōn* before Nonnus (who uses the same phrase as 3.79 too) is Empedocles, whose name might be heard in the new word *empedokuklon*. Empedocles describes a world of clashing forces and instability made up of the combining and separating elements.[101] Things come to be (*gignontai*), Empedocles declares, there is *no* 'stable life'. Rather, the elements 'never cease their continuous interchange, to the extent that they always are in a cycle, changeless'. The life (*aiōn*) of the elements is echoed in the 'always are' (*aien easin*) – an etymological pun – that is ceaseless, cyclical change.[102] If Empedocles is evoked by the coinage *empedokuklon* and the direct echo *empedos aiōn*, it is to emphasize Christianity's rejection of his paradigmatic materialism with

---

[99] Hernández de la Fuente (2014) 231–2. Circular time is common in imperial epic, as discussed by Greensmith (2020) for Quintus.

[100] As Agosti (2003) 454 *ad loc.* notes, Gregory of Nazianzus (*Carm.* II.2.7.179) has the phrase *theos empedomuthos*. This is the only prior extant use of the four compound terms.

[101] B26. 9–12 DK. On Empedocles in Christian polemic, see Mansfeld (1992) 209–31.

[102] For this etymology see Aristotle, *De cael.* 279a; Degani (1961) 29–35.

its arbitrary collisions and combinations of elements, his cycles without the Providence of the Christian god.[103] If rewriting scripture is based on the dual desires, *durus uterque deus*, of recognizing the need for fixity in scripture and recognizing the fluid, fissile potential of its language, Nonnus performs this tension in his very language of fixity. Time, for Nonnus, is *empedos*, a fixed grounding for his theological certainty about human placement in time – expressed in his novel compound adjectives, the swirling combination and separation of new linguistic elements.

The personification of *Aiōn* as a charioteer continues throughout the *Paraphrase*. So, too, does Nonnus' insistent qualification of the idea of time that such personification allows. John writes (6.35), 'He will never be thirsty'. Nonnus writes (6.146) 'he will never be thirsty' – but then glosses 'never' with 'as long as Aion, curved, crawling, broad-bearded passes the boundless (*atermona*) turning post'. 'Broad-bearded' might suggest an anthropomorphic representation of a god;[104] but 'curved', like the expression *empedokuklos*, suggests a concept of an embracing topography of time.[105] Yet most striking is the adjective *atermona*. A 'turning-post' is the very definition of a boundary, the end of the track. For Aion the charioteer, however, this boundary is boundless – a paradox that not only captures the strangeness of Eternity as a chariot racer, but links his course to the theology invoked from the opening lines of the poem.[106] Similarly, in the narrative of the miracle of the blind man, the word 'never' is also glossed by the image of the charioteer Aion, who never brought such a fellow 'into the coeval world', *hēliki kosmōi* (9.8). 'Coeval', once again, opens a vista of cosmological time, insisting that Jesus' miracle requires such a frame. The personification of Aion is a particularly telling example of how inadequate it is to describe Nonnus' classicizing expressivity either as 'decorative' or as a sign that classicizing is no more than the unmarked discourse of the educated writer of antiquity. Nonnus appropriates and

---

[103] For a surprising use of Empedocles in contrast to a Christian model see *AP* 8.28 where Gregory of Nazianzus compares his saintly mother's death to that of Empedocles.

[104] Franchi (2013) 146–8 *ad* 147. Spanoudakis (2016) 610.

[105] On the Neo-Platonic curve of time in Nonnus see Hernández de la Fuente (2014) 246–8, and on this passage Franchi (2013) 436 *ad* 146.

[106] This point is not contradicted by *Dion*. 19.280 where the same phrase, with the participle *ameibōn* and the adverb *stephanēdon*, is used to capture the movement of a dancer spinning continuously in a circle around the same spot. At *Dion*. 38.250, the sun travels a 'boundless circle round the Zodiac turning-post'. Spanoudakis (2016) 610 calls this language 'a formalistic embellishment', but sees Aion nonetheless as 'a construct of achronous God bound with creation'. The second expression is far better than the first.

redrafts the very tropes of eternity, within his Christianizing conceptualization of God's time.

The reward of eternal life for the disciples who believe in Jesus' message, is central to the narrative of John's Gospel, and to Nonnus' image of time in the *Paraphrase*. As early as Book 5, Jesus lays out the decisive moral choice. For those who do not believe, there will be a swift judgement. But the person who believes the *marturon empedomuthon*, 'the witness fixed in utterance' (5.89), there will be no judgement. Instead there is this promise (5.92–9):

> ἀλλ' ἐπὶ κείνην
> ζωὴν ἀμβροσίην, τὴν οὐ χρόνος οἶδεν ὀλέσσαι,
> ἵξεται ἐκ θανάτοιο μετάτροπος· ἀπροϊδὴς γὰρ
> μαῖα παλιγγενέων μερόπων νεκυοσσόος ὥρη
> ἵξεται ὀψιτέλεστος, ἀναυδέες ὁππότε νεκροὶ
> αὖτις ἀναζήσουσιν ἀνοστήτων ἀπὸ κόλπων,
> πάντες ἀλεξιμόροιο μιῆς ἀΐοντες ἰωῆς
> παιδὸς τηλυγέτοιο φερεζώοιο τοκῆος

> But to that
> ambrosial life, which time does not know how to destroy,
> he will come, transformed from death. For unforeseen,
> a corpse-saving hour, a midwife of mortals born long ago,
> will come, late-fulfilled, when the speechless corpses
> will come back to life again from the embrace of no return,
> all listening to the single fate-averting voice
> of the beloved child of the parent who brings life.

The adjective *ambrosios* is closely associated with the divine world of Homeric epic. 'Ambrosial' is very rarely used in Greek without such connotations,[107] but, specifies Nonnus, Time – *Chronos* – does not know how to destroy this 'divine life'.[108] It is a commonplace of earlier Greek writing that Time is *pandamator*, 'conqueror of all'. So Nonnus himself writes later in the poem, where Jesus declares that he is willing to give his life for his flock (10.61–4): 'No law from my father will take my life from me, not creeping time (*chronos*) that conquers all (*pandamator*), nor unconquerable (*adamastos*) necessity, fixed in its plotting (*empedometis*), but, self-commanded, I lay down gladly a life that wills the same'. Time is ironically called *pandamator*, and necessity, with equal lack of purchase, is

---

[107] In Nonnus it is especially associated with Jesus or John the Baptist, and, as Agosti (2003) 457–8 *ad* 93 notes, is a good example of Christian appropriation of Homeric diction.
[108] Agosti (2003) 456–7 *ad* 93 usefully lists variations of the phrasing for 'immortal life'.

called 'unconquerable', since at this moment Jesus is proclaiming his triumphant conquering of the necessity of death, time's end.[109] Necessity is called 'fixed in *metis*', the sort of plotting that is traditionally anything but 'fixed', but will turn out to reveal another fixity altogether, namely, the fixity that the language of *empedos* has constructed around Jesus and his promise (*mētis* is not a word that occurs in the New Testament or in Patristic Greek; it speaks of another time).[110] Jesus repeats the same phrase 'that time does not know how to destroy' also in Book 10, also applied to the life to come (10.35), but he adds a surprising phrase that shows a very Nonnian reading of John's sense of time. John writes (10.10): 'I have come so that they may have life and have it in abundance (*kai perisson*)'; Nonnus writes (10.36) 'or have it in superior abundance (*ēe perisson huperteron*)'. Adding a comparative ('superior', 'higher', 'longer') to an expression of excess ('abundance', 'more than required') and changing John's 'and' to 'or', creates a theological hyperbole.[111] What alternative can be more than eternal life? Nonnus' linguistic exuberance – his excessive language of excess – in striving to capture the enormity of eternity, has Jesus enter something of a conceptual morasse. In Book 5, too, Jesus proclaims his triumph over time and its link to death, which is also a triumph over the standard language of the Hellenic tradition. Thus, the believer will come from death *metatropos*. *Metatropos* is a turn that is a change (hence my translation of 'transformation'). I could almost have written 'conversion'. Jesus demands a *metatropon ēthos* in his believers, 'a change of life and character' (3.83;[112] 6.208), a change predicated on this flight from death (17.52; 13.3).[113] It is the change of moral – religious – affiliation that is being demanded, a change which will end in this journey away from death, the reversal of time's arrow.

John's language is far simpler (it will be no surprise by now to learn). 'He has passed from death to life' (the basis of the paraphrase just discussed) is followed by 'the hour is coming and now is, when the dead will hear the voice of the Son of God, and those who will listen will live' (5.25). I have translated Nestle-Aland's standard text of John, which is here different from the text given in Scheindler's edition of Nonnus: Scheindler prints

---

[109] At *Dion.* 34.109, a delusional lover misled by a deceptive dream calls a maiden's cheeks a 'meadow which time does not know how to wither (*marainein*)', in contrast to the regular erotic discourse of the fading of flowers and beauty.

[110] For *mētis* as archetypally fluid and flexible, see Detienne and Vernant (1974).

[111] Simelidis (2016) 295 notes a motivated change of 'and' to 'or' at 2.21.

[112] Franchi (2016) 256 misleadingly translates it here as 'variable', and connects it to Dionysiac imagery.

[113] In *Dion.* 12.139, the verb *metatrepein* is used for the reversal of the thread of fate to allow the rebirth of Ampelous as a vine, on which see Shorrock (2016) 592.

future rather than perfect, 'will pass', and leaves out 'and now is'. This may indeed have been the text Nonnus read. At 4.111, however, the previous occasion that the (full) phrase occurs, Nonnus adds ἄγχι, 'and has nearly come', as he does at 5.108 (where John does not have 'and it is now here' anyway). Some centuries after Paul, it is harder just to say the 'and it is now here'. The relation between 'now' and 'waiting' and what is near and what will come has become thoroughly theologized. Nonnus' redrafting of John's language of time reflects this shift. We can note that Nonnus repeats *angchi* compound adjectives 55 times in the *Paraphrase*. What is 'nigh' obsessively echoes in his language. How close is fulfilment?

The connection between the believer's immortal life and the resurrection of the dead is asserted in Nonnus by the repetition of the verb *hixetai* (94/96): the faithful human 'will come' to ambrosial life as the hour of resurrection 'will come'. This 'corpse-saving' hour is described first as *aproïdēs* 'unforeseen'. This is a riveting moment. In the previous chapter, I discussed how the fascination of narrative and narrative theory with 'the sudden', 'the unexpected' becomes a leap of faith, a religious moment of transformation – or, as with Gregory of Nazianzus' poetics, a resistance to such suddenness in the name of slow reading and gradual change. In the Gospel, there is a foundational tension between Jesus' knowledge of who he is and what will happen to him, and the disciples' uncertainty – thematized by Peter's denial or Thomas' doubt – as well as the slow understanding of the crowds or their refusal to recognize Jesus at all. 'The unforeseen' becomes a narrative node for such a tension. From the beginning (1.28), we are told that the Logos came *aproïdēs*, unforeseen, to an unbelieving world. Later (7.167), we are told that unthinking men undertook to oppress Jesus, because they were *aproïdeis*, 'unable to foresee' (the adjective can have both the active and passive sense). When Jesus escaped from the temple *akikhētos*, 'untouched', no-one knew what he was doing: he was *aproïdēs*, beyond the foresight of his human enemies (8.193). The Jews who are willing to believe, but are too frightened of the priests to speak out, are said to have a faith that is hiding *aproïdēs* 'unforeseen' – unforeseen by the priests, misunderstood by themselves, with consequences they cannot fathom (12.173). In contrast, Joseph of Arimathea became a disciple *aproïdēs* (19.194), without letting anyone perceive his conversion – escaping the attention of the crowds around him. Two final surprises sum up this language of failed anticipation, concealed intention and misplaced consequences. After the crucifixion, the disciples are eating together, and doubting Thomas is singled out by name.[114]

---

[114] On the play with the double name of Twin Thomas see Whitby (2007) 203–5.

At that moment, Jesus appears *aproïdes*, and overcomes Thomas' doubts (20.118).[115] Jesus' arrival is no surprise to Nonnus' readers who know the story; John does not mention 'surprise' in his account. Nonnus, by adding the specific narratological indication, is not only designedly reminding us that this *is* a surprise, but also highlighting the role of surprise in Jesus' impact on a person (what Augustine would theorize as grace, whose moment is overwhelmingly surprising even when anticipated). Nonnus' addition draws attention to the theological implications of narrative's *thauma*. By contrast to the episode with Thomas, Jesus before his death warns that Satan – 'the arrogant ruler of the ever-flowing universe' – is coming *aproïdēs*, unforeseen by the disciples and misrecognized by the unthinking world (14.120–1). Jesus thus foretells the future failure of human anticipation and understanding. It is absolutely archetypal of Nonnus' narrative language that this prophecy is called an 'inspired voice before its hour (*proōrion*)', and the coming time of fulfilment of the oracle is described with 'if twisting, unstable time, as it crawls along, fulfils this' (14.115–16). As the threat of the 'unforeseen' Satan is proclaimed, we are reminded volubly of the unfixed, disorderly unfurling of time he rules in and through. At significant junctures of the narrative, the repeated use of the word *aproïdeē* links narrative surprise with a theology that contrasts human ignorance and divine knowledge.

The Gospels each have miracles at their heart: the unexpected, the sudden. Miracles are designed to transform the inner lives of their observers: amazement, shock, awe are expected reactions. If Herodotus promises *thōmata* for his history, and Aristotle insists that *thauma* is the beginning of the philosophical pursuit of knowledge, the Gospels place *thaumata* – miracles – and the response to them – *thauma* – at the centre of its narrative of revelation and belief. Nonnus makes the dynamic of the unforeseen central to his narrative dynamics: who can anticipate what? Who fails to see what should be seen? Yet, as Kierkegaard knew all too well, the abyss of narrative surprise also opens the leap of faith. The 'unforeseen hour' anticipated by Jesus is part of a persuasive story to recognize and accept the promise of an unchanging time of 'ambrosial life'. The turn from death – *metatropos* – goes hand in hand with the 'unforeseen'. Nonnus' paraphrastic poetry displays a *metatropic* poetics that transforms John's promise of a life to come into a fully theological Christian discourse of time and narrative.

The 'corpse-saving hour' that is to come is also termed the 'midwife (*maia*) of mortals born long ago'[116] and 'late-fulfilled' (*opsiteleston*).

---

[115] Whitby (2007) 205.
[116] On *palaigeneōn* see Agosti (2003) *ad loc.*; on its use in *Dion.* 2.650 see Goldhill (2020) 121–2.

'Midwife' suggests, of course, the imagery of rebirth central to Christianity, but it specifically recalls here Nicodemus' confusion in Book 3. Jesus tells Nicodemus that a human who wants to see the 'eternal (*aiōnion*) kingdom' must be 'born a second time' (3.17–18). He replies: 'How can an old man, with white hair, have another, late-fulfilled (*opsiteleston*)[117] travail of birth? Surely he cannot, without a father, enter through the swollen lap (*kolpos*) of his ancient mother into her pregnant belly to see the groaning rite of a labour that is a return again (*palinnostos*)?'.[118] The rebirth of conversion is now expanded to the resurrection of the dead; shared promises, a link encouraged by the shared language – *opsiteleston, kolpos, nostos, palin* – between these two passages. But *maia* also recalls Plato's famous use of the term for how philosophy in the form of Socrates can bring new ideas to life.[119] Jesus' teaching appropriates and transforms the privileged language of the philosophers of the past. This metatropic poetics, rewriting both Nonnus' own earlier expressions of rebirth, and Plato's language of the birthing of ideas, is fully in service of his Christian theological agenda.

The final phrase of Jesus' promise reveals Nonnus' close reading of the syntax of John. John writes that the corpses will hear 'the voice of the son of god' – three genitives: *tēs phōnēs tou huiou tou theou* (5.25). Nonnus specifies that what is heard is a single voice (*miēs iōēs*). This is not the 'single voice' of Christianity that Ps-Apollinaris celebrated, but anticipates the two genitives that follow: 'of the beloved[120] child of the parent who brings life'. The potential for ambivalence that a genitive dependent on a genitive can bring in Greek – which genitive is dependent on which? – is mobilized here pointedly. There is one voice of Father and Son, Son and Father: hence the emphatic addition of *miēs*, 'single'. The adjective *pherezōoio*, 'life-bringing', could go with either noun, child or parent – and up to this point it has been the son who has promised *zōē*, 'life' (93). By the end of the sentence it is clear enough that it qualifies the Father, but marks the shared ordinance of Father and Son, their consubstantial nature, as the Nicene creed asserts. Nonnus' explosive polyphony seeks yet to promote the one voice of truth, fixed.

The relation between waiting and the moment, and the *metatropic* poetics of paraphrase, reach a remarkable climax when Jesus predicts his resurrection to the confused disciples in Book 16. The prophecy itself is surrounded by expressions of changing language. Jesus promises he will

[117] A Homeric *unicum* (*Il.* 2.325) of an oracle whose fame 'will never die'.
[118] For links between this passage, the *Dionysiaca*, and Quintus of Smyrna, see Shorrock (2016) 582.
[119] See Burnyeat (1977); Pender (1992); Sheffield (2001).
[120] On the sense of this Homeric expression see Agosti (2003) 465 *ad loc.*

utter '*heterotropa* (turned otherwise) paths of song' (16.95), which the disciples struggle to paraphrase as not the 'twisted mysteries of utterances *paratropeōn* (turned aside)' which need to be read with 'another voice' (*heterēs phōnēs*) (16.111–12). The disciples express their confused reaction 'by concealing their utterance, pregnant with voice, travelling close to the tongue, a contest with silence' (16.54–6) – this a wonderfully evocative version of John's 'they said to one another'. Where John pictures the disciples turning to one another in doubt, Nonnus depicts their thoughts and unsaid words in contorted tension. But Jesus' prophecy itself transforms John's hauntingly simple expression, 'A little while, you will see me no more; again, in a little while, you will see me' (16.16) into an incantatory performance. And it does so in a quite extraordinary way. Over eleven lines (50–61), Nonnus repeats and repeats the same terms in multiple forms. 'There is', he says 'still a small time' (βαιὸς ἔτι χρόνος, 50), 'you will no longer (οὐκέτι) see me' (51); 'there is left still a small spiralling time' (εἰσέτι βαιὸς ἕλιξ χρόνος, 52); 'know that there is still a small | a small time still left' (ἔτι βαιός | βαιὸς ἔτι χρόνος, 57–8); 'you will no longer (οὐκέτι, 59) see me'; 'there is left still a small spiralling time' (εἰσέτι βαιὸς ἕλιξ χρόνος 60). 'Again (*empalin*) you will see me' (53); 'again (*empalin*) you will see me' (61). In this shifting, paraphrastic repetitiveness the adverbs *eti*, *eiseti*, *ouketi*, 'still', 'no longer', are repeated six times, *baios*, 'small', five times; time itself, *chronos*, is repeated four times, twice with *helix*, 'spiralling', and always with 'small'. The chiasmus of enjambment is arresting – *eti baios* | *baios eti*, 'still small, | small still' (which comes first, the smallness or the still, brevity or waiting? How long, O Lord, how long . . . ?). Homer's verse is celebrated for its repetitions and formulae. But here this Homeric *lexis* has become something else, transformed, like the simple, clarifying repetition in John, into a profusion of spiralling language, as time is repeated time and again to try to capture the prophecy of the future, which, for the faithful, is changing time, their sense of time, for ever. The incantatory repetition is an anticipatory cry to bring forth the short measure of transformative time. How long, how long . . .

After such a dizzying swirl of words, Jesus asks the disciples (16.65–7), 'why do you seek each other in neighbouring voices, if I said that after a small while (*baion*) I will pass from your sight, and again still in a small while (*palin eiseti baion*) you will see me revealed?'. Again, Jesus' recapitulation repeats the language of return and repetition, paraphrasing his own words of resurrection, performatively; again 'small' is repeated, twice. (Is there any other passage of ancient poetry where the same adjective is repeated seven times in sixteen lines? The hour has become emphatically its *scale*, the contrast between the enormity of the event and the brevity of its time.) Did they not get it (though their earnest bafflement at such impossibly fracturing language is understandable)? Jesus' question prompts a further prophecy (16.68): 'I say to you an oath that is fixed (*empedomuthon*), amen, amen'. The words may swirl but the oath – God's word – is fixed, *empedon*. At this moment of declaration, as Jesus announces that the disciples must become the apostles, the fixity of the promise is again contrasted not just with the explosive polyphony of language but also with the sheer material resounding of paraphrastic repetition, its intricate, bewildering clatter of sound.

The *form* of Nonnus' poetry violently redrafts John's idea of the small moment and waiting for change. Jesus' time is coming, the hour that will transform time: the language of time, the smallness of the moment, is made to sound out again and again – taking more and more time – as the narrative rests in the small moment, still, awaiting, still, the moment of transformation to come, aware of its enormity. As we read and re-read Nonnus, as his language swells and changes the language of John, we participate in a continuing reflective process of *conversio*, a turning to God, a turning of John, that allows no completion – the enactment of the *achronos*, the timelessness, with which the poem starts. This is an astounding demonstration of how the performative power of Nonnus' paraphrastic poetics dramatizes the transformation of Christian time.

*

Paraphrase is the metatropic genre *par excellence*. Nonnus takes up John's Gospel, which, significantly, is the Gospel from the beginning most engaged with Hellenic intellectual tradition and most focused on time and eschatology, and rewrites it, overlaying its *koinē* Greek prose with the different affordances of hexameter verse. In so doing, he rewrites the culture of Christianity as an elite, educated, *theological* narrative. Both the form of his poetic work, and his intellectually challenging discourse, with its rebarbative vocabulary and syntax, demand a theologically educated and culturally sophisticated reader. The exuberant poetic extremism of Nonnus is in service of an agenda, to make the Gospel *more Christian*, to reveal its conformity with a post-Nicene settlement. This poetics of paraphrase is to be located thus not only within an aesthetic contrast with cento and translation, but also within a contemporary politics of language that has placed so much theological emphasis on the fixity of tradition and dangerous innovation at the level of the word. Within this process of rewriting, the language of time and its narrative expression are articulated into a newly theological expression. Although, in line with John's focus, only a few passages in the *Paraphrase* talk about time itself, each of these passages demonstrates an explosive semantic expansion of John's temporal discourse. Timelessness, the end of death, the violence of rupture, the normativity of synchronicity, waiting, the smallness/enormity of the moment, the biology and narrative of a life-story, even making the figure of Eternity visible – our building blocks of the understanding of time – are integral to this transformation of the narrative of John in Nonnus' metatropic poetics. From its first word onwards, Nonnus' *Paraphrase* provides a wildly experimental but brilliantly revealing example of how the literature of late antiquity invents Christian temporality.

CHAPTER 12

The Eternal Return
Nonnus' Dionysiaca

Let us return to eternity, which in truth we have not ever fully left since the opening chapter.

Book 7 of Nonnus' other epic, the *Dionysiaca*, portrays Eternity as an instrumental character in the narrative, which, as we will see, is only one of many ways that time becomes an integral part of Nonnus' writing. Book 7 marks another beginning too, and, if Eternity becomes a character for the *Dionysiaca*, going back to the beginning and starting again is a thematic obsession for this swirling, twisting, polymorphic *and* perverse storytelling. At this point in the narrative of the *Dionysiaca*, Zeus has just relented from destroying the whole cosmos with an apocalyptic flood. Typically for Nonnus, and again we will return to this theme, because Nonnus' imbricated and entangled narrative constantly requires such gestures of return, the storm from God is expressed first in *astrological* terms, as the Sun God drives his chariot into the quarter of the Lion, and the Moon rides down Cancer, the Crab (6.233–6). But – and do I need to repeat that this again will become part of our discussion later? – the flood becomes an opportunity for a wonderfully baroque whirlpool of myth. Earlier, drier narratives of a world-threatening flood from both the classical tradition and Jewish or Christian narratives of Noah, are overwhelmed by Nonnus with a full influx of end-of-the-world horror, twisted into bizarre *Witz*, so that a whale can meet a lioness in her mountain den, while bloated bodies pass; and, what is more, the flood also becomes an opportunity for figures of the mythic repertoire to float past each other – dramatically embodying the familiar epic use of the ocean's roar or the flow of water for the tradition of song. So, Pan, looking for Echo – who else in the sea of song? – asks the sea nymph Galateia if she is looking for her Cyclops' song (*aoidē*, 6.303, the classic marker of hexameter poetry) – the song, that is, of her lover, Polyphemus, most famously composed by Theocritus, but echoed by others since. The Nile meets the river Alpheius, who laments he cannot find the spring Arethusa, *his* lover, always concealing herself – but no doubt in a flood river gods do lose their boundaries if not their identities. Alpheius also encourages Pyramus to hunt for Thisbe lest Zeus find her first, and he takes some small comfort for his own loss in the fact that Aphrodite has lost Adonis: further familiar love-stories from the erotic novels or the poetry of Ovid. At that point, Deucalion too 'crossed the crowning waters, a sailor unattainable (*akikhētos*) on a skyborne voyage' (6.367–8). Deucalion in traditional Greek myths is the Noah-like figure who survives the inundation with his wife to found the human race again by throwing stones over their shoulders into the earth, stones which become people, a new beginning for humans. His story has been briefly told in Book 3 already (3.209–14). Here, it is as if Deucalion has floated in from another myth, and, far from ending on the stony ground, he is not only still floating, but raised into the sky, 'unattainable' – that marker of transcendence, Nonnus' much repeated Homeric *hapax*, that we saw so strikingly used in the *Paraphrase*. Deucalion is travelling out of reach, founding nothing ... From this wild, self-conscious and funny mythic mélange, it will immediately be clear why those who are committed to the paradigms of classical aesthetics find it hard to appreciate Nonnus – Matthew Arnold on Dover Beach heard Sophocles not Nonnus in the roar of the ocean – and why those who love Nonnus do so.

At this point, however, Nonnus concludes, the 'cosmos would have become no cosmos', *kosmos akosmos*; and, had not Zeus relented, 'Aion (Eternity) who nourishes all, would have dissolved the unsown concordance of humanity', Καί νύ κεν ἀνδρῶν | ἄσπορον ἀρμονίην ἀνελύσατο πάντροφος Αἰών (6.371–2). In the *Paraphrase*, the figure of Aion, it will be recalled, appeared as a personification, a surprising charioteer of time in a Christian poem, although his representation in the art of late antiquity suggests at least an imagistic continuity with his broad role as a divine force from Hellenistic Greek and Roman culture onwards.[1] It might seem here too as though 'Aion' is less of an agent than an expression of the magnitude of the threat to the order of things. The phrasing, however, is not the familiar *topos* of the race track, but powerfully evocative. Aion 'nourishes all', a beneficent and supportive framer of existence, who in this counter-factual case, would have had to dissolve the *harmonia* that binds men, making the *cosmos* without seed, without the future of procreation. Aion is not just 'eternity' but embodies 'the vital principle that governs the cosmos'.[2] *Harmonia* is also the name of the bride of Cadmus, whose

---

[1] Levi (1944). For textual continuities, Nock (1935); Zuntz (1992); Foucher (1996); Keizer (1999).
[2] Vian (1993) 48.

song had lulled the monstrous Typhon into destruction and thus saved Zeus as king of the gods and ruler of the cosmos, and who, as the father of Semele, is destined to be the grandfather of the not yet born Dionysus. Had this flood continued, the genealogy that gives the world Dionysus would not have been fulfilled, and the epic narrative would have been stalled; it would not have reached its dénouement, its (*ana*)*lusis. Harmonia* not only evokes the ancient anthropology that makes social ties the essence of culture, threatened by this flood, but also imbricates the closure of the flood narrative into the genealogy of the epic's hero.

It is significant, then, that the language of Aion's appearance here at the threatened end of the order of things is recalled at the beginning of Book 7, an episode which narrates how Aion supplicates Zeus to allow the invention of wine to make the world a happier place.[3] His observation of the flood is an anticipation of his involvement in the birth of Dionysus, and the world-changing innovation of wine. His passing appearance in Book 6 prepares the reader for his intervention in the story of the cosmos in Book 7.

The book opens with a summary of how the world has progressed since the flood (7.1–7):

ἤδη δ' ἀενάοιο βίου παλιναυξέι καρπῷ
ἄρσενα θηλυτέρῃ γόνιμον σπόρον αὔλακι μίξας
ἄσπορον ἤροσε κόσμον Ἔρως, φιλότητος ἀροτρεύς.
καὶ Φύσις ἐρρίζωτο, τιθηνήτειρα γενέθλης,
καὶ χθονὶ πῦρ κεράσασα καὶ ἠέρι σύμπλοκον ὕδωρ
ἀνδρομέην μόρφωσε γονὴν τετράζυγι δεσμῷ.

Already now Eros, with regrowing fruit of ever-flowing life,
Had mixed male generative seed in the female furrow,
And ploughed the unsown cosmos, Eros, the ploughman of love.
Nature, the nurse of generation, was rooted;
And she mixed fire with earth, and water woven with air,
And in the fourfold bond formed the human race.

Now the 'unsown cosmos' is productive, nature is 'rooted' rather than floating and disrupted, and humans are linked in love, rather than swimming after lost lovers. The scene after the flood reverses the threat of the flood. The language of sowing, ploughing and furrows is absolutely traditional, going back to the ritual of the wedding ceremony in fifth-century

---

[3] Foucher (1996) 10 notes that Aion appears just below Theos, the supreme being, in *Corpus Hermeticum* 11.2, which he says is 'doubtless the reason why he plays an essential role' here with Zeus. 'Doubtless' is a little strong.

Athens, and earlier too, as we saw, in Homer's 'generations of men' and the language of the Homeric Hymn to Aphrodite.[4] But what makes this introduction so remarkable is its rich layering of philosophical language, language specifically to do with the creation of the world. If Empedocles was a haunting presence in the *Paraphrase*, here again his theory of matter runs through Nonnus' Greek.[5] In Empedocles, Eros is a founding principle of how the elements combine to make things (as in the form of Phanes, he is for Orphic texts too).[6] Empedocles calls the elements *rhizōmata*, 'roots', and here nature is 'rooted' (*errizōto*), and the human race is made out of the 'fourfold' mixing of the elements of fire, earth, water and air. No Deucalion here. This is not the first time that Phusis, Nature, has been called on to regenerate the earth in Nonnus. In Book 2, after the battle between Zeus and Typhon has uprooted mountains and destroyed the fields, Nature heals the world (2.650–4):

> καὶ ταμίη κόσμοιο, παλιγγενέος Φύσις ὕλης
> ῥηγνυμένης κενεῶνα κεχηνότα πῆξεν ἀρούρης,
> νησαίους δὲ τένοντας ἀποτμηγμέντας ἐναύλων
> ἁρμονίης ἀλύτοιο πάλιν σφρηγίσσατο δεσμῷ.

> The steward of the universe, Nature, made of regenerative matter,
> Fixed the gaping flank of the broken field-lands,
> And sealed with the bond of indissoluble harmony
> The island cliffs broken from their beds.

Nature is an organizing principle of the cosmos, its steward, and is composed 'of regenerative matter'.[7] This phrase cues the philosophical argument, central to the polemics surrounding Genesis' account of creation, about whether matter is eternal or whether creation *ex nihilo* is possible.[8] It was an argument that set Platonists against Christians; it seeded Augustine's reflections on God's time, and, in contrast, was one of the sticking points that Synesius expressed when he agreed to become a bishop: as a philosopher, he could not countenance *creatio ex nihilo*.[9] Regenerative matter, however, seems to look towards Christianity's promise of eternal life, the resurrection of the body. Nature here, in anticipation of the flood's disruption of Book 6 and Eros' regenerative work at the

---

[4] See above, pp. 25–7 and 31–4.

[5] Faulkner (2017) 108 with n. 28 does not consider this passage in his unduly cautious comment 'Nonnus may well have known Empedocles'.

[6] For all things Orphic see Bernabé and Casadésus eds. (2008); and, more specifically, García-Gasco Villarubia (2008); Otlewska-Jung (2014); Bernabé and García-Gasco (2016).

[7] On the materiality of this expression see Goldhill (2020) 119–21.   [8] See above p. 28 n. 26.

[9] See *Ep.* 105 with Bregman (1982).

beginning of Book 7, guarantees an 'indissoluable *harmonia*' – the flood's threat was precisely 'dissolving *harmonia*' – which is 'sealed with a bond (*desmos*)' – as Nature 'formed the human race with a bond' (*desmos*). Nonnus knows his Plato, too. In *Timaeus*, a text central to the history of thinking about time and creation, as we saw, and especially to Christian Neo-Platonism, Plato describes how the universe was created by the Demiurge, perfect and ageless, out of the four elements, combined in harmony, by bonds (32a–c).[10]

Yet the opening line of Book 7 echoes with another philosophical possibility too. *Palinauxei*, which I translated as 'regrowing' ('renascent', 'increasing again'), is a word that Nonnus seems to have coined, and he uses it nine times in the *Paraphrase*. There it refers to Jesus' honour; the miracle of the loaves ('the feast increasing again'); the circle of time; Eternity (*Aion*) itself; but, most strikingly, as Gigli Piccardi notes, at the beginning of Book 15, Jesus announces, 'I am the vine of life in the regrowing Universe' (*palinauxei kosmōi*) (*Par.* 15.1), and a few lines later calls himself 'the regrowing plant (*palinauxei thamnōi*, 8), and encourages the disciples to stay together in faith 'to swell the fruit' (*auxein karpon*, 14).[11] That is, *palinauxēs* is associated especially with the power of Jesus and especially his power over time, the promise of the life to come – a world view, imbued with a particular sense of time. *Aenaos*, which I translated 'ever-flowing', is another common word in the *Paraphrase* and the *Dionysiaca*, an ambiguous term which 'combines the ideas of "perpetual flux" and "eternity"',[12] and which, from its first use in the prologue applied to God himself (1.6), is associated especially with the 'ever-lasting' life of the Christian promise, and the constant flow of the cosmos – interconnected theological ideas.[13] The introductory phrase of Book 7 of the *Dionysiaca*, 'The re-growing fruit of ever-flowing life', is a paraphrase of the language of the *Paraphrase* (or the *Paraphrase* a paraphrase of the *Dionysiaca*), at the moment when we are about to discuss how Eternity asks for the birth of Dionysus, and as we read of the beginnings of the human race. The introduction to Eternity's supplication is replete with *both* Empedoclean, Platonic, philosophical expressivity, *and* echoes of Christian eschatology, here at this significant juncture of beginning. This beginning combines its

---

[10] See pp. 165–9 above.   [11] Gigli Piccardi (2003) *ad loc.*   [12] Vian (1993) 48, my translation.
[13] 1.6; 4.69; 3.119 (of the Jordan; baptismal water); 4.121; 6.57; 6.217; 8.10 (of Jesus' words); 11.15; 12.199; 14.121; 16.35. F. D. Maurice was sacked from his role as professor from King's College, London, by Evangelical authorities for observing that the ambiguity of the term *aenaos* could mean that everlasting punishment in Hell might be tempered by Jesus' mercy: see Morris (2005).

echoes of different, philosophical accounts of the origins of the world, matter, and time.

Nonnus, as we have noted, does not explicitly refer to Christianity or the Christian god in the *Dionysiaca*. But he is writing from within a Christian community and for a community familiar with Christian language and argument. Especially when Christ is described in the language of the vine, as in Book 15 of the *Paraphrase*, an extended image beyond even its marked use in the Gospel (itself an echo of Jeremiah's prophecies), or when the *Dionysiaca* appears to adopt such Christian language in its description of Dionysus and his exploits, these cross-echoes between the *Paraphrase* and the *Dionysiaca* have repeatedly vexed modern critics, because they pose in the most sharp form the general question of the purity of either Nonnus' Christianity or his classicism.[14] The difficulty is exacerbated by the fact that it is not known which epic was written first, or whether indeed they were composed at the same time. Yet the hybridity that these opening lines of Book 7 embody, seems to capture something integral to Nonnus' poetic discourse. It is not just that he represents a culture where the tradition of Greek learning continues to play a formative role in the performance of elite self-awareness. It is rather that Nonnus revels in interweaving different narratives of the beginning of the human race, allowing echoes of these different paradigms to pose a question about their distinctiveness and mixing – as when Deucalion floats by Pyramus, a meeting which makes no chronological sense, but allows a trace of another mythic world to impinge on the scene he has created. Critics, that is, are right to be vexed, because there is a provocation in such a style of writing. This provocation has repeatedly resulted in critics laughing with or at Nonnus; declaring the tension between Christian and pagan languages trivial or insisting on religious rivalry between Dionysus and Jesus; wondering how theologically or even aesthetically sophisticated Nonnus should be taken to be: *performing*, that is, *their reaction* to the provocation. In short, much as we have come to realize that the voices of Virgil's *Aeneid* require a reader's political negotiation of the imperial epic's narrative of the coming to be of Rome, a negotiation often performed in assertive political readings of the poem, so the *Dionysiaca*'s story of the coming to be of Dionysus, with the invention of wine and the mythic repertoire associated with the god's transformative power, sets its readers to negotiate the hybridity of tradition, their place in cultural normativity.

---

[14] For modern views with discussion of earlier stances see Shorrock (2011), (2014), (2016); Spanoudakis (2016); Simelidis (2016); Sieber (2016) all with further bibliography.

It is thus significantly at this layered moment of a new beginning for humans that we are to meet Aion, Eternity. Aion, as he observed the flood has been observing the grim life of men, which 'begins in toil and does not cease from care' (7.8) – an echo of God's promise to Adam at the fall? – and which needs wine to discover joy, celebration, relaxation, release from pain. He will supplicate Zeus to allow the birth of Dionysus and the consequent invention of wine (an invention of what is thus already known). In the *Iliad*, the narrative takes a decisive turn when Thetis, the mother of Achilles, supplicates Zeus, a celebrated scene which in the imaginary of art history inspired Pheidias' monumental statue of Zeus at Olympia.[15] The supplication here too marks a determinative moment in the narrative – the rebirth of Dionysus – and is equally portentous in its descriptive language: it makes Aion an instrumental force in the direction of the narrative. The description of Aion, consequently, may be brief but is telling. He is, first, *suntrophos* (10), 'nourished/nourishing alongside' and *poikilomorphos* (23), 'variegated in form', 'shape-shifting'. *Suntrophos* implies that Aion is coeval with Zeus, or perhaps Nature[16] – there is no time for the gods without the time of Eternity. *Poikilomorphos* opens a vista, however, onto Nonnus' poetics. It is a term applied to Aion in the *Paraphrase* too (9.154): the Gospel phrase 'for ever' (*ek tou aiōnos*) is paraphrased with ἐξότε ποικιλόμορφος ἀέξετο πάντροφος Αἰών, 'from when Eternity, variegated in form, all-nourishing, increases', a line which reworks precisely the vocabulary we have been tracing ([*palin*]*auxein*, *trophos*, *poikilomorphos*, *Aion*) – or, as we must say, alternatively, which is reworked in the *Dionysiaca*: the view of Eternity is shared between the discourse of the *Paraphrase* and the discourse of the *Dionysiaca*, a time-frame for both classicizing and Christianizing vectors of Nonnus' language. *Poikilomorphos*, however, also summons the presiding muse of the poetics of the *Dionysiaca*, Proteus. We could indeed with Gigli Piccardi translate it 'proteiforme', 'of protean-form'. Proteus is the god invoked at the beginning of the *Dionysiaca*, with a call for *poikilon eidos, poikilon humnon* (1.15), 'variegated form, variegated song'. Dionysus is the god associated with making other, the god of transformation (theatre, alcohol, boundaries of identity), and, as many critics have begun to develop in recent years, the Dionysiac poetics announced in this prologue of the *Dionysiaca* signals both the narrative form of the epic, where, as we have already begun to see, mythic narratives blend and transform into versions of each other, and the

---

[15] Platt (2011) 293–333.
[16] The ambiguity is discussed in Gigli Piccardi (2003) *ad loc.*, which she rightly does not resolve.

language of the poem, which, with its long, twisty phrases and strangely accumulating compound adjectives, transforms the reader's perspective in the journey through a single sentence. Aion is an agent of change, the embodiment of shifting form – the perfect frame for the transformation that is Dionysus' story and style of storytelling – against which the proclamation of the *empedon*, the fixity of Jesus' message, echoes. And here at the beginning of Book 7, Aion, introduced as *poikilomorphos*, asks for a transformation in the world.

Aion is also introduced with 'holding the key of generation (*genethlēs*)' (23) and 'stretching out his boundless (*atermona*) hand, the old man, shepherd of ever-flowing life' (28). Eros may be the 'nurse of generation' (*genethlēs*) (7.4), but Eternity links the flow of time to the life of humans that is the passing on of the generations – a fundamental idea we saw in Plato's *Laws*: eternity for humans is the continuation of the generations, through time.[17] Such an introduction significantly announces Aion's speech. For he outlines how he has observed that the flood and warfare have made men's lives painfully short, and asks therefore for some other god to control 'the course (*dromon*) of my years' (7.39). The charioteer imagery of the *Paraphrase* is taking on a new contour. Aion wants to change how lives run. Aion marks his pity for humans who are buffeted by the twin miseries of old age, when man 'walks with a foot too many' (43) (a phrase to delight Freud in his reading of Oedipus), or early death 'which dissolves the life-bearing hawsers of indissoluble union' (47) (echoing again his observation of the threat of the flood to dissolve harmony, and destroy the possibility of generation). Aion explains human time to God. Human lives are not merely short and insignificant, as Apollo in Homer insisted, but broken by failed fulfilment and marked by pain. And he proposes a 'cure' (*pharmakon*, 56) that 'saves life'. Wine can dissipate the cares of the universe. Zeus with much thought agrees. 'The primordial (*archegonos*) cosmos will be in pain', he declares 'until I bear one child'. Again, within a Christian environment it is hard not to hear at least an echo of 'God gave his only begotten son' (John 3.16, *Paraphrase* 3.82), so that the faithful might have 'everlasting (*aiōnion*) life'. Eternity asks Zeus to change the primordial universe – the philosophical language of the opening of the book has its purchase now – by bringing men the transformation that Dionysus promises.

This representation of Aion in Books 6 and 7 grounds the later depictions of the god. The second half of the *Dionysiaca* – Book 25 opens with

---

[17] See above, pp. 45–7.

a second prologue – builds up the scale of the Indian war to come by announcing that 'Aion never witnessed such a war' (25.23). As Aion witnessed the flood and the miseries of primordial humanity, now he is called on to mark the magnitude of Dionysus' triumph. So too, he was there observing at the foundation of Beirut, the earliest city in the world (41.84).[18] Elsewhere, Aion appears only in the speech of characters rather than the narrator. In Book 12, the Season (*hōra*) of the Vintage asks the Sun which god will have the privilege of overseeing the growth of the vine (12.23) – reprising Aion's role in asking Zeus for Dionysus to come for precisely this privilege. In Book 24, Leukos sings a song for the resting warriors of Dionysus which includes the light-hearted tale of Aphrodite learning to weave and hence giving up her usual role in fertility. Aion, called 'charioteer of life', grieves that the *harmonia* of wedded life is disrupted and useless (24.265–7). The language of the race track – the cliché revitalized – is now linked to the cycle of marriage and childbirth – as the threat to *harmonia* recalls Aion's role preserving human intercourse in the poem earlier. Hermes tells Dionysus how Aion had observed Phaethon's fiery self-destruction – another witnessing of a major cosmic event (38.90–5); and finally, Heracles tells Dionysus how Aion observed the birth of autochthonous people (40.430–1).

The representation of the figure of Aion is used thus to witness, and, in Book 7, to intervene in the great course of cosmic events, and to mark the course of human life as a pattern of generation, understood as the eternal cycle of procreation over time (the two senses of generation), which, through the language of *harmonia* and union, also links Eternity's directorial role over life to the genealogy and life-story of the epic's hero, Dionysus. The repeated representation of Aion as a charioteer of time in the *Paraphrase* is enriched by the recognition of how the cliché of time's cycle is thus integral to understanding human time. The shared language between the *Paraphrase* and the *Dionysiaca* creates a shared framework – with an intense link between Aion and the Christian promise of eternal life in the *Paraphrase*, and an equally intense insistence on the pattern of generation and constant change in the *Dionysiaca*. Yet what is most striking is the shift in epic discourse. For Achilles in the *Iliad*, the promise is of 'unfading fame', as Odysseus, too, in the *Odyssey* can assert his *kleos* against the roll-call of the dead heroes. Immortality is predicated on the undying power of song to record and memorialize a hero's life. Now in both the *Paraphrase* and

---

[18] On Aion as observer see Vian (1993) 47.

the *Dionysiaca*, the framework is Eternity, Eternity as a concept, a concept marked by a long tradition of philosophical reflection and, above all, theology. Both the *poikilos humnos* the 'shifting song', of the *Dionysiaca*, and the *empedos muthos*, 'the fixed word' of Jesus in the *Paraphrase*, are set within Eternity's scope. Poetry is being rethought through God's time.

*

The two most insistent problems that haunt eternity as a concept are the interrelated anxieties about origins or beginnings, on the one hand, a concern we have already discussed with Augustine's repeated reflections on creation and God's time; and genealogy, on the other, human continuity in and against eternity – anxieties exacerbated by the theological debates between the councils of Nicaea and Chalcedon about the temporal and hierarchical relations between Father and Son in the language of the Trinity. The eternal problems: how do things start and where do we come from? Nonnus writes with extraordinary panache about both genealogy and beginnings. The narrative of the *Dionysiaca* repeatedly returns to multiplied stories of origin, plays with the chronology of stories of descent, and through prophecy and retrospect makes the time of its narrative a mythic 'always already'.

In Book 3, for example, Cadmus, Dionysus' grandfather, is asked his lineage, and pours forth 'like a fountain a line of ever-flowing stories' (*aenaōn muthōn*) (3.246). Cadmus' answer is indeed one of Nonnus' mythic narratives that keep flowing into each other throughout the epic. He begins by rehearsing the famous answer of Glaucus in the *Iliad* to the same question about lineage. One point of origin for *muthoi* is always Homer. 'I liken the generation of swift-fated men to leaves', Cadmus declares. Homer's Glaucus said that 'the generation of leaves and men are alike'; Cadmus now, with typical upping of self-consciousness, marks his formation of a simile: 'I liken' . . .'. 'One generation', he continues, 'rides life's course and is conquered' – the charioteer of time, again – 'another flourishes, to yield to another', which leads to his conclusion (3.255–6):

ἐπεὶ παλινάγρετος ἕρπων
εἰς νέον ἐκ πολιοῖο ῥέει μορφούμενος Αἰών

Since, turning back as it goes,
Changing form from old to young, flows Eternity

Palinagretos is another term that appears only once in Homer – Zeus' nod in answer to Thetis' supplication, which cannot be 'turned back' – but, like *akikhētos*, it is repeatedly reused by Nonnus (8 times in the *Paraphrase* and 22 times in the *Dionysiaca*). The etymology seems to have been understood from *egeirō* 'to awaken', and it is most pointedly used not just for repeated actions but in the context of resurrection – including the story of Lazarus in the *Paraphrase* and Tylus and Zagreus in the *Dionysiaca*, used in each case with *archē*, to capture the 'paradoxical miracle' of beginning life again.[19] I translate it 'turning back' here, but it is used by Cyril too for the return to life, or, metaphorically, of light to the world, or even of Christ's soul returning to him.[20] This repetition is not a trivialization, as Spanoudakis suggests,[21] as if its use were no more than a verbal tic. In fact, the example he disparages (11.47), emphasized by the verb *egeirō* and the adjective *anēgretos* in the preceding lines, announces Jesus' intention to raise Lazarus and is pointedly repeated at the climax of the resurrection (11.164) – thus overlapping verbal repetition, ring composition and resurrection as models of return. Both the simple word *palin* and compound adjectives with *palin* as a prefix, come again and again in Nonnus – there are 14 different such compound adjectives in the *Paraphrase*, used fully 69 times, and the simple *palin* itself occurs another 41 times. *Palin* sounds like a knell through the *Paraphrase*: it is the sound of paraphrastic poetics. The *Paraphrase*, whose *raison d'être* is *re*writing the Gospel, making it speak *again*, is also about Jesus' *re*peated assertion of his message, and his *re*turn to heaven, and the *re*cognition that Jesus' incarnation is the *re*direction of world history, a look back to the first man and the possibility of undoing the primal scene of human error. The repetition of the word *palin*, which marks repetition, return, reversal, resurrection, is both the materiality and the thematics of the *Paraphrase*.

In the *Dionysiaca*, the first adjective applied to Dionysus is 'twice-born'; and, as in all epics since Homer, the hero's first epithet is a definitional moment. As we will see, Dionysus is the reborn god (in more than one way); he is to rise back to his father in heaven; his actions – planting vines, changing landscapes, defeating his enemies with plant-weapons, raping women – are repeated again and again in the epic, not in aesthetic homage to Homeric 'type scenes', but because this is, first of all, how the epiphany of the god is constructed and recognized. As the opening speech of Euripides' *Bacchae* shows, Dionysus travels from town to town establishing

---

[19] 'Paradoxical miracle' is taken from Greensmith (forthcoming) discussed below, pp. 310–12.
[20] Resurrection language: *Paraphrase* 5.82; 6.162; Christ's soul *Paraphrase* 10.61. Cyril, *In xii proph.* I.275.20. For the etymology, see the *schema etymologicum* at Ps-Apoll. *Met. Psalm.* 22.4 *palinagreton ētor egeirei*, and, most saliently, Nonnus *Paraphrase* 11. 41; 45; 47.
[21] Spanoudakis (2014a) 194 *ad* 47c.

his rituals across the world, again and again. But, more importantly, it is because Dionysus as a divine figure is especially associated with change, with turning things upside down and back again, ecstasy and return from otherness, even with bringing back to life, that the language of reversal, turning back again, becomes a thematic resource of the *Dionysiaca* too. Indeed, as we will see in the final section of this chapter, mythic figures are repeatedly constituted as *tupoi*, 'types', of each other, 'another Agave', 'another Diomedes', even 'another Zeus', 'another Dionysus'. As in the *Paraphrase*, so in the *Dionysiaca*, the language of *palin* is integral to the conceptualization of narrative time.

No surprise, then, that here it is precisely *Aion*, Eternity, that Cadmus describes as 'turning back'. As time flows, it changes its own form 'from grey age to the new/young'. *Aion* is an old man, but here it is imagined that even Eternity, it seems, refreshes itself and becomes something *neos*. (Does this echo a Christian promise of the 'new age'?) It is an image that returns towards the end of the poem, in a scene we shall come to shortly, where Aion is portrayed as shedding skin like a snake and becoming youthful again (*empalin*), washed as he is in the purifying delights of Roman law (sic) (41.179–82). Here, Cadmus frames his genealogy, usually the most linear form of time, within a life-course (*biou dromon*) that is ridden down or flourishes, but finally twists back on itself.

How this portrayal of time twisting back on itself might form a genealogical narrative and make a determination of origin difficult is vividly demonstrated by an earlier passage in Cadmus' journey of Book 3. As Cadmus comes to Electra's palace and looks amazed at the statues in her garden, Emathion, Electra's son, is riding from the market place. Emathion, we are told, has a brother Dardanus, who was fathered by Zeus and nursed by Dikē ('Justice'), 'when the Seasons (*hōrai*) ran to the house of Queen Electra, bringing the sceptre of Zeus and the robes of Time (*chronos*) and the staff of Olympus, as prophets of the indissoluble power of the Romans (*Ausonieōn*)' (3.196–9). As we go back to the birth of Dardanus, we go forward to the Roman empire, which, like the *harmonia* brought by *Aiōn*, is an 'indissoluble' (*alutos*) rule – everlasting. Unlike the Greek novels of antiquity or an epic such as Quintus Smyrnaeus' *Posthomerica*, Nonnus mentions Rome explicitly. Augustus too will have his prophecy (41.388). The *Hōrai*, Seasons, are also characters in the epic who will return, and who here are the nurses of Dardanus, and they bring 'the robes of Time'. The robe of time is not such an obvious symbol of power as the sceptre and staff of authority. It places the narrative of genealogy and power within the scope of a broadly conceptualized temporal frame. Indeed, Dardanus, continues Nonnus, came to the 'flower of regrowing (*palinauxēs*, again) youth' (3.201) when the third flood threatened the foundations of the cosmos. The first flood was in the time of Ogygos (205). This rather shadowy primordial figure gives his name to the earliest period of human life in Greek and Latin mythography; the second flood destroyed all humans except for Deucalion and Pyrrha, the story also evoked in Book 6, when Deucalion again floats through a new flood (209–14). The third flood lifted Dardanus away from his family (and he plays no further part in the *Dionysiaca*). To introduce Emathion, Electra's son, we are given a birth-story and genealogy of his brother, which takes us back to two different flood stories at the origins of history – the genealogy of the human race – and forward to the everlasting power of Rome (itself a charged and precarious history in the fifth century after the sack of Rome). As Cadmus prepares to meet Harmonia, the temporal scope of the narrative is stretched, back and forward, into eternity.

The final story of Dionysus in Asia, before he returns to Europe and the familiar mythography of Pentheus and Perseus, constructs the most complex narrative of aetiology, genealogy and the search for origins. Books 41–3 tell of the birth of the beautiful Beroe, how Dionysus and Poseidon both desired her, and how they fought, until Poseidon won and married her. Beroe is the patron divinity who gave her name to the city of Beirut (*Berutos*), and Book 41 opens with an extended description and eulogy of the city.[22] This eroticized ecphrasis leads, however, into a remarkable passage about the earliest inhabitants of the place, opening with a single sentence that stretches over fifteen lines of verse (41.51–65), which describe these earliest inhabitants' autochthonous birth. These people are 'of the same age as dawn' (51); Nature brought them forth 'by some system' (*tini thesmōi*) which did not involve marriage, a father, childbirth, mother (52–3), but – as in the beginning of Book 7 – by the combination of atoms in a 'fourfold bond' (*tetrazugi desmōi*, 54, as in 7.6) which allowed the 'unsown (*asporos*) mud' to form a living generation to which Nature gave a 'perfected form' (*eidos telesphoron*, 58). This perfected shape is contrasted to the Athenian story of autochthony, when Hephaestus, failing to catch Athena, ejaculated onto the ground, and sired Erechtheus (as he is named here), a half-man, half-snake. In Beroe, the autochthonous race is 'the image of the gods' (*indalma theōn*, 65), the 'first appearance' (*protophanēs*, 66) of humans, the golden race. Again in the *Dionysiaca*, we are taken back to the beginning of time. The autochthonous race of Beroe may seem

---

[22] Chuvin (1991) 196–221.

unfamiliar to Greek mythic tradition, but the story is immediately linked into the Hesiodic theogony. For Beroe, asserts Nonnus (66–76), was a city founded by Cronos at the time when he was siring and eating his children (the Hesiodic theogony). 'Zeus', Nonnus reminds us, 'was then a young fellow, a baby still, I guess (*pou*)'. Before even the battle with the Titans which established Zeus' authority, 'the city of Beirut was there' (83). And the framework is once again made plain (83–4):

ἦν ἅμα γαίῃ
πρωτοφανὴς ἐνόησεν ὁμήλικα σύμφυτος Αἰών.

Eternity [*Aiōn*] first-appearing
born together with the earth observed the city his coeval

Eternity is the first to appear (*prōtophanēs*, as with the golden age [66]) but is born together with the earth – a cosmogony where matter and time coexist from the start – but Beroe is already there at this primordial moment, although it was built by Cronos. This baffling, entangled temporality of beginning is immediately compared to other mythic claims for priority. Tarsus did not exist at this point (65), nor Thebes, nor Sardis 'coeval with the Sun' (88). Nor was there the 'race of men' (*genos andrōn* – the term familiar in mythic accounts of the origins of humans): a city without humans, then … Nor did Arcadia *proselēnos* exist. *Proselēnos*, which is etymologized as 'before the moon', is a standard term to express the extreme antiquity of Arcadia in the Greek mythic imagination.[23] Nonnus here plays out this etymology: Beroe, he claims, is older than Phaethon, from whom Selene, the moon, gets her light. Beroe, unlike Arcadia, is literally 'before the moon'. Before the heavenly bodies shone, then, Beroe was the first to dispel the cone of darkness, and 'disperse the dark covering of Chaos'. Chaos, in Hesiod, is the primordial chasm: the beginning – the before – of everything: another echo of another distant, archaic didactic epic.[24] Here, the dark of Chaos is illuminated by the city of Beirut. Beirut, in this fantastical search for origins, is before every other first, a city before humans, before the sun, before chaos …

The narrative of Beroe takes place 'under the sign of origins', as Chuvin writes.[25] It is also haunted by another set of stories, namely, the bizarre cosmogony written by Philo of Byblos in the second century, which we know primarily through the lens of Eusebius' *Praeparatio Evangelica*, who is citing it from Porphyry's *Against the Christians*, a contemporary anti-Christian

---

[23] See the scholion to Ap. Rhod. 4.264; Dueck (2020).
[24] Faulkner (2017); see also Bajoni (2003).   [25] Chuvin (1991) 212.

polemic: a story or origins, that is, mediated through the appropriative strategy of a Christian search for prescient intimations in the past of what the world truly is – and a polemic about Christian truth concerning creation.[26] Philo claims to be translating into Greek the Phoenician cosmogony of Sanchuniathon of Beirut, which was itself a version of the account passed down by Taautos, a primordial figure assimilated to the Egyptian Thoth – a fivefold metaphrastic recession of voices (Eusebius, Porphyry, Philo, Sanchuniathon, Taautos), a story in search of an author. Eusebius is dismissive because Philo offers both physical, material explanations of creation, and a Euhemeristic account of divinities that supposes gods were once extraordinary men – a long-established Greek theory that was particularly dismaying to Christians struggling with the Christology of a *theos anēr*. 'Downright atheism!', snorts Eusebius. Sanchuniathon's cosmogony begins with thickening air, a form of boundless chaos, which leads to a mixture (*sunkrasis*) called Desire (*Pothos*), which led to a primordial mud (*Mōt* in Phoenician or *ilus* in Greek), from which all was seeded, including the celestial bodies. From wind and night are born the first humans, who are called Aion and Protogonos. Their children were Genos and Genea ('Race' and 'Generation'). The Greeks in their ignorance misunderstood this story, says Philo, 'because of the ambiguity of translation'. 'Eternity', Aion, is a mistranslated Phoenician word for ... the origin of humans. In Nonnus' story of Beirut too, 'mud', (*ilus*, 56) is the primal goo;[27] nature rather than a divinity is the cause of the generation, which depends on a mixing; Aion is the 'first-appeared'. Cronos too appears later, as in Philo, who pictures an awful, violent version of the so-called Golden Age. Nonnus' Beirut has been shaped by a Greek translation of a Phoenician cosmology, a provocative text provocatively reappropriated.[28] This exotically ludicrous account, comments Eusebius with dry dismissiveness, 'was approved as true by Porphyry' – the Neo-Platonic philosopher from Tyre, the Phoenician city nearest to Beirut, and Dionysus' grandfather's hometown. Philo, as he enters a polemic of quotation and counter-quotation in Eusebius, lurks behind Nonnus' account of Beirut, a local tale of universal history, a hidden quotation, another origin for the multiplying origin stories.

This exuberant, drunken excess of origin stories, which are so hard to reconcile into anything resembling linear time, is far from over, however. For Aphrodite, it turns out, did not go first to Cyprus when she was born in

---

[26] Eus. *Praep. Evang.* 9–10: see P. Johnson (2006), (2014).
[27] Cf. Ap. Rhod. 4.676 for an Empedocles-influenced account of similar primal goo.
[28] Chuvin (1991) 221–2.

the foam of the sea out of the castrated genitals of Ouranos. She did not go to Paphos, Byblos, Colias or even Cythera (all famous sites of her worship) (87–109). Rather – as you may by now have guessed – her first landing was at Beirut. This extraordinary transformation of what was such an established literary and artistic and cultic tradition is duly marked in passing by the comment that the other story is 'a lie told by Cypriots', a blithe and traditional marker of mythographic infighting.[29] The goddess calmly swims through the 'god-bearing water' (*theētokon hudōr*, 112) towards the shore. Granted the *casus belli* for the Council of Chalcedon was the applicability of the term *theotokos* to Mary, it is hard not to be provoked by this adjective here, as if we are being asked to set Christian and pagan theological vocabulary against each other.[30] The heady description of her arrival from the sea, full of roses and soft breezes, also includes, however, the birth of Eros. His spontaneous generation, as soon as the goddess was seen on the shore, is a different story of his birth from elsewhere in the *Dionysiaca*, where he is the child of Hephaestus and Aphrodite (5.138–44). Eros, like Dionysus, has multiple forms, multiple births, multiple stories in the *Dionysiaca*. He is described here as a cosmic force, as at the beginning of Book 7 (41.129–30):

γονῆς πρωτόσπορον ἀρχήν,
ἁρμονίης κόσμοιο φερέσβιον ἡνιοχῆα

First sown origin of generation
Life-bringing charioteer of the harmony of the cosmos.

Eros, as in Empedocles (and in Orphic cosmogonies, where he is named Phanes)[31] is the primal force of generation, here invested with the iconography of Aion, a charioteer on the course of life, committed to the harmony of the world. Another authoritative origin story is woven into Beroe's narrative. Eros immediately and exuberantly breast-feeds on the bosom of the goddess of sexual gratification.

This wild sensual tale leads the narrator to burst into an amazing, ecstatic hymnic address to Beirut (143–54): 'Root of life, Beroe, nurse of cities, boast of rulers, first-appearing, same-sown as Eternity (*Aiōnos homosporos*), cotemporal (*sunchronos*) with the cosmos, seat of Hermes, plain of Justice, town of laws', and so on for eleven lines of chanted vocatives, ending – extraordinarily – with the triumphant announcement of the birth

---

[29] Faulkner (2017) sets this within a didactic tradition back to Aratus especially.
[30] Shorrock (2011) 61–2 is much more convincing than Sieber (2017) on the pointedness of the term.
[31] See n. 6 above.

of Beroe to her parents Ocean and Tethys. The city has (finally) become the nymph after whom the city is named, and who Dionysus and Poseidon will fight over in the next two books of the epic. The genealogy which starts with time before time, the very origin of things, and claimed for itself the birth of Aphrodite and Eros, has somehow come right up to the point of the narrative present. As the poet becomes increasingly heated, till he explodes into a hymn of eulogy, the stories become increasingly hard to tie into a coherent chronological schema, as tales of origin multiply and compete, and familiar stories are transformed into new paradigms. In the prologue to the *Dionysiaca*, the poet called for the accoutrements of Dionysus to sing his Dionysiac song, and here the poetry indeed seems increasingly Dionysiac as stories swirl and blend into each other and transform, until the poet himself is overwhelmed and breaks into an ecstatic outpouring towards the city – which metamorphoses into the announcement of the nymph's birth. This is Nonnus' Dionysiac poetics dramatically on display.

And yet, with a startling and very funny *coup de théâtre*, at this very moment of ecstatic outpouring, Nonnus stops his narrative dead in its tracks and states 'But there is a younger story (*phatis*)' (155) – and the narrative which had reached such a pitch simply takes another track.[32] Now, Beroe is the daughter of Aphrodite and Adonis (157), a quite different genealogy, and as her birth approaches, Hermes arrives with a *Latin* letter, a 'herald of things to come' (160).[33] Latin poets regularly recognize earlier Greek models, and Ovid is happy to add a Greek explanatory account of a ritual to supplement his Roman understanding. But Greek writers only very rarely acknowledge Latin authorities explicitly, and do not readily admit even to knowing Latin well.[34] It is marked, then, that Hermes, a Greek god, arrives with a Latin message. It would seem that after the first, Dionysiac splurge of origin stories in the deepest past, a real abyss of time, now we are to enter the historical time of the Roman world. Indeed, Beroe's birth is also aided by Themis, the goddess of order and justice, who has Solon's laws in hand, and Hermes, the 'male midwife', is there because he is *dikaspolon* – a Judge. Aion – again present and observing – 'coeval with Ocean', Beroe's father, swaddles the baby with 'the robes of Justice', *pepla Dikēs* (179). The baby Beroe is nursed by Astraia, who pours statutes into her mouth with the milk, as the baby burbles laws in response (215–17). All Beroe's drinks are brought from

---

[32] Faulkner (2017); it has often been discussed if this second story is an invention of Nonnus: Accorinti (2004) *ad loc.*
[33] See Lightfoot (2014) 41 who calls this 'a favourite usage' and lists parallels.
[34] See Goldhill (forthcoming a) with bibliography and further discussion.

inspirational springs to foster her eloquence. Beirut is a famous centre of Roman law, and this new birth story, its new genealogy, is building towards a celebration of this Roman foundation – a very different historical model. As we have more than one Dionysus, more than one Eros, now we have more than one Beroe.

The girl's birth is greeted with a cosmic joy that turns the prophecies of Isaiah into a Dionysiac scene, where the lion gently kisses the bull on the neck, the wolf with playful cries kisses the sheep, the hound dances a sprightly jig with the boar, a calf licks a lioness ... Aphrodite too is thrilled with the child, and decides to build a town to celebrate her daughter and the Law she embodies. So she tours in her imagination cities with a claim to fame and age: Mycenae, Thebes (named for the primeval (*archegonos*, 270) city of the same name), and decides to compete with Athens, Athene's city, celebrated for its legal history.[35] So she visits the house of Harmonia to ask for advice. This Harmonia appears to be a goddess, rather than Cadmus' wife – a second Harmonia to add to our list of repeated doubled figures ('I seem to see two cities, two suns', is the paradigmatic sign of Dionysiac possession). Harmonia, with her care for order – the justice that is harmony – is asked which city should house the Law her daughter brings. Harmonia reveals that she has seven tablets, each named after a planet, which contain 'prophecies of the cosmos'.[36] The world's history is inscribed on these tablets, and if Aphrodite consults the tablet of Cronos she will learn what she needs to know. She will find out if Arcadia or any other place can make such a claim to be the oldest city, and thus have the right to house the process of Law.

The oracle is not only clear but also close to the claims of the first genealogy of Beroe (41.364–7):

πρωτοφανὴς Βερόη πέλε σύγχρονος ἥλικι κόσμῳ,
νύμφης ὀψιγόνοιο φερώνυμος, ἣν μετανάσται
υἱέες Αὐσονίων, ὑπατήια φέγγεα Ῥώμης,
Βηρυτὸν καλέσουσιν, ἐπεὶ Λιβάνῳ πέλε γείτων.

Beroe was first-appearing, cotemporal with the cosmos her coeval,
She has the name of the nymph born later, which the colonizing
Sons of the Ausonians, consular lights of Rome,
Will call Berytos, since it is a neighbour to Lebanon.

---

[35] The D Scholia to *Il.* 18.491 states that Athens was the first city to be created: the claim of Beirut to be first may echo also against that tradition.
[36] Vian (1993); Accorinti (2004) 162–4.

With the same vocabulary as before, Beroe is 'first-appearing', 'coeval with the cosmos', but – with more clarity – named after a descendant (not a problem in Nonnus' 'preposterous poetics'), and renamed by Romans.[37] Beroe, then, will house Beroe. Yet as Aphrodite consults the tablets and finds lists of other 'first founders', she also sees an oracle of Beirut's future, written in many verses of Greek poetry (unlike Hermes' Latin letter) (389–97). When Augustus comes to power, he will found the school of Roman law at Beirut which will become a bulwark of social justice in the world. The Greek poem predicts the triumph of Roman law and order. The city of Beirut was in fact destroyed in 140 BCE and only slowly recovered to become a city of the Roman empire and centre of Roman law thanks to Augustus' intervention. The two oracles juxtapose the eternity of Beirut and its modern history of change and redevelopment. Aphrodite returns to set her son Eros to incite Poseidon and Dionysus to fight over her daughter . . . and the narrative continues with Beroe reshaped as the object of divine lust rather than the harbinger of Law.

The narrative of Beroe in Book 41 – on which more could certainly be said – is an arresting example of Nonnus' poetics, and of what happens to time in his narrative. As befits the celebration of a law school, different accounts are set against each other for the judgement of the reader. As befits Dionysus' inspiration, the first account swirls with increasingly exuberant assertions about the beginning of time, mixing different authorities and different claims about the primordial, transforming familiar stories into wild alternative imaginings – and it ends with bursting into an invocation as the poet appears overwhelmed by the sheer marvel of Beirut, which by the end of the invocation has become transformed into a nymph born to the Ocean. But equally befitting a Dionysiac poetics, the return from such ecstasy is dramatized with a second story, with another, wholly different genealogy – and the juxtaposition, with its alternative birth stories, is itself paradigmatic of the ebullient excess of the mythic repertoire in the *Dionysiaca*. In this second story, Greek and Latin prophecies combine to create a historical account of what is to happen, and unlike the misty past of the first story, the tale of Beirut is firmly located in the political history of Rome and Augustus. In contrast to linear aetiology, the city is named after a nymph who will be born generations later, and, as the city's name will be changed by Romans, its role will be redefined by a future intervention by Augustus, anticipated at the nymph's birth by Hermes with a Latin letter,

---

[37] In what remains for us an incomprehensible etymology: see Accorinti (1995/6). On the contrast between this passage and Quintus Smyrnaeus in terms of cultural politics see Hadjittofi (2011).

as if Rome and its laws were there at the origin of time. In the narrative of the *Dionysiaca*, the search for origins and the descent of genealogy are transformed into a kaleidoscopic mélange of competing and conflicting stories of how things start, and a designedly entangling challenge to the linearity of time's narrative.

*

This story of Beroe has been anticipated in the *Dionysiaca* especially by two, further scenes that embody Nonnus' fascination with time, prophecy and the twisting unfurling of continually entangled narrative.[38] The first is in the previous book. Dionysus visits Tyre, his grandfather's birthplace, and as with Beroe, we have an ecphrasis of the city, focalized now through the god's wondering gaze.[39] The maritime city is described in one of Nonnus' rare similes as like a girl swimming – the picture is, of course, carefully eroticized[40] – stretching her arms, whitening her body in the foam, with her feet on the earth, while Poseidon embraces her neck with a splashing arm – anticipating Aphrodite's swim towards shore in the following book. Dionysus views the town, expresses his amazement, and then offers a prayer in the temple of Heracles Astrochiton. Within the arena of Greek religion, the assimilation of the Phoenician god Melikart with Heracles and then the Sun is a familiar slide.[41] So Dionysus addresses the god as 'Astrochiton Heracles, lord of fire, prince of the cosmos, Sun' (40.369–70). The Sun is invoked as driving his chariot in twisting circles (*helikēdon*): 'twisting the son of Time, the year, with its twelve months, you drive circle after circle. From your chariot flows Eternity (*Aiōn*), taking its shape in old age and youth'. *Aion* in Homer, as we saw, can mean a 'life-time', and Accorinti translates Aion here reasonably as *Vita*, 'Life'; or perhaps, with Vian, 'the vital principle' that governs not just the universe but an individual's life-course.[42] This invocation of the Sun's power in the cosmos continues with his birthing of the light of the moon, the sequence of the seasons, the contrast of night and day, and the weather. In Dionysus' prayer, the Sun is taken as a central force to understand the flow of all levels of calendrical time – year, seasons, night and day – and the weather, and the experience of a life-time from youth to old age. There is a recognition of the full panoply of Time's robes, conceptualized through the heaven's

---

[38] Vian (1993), who calls these scenes 'cosmic preludes', has been especially influential.

[39] Chuvin (1991) 224–54.

[40] Chuvin (1991) 226–7 notes the influence of Achilles Tatius throughout the description.

[41] See Fauth (1995) 165–84. Fauth starts his chapter with a 13-line sentence: writing about Nonnus affects style. See also Chuvin (1991) 233–9.

[42] Accorinti (2004) *ad loc.* Vian (1993) 48.

order: this is the grounding of the narrative of the god's ascent to Olympus. The grandeur of the address is part and parcel of the grandeur of the vision of time and the heavenly bodies as the necessary frame of the *Dionysiaca*. Beyond even the *Aeneid*'s recognition of Roman *imperium sine fine* as an expression of divine providence and necessary order,[43] Nonnus makes cosmic order and cosmic time figures in Dionysus' narrative. The *Dionysiaca* parades how time has become a subject of epic in late antiquity.

This hymn to the Sun is especially evocative in the religious context of late antiquity.[44] The *Dionysiaca* has already 'invented' the hymn, when the character Hymnos is murdered for his forlorn passion, and lamented with a pastoral *hymnos* by nature and its divinities (15.395–422). But the worship of the Sun was central in the tensions and battles between Christianity and paganism, not least through the emperor Julian, who wrote a long piece, which he called a (prose) *humnos*, 'On the Sovereign Sun', which mixes Neo-Platonism, ethnography and religious history to set the Sun as a central and dominant force in all religion.[45] Orphic Hymn 8 is to the Sun and demonstrates the place of Helios in liturgy and Bacchic mystery cult.[46] Most saliently, however, Nonnus' contemporary Proclus wrote a 'Hymn to the Sun', which begins with an address to the 'king of fire', as Dionysus begins here, 'lord of fire'. This image goes back to Plato's *Timaeus*, and is found too in the Chaldaean Oracles.[47] There are many shared terms and expressions between the language of Proclus and the discourse we have been tracing in the *Dionysiaca*.[48] In Proclus, the Sun is even hymned as 'the father of Dionysus', which associates the Sun and Zeus, and Dionysus is consequently also associated with Attis and Adonis as figures who come back from death, and thus stand as symbols for the journey of the soul. The Sun itself, in a phrase likely to catch the eye of a Christian reader, is the creation of the 'ineffable (*aporrhētos*) deity'. Three examples, then, not just of non-Christian but of stridently anti-Christian hymns to the Sun, to which we could add the huge religious debate in Book 1 of Macrobius' *Saturnalia* which relates every god of the Greek and

---

[43] Hardie (1986).

[44] See Agosti (2015) on the importance of hymns in late antiquity. He surprisingly does not discuss this passage, however.

[45] See, on Julian, Smith (1995) 114–78; in general, see Fauth (1995), who includes the evidence from magic papyri for the role of the Sun in spells and mystery cults.

[46] Graf (2009).

[47] See Van den Berg (2001) 152–5 for full discussion. Proclus' commentary on Plato's *Timaeus* 3.141f Diehl; 3.9.8–27 Baltzly provides the theory and quotes the oracles. See also the extended discussion of the sun in Book 4.

[48] Saffrey (1984) discusses the richness of Proclus' language in the hymn, and especially its Platonism.

Roman pantheon to worship of the sun, in a bravura mixture of scholarship and literary mythography (*Sat.* 17–23), a parade of learning in Latin and Greek which includes an Orphic hymn that links Phanes to Dionysus to the Sun (18).[49] For Nonnus, the Christian poet, to write a hymn to the sun may, then, be a charged moment, even in an epic about Dionysus. Here with Dionysus' prayer, as god prays to god, both the shift between Melikart, Heracles and Helios, and the intense, culturally resonant language of the hymn to the Sun, constructs a moment where the boundaries between Neo-Platonism, Orphism, and the Christianity of Nonnus are especially fluid. Nonnus' hymn may be 'eccentric and hyperbolic' – Dionysiac? – in its combination of 'religious, mystical, magical, astronomical, or, more precisely, astrological, and philosophical-cosmological' elements;[50] but what is distinctive, however, is that although Proclus calls the Sun, in a list of vocatives, the 'father of Time (*chronos*)' (8.13), in contrast with such anti-Christian hymns it is only in Dionysus' prayer that the Sun's power is articulated through so many levels of human and cosmic temporality.[51] The hymn to the Sun becomes another occasion to envision the role of time in the god's story.

Heracles answers Dionysus' hymnic invocation and appears in epiphany to him, and, in answer to the god's request, offers two long stories, the first about the foundation of the city, the second about the foundation of its fountains, two more origin narratives. The first begins with yet another story of autochthonous generation of an original population, an anticipation of the first genealogical myth of Beroe, in language that has by now become familiar to this discussion. 'Eternity (*Aion*), same-sown, saw them alone as coevals of the ever-flowing cosmos' (40.430–1). Again, to talk of the origin of humans is to talk of eternity and the everlasting cosmos; it is a necessary frame. Second, he explains that the fountains were once water nymphs who wished to avoid sex, but Eros – with a mythical excursus on the many rivers and fountains who have fallen in love, including Okeanos and Tethys, the parents of Beroe, Galateia and Polyphemus, and Arethusa and Alpheius, pairs we saw in the flood of Book 6 – forces the water nymphs to join with 'sons of the soil'. These relationships are, concludes Heracles, 'the divine blood of your generation' (40.573). That is, Heracles' first story takes Dionysus' genealogy back through Tyre, his grandfather's city, to an autochthonous race at the beginning of time, back to the earth itself; his second story takes Dionysus' blood back to divine water nymphs

[49] Chuvin (1991) 231–2.   [50] Fauth (1995) 165, my translation.
[51] Proclus' commentary on *Timaeus*, especially Book 4, reveals a far deeper background, of course.

(anticipating his own sexual exploits in the *Dionysiaca*) and another set of primal ancestors (*archegonoi*, 40.538). Dionysus comes now from earth and water, another beginning, after his birth in fire. 'Twice-born' Dionysus multiplies the narratives of his genealogy, how this history in-forms him through time.

The second scene that anticipates the narrative of Beroe in an especially striking manner comes in Book 12 – and involves the first visit to the prophetic tablets of Harmonia, the tablets at the house of the Sun.[52] At the end of Book 11 and the beginning of Book 12, the Horai, the 'Seasons', are figures of the plot, as was Aion in Book 7. They are the daughters of the Year and the Sun, maidens who dance and sing, and they visit their father's palace to ask about the coming of the vine, which will grow, be harvested and make wine under their ordering of time. When they arrive, the Sun has just dismounted from his chariot with its sweaty horses, and they are greeted by the twelve Hours, also named Horai, who are called the 'daughters of Time' (*Chronos*) (12.15). Nonnus revels in the confusions of myth's nomenclature, and in the doubling it allows. *Hōra* means both 'season' and 'hour' in Greek, so Nonnus has one set of *Hōrai* meet the other *Hōrai*, juxtaposing rather than disambiguating the two senses. With the Sun, the Seasons, and the Hours who are the daughters of Chronos, gathered to talk about Aion's gift of the honour of the vintage, Nonnus dramatizes time as an extended family – a *divina comedia* of time.

In answer to his daughter's request – she is, as a Season should, 'circling' round him (31) – the Sun directs her to the tablets of Harmonia, which, the narrator indicates, contain 'all oracles in one place', inscribed by 'the prophetic hand of Phanes the first born (*prōtogonos*)' (12.33–4). Phanes, assimilated to Eros, is the primal figure of Orphic cosmogony – yet another explicit claim to the origin of things. The first tablet is described as *atermonos hēlika kosmou*, 'the same age as the boundless universe' (12.43). As with 'beginning again' there is something of an oxymoron in the claim to be the same age as a universe without limit – without a beginning or end. So, at 9.140, Hermes, in order to save the infant Dionysus from Hera's wrath, takes on 'the boundless (*atermona*) shape of first-born (*prōtogonos*) Phanes' – an even sharper paradox, since it is hard indeed to imagine a distinctive shape (*morphē*) with no boundaries. The first tablet contains the story of Ophion, who ruled Olympus before Cronos, another story of a time before – and the tablets of Book 41 were written by Ophion (not Phanes – though Phanes and Ophion are sometimes 'associated' as primal

---

[52] The fullest commentary on this passage is still Stegemann (1930) 128–72.

figures). As ever with Nonnus, the beginning – the point of origin – dissolves into paradox, uncertainty, multiplicity. Yet the existence of these oracles at the beginning of time, which reveal all of coming history, is the perfect image for the 'always already' of Nonnus' mythic narration.

The Season views the tablets of Harmonia to find an answer to her question, and, as she reads, discovers a potted history written by 'the primal (*archegonos*) mind, variegated in tales (*poikilomuthos*)' (68). To call Phanes 'the primal mind' is to emphasize the philosophical, Orphic roots of the figure,[53] but *poikilia* is a key sign of Nonnian poetics, here placed at the very beginning of the world, a founding principle: the oracles are indeed *polutropa* (66), a word which both encodes the Odyssean slipperiness of Nonnian poetics and performs it in its allusivity. These oracles may be fixed from the beginning of time but, like Odysseus, they also pose a trap for interpreters.[54] On the second tablet, indeed, the Season sees 'how the pine-tree gave birth to the human race' (56) – yet another, variegated origin story for humanity, which nonetheless turns into a further telling of Deucalion and the flood (59–64). In contrast to the story of Genesis, which may tell the story of creation twice, but which locates the beginning of the human race firmly in Eden and firmly brought about by the willed creative force of God, the *Dionysiaca* produces multiple sites, multiple methods, multiple histories for the origin of humans, a multiplicity combined with conflicting, entangled genealogies and hazy, uncertain processes of coming to be. The contrast between what we can call Dionysiac history and biblical history is stark and telling. Unconcerned, the Season reads on, until she reaches some astrological signs, under which she discovers the promise of the vine, as the special province of Dionysus. She can then depart happy with the future to come.

This first visit to the tablets of Harmonia at the palace of the Sun anticipates the second viewing of the prophetic tablets in the Beroe episode at the palace of Harmonia. In both cases, prophecy plays a determinative role in the narrative (as so often in epic). In both cases, however, the tablets also record written prophecies from and about the primal moments of humanity, prophecies that lay down all the future for always. Both passages recognize, manipulate and happily play with the paradoxes of temporality and narrative established by the search for a beginning, a 'first-born' moment. Yet there is a further doubling that opens a particularly difficult area of Nonnus' representation of time. For in both passages the zodiac is woven into the very structure of the prophetic tablets. Indeed, Harmonia's

---

[53] Gigli Piccardi (2009).   [54] Fincher (2017) 123–4 – a little heavy handed.

seven tablets are said to stand each for one of the seven planets (41.341), just as the tablets of Book 12 are named after the signs of the zodiac. What does astrology mean to Nonnus?

I write astrology but I could say astronomy, because the modern distinction between observation of the astral bodies and the use of astral bodies for prediction of the future are constantly blurred in antiquity – to the degree that one standard term for the person who engages in such work is *mathēmatikos*, which signifies the science of mathematical calculation at the shared base of what modernity would insist are separate subjects.[55] From at least Hesiod's *Works and Days* onwards in Greek culture, the movement of the stars, as well as the moon and sun, are fundamental not just for telling the hours of the day, but also for marking the seasons and thus the pattern of the agricultural year. The stars predict the weather, and this sort of general ('catholic') astrology leads to – and is sometimes in antiquity contrasted with – personal horoscopes. The slide from using the stars to predict the seasonal weather into using the stars to predict the changes of a human life stands at the heart of late antique polemic around astrology.

The history of the field is long and complex, certainly.[56] Already in the fifth century BCE through the public performances of drama, imaging the cosmos through astrology is an integral element of the interconnection of prediction, inescapable fate, order and disorder that runs through the overdetermined narratives of tragedy – just as playing with the calendar, based on lunar observation, in order to avoid the due date of a debt can become a joke in Aristophanes. The chorus of stars and chorus on stage mirror each other's circular dances.[57] The detailed mathematics of personal astrology, however, as well as the science of astronomy, seem to have developed with particular intensity as practices in the Hellenistic world, and continued through to late antiquity.[58] There are 'thousands of pages (mostly Late Antique manuals, didactic poems, and papyrus horoscopes) . . . written in Greek',[59] two major technical works in Latin,

---

[55] See Kennedy (2011).
[56] See, for examples: Hegedus (2007) (Christian); Reed (2004); von Stuckrad (2000); Leich (2006) (Jewish tradition); Rudolf (2014) (Aramaic); Jacobus (2020) (brief summary of Dead Sea Scrolls); Cramer (1954); Barton (1994a), (1994b); Gee (2000); Volk (2009) (Greek and Roman); Heilen and Osnabrück (2016); Oestermann, Rutkin and von Stuckrad eds. (2005) (horoscopes). For the standard older discussions, see Bouché-Leclercq (1899); Cumont (1912); Festugière (1950); Gundel and Gundel (1966). For Nonnus, see Stegemann (1930).
[57] Gagné (2019); Csapo (2008) (well criticized for its Orphic turn by Gagné); Hannah (2002); and the extraordinary Miller (1986).
[58] Volk (2009) 67–75; Lightfoot (2020).
[59] Heilen and Greenbaum (2016) 123. For the works of Ps-Manetho, see now Lightfoot (2020).

Manilius and Firmicus Maternus (both heavily indebted to Greek material), and innumerable poetic references and discussions in Greek and Latin
writing – to the degree that Quintilian (*Inst. Or.* 1.4.4) insists that you
cannot understand poetry without some knowledge of astrology, since it is
so common for poets to use astrology to differentiate time.

Quintilian (*Inst. Or.* 10.1.55) also finds the popularity of Aratus baffling – to
the orator he is dull and his story has no action – but Aratus' Greek
hexameter poem *Phainomena* was indeed hugely influential on the Latin
poets of the Republic in particular, and his work is especially telling for
Nonnus too. Aratus, 'master word-smith and amateur astronomer',[60]
offered a highly sophisticated poetic account of the moving map of the
stars, which encapsulates a broadly Stoic image of the universe as an ordered
cosmos.[61] Echoing Hesiod, Aratus lays down as a principle that god 'fixed
signs (*sēmata*)' in heaven; and 'comprehensively reveals the signs' (772).
These signs are so that 'the being of everything is fixed (*empeda*)' (13) – the
establishment of order demands regularity and certainty (and it will be
remembered how important the term *empeda* is in Nonnus' *Paraphrase*, in
contrast with the exuberant world of Dionysiac change). This sense of order
is temporal. In a remarkable programmatic phrase, the multiform travelling
stars are said to be in motion πάντ' ἤματα συνεχὲς αἰεί, 'all the days,
continuously, always' (20). Three ways to express the span of time, and
the inevitable connection between astrology and temporality: that is,
duration – each moment of each day – *sequence*, and a sense of *eternity*,
stretching forwards and backwards. Aratus parades the constancy and
certainty of his vision – words for 'all' and 'ever' are repeated through his
text from the opening line, where Zeus is declared to be always celebrated, or, to be precise, with a double negative, to be 'never unspoken',
oudepote ... arrhēton (1–2), 'never "ineffable"'. The ineffability of the
Logos that is so important to Nonnus' Christianity is also a contradiction
of this certainty of Aratus' theologically grounded science. So, in the final
two lines of the poem, with a ring composition that mirrors the circling of
the stars, Aratus promises that if you observe 'all these things together' –
his comprehensive science of the comprehensive universe – you would
never (*oudepote*) be uninformed (1153–4). *Arrhētos* is also a pun on the
author's name *Aratos*. Aratus' text revels in puns, etymologies and acros-
tics, including an acrostic of the word *leptē*, which means 'sophisticated'
in the poetics of the era, and also may recall the term *lepton*, a technical
term for a 'minute' in astrology – a triply self-reflexive game with *sēmata*.

---

[60] Gee (2001) 534.   [61] See Gee (2000); Hunter (1995); Fantuzzi and Hunter (2004) 224–45.

The 'signs' revealed by god are expressed in the signs of Aratus' language which also dance to hidden and revealed tunes.

No surprise, then, that Ovid, most self-consciously sophisticated of poets, and especially in his *Fasti*, repeatedly turns to Aratus. The *Fasti* is announced from its beginning as a poem of time (*tempora* is its first word), a poem which sings 'the signs (*signa*) that rise and fall under the earth' (2) – astrology as the measure of time.[62] Yet Ovid, unlike Aratus, happily imagines a drunk observer of the stars blurrily missing the signs (6.785–90), and his observer of the heavens is 'subjective and problematic'[63] – blithe in error and accuracy – a stance that epitomizes the 'indeterminacy and uncertainty of knowledge'[64] in Ovidian explanations of the world. The god Janus, after whom January is named, paradigmatically confesses '*me Chaos antiqui (nam sum res prisca) vocabunt*', 'The ancients, for I am an old thing, called me Chaos'. The beginning of Hesiodic time turns out now to be a talkative, misnamed Roman god (who faces two ways). There is always a profound instability in Ovid's shifting stories, and not just in the *Metamorphoses*. If such instability, set against the imperial claims of *imperium sine fine*, has a political snideness, it is all the more marked when the subject is time. Caesar's calendar embodied the emperor's 'control over time'.[65] The state, as we have seen, organized and regulated the time of its subjects. The coincidence of the celestial and the terrestrial calendar now meant that reading the stars had less necessity, on land at least (navigating at sea still required astrology's accuracy). With the new accuracy of Caesar's reordered calendar, the date now matched the season, predictably. In 11 CE, astrologers were exiled from Rome and personal astrology, especially casting the horoscope of the emperor, banned.[66] Yet Augustus allowed his zodiacal sign to become a symbol of his fated power, and the Hellenistic practice of catasterism, epitomized by Berenice's Lock, celebrated by Callimachus and then in translation by Catullus, took imperial power into the map of the heavens.[67] If personal horoscopes were instruments of power, dynamic in their manipulation,[68] so too the imaging of the cosmos was a framing of imperial authority, in and against which Ovid's *Fasti* is shaped. Astrology becomes a factor in the politicization of time.

---

[62] Gee (2000).   [63] Kimpton (2014) 36.

[64] Schiesaro (2002) 74. Herbert-Brown (2002) suggestively argues that the *absence* of predictive horoscopic astrology in the *Fasti* is a politically loaded strategy.

[65] Kimpton (2014) 45; or Volk (2009) 131. See, more fully, Feeney (2007).

[66] Ripat (2011) (with full bibliography) who nuances the seminal Cramer (1954).

[67] Gee (2000) 154–87; Volk (2009) 127–73.

[68] Barton (1994a), (1994b); Heilen (2005); Volk (2009) 127–73.

Or for the designed refusal of the politicization of time. Marcus Argentarius is probably a contemporary of Ovid's, and one of his best-known epigrams manipulates the analogy between the macrocosm of the universe and the microcosm of a human life with an elegant melancholy, which recognizes the tension between the observation of the heavens and the predictive authority of astrology with an amused disavowal of responsibility or duty (*AP* 9.270). 'I am on the razzle', it begins (*kōmazō*), 'and I am staring at the golden chorus of the evening stars'. Like Ovid's drunken astronomer, Marcus Argentarius has stars dancing before his eyes. 'Nor do I step heavily on the gossip of others'[69] – amid the partying, he disturbs nobody, his eyes elsewhere. 'I have crowned the dark hair of my head with shaking flowers, and with my musician's hands I have struck the strings.' The poet is dressed for the party and performing. He concludes:

καὶ τάδε δρῶν εὔκοσμον ἔχω βίον· οὐδὲ γὰρ αὐτὸς
κόσμος ἄνευθε λύρης ἔπλετο καὶ στεφάνου.

When I act like this, my life is a model of good order. For not even
The cosmos itself moves without the lyre and crown.

As if rehearsing the moral injunctions of sympotic poetry that require an orderly display of drinking and singing – a demand observed, as here, mainly in its transgression – Marcus declares his life is 'well-ordered': *eukosmos* is a well-worn political and social admonition. But the term sets up the final joke. The *cosmos* – in its other sense of the universe – has its Lyre and Crown, that is, the constellations so-named in the night sky, the stars he was watching at the poem's opening: the heavens provide a determinative analogy for the poet's life on earth. The belief that the heavens are a model of god's order, and that the stars determine a man's life, become here an amused defence of his partying. The refusal to be serious about the seriousness of good order is also a political stance, of course, and this lovely little poem exemplifies how there is more than one way to use astrology to illumine a life-time. As the poem sets the poet against the stars and his companions, so it sets the epigrammatic moment against the eternity of the universe.

Marcus Argentarius' melancholic wit becomes full-throated parody for the second-century satirist Lucian, whose *On Astrology*, written with his typical flair and wit, performs a mock eulogy in Ionic Greek – as if it were a defence of the science by a trendy sophist of old, now hopelessly out of

---

[69] The text is difficult here: see Gagné (2019) for the most recent discussion.

date. He writes an imaginary history of divination by the stars from its invention by Ethiopians, and turns a series of familiar myths into an increasingly baroque, rationalizing anthropology of astrology. So, Orpheus and his music are no more than a pointer towards the Lyre constellation, the animals who listened to the singer, the animal signs of the zodiac; even Aeneas' birth from Venus, Rome's foundation myth, is explained as the good-looking Aeneas being born 'under Venus', a birth sign misremembered and mythologized. For Lucian, both astrology's status as a science and its very claims to historical authority are already an open target for his intellectual satire. The Ionic Greek – old-style science – is the mocking voice of Lucian's snide takedown of astrology's pretentions to antiquity.

The tradition of astrological writing, distinctive from the use of astrology in mundane circumstances of doubt and anxiety, is established *as* a literary tradition.[70] Hipparchus explains that Aratus had turned a prose treatise of Eudoxus into verse, and Eudoxus was a pupil of Plato's, which takes us back to the *Timaeus* and its account of the birth of time with the sun, the moon and stars. Aratus in turn is translated into Latin by Cicero, and by Germanicus, the emperor in waiting, and, a third time, by Ovid, the poet exiled by the emperor (and three others, too).[71] But both the practice of consulting horoscopes and this literary delight in the sky's *sēmata* or *signa* proved dismaying to Christian normative writers, intent on projecting a new vision of time and the cosmos, as it was contested by earlier philosophers too.[72] This angry and rhetorically potent Christian engagement with astrology provides a fundamental framework for Nonnus' poetry.[73]

Astrology in the sense of the observation of the heavens can certainly play a role in a Christian perspective on the universe, because of the argument from design, which runs on into the nineteenth century (and beyond). The orderly movement of the stars, sun and moon is a sign of God's creation, God's direction. So Clement of Rome, from as early as the first century, writes 'The sun and moon, with the companies of the stars, roll on in harmony according to His command, within their prescribed

---

[70] For the awkward literary status of a text such as Ps-Manetho, neither Anubion nor Aratus, see Lightfoot (2020), who tries to evaluate the practical use of its 'solemn pedestrianism' (183) and 'sclerotic catalogues' (195).

[71] Varro of Atax (first century); Avienius (fourth century); anonymous (eighth century) see Volk (2009) 28.

[72] See Long (1982); his Christian sources are no more than Augustine, *Civ. Dei* 5, however.

[73] On Christian astrology Hegedus (2007) is seminal.

limits, and without any deviation.'[74] In this statement, Clement is close to the standard principle of Stoic philosophy, the elite educational *lingua franca* of the Roman empire, that also saw proof of the divine *logos* in the order of the celestial bodies. This sense of celestial order can include the role of the heavenly bodies in predicting the seasons and the flow of time. Ps-Clement's *Recognitions* gives an account of creation which states that the stars were placed in the sky so that 'they might be for an indication of things past, present, and future. For they were made for signs of seasons and of days, which, although they are seen indeed by all, are understood only by the learned and intelligent', *ab eruditis et intelligentibus*, as Rufinus' translation has it (1.28.2). Astrology is here also a science, and even Gregory of Nazianzus praises his brother Caesarius for his skill in this area of elite knowledge.[75] The roots of this religious recognition of the science of astrology are found in Genesis, where Abraham is instructed by God to 'observe the stars and count them, if you are able' (15.5). Artapanus, the Hellenistic Jewish historian, with the typical projection of Jewish cultural precedence that we have seen in the Letter of Aristeas, concludes that Abraham learned astrology and taught it to the Egyptian kings, who are usually assumed to be the first authorities in the field.[76] Josephus with a more complex apologetics tells the same story, but now within more conflicted Greek and Roman attitudes towards astrology and its foreignness.[77] Eusebius lists other sources that Abraham taught the Pharaoh, but takes Enoch to be the inventor of astrology.[78] Origen, with his more philosophical theology, firmly distinguishes between the stars as signs and stars as causes – separating, that is, what modernity would call astronomy from astrology – but he, too, in a remarkable and beautiful poetic expression, describes 'the stars dancing in the heavens for the salvation of everything', ἐν οὐράνῳ σωτηρίως τῷ παντὶ χορευόντων ἀστέρων.[79] Wonder at the order of things leads even Origen to see in the stars a dancing sign of salvation.

Yet reaction against astrology as a science of prediction produces a complex polemic among Christian writers. Augustine is a particularly fascinating case. In the *Confessions* (4.3.4–6), he confesses that he used to frequent astrologers, and begs God's mercy for such behaviour. It was, he

---

[74] Clement of Rome, *Epistle* 1.20.   [75] *Or.* 9.9 – also with the argument from design.
[76] Eus. *Praep. Evang.* 9.17.2–9.
[77] See Josephus, *AJ* 1.155–6 with the excellent discussion in Reed (2004).
[78] Eus. *Praep. Evang.* 9.17.8 on which see Charlesworth (1977). See also Clem. Alex. *Misc.* 124.
[79] Origen, *On Prayer* 7. On Origen, see Hegedus (2007) 329–38; Hall (2020), Hall (forthcoming), especially 92–9.

explains, because the astrologers did not use sacrifices, as did Roman practitioners of divination, nor did they pray to any spirits. The lure was the science: he calls the experts *mathematici*, and the objectivity of numbers was crucial to the astrologer's self-representation.[80] He was taught by his friends to dismiss such personal horoscopes. Astrology, like sex, must be rejected on his route to orthodoxy.[81] It is a rejection that he comes back to again and again, not just in the *Confessions*,[82] but throughout his career, with increasing aggression, at greatest length in the first seven chapters of Book 5 of the *City of God*. He sarcastically rebuts astrologers who use Jesus' expression from the Gospel of John *nondum venit hora mea*, 'My hour has not come', as if it justified a horoscope for Christ. Rather, to be 'a slave of sin', a phrase from the Gospel of John that we discussed with the *Paraphrase*, is explained by Augustine as 'a slave of Venus or Mars' – that is, going to an astronomer is an act of self-slavery; or, as he puts it most forcefully elsewhere, consulting an astrologer is like paying for one's own spiritual death.[83] Augustine repeats the same arguments against astrology again and again. How can it be that twins have such different lives when they are born at the same time (an argument he repeats no fewer than seven times)?[84] If the astrologer retorts that the small difference between birth times for twins determines the difference, Augustine counters with the impossibility of making any such accurate measurements. Above all, however, the determinism of astrology – that a man's life is fixed from the moment of his birth – is hated as a denial of free will, the possibility of moral choice, the transformative efficacy of prayer, the very dignity of a Christian life. 'The stars', like fate, become an excuse to hide a person's moral culpability.[85] The sheer repetition of Augustine's complaints and the forcefulness with which he makes them indicate the threat of astrology both to Christianity as a practice and to Augustine himself – he needs to enforce the distinction between astrology's predictions and his own interests in unwilled grace or in his practice of opening a book to find an omen in a verse, as he separates his new self from his old. It is worth questioning why in the myriad modern discussions of Augustine on temporality so little attention has been paid to this repeated – obsessive – rejection of a model of understanding time and the unfurling of a life. It is hard, it seems, for

---

[80] Kennedy (2011).   [81] Ferrari (1977).
[82] *Conf.* 7.6.3. Augustine is discussed in Hegedus (2007) 43–84.   [83] *Enarr. in Ps.* 140.9.
[84] *Civ. Dei* 5.1–7; *83 Diverse Questions* 45.2; *Diverse Questions to Simplicianus* 1.2.3; *On Christian Teaching* 2.22.33–4; *Literal Commentary on Genesis* 2.17.36; *Confessions* 6.10; *Against Two Letters of the Pelagians* 2.14–16 – data taken from Hegedus (2007) 74 n. 80.
[85] *Enarr. in Ps.* 140.9.

contemporary critics to take astrology as seriously as Augustine does, even when discussing Augustine on time.

Cyril of Alexandria and Gregory of Nazianzus, both much read by Nonnus,[86] reiterate the argument against astrology that it denies free will and thus moral action, but Gregory also points up another difficulty. Did not the Magi predict the birth of Jesus by astrology? Did these wise men of the East not follow a star, precisely? 'A star will arise out of Jacob' (Num. 24.17) provided a ready prophetic source, but this did not remove the problem.[87] Ignatius in his *Letter to the Ephesians* asks 'How was [Christ] revealed to the aeons?' – *aiōsin*: 'eternity' again is at stake – 'A star shone in heaven, brighter than all the stars, and its light was ineffable (*aneklalēton*), and its novelty (*kainotēs*) caused astonishment. All the other stars together with the sun and moon became a chorus for the star, and it outshone them with its light' (19.2). The star that heralds Jesus is a sign of the majesty of Christ, but also typologically evokes Joseph, in whose dream the stars and sun and moon bow down to him. Yet the result of this star's appearance is that 'From this moment all magic (*mageia*) was destroyed and every bond of evil disappeared' (19.3). The Magi come to end the power of magic, and the end of evil is explained by Ignatius as the promise of the new life, a beginning (*arkhē*) prepared by God. Tertullian puts the same case more bluntly: the magic of astrology was 'allowed until the Gospel', *usque ad evangelicum concessa*, but then becomes a sign of evil.[88] Theodotus declares that this new star destroyed *tēn palaian astrothesian*, 'the old order of constellations'.[89] It was, Gregory announces in a poem we will come back to in the next chapter, 'not the sort of star dealt with by astrologers' – Christian astrology has to supersede the science of pagans.[90] The star is not just a sign, but a redesign of the heavens.

Hippolytus' *Refutation of All Heresies* has almost a whole book dedicated to astrological heresies, which include (4.46–50) a group of heretics known as the allegorizers of Aratus – Christians who use Aratus' poem as a guide but assimilate it to Christianity though an allegorical technique. So, these heretics take the Lyre and the Crown, the constellations that Marcus Argentarius brought to bear on his life, to stand for the divine law and the crown achieved by those who follow the divine law. (So if you look at the stars you can see the (constellation of the) kneeling man, Engonasin, who is associated with Adam, kneeling before the Divine Law (Lyra), that

[86] Simelidis (2016) with bibliography.
[87] Von Stuckrad (2000) 555–86; Heilen (2015); Hannah (2015).   [88] *On Idolatry* 9.4.
[89] Casey (1934) 86.650.   [90] *Poem. Arcan.* 5.

is, confessing his sins, and reaching with his hand towards the Crown (Corona): the allegory is not sophisticated, at least in the text of its refutation.) As Tertullian humiliates a Christian convert who wishes to keep up his profession of astrologer,[91] and as Augustine talks of how he himself had been seduced by the mathematics of astrological prediction, and as artistic representations, including famous examples in synagogues,[92] figure the zodiac prominently, so Hippolytus' huge exposition of astrological heresies indicates the anxiety that the continuing purchase of astrology on the community produces among Christian ideologues.

One figure who seems amusedly comfortable in his Christian superiority to astrology is the little discussed Zeno of Verona, a bishop from the 360s, who despite his Greek name writes in Latin, at around the same time as Gregory of Nazianzus.[93] Some ninety-two of his sermons ('Tractates') survive, mostly about Easter, often bitterly dismissive of Jews, but one speech is addressed to a group of recent converts to Christianity.[94] Zeno takes the idea of rebirth literally. Unlike the very recent past, these converts are now 'pure infants free from all guilt'. With a pleasing wit, he imagines that they must be fascinated to know, then, since they are so many, and from such a diverse background, what their 'natal constellation' (*genitura*) is or 'under what sign' (*quo signo*) they are born. And so, as if they were just kids (*parvulis*), he offers to reveal their 'sacred horoscope', that is to say, their 'birth-chart' (*genesis* – like *genitura* and *signum*, a technical term in astrology). Zeno thus starts with a bravura string of smart rewrites of astrological expectations: 'It was not Aries but the Lamb, who received you' – that is, their natal sign is not the Ram of astrology, but the Lamb of Christian symbolism. Each zodiacal sign becomes the opportunity for a supersessionist expansion of the theme of rebirth. Tertullian in *De spectaculis* had made this sort of reading a familiar part of Christian interpretation, with his characteristically overheated rhetoric: 'You want blood?', he demands of those who like gladiatorial games, 'Take the blood of Christ!'.[95] Zeno, in a more gentle but no less triumphalist manner, demands we read the zodiac now through Christian eyes, to celebrate a Christian view of a new type of birth, a new time.

---

[91] *On Idolatry* 9.1.

[92] See Hachlili (2002) for bibliography and discussion; also Levine (2012). *Targum Pseudo-Jonathan* surprisingly explicitly allows mosaics in synagogues provided no obeisance is made: Charlesworth (1977).

[93] See Hegedus (2007) 353–70; and McEachnie (2018) for basic introduction and bibliography to this obscure figure.

[94] 1.38 (Löfstedt).   [95] *De spect.* 29.

The strength and scope of Christian polemic is highlighted by the contrast with Jewish responses.[96] We have already seen how Jewish writers – from Philo and Artapanus to Josephus and beyond, including the Talmud – took Abraham as the inventor of astrology, from which, according to *Jubilees*, he deduced the principle of monotheism.[97] Philo is clear that worshipping the celestial bodies is a gentile error, but he too takes astrology as a normal part of cultural tradition,[98] and there is striking evidence in the Qumran documents and in *Jubilees* of active Jewish involvement with astrology.[99] The *locus classicus* for rabbinical discussion is tractate *bShabbat* 156a–b, which is the closest parallel to the Christian theological debates.[100] This passage is a carefully edited *sugya* which recognizes horoscopes and the influence of the planets on individual lives, but which also insists that a person can avert a bad fate by moral action, thanks to divine intervention. It concludes that there is no *mazel* (luck/fate) for Israel. The argument, that is, not only recognizes the power of astrological horoscopes, especially over the lives of gentiles, but also allows the necessary place of the free will of moral choice. Within the overall principle of Talmudic editing 'to systematize and synthesize earlier halakhic traditions to eliminate contradictions and provide more general and abstract formulations', this passage seems to bring together what might seem contrasting vectors in a conclusion that is 'a type of compromise',[101] and an apologetic: the power of a commitment to halachic living, according to rabbinic regulation, could avert the evil decree of fate.

In contrast to Roman ambivalence, then, which criminalized personal horoscopes, while avidly reading astronomical literature, and in contrast to this cautious Jewish assimilation, Christian ideological objections to astrology stand out all the more clearly. For Christian theologians, the normativity embedded in the invention of Christian time required the excision of astrology. Yet, for all this concerted effort of theological theory, here is one area where even Christian communities failed to follow. Into the Renaissance, astrology in the hands of figures such as Ficino, Melanchthon and Kepler – against the rage of Luther and Calvin – remained part of serious attempts to gain knowledge of the future: where

---

[96] Fullest discussions in von Stuckrad (2000); Leich (2006).
[97] See Reed (2004); Rubenstein (2007).   [98] Taylor and Hay (2012), with further bibliography.
[99] Dimant (2014) 489–97, with further bibliography; Leich (2006). Von Stuckrad (2015) discusses Jewish political uses of astrology.
[100] See Rubenstein (2007), which is sounder than Gardner (2008); von Stuckrad (2000) 460–80.
[101] Rubenstein (2007) 139.

the Christian insistence on the reinvention of time failed in practice to change the fascination with the stars.

*

Nonnus, therefore, writes from within a deep history of a continuing practice of astrology, a continuing representation of the zodiac even in religious surroundings, an extended literary history of the use of astrology especially in epic, and a very active ongoing debate about 'fate', 'the stars' and prediction within the Christian community, including in Cyril, and the Cappadocian fathers, whom he seems to have read most avidly. It is a frame of cultural expectation that nonetheless makes it extremely hard to evaluate the impact of Nonnus' repeated use of astrology within the *Dionysiaca*.

Nonnus is the only ancient literary writer, for example, to describe at length a scene of casting an individual horoscope, a *genesis*.[102] At the beginning of Book 6, Demeter becomes terrified for her daughter Persephone. The girl's beauty, she recognizes, has inflamed the desire of the divinities, and the goddess is frightened about which god might consequently try to assault her; she is especially concerned about the lame Hephaestus (6.14). So, Demeter goes to consult Astraios, 'the god of prophecy', *daimonos ompheontos* (16). Astraios is the father of the winds in Hesiod (*Theog.* 375–80), but in Aratus, the father of the stars (as his name suggests), and hence the god for astrology. Astraios is found with a table covered in dark dust in which he had described a circle with a square inside it with a pointed metal tool, and next to it an equilateral triangle (19–23). He is engaged in the mathematics or symbolics of astrology. After dinner and a dance, Astraios turns to the matter in hand. He gets the *genethlia metra*, 'the numbers of her birth', the time (*chronos*) and hour (*hōra*) of her birth (*archegonos*), and does a calculation with his fingers (59–60). The vocabulary for this initial calculation includes several terms that we have seen to be highlighted in the *Paraphrase* and *Dionysiaca*: his fingers 'move back and forth' (*metatropa*); the number 'recurs again' (*palinnostoio*); the number is a moving 'circle' (*kuklon*) (61–3). The returning circles that we have seen associated in both poems with *Aiōn*, 'Eternity', here are linked with the mathematics that grounds the girl's fate. The suggestion might be that the inevitable flow of time and the inevitable flow of fate are connected by this mathematics of circles.

---

[102] Since Stegemann (1930) 88–100, scholars have concentrated on how detailed or sophisticated the science is – it is clear that it is not especially technical – but have not discussed adequately the striking novelty of the subject itself.

Astraios, armed with these numbers, has a servant bring him an astrological globe, (*eukuklon sphairan*), a device that fascinated ancient science since its invention, ascribed to Pythagoras, and its development by Archimedes. He turns the circle of the Zodiac (*kuklon* again (68)) on its pivot, and consults the 'moving and fixed stars' (69). The 'counterfeit sky turns' (*kukloumenos*) and 'bends its course around the boundless turning post' (*atermoni nussēi*, 71). Again, the language closely echoes the description of Aion, the charioteer, in the *Paraphrase*. Astraios observes an eclipse on his model, and especially focuses on Ares – and Aphrodite, as the question is one of sexual transgression – and finds them linked (75–85). But he also notices the rising of Spica, the Ear of Corn. Thus, with his 'prophetic voice'[103] he predicts both that a deceitful half-monster will ravage Demeter's daughter, and that Demeter will be celebrated for the bounty of crops.

The scene is remarkable not just because of the grandly exotic scene of performed astrology, but also because the process involves divinities and a personal horoscope. On the one hand, in the tradition of ancient epic, with its multiform representation of gods receiving and giving prophecies, and recognizing and acceding to fate, there is no other example of a god behaving like an anxious mother turning to such a science for comfort or knowledge. In the same way, only in the *Dionysiaca* in ancient literature does a god have dreams – deceitful, encouraging, confusing prophecies of the action to follow – like a human.[104] This is not how the gods of epic usually – generically – conduct themselves. On the other hand, while the epic tradition is happy to imagine the cosmos as an ordered whole as part of its politics and teleology, there is no other scene where any individual turns not to divine prophecy but to the far dodgier practice of astrology to gain access to the future. What makes the scene so hard to evaluate, therefore, is the particular combination of exotic grandeur and conceptual bathos. Magic can have its own grandeur and terror, certainly, with Medea, say, or Circe. The attempts of humans to control the world through magic can also play with the bathos of failed and self-deceptive hopes, as with Theocritus' portrayal of Simaetha, the girl deserted by her seducer (*Idyll* 2). Nonnus takes the scene of the anxious and fearful Demeter from the Homeric *Hymn to Demeter*, where she hunts for her stolen daughter in desperate silence and grief, and builds from it a moment

---

[103] See Lightfoot (2017).

[104] Dionysus at 18.169ff.; and Ares at 29.238ff. Plutarch (?) *De facie in orbe lunae* 942a–b in a remarkable image depicts Cronos, bound for ever by Zeus in sleep, dreaming of Zeus' plans, but this rarity is not parallel to Dionysus' human-like dreams.

where the awesomeness of the *divina comedia* – the very aetiology of the seasonal world – is freighted with the less than grand associations of personal horoscopes.

This moment tests the comfortableness of modern critical assumptions of the ease of assimilation between the classical tradition and Christian normativity. Nonnus portrays a scene which draws on ancient epic but which is also unparalleled in its subject and treatment. He is also depicting a subject that Christian authorities have repeatedly railed against, but which in some form continued in society, both in practice and in material culture's representations. Is Nonnus' portrayal a way of marking the poet's or reader's distance from the old gods, gods who need such human measures? Is the display of the practice of casting a horoscope to be taken as the exoticism of the other, a moment of fascinated voyeurism for the sophisticated reader? Is the scene a version of the delight Hellenistic authors could also take in imagining the gods engaged in more trivial human activity, as Eros is often portrayed playing with his friends and cheating – also, needless to say, an image of desire's delusions? Or is it no less or no more concerning than the other tales of divine activity in the poem, where, for example, in scenes that are especially unpalatable for a modern reader, the celebrated hero of the poem, Dionysus, is happy to rape young females, who are stridently committed to their virginity? What, in short, does – should – astrology mean to Nonnus?

This provocation, this challenge seems likely to be experienced by ancient readers too. If the *Paraphrase*, by its very form, entails a triangulated reading between the Gospel, commentary and the new poem in a way which constitutes a transformative cultural intervention, the *Dionysiaca*, a poem dedicated to the transformative god *par excellence*, repeatedly constructs scenes that pick up and transform the stories and imagery of antiquity, its own cultural intervention: how is this past to signify now? This metatropic discourse tests its reader's engagement. Jane Lightfoot, rightly suspicious of syncretism, accommodation, or resistance as adequate models for Nonnus' writing, asks whether 'despite rhetorical elaboration, Nonnus has in the case of prophecy at least, remained essentially true to the postulates of each cultural whole',[105] that is, true to paganism in the *Dionysiaca* and to Christianity in the *Paraphrase*. Yet, not only is this scene of casting a horoscope designedly challenging both to the epic tradition and to Christian expectations, but also for fifth-century Christianity and for the tradition of pagan culture viewed from the fifth

[105] Lightfoot (2016) 643. See also Lightfoot (2014).

century, the integrity of a 'cultural whole' is fragmented into contested arenas of normativity and behaviour.[106] It is because a scene of casting a personal horoscope for a god does not fit easily into the cultural tradition of epic or of Christianity that it poses a question to its readers – a multi-layered question which includes reflections about how a life-time is determined and about the function of prediction – about, that is, the experience of temporality.

The figures of the Zodiac have been present in the *Dionysiaca* as active characters, like Aion, since the opening book. Epics begin with a scene of chaos that needs resolution: the breakdown of civility between Achilles and Agamemnon; the household of Odysseus in turmoil; the storm as a bar to Aeneas' mission of foundation. In the *Dionysiaca*, where a new god requires the reordering of the cosmos, the initial scene of chaos is a battle of the cosmos, where Typhon attacks the very structure of things. In this opening conflict, the signs of the Zodiac including the sun and the moon, are fully involved as combatants. If the celestial bodies in their regular movements are the very signs of order, disorder is expressed at the grandest level when the heavenly signs are violently disrupted.[107] At the start of the fight, the number seven is especially emphasized in Nonnus' astrological numerology. 'Itself answering the seven-zoned sky, the seven-mouthed echo from the mouths of the Pleiads of equal number raised the war-cry, and the planets smashed out a noise of equal measure' (1.240–3). The parallel between the seven zones of the heavens, the seven Pleiades and the seven planets is reinforced by the language of echoing and answering, and the repeated insistence precisely on the equality of number (*isērithmōn, isometron*). This parallelism is also seen at the beginning of the second half of the *Dionysiaca* on the Shield of Dionysus, introduced as 'the starry (*asteroessa*) shield of the sky' (25.352).[108] The first and framing scene described on the shield is the 'seven zones' (25.396) of the heavens; the second is the 'seven gates' (416) of Thebes, which are played into existence by the 'seven-stringed' lyre (428). The shield itself is come to end the 'seven years' (397) of the war against the Indians. By this numerological repetition of the number seven, the narrative time of the war is linked to the mythic time and space of Thebes, home of so many Dionysiac stories, which is made parallel to the cosmological time of the zodiac in this programmatic

---

[106] For 'integrity' as the relevant but misguided normative term, see Lightfoot (2017) 155.

[107] Komorowska (2004) – though her final suggestion that this represents the regrettable violent triumph of Christianity over paganism is too tendentious.

[108] A much-discussed ecphrasis: see Spanoudakis (2014b) with bibliography; Hernández de la Fuente (2011); Miguélez Cavero (2008) 297–300.

ecphrasis. The analogy between the heavens and life on earth, as ordered and mathematically regulated patterns, and the embodiment of this analogy in the narrative of the epic, is foregrounded in the programmatic image of Dionysus' shield. Astrology is invoked to underline how time's order and narrative's order are mutually implicative, equally fated.

It is not by chance, therefore, that the Seasons go to the House of the Sun at the 'fated time' (*chronos memormenos* 11.520) to consult the prophetic tablets of Harmonia. Prophecy is located in the domain of the central figure of the celestial world, the Sun, central on the shield of Dionysus. The tablets are marked by zodiacal signs – the third tablet which predicts Dionysus' patronage of the grape-harvest is under the name of Leo and Virgo (12.39). Similarly, in Book 41, as we have already mentioned, we are told that there are seven tablets, one each named for the seven planets (41.341). Each tablet is engraved in red, by Ophion or Phanes, who is a primal god in Orphic myth, and each contains all the prophecies of what will happen. In Book 12 the oracles that are emphasized concern not only Dionysus, but also a range of mythic stories including Philomela, Niobe, Pyramus and Thisbe, who appear elsewhere in the narrative of the *Dionysiaca*; in Book 41, as we have seen, the prophecies go up to Solon and to Augustus, a political and historical narrative of law-giving. At crucial junctures of the narrative, the prophecies that stimulate or announce the action to come are presented within a set of other prophecies, other stories. Each prophecy marks the interlinked and fated network of foretold actions.

So in Book 7, after Aion has supplicated Zeus to allow the birth of Dionysus, Eros goes on to shoot his arrow of desire into the king of the gods in order that he will pursue Semele to procreate Dionysus in fulfilment of his promise. Eros is described immediately as *sophos autodidaktos aiōna nomeuōn*, 'Clever, self-taught manager of *aion*'. Gigli Piccardi and Vian both translate the final phrase as 'shepherd of life'.[109] It is hard, however, immediately after a scene dominated by the personified figure of Aion, to take *aion* in this unspecific and general sense. Especially when the first two words of the line not only invoke a long history of the representation of love as a smart rhetorician who makes lovers eloquent, but also encourage sophisticated reading. Eros, 'Desire', rather, is here the motivational force in how the long unfurling of eternity is shaped. Desire 'organizes eternity' (where the slide between the personified figure, who has asked for Dionysus' birth, and the abstract, temporal category recalibrated

---

[109] Gigli Piccardi (2003) *ad loc.*; Vian (1993).

by the birth of Dionysus, is significant). Hence Eros goes immediately to the gates of 'First-born Chaos'. Eros – often taken as a primeval force in his own right – goes back to the beginnings of (Hesiodic) time, where he has twelve arrows in a quiver, each inscribed with a foretold name of one of Zeus' mortal lovers. They run from Io all the way to Olympias, the mother of Alexander the Great – from the mythic world of the oracles of Book 12 to the historical world of the oracles of Book 41. From Io to Alexander, time's narrative is ordered by Eros' arrows.

The arrows of desire, it would seem, drive time forward in a sequence of twelve to match the months of the year. Yet for all these indications of order and regulation that the figures of astrology provide to the discourse of the *Dionysiaca*, there is still the destabilizing effect of the multiple points of origin and the multiplying and transformative stories of myth. Eros here may be the 'organizing driver of eternity' but his status as primeval, originary force is shared with many other claims to be the first born, including Chaos, Aion, Ophion, Phanes, Beroe ... This unresolved tension between the always already of foretold mythic inevitability, and the multiplying, contrasting and even chaotic stories of origin, or the contingent, mutating, stories of myth, runs throughout Nonnus' narrative.

So what does astrology mean to Nonnus? The extended use of astrological discourse in the *Dionysiaca* first of all suggests, then, that time is ordered. The regular movement of the stars, the sun, the moon constructs a model for a determined cosmos. The language of *kukloi*, 'circles', associated with Aion and other figures of time, therefore links the representation of time as a figure into the model of the cosmos as a regulated pattern of movement, which is itself a principle of Stoic philosophy, but also expressed through echoes of Orphic cosmogony in particular. Dionysus may be the god of transformation, and the narrative may relate Dionysus' ascent to heaven and the changes he brings to human life, but the temporal framework is one of threatened but maintained cosmic order. Second, however, the use of astrological discourse to predict the future, tied as it is into other forms of prophecy by the narrator's voice or by internal figures of the narrative, helps inform the narrative itself as determined, the always already of myth. As we will see in the final section of this chapter, the notion that stories are not merely determined but repeat each other is integral to the typological redrafting of the narrative. Yet, thirdly, despite this overdeterminism of astrology, Nonnus sets such predictive certainty in tension with multiple versions of stories, stories that are transformed into new and surprising accounts, and multiplying points of origins, multiplying claims to be the 'first-born'. The swirl of narratives, blending into each other, without regard for chronological order, repeatedly threaten the directed purposiveness of the narrative, extending and twisting the story round again on itself. If the *Odyssey* is the archetypal narrative of *nostos*, the *Dionysiaca* tells its story under the sign of *palinnostos*, 'turning back on itself'.

Nonnus' astrology, then, is integral to his temporal discourse and to how he explores the inevitability and contingency of his storytelling. But his astrology remains hard to relate simply to the dominant normative discourses of fifth-century intellectual life. Astrology is dismissed by authoritative religious intellectuals as a travesty of moral judgement and a false science, if it is used for the personal predictions of horoscopes – and here it is a pagan god who uses astrology; but the science of observing the heavens is respected, not least when it testifies to God's order in creation. And it is clear that the use of astrology never stops in the Christian community. Whether or how the pattern of the stars should be part of a Christian understanding of time's order remains a fight between intellectual authorities and their communities for centuries to come – much as predestination, original sin and moral responsibility remain an integral part of Christian polemics. Is the experience of a life-time as it must be or as it can be? Nonnus' astrology does not fit easily either into the generic expectations of epic – how can an immortal god need a horoscope? – or into the normative expectations of Christian narrative – no-one should use a horoscope! Perhaps Nonnus' bold representation of astrology is best seen as something of a provocation, then – a potential flashpoint in the border wars between paganism and Christianity. What astrology means to Nonnus, that is, is not to be answered with a simple declaration, but recognized as an ongoing question, a question not just about *how* Christian or *how* pagan Nonnus' poem is taken to be, but also about how determined and how fluid are the narratives of past and future time.

*

The presence of the future in the past is strikingly on display when Zeus, as king of the gods, summons his supporters to aid Dionysus in his battle against Deriades and the Indians. Zeus' rhetoric echoes with the *epitaphioi* of classical Athens, the funeral speeches over the war dead that repeatedly laid out the importance of past military exploits for the soldiers of today (as we saw with Demosthenes' version).[110] *Mnōeo*, he repeats (27.254, 263, 285), 'remember', 'remember', 'remember' – an injunction that also draws

---

[110] See above, pp. 142–3.

attention to the literary tradition he is invoking. But the gods are being asked to remember what has not yet happened. Athene is asked to repay a favour to Icarius, the first in Athens' territory to receive wine, who in the chronology of the mythic narrative of the epic is not met until some twenty books later (47.34ff.). What is more, Athene is also asked to recognize the history of centuries *after* the time of the poem. She is asked by her father to preserve Pan, 'the future helper in Attic battle; preserve the preserver of shaken Marathon, the killer of Medes' (27.299–300). The *epitaphios* paradigmatically used Marathon as a spur to encourage Athenian virtue, so here Zeus uses Marathon and Pan's role in it as a spur for Athene to fight and as an example of glory – but for the future. This typology allows, indeed insists on, the reversibility of chronology and a spreading of exemplarity, which melds story into story as exemplars or contrasts of each other.

So Aiacos, the grandfather of Achilles, has his *aristeia*, where he fights, the narrator announces, *ouk atheei*, 'not without god' (22.384). This is a significant introduction. For Aiacos is described immediately as fighting in a river, 'as he is the father of Peleus', a watery battle which 'foretells the half-finished (*hēmiteleston*) fight to come for Achilles by the flow of the river Kamandros; the clash of the grandfather prophesied the clash of the grandson' (22.386–9). Odysseus at the end of the *Odyssey*, as the fight with the suitors' relatives erupts, stands shoulder to shoulder with Telemachus and Laertes, and Laertes notes with joy that the three generations of single sons are competing in valour, an image of the achieved patriarchal triumph of the household. In the *Iliad*, Priam appeals to Achilles to recall his father, as Odysseus in the *Odyssey* is asked by Achilles to tell him about Neoptolemus, his son. As Odysseus provides the iconic image of the household saved, Achilles' young death represents a negative picture of normative generational succession. Here, however, we have three generations of the family of Achilles, but only in the text's representational strategy: the here and now of the narrative is replete with the literature of the past that is yet to take place in the time of the epic and thus three generations stand, as it were, shoulder to shoulder in the river of poetry. A hero's battle is made to foretell (*prothespizein*) and prophesy (*manteuesthai*) the generations to come. Where Pindar is happy to imagine that the triumph of a victor in the games embodies the breeding and inheritance of the generations of past Aeacids, Nonnus reverses the flow of time, and has the grandfather embody the 'prophecy' of generations yet unborn.

Achilles' fight in the *Iliad* will be with the god of the river (as his father Peleus 'fights' with Thetis, a watery god, in his tempestuous marriage: hence the loading of *ouk atheei*), and the fight is termed here *hēmiteleston*, 'half-finished', 'half-fulfilled', a typically Nonnian word, freighted with implication. On the one hand, it invokes the battle in the *Iliad* where Achilles cannot defeat the river and has to pray to the gods for help: it is a battle that does not end in death or defeat. Hence, 'half-finished'. On the other hand, the river here does not complain of being choked with bodies, nor does the river fight with Aiacus: the fight of Achilles only half fulfils this prequel. When the river does rebel, it is against Dionysus not Aiacus, and threatens a cosmic disruption which requires Zeus' intervention to quell it (23.162–24.67) – where the river begs for mercy from the flame-throwing Dionysus. The 'burning star' Achilles was swamped by the river; Dionysus' flames, and his rhetoric, threaten to scorch the river into submission. The complaint that does arise here is rather from a Naiad who begs Aiacus not to spread so much blood in the waters. She appeals to him as the son of Aigina: 'For your Aigina I hear was the daughter of a river' (22.396). 'I hear', *akouō*, as often, marks the inheritance of the mythic tradition, as the scene deepens its grounding in the stories of the past, and lengthens Aiacus' genealogy backwards as well as forwards (his grandfather, the river Asopus, has been represented on Aiacus' shield and will appear again in the narrative). This Naiad will not fight but flees to another pure (*akēraton*) water, the sea, where Thetis will receive her (23.399–400). The reference to Thetis is pregnant. Thetis will become the daughter-in-law of Aiacus, the mother of Achilles; she has already received Dionysus in her waters. As the book ends, the Nymph's language evokes the interconnected stories of past, present and future. Each fight is seen as layered with fights past and fights to come, images and foretellings of each other.

It is a fascinating background to this half-fulfilled shadow of Achilles' future that it may perhaps also recall both theoretical writing on typology and an earlier cento, which embodies the possibilities of how past verse can be reformed into a Christian present. On the one hand, Cyprian, writing about the practice of typology, declares – directly enough – that any mention of water in scripture can represent a baptism (*Ep.* 63.8.1–9.1), and such baptismal imagery is prevalent in early Christian texts and art.[111] By the later fourth century 'these typological tropes were commonplace'.[112] On the other hand – and more pertinently – Eudocia, the empress who was exiled to Jerusalem, and herself baptized in the Jordan, when she turns in her cento to represent the baptism of Jesus, does so precisely by reutilizing lines from Achilles' struggle with the Scamander. Her depiction (447–61) includes three different lines from *Iliad* 21 – where Achilles steps into the

---

[111] Jensen (2012), especially 149–60.   [112] Jensen (2011) 35.

river (21.8), the river hides the Trojan bodies (21.239) and Hephaestus stops burning the river (21.382) – as well as two other lines about the Scamander from the epic (14.433; 16.679). The lines are (now) full of 'symbolic descriptions that would resonate for a Christian audience', especially of a soteriological tenor.[113] As Achilles in the Scamander can be rewritten as a baptismal *tupos*, so here Achilles' grandfather becomes a figure to image the buried future present in the past.

The language of *tupos* occurs frequently in the *Dionysiaca*. Cadmus founds the city of Egyptian Thebes as 'an intricate earthly type, equal to Olympus', *poikilon chronion tupon* (5.87), evocative language for a Christian who knows Augustine's *City of God*. Blood on an altar is the 'type' of wine (11.93) – again evocative – as wine is also the 'the earthly type (*chronion tupon*) of heavenly ambrosia' (12.159). When a Lydian warrior is richly apparelled he is 'a type of Glaucus', the Homeric warrior famously adorned in gold armour, where the literary echo is also a foretelling of a time to come (22.147) – a scene famous for the swapping of genealogies as much as the exchange of armour. Characters are models of each other, and even genealogy can become *palinnostos*, 'twisted back on itself'. When Cadmus explains his genealogy, he starts from Zeus' pursuit of Io. Io gives birth to Epaphus; Epaphus gave birth to Libya, and Libya gave birth to Belus, 'Libyan Zeus', that is, Zeus Ammon whose prophetic shrine is in the Libyan desert by Asbystes (3.291–3). Zeus is the grandfather of . . . Zeus.

This typological morphing of identities is nowhere more pertinent than with Dionysus and his precursor Zagreus. Zagreus is the baby of Zeus and Persephone, whose horoscope was cast in Book 6. His life is short, as he is ripped apart by Titans, motivated by the jealous Hera. But Zagreus also becomes Dionysus. The description of this process is quite extraordinary (6.172–6):

Ταρταρίη Τιτῆνες ἐδηλήσαντο μαχαίρῃ
ἀντιτύπῳ νόθον εἶδος ὀπιπεύοντα κατόπτρῳ.
ἔνθα διχαζομένων μελέων Τιτῆνι σιδήρῳ
τέρμα βίου Διόνυσος ἔχων παλινάγρετον ἀρχὴν
ἀλλοφυὴς μορφοῦτο πολυσπερὲς εἶδος ἀμείβων, . . .

The Titans destroyed him with a hellish dagger
As he contemplated his counterfeit form in a reflective mirror.
There and then as his limbs were split with the Titans' iron,
The end of his life Dionysus had as a returning beginning,

---

[113] Lefteratou (forthcoming) ; she also notes that Tertullian's *De baptismo* includes a string of 'types' from scriptural sources.

He changed shape into another nature, and transformed into many forms . . .

It is unclear what the myth evoked by the scene of looking in a mirror is, but it is salient that this introduction projects the idea that the form (*eidos*) of the baby is *nothos*, 'counterfeit', 'made up', and that the mirror is a revelation of the *antitupos*, 'the counter-type'. For as Zagreus is 'split', Nonnus stunningly transforms the subject of the sentence from the expected 'Zagreus' to 'Dionysus': it performs the transformation in a word. In Zagreus' end is Dionysus' beginning. It is a beginning which is *palinagretos*, which, as Emma Greensmith has sharply analysed, is a paradox of genealogical rebirth, which stands against the frequently repeated story of Dionysus' birth from the thigh of Zeus. Zagreus/ Dionysus, to avoid the Titans, changes shape (*morphouto*) and form (*eidos*). Dionysus is not yet born but comes into being as the subject of change. In the following lines he becomes his father Zeus, as a tricky young man (177), then his grandfather Cronos (178), weighed down with age – as if, as with Aeacus, three generations are present in one body. He becomes a lion, a bull, a serpent too – but to no avail, as the Titans finally 'slaughtered the bull-shaped Dionysus' (205). Zagreus has become Dionysus and is killed as such. Zeus is furious that the 'earlier (*proteros*)[114] Dionysus had been slaughtered' (206), and it is his rage at the murder that leads to the flood with which we began this chapter.

Greensmith has demonstrated with great acumen how this story grounds a set of images of resurrection which play the Dionysus story off against the Gospel narratives, most obviously with the long tale of Tylus on the shield of Dionysus, and with Ampelus, Dionysus' beloved, who comes back as the vine (*ampelos*) – which, as we have suggested, links his miracle of rebirth further to the language of Jesus as vine in the *Paraphrase*.[115] The same language – the *terma biou* leading to *palinagreton archēn* – also notably appears in the resurrection of Lazarus in the *Paraphrase* (11.164) – again suggesting that these figures of reincarnation are versions of each other, morphing through the different narratives, which are themselves versions of other poetic and religious accounts: 'Nonnus creates a "hybrid of hybrids." Dubious Orphism and illicit Christianity, Homeric lexica and Callimachean allusion: Zagreus contains within him multiple different parts.'[116] The 'counterfeit form' of Zagreus has thus also been taken as an

---

[114] It is not clear why Gigli Piccardi, like most scholars, translates *proteros* as 'first'.
[115] Greensmith (forthcoming), building on Spanoudakis (2013) and Kroll (2016). See also Bernabé (2008).
[116] Greensmith (forthcoming).

image of Nonnian poetics.[117] Yet for my purposes here, what is most pertinent is the transformative nature of the poem's central divine hero, who not only changes shape and changes the world around him, but whose very nature has its 'before' in Zagreus and, by the end, has also a future in the figure of Iacchus. Iacchus is the child that results from Dionysus' rape of drunken Aura and her horrific torture by Artemis. The boy is called Bacchus after his father, and Iacchus is honoured as a god in Attica with Zagreus and Dionysus. 'They established sacrifices for late born Lyaios and first-born (*archegonos*) Dionysus and sang out a new hymn to Iacchus third' (48.964–5) – that is, to Lyaios, who is Dionysus, to Zagreus who is called Dionysus, and third to Iacchus, who is also Bacchus. 'The citizens beat out their choral dance late-fulfilled in honour of Zagreus with Bromius and Iacchus' (48.967–8). The three gods, with shifting names, are interconnected in cult – versions of each other.

A poetics of typology is one frame for understanding Nonnus' use of analogy, proleptic and retrospective figuration – how stories blend into each other, offer different versions of each other, anticipate the future and fulfil the past. We have already discussed typology as a contested mode of interpretation among Christian theologians and translators.[118] We can now see how typology as a mode of thinking can shape a mythological narrative, providing a structure of comprehension and expressiveness for its storytelling. The lack of stable, linear narration, as stories and genealogies turn back on themselves, and the apparent confusion or overlap of figures in the poem – stylistic features that have often dismayed classical critics – are features of this poetics of typology. For Nonnus at least, this is one way that a specifically Christian mode of thinking has informed the epic's narrative temporality.

Both the *Paraphrase* and the *Dionysiaca*, for all their evident differences of tone, subject matter, and perspective on the relation of man and god, share some fascinating trajectories. Both are focused on transformation, the transformation of the cosmos, the transformation of individuals, and the transformation of poetic language and poetic form – what I have termed metatropic poetics. Both make interventions in the culture of Christianity: the *Paraphrase* by redrafting the *koinē* of the Gospels into the sophisticated, poetic theology of a post-Chalcedon Christian thinking; the *Dionysiaca* by redrafting the myths of pagan tradition into a new, provocative vision of what the pagan past can say to the educated reader. Both thus are committed to changing their readers in their cultural

---

[117] Greensmith (forthcoming), criticizing Shorrock (2001) 116–21.   [118] See above, pp. 107–12.

embedding. Both poems inhabit the narrative space of a not yet fulfilled religious vision – stories of becoming recognized as (a) God – expressed from within the 'always already' of an achieved system. The fascination with the language of fulfilment – half-fulfilment, non-fulfilment – in both poems mirrors the reader's narrative journey towards this foretold end. In both epics, the language of temporality – and its embodiment in narrative – undergoes its own transformation, not least by the newly conceptualized obsession with eternity, God's time. The vision of time in John's Gospel is paraphrased into a more expansive and nuanced theological apparatus, just as in the *Dionysiaca* the teleology and foundational order of epic tradition is framed both by an astrology, a cosmic patterning, a typological poetics, *and* by a multiplication of competing stories of origin, a twisting, swirling overlap of myths, *and* an explosive semantics of temporal language, pushing poetic expression to a limit. Eternity through the figuration of Aion, origin stories, and the technology of time as expressed through astrology – crucial elements of time's redrafting by Christianity – are flamboyantly imaged through Nonnus' metatropic, Dionysiac poetics. It is unnecessary (if tidy) to distinguish the two poems as representing two poetic stances – 'the poet of the Muses' and the 'poet of Christ':[119] in both poems, Nonnus provides the most vivid testimony of what the Christian invention of time can do to the narrative of epic.

---

[119] Shorrock (2011), who is more open to the overlap between such stances than some who have adopted his critical vocabulary.

CHAPTER 13

Regulation Time
Gregory's Christmas Day

Gregory of Nazianzus also had stars in his eyes. Some seventy years before Nonnus depicted Demeter having a horoscope cast for Persephone, in both sermons and theological poetry Gregory was setting the agenda for a Christian understanding of astrology. It was a project that deeply influenced the discourse of Nonnus' epics, not just on the stars but also on temporality. For, in Gregory's hands, astrology opened a broad vista onto the theological conflicts of the post-Nicaean era, and the active heritage of Greek learning in Christian culture. The collection of hexameter poems known as the *Aporrhēta* in Greek (and *Poemata Arcana* in Latin) – the 'ineffable', or 'secret' matters – takes on astrology as a subject for debate along with such weighty topics as the Cosmos or the Soul, which gives an immediate indication of astrology's importance in Gregory's thinking; but it is first in one of his Theological Orations, the sermons which were one primary cause of Gregory being known in Byzantium simply as 'The Theologian',[1] that we will trace the significance of looking at the heavens for Gregory's understanding of how a Christian inhabits time. How should the regularity of the movement of the celestial bodies be related to the regulation of a Christian life? This extended discussion of the circles of time will lead, however, to another of Gregory's sermons, delivered on a Christmas day, about how to celebrate Christmas, the moment when, thanks to the star that led the Magi to the manger, astrology was, declares Gregory, defeated. To understand Christmas, to *experience* Christmas, Gregory explains to his congregation, is to see yourself in the framework of a Christian comprehension of temporality. It is towards this fascinating recognition that celebration requires a theology of time that this chapter travels.

Gregory of Nazianzus, like Gregory of Nyssa and Basil, is insistent, confident indeed, in the inability of human language to express what God

---

[1] A title probably stemming from Theodoret at the Council of Chalcedon and one shared with the author of John's Gospel: see Langworthy (2019).

is. His Second Theological Oration (*Or.* 28) is one of the greatest statements of the limitations as well as the soaring hopes of this apophatic theology. Its opening image is an extended allegory, based on the revelation in Exodus, of the theologian climbing Mount Sinai, leaving behind the material world of the Israelites at the foot of the mountain, to enter the cloud and converse with God. But even as the theologian confidently enters the cloud, he is faced by the necessity of seeing only God's back from behind a cleft in the rock: he can only see *gnōrismata* 'tokens', which are 'like shadows of the sun on water', 'images for our corrupt sight', *eikones tais sathrais opsesin* (3.15–16). It is an image he shares with Gregory of Nyssa, who also makes Moses the type of the striving and incomplete knowledge of the theologian.[2] Yet Gregory of Nazianzus extends this idea into a detailed epistemology of fleshly and conceptual failure of knowledge which itself becomes part of a statement of faith.

Gregory 'runs to grasp', *katalēpsomenos* (3.2), the ungraspable essence of divinity. As we saw with Nonnus' *Paraphrase*, and Cyril's commentary on John, the ungraspability and ineffability of the divine becomes a watchword of the nature of God – (*kata*)*lambanein* in its many forms is repeated throughout Gregory's sermon as the impossible hope of human understanding.[3] This unfulfilled striving takes two trajectories. On one level, the failure is a question of language and thought.[4] It may be hard to conceptualize (*noēsai*) God, but it is impossible to express (*phrasai*) God (4.1) – though Gregory will quickly qualify this with actually 'it is more impossible to conceptualize him' in the first place, as any concept can be at least darkly expressed in language. The tension dramatized here is played out in multiple ways through Gregory's sermon. God is uncircumscribable – infinite – yet is not 'comprehension', *katalēpsis*, precisely a form of circumscription? (10.21–2). Any adjective indeed will miss the point. Yes, God can be called incorporeal, but 'this term "incorporeal" neither sets forth nor contains his essence, any more than "unbegotten" (*agennēton*),[5] or "without origin" (*anarchon*), or "unchanging" (*analoiōton*) or "incorruptible" (*aphtharton*) or anything else said on the subject of God or about God' (9.4–7). Such negatives give no access to the positive essence of divinity. It is, as Gregory sharply puts it, like answering the question 'what is five plus five?'

---

[2] See Greg. Nyss. *Vita Mos.* 2.152–66, and on it, Frank (2000b) 86–96 with general background of Vassilopoulou and Clark (2009).

[3] See Beeley (2008) 80–91; Ayres (2004) 282–301: 'Pro-Nicenes universally assert that God's nature or essence is incomprehensible' (282).

[4] Meinel (2009) 75–8.

[5] The definition of God given by Eunomian Arians: see Norris (1991) 53–68.

with the answer 'not six, not seven'. Yet each positive term in turn captures only a part of God, hence the definitional process is itself infinite, and continually partial. Again and again, Gregory reverts to the failures of human conceptualization and the insufficiencies of human language to grasp God. As Basil in his *Contra Eunomium* also shows, this insistence on ineffability is a polemical rejection of Eunomius' Neo-Arianism, which confidently asserted the knowability of the divine.[6] *Ta aporrhēta*, what cannot be said, is integral to Christian comprehension. As Augustine famously declares (*Serm.* 117): 'If you comprehended something, it is not God'.

At another level, this failure is because of humanity's very bodily existence. Even for the lofty lovers of God 'this darkness' and 'this fleshly thickness (*pachu sarkion*) blocks comprehension of the truth' (5.13). 'This fleshly thickness' (*pachu sarkion*) is like the cloud that came between the Israelites and God, a sign of what Jeremiah calls 'the bonds of earth' (Lam. 3.34).[7] Gregory seems to have been instrumental in the development of the language of *pachu*, 'thickness', 'fatness', which barely occurs in the Bible, to express how the physicality of human existence is itself a bar to reaching towards the spiritual.[8] As humans are marked by this thickness, God, the incorporeal, is *exō tēs pachutētos*, 'outside thickness/fleshliness'.[9] As the more extreme ascetics mortified the flesh to transcend their fallen state, so Gregory constantly reminds his listeners that the condition of humanity requires a constant redemptive effort to recognize and attempt to transcend their necessary materiality.

One consequence of humanity's fleshliness is that human sight, flawed as it is, fixes on the physical. As the Psalms say (Ps. 8.4), 'I will consider the heavens, the work of your fingers, the moon and the stars' (5.13). At best, this vision leads to a recognition of God's authorship. Gregory lyrically declares the argument from design:

> that God exists, our sight is the teacher along with the law of nature. Our sight, because it falls on visible objects, that are beautifully stable and moving on their way, and, so I may say, immovably moving and revolving; the law of nature, because it reasons back to the author (*archegonon*) from what is seen and their order. (6.1–7)

---

[6] Beeley (2008) 91–101; Norris (1991) 53–68.

[7] The phrase in Lamentations, *asireh eretz*, means 'prisoners of the land' in a military context, but is turned by Gregory into an image of 'earth-bound' humanity, which, as we will see, is typical of his discourse. The Septuagint has *desmious gēs*, which may help Gregory's understanding.

[8] As discussed in Goldhill and Greensmith (2020), this specific language looks back to Origen. See also Beeley (2008) 81–4; and on the Spiritual 154–85; and on Origen, *passim*.

[9] *Or.* 40.45. Hence, in the incarnation what is taken on is precisely *pachutēs*: *Or.* 29.19.

The wonder of contemplating the heavens with their regularity of motion, for Gregory, is a foundational moment. But it is dangerous to stay in the moment of awe. This is the cause, declares Gregory, that some people think that the sun, the moon and the stars 'rule everything according to the quality and quantity of their movement' (14.1–2). To hymn the sun or to allow astrology's determinative power is an epistemological failure based on a misguided, corrupted human materiality.

Gregory circles back to his attack on astrology in the sermon's concluding paragraphs. He has led the audience through the multiform evidences of the natural world, from peacocks to mountains, and now reaches the heavens. He demands now that Faith, *Pistis*, rather than Reason, *Logos*, is to be their guide, because in going beyond mere Reason, they can transcend their completely earth-bound nature (28.41–4).[10] Aiming to grasp the heavens, that is, is both a search for the spiritual, and a corollary resistance to the earthly, which includes both the terrestrial world left behind in turning to 'consider the heavens', and the materiality of the human condition, the basis of their ignorance, 'the bonds of earth'. With such a programmatic introduction, Gregory's argument is an intensely charged, rhetorically flamboyant dismissal of astrological science. 'Who gave the sky its circular movement, who ordered the stars?', he begins. 'Before these questions, you lofty thinker (*ho meteōros*), who is ignorant of what is at your feet, and cannot measure yourself, can you tell me what the sky and stars are?'. This aggressive opening flourish combines the moral and intellectual failures of the scientist with the inability to answer the most basic questions about the subject of his science.[11] With a wonderful dismissiveness Gregory grants the scientist his understanding of orbits, and periods, and waxings and wanings, settings and rising, and degrees and minutes – the full panoply of the astrologer's technical vocabulary – only to conclude 'This is no comprehension (*katalēpsis*) of reality, but an observation of some movement'. Gregory takes the very essence of astrological science, its mathematical observation, and disparages its mere empiricism ('*some* movement' is an especially nice sniffiness): understanding reality is a different procedure, which the Christian theologian is arrogating to his own knowledge, theology.

---

[10] On faith see *Or.* 6.7; 14.33; 22.11 and especially the discussion in 32.23–7. Norris (1991) 126–9 makes the appeal to faith central. Florensky (1997) 41 describes Nicaea as where 'rationality was given a death blow' – an account criticized by Ayres (2004). On Gregory's dependence on Aristotle, see Norris (1991) 17–39 (and e.g. on Clement see Clark (1977)).

[11] On the figure of the astrologer, see Hübner (2020).

The astrologers, continues Gregory, have firmed up their measurements by collecting together many such observations. This generalizing they 'consider an explanation (*logos*), and name a science (*epistēmē*)' (29.11–12). But, he rounds on them, 'If you want our admiration, tell us what is the cause of this order, this movement?'. This demanding question opens into a sort of hymn to the sun, a beacon-fire to the whole world, a leader of the chorus of heavenly bodies – a hymn of praise which pivots, surprisingly, around a quotation from Plato – called 'one of the pagans (*allotriōn*)' – that 'the sun has the same position among perceptible objects as God does among intelligible things'.[12] But, repeats Gregory as his speech becomes more incantatory, 'What gave the sun its movement from the beginning? What is it which keeps him always moving, and circling round, and fixed in order (*logōi*) and unmoving, in reality tireless, and life-bearing and life-giving, and all the things which the poets hymn with reason (*kata logon*)' (30.5–8). As he gets more poetical, he self-consciously notes his adoption of poetic form. The obvious and required answer to Gregory's questions is repeatedly left for the audience to fill in – we are to be complicit with his disparagement – until with a final scornful turn he asks whether all this measuring and naming is in order that 'I should trust you, [the astrologer,] when through this measurement you weave/plot our lives and arm creation against the creator?'[13] As Augustine reverts again and again to the act of creation to understand human temporality and our bafflement, so Gregory turns the very event of creation against astronomy's claim to regulate time and life.

Thus he begins the final paragraph of the speech, self-consciously marking the transition to his conclusion, with 'What do you say? Will we stop our speech here, with materiality and the visible?' (31.1–2). Rather – no surprise – he concludes with a typological reading (*antitupon*, 31.3) of the Tabernacle, and his desire to go beyond the veil, and perceive the world which is made up of visible and invisible, material and spiritual, all a product of God's will, God's creative force, which is the only true object of worship, for man and angels alike. As he began with the image of Moses climbing Mount Sinai as a model for the theologian's striving, so he ends with Moses' tabernacle as an image of the veiled immaterial truth which is the object of the theologian's striving.

---

[12] A paraphrase of Plato, *Rep.* 6.508c – where the subject is the Good rather than God.

[13] The phrase *plekonti ta hēmetera*, which I translated as 'weave our lives' (Gallay (1978) translates 'tu règles notre sort': 'you rule our fate') is best understood through the 'composition of things' as is evidenced in *Poem. Arc.* 4.21; 4.36; 4.40.

This Second Theological Oration of Gregory is titled 'On Theology', and unlike the First Theological Oration, which is an attack on Eunomius, and the third and fourth which also criticize Arian and other misunderstandings of the Son, this speech is aimed not so much at infighting between Christian thinkers, as at promoting a theological view of the world, a Christian *Sophia*. In particular, its discussion of the essential theological question of the essence of God determines that human failures of comprehension stem from the materiality of language and the materiality of humanity's fleshly, earth-bound, physicality, an argument which does tie this sermon together with the other, anti-Neo-Arian diatribes. Flesh is an epistemological issue. Astrology, as a science that is competitive with theology, is therefore also stridently denigrated because its claim to measurement is no more than a sort of material empiricism, and, as such, not only ignores the primary question of creation and causality, but also, unlike theology, cannot lead humans to transcend the earth-bound limitations of their knowledge.

We saw in the last chapter that astrology, regularly contested as a science as well as regularly adopted as a form of prediction as well as description, was particularly worrisome to Christian thinkers. For Augustine, whose strenuous denials of astrology also constantly remind us of his former commitment to it, astrology was to be rejected both on moral grounds because it denied free will, and on scientific grounds because it could not defend its claims to predictive rigour. Gregory of Nyssa in his *On Fate* makes these same cases in his dialogical polemic against a pagan *mathematicus*. He describes with a certain rhetorical flair the collaborative work of building a ship – from the wood cutters to the nail-hammers – and the formation of a crew of sailors, and then points out that if the ship goes down, it would seem that a whole team of people – and objects – with quite different birth dates and thus horoscopic predictions nonetheless meet the same fate. When – at what precise point? – did the influence of heaven become instrumental in such a multiform collaborative project? As we will see, Gregory of Nazianzus is fully aware of such arguments, which he too can marshal when required.[14] Yet in this oration he is making a different sort of claim (one not recognized by Hegedus' seminal study).[15] Gregory wants us to think about the order and regularity of the heavens differently.

---

[14] It is striking that Bowen and Rochberg eds. (2020), despite an advertised date range of 250 BCE to 750 CE, including a chapter on Christian astrology (Denzey Lewis (2020)), and 750 pages of exposition, do not discuss Augustine, Gregory of Nyssa, Gregory of Nazianzus, Nonnus and several others discussed here.

[15] Hegedus (2007). Gregory of Nazianzus gets short shrift in what is now justly the standard study.

It is necessary to see in the sun, the moon and the stars the hand of God, and he repeats his happy paradoxical determination that theirs is a fixed movement, a stable set of changes. Yet such a recognition of the relation between the celestial bodies and the flow of time is salient only if it allows the Christian to move beyond the material, beyond such empirical, physical sightings, and recognize the limits of human, earth-bound scientific knowledge, the necessity to strive for the spiritual, and, above all, to grasp, as best as possible, the will and majesty of God in and through creation – which is a matter of faith not measurement. 'The textual blank left by paradox is filled by faith'.[16] Where Augustine suggests that accurate measurement is an impossibility for astrology, as if with accurate measurement things might be otherwise, Gregory is clear that the desire for accurate measurement is itself a misplaced acquiescence in the fleshly thickness of human knowing, its failure to grasp what reality truly is. 'How will you ever grasp (*hupolēpsēi*) the divine, if you put all your trust (*pisteueis*) in the method of logical analysis?' (7.1–3). Gregory's carefully reasoned – and occasionally poetically evocative – plea for faith rather than reason to be the guiding principle of a Christian life thereby constructs astrology as the iconic counter-science to theology.

*

Gregory returns to astrology in his poems known as the *Aporrhēta*, 'Ineffable Matters' (a title used at least since the Byzantine manuscript which includes the poems and a commentary or paraphrase of them by one Nicetas David Paphlagon).[17] These eight poems seem to have been composed as a set.[18] They are in hexameters and announce themselves as if they are hymns: 'First we will sing (*aeisomen*) the Son', (2.1); 'Sing (*aeide*) praise of the Spirit' (3.1); 'let us hymn (*humneiōmen*) creation' (4.1), a 'framing of the poems as "hymnic prayer"', therefore, which contributes to the 'validity' and performativity of the poems' 'didactic content'.[19] As with the Theological Orations, the poetry's polemical engagement with Eunomian Neo-Arianism leads to a broad range of critical argument – from epistemology

---

[16] Meinel (2009) 89.

[17] Christos Simelidis has pointed out to me (*per litteras*) that the *Life of St Gregory of Akragas* from the 8th/9th century is the first reference to the title – in an extraordinary tale where a visiting monk is asked to read and interpret 'The Theologian's *aporrhēta* poems' as a test of his intellectual excellence.

[18] See Norris (1991). Gregory was the first to publish an edited collection of his own letters in Greek, and there is an argument to be made that Book 8 of the *Palatine Anthology* is also an integrated volume: Goldhill and Greensmith (2020). He curated his output and his self-representation more purposively than many others.

[19] Meinel (2009) 86, who rightly criticizes Keydel's description of them as failed didactic epic. Faulkner (2010) cautiously nuances the claim of Daley (2006) 29 that the hymns imitate Homeric hymns.

to theology to astrology – but the form of hexameter also marks Gregory's continuing contribution to the transformation of Christian elite culture. It is not so much that different audiences are projected for the orations and the poetry: in both cases, there is a staging of performance, sermon or hymnic, and an embrace of a complicit congregation along with a corollary dismissal of wrong-thinking opposition. But this poetry engages with the tradition of Greek cultural heritage differently – especially hexameter poetry with its Homeric and religious associations – and allows a different expressivity, which, as we will see, seems also to have been formative for Nonnus.

The fifth poem is titled *Peri Pronoias*, 'On Providence'. Its first word is ὧδε, 'Thus', which both looks back to the previous poem on the Cosmos and forward to the summary description of the creation of the world which opens this poem: an internal marker of the collection as a collection, its continuity. The Cosmos, declares Gregory, was set on its foundations by 'the infinite great mind', and 'with an initial impetus (*rhipē*), as a top is urged into whirling by a whip, it was set in motion by the great unmoving design (*logoisin*)'. As with the Theological Oration, the paradox of 'moving according to an unmoving sequence' grounds the description of the celestial bodies. There can be, he argues, nothing 'automatic' – in the strong sense of 'self-moving' – about such an order, any more than a house, a ship, a chariot – the familiar examples of the argument from design – could be built without hands (Nonnus' fascination with the self-moving in the *Dionysiaca* has this precedent). Nor could the cosmos have existed for so long if it were *anarchos* (12). *Anarchos* here means, it seems, without a director, or without an originating principle (*archē*), as it is immediately glossed by the image of a chorus without a leader (*anhēgemoneutos* (13)). But as our reading of the collection continues, it will become clear that the term can also mean 'without an origin' (*archē*). There is an argument, of course, whether the opening words of the Gospel of John mean in the beginning in a temporal sense, or in the beginning as an originating principle. It will turn out to be crucial to Gregory's understanding that there is no temporal beginning but there absolutely is an originary moment of creation: the language of *archē* and its negation marks this tension.[20]

This description of the celestial order leads directly into the opening assault on the astrologer who thinks the stars are the 'guides (*hēgemonēas*) of our birth and, at the same time, all our life' (15–16). His opening attack is

---

[20] On the importance of *archē* in Gregory's concept of God the Father see Beeley (2008) 204–17; on the influence of Origen on this see Daley (2012) 8–10.

allusive. 'What other heaven do you turn for the very stars? Again, to whom do you ascribe the other heaven, piling leaders on leaders?' (17–18). This appears to pose the question: as individual stars affect individual lives, what is the principle that controls this disparate system? If there is no single force – God – behind the system, how do you stop multiplying agents of influence? This opening quickly translates into the standard argument that many people with the same moment of birth have different fates (19–23), and groups of different people have a common fate. Either the stars are in conflict or some superior power must be mixing these different fates together. Thus, he concludes, with blunt certainty (33): 'For either God is the director or the stars are the directors (*hēgemonēes*)'. The allusive opening is thus in fact foundational. Gregory takes what is often expressed as a common-sense observation – how can horoscopes be predictive if different people born at the same time have different fates, or people born at different times suffer the same fate? – and turns it into an argument about the foundation of the order of the cosmos. Which is also a statement of Christian knowledge, a confession: 'But I know this. God governs all these things' (33). And in lines that strikingly look forward to Nonnus (36–7):

> τοῖς μὲν ἔδωκεν
> ἁρμονίην τε δρόμον τε διαρκέα ἔμπεδον αἰεί,
> τοῖς δὲ βίον στρεπτόν τε καὶ εἴδεα πολλὰ φέροντα·

> To the celestial bodies he granted
> Harmony, and a fixed course lasting for ever;
> To the lower world, a life of change with its multiplicity of forms.

As in Nonnus, 'Harmony', a quality of the natural word, is distinguished by a pattern that is established and certain, *empedon*, while human existence is marked by change, and the *poikilia* of shifting forms. Where Nonnus allows the poetics of metatropic, polytropic transformation to stand in tension with the establishment of order through the *empedos* Word of Jesus in the *Paraphrase* or the pantheon of Olympus in the *Dionysiaca* – a tension also enacted in the different scopes of the two epics, Dionysiac change versus Christian transformation – Gregory contrasts God's order with human changeability, and sets human uncertainty as the necessary mediation between the fixity of God and the unfurling of a human life over time.

> Some things God has revealed to us; the rest he preserves in the crypt (*keuthmōsi*) of his wisdom. He wants to prove empty the boast of mortals.

> Some he has established in the here and now; others he will meet in the later days: the farmer cuts down everything in season (*hōria*). (39–42)

Human epistemological doubt is a function of the difference between God's time and human time. Humans can only have a partial and incomplete knowing, because God's *sophia* takes shape over the infinity of God's time, and humans can only wait for 'the later days'. There is a certain violence in this alienation from understanding: God the farmer 'cuts down', *keirei*, but does so according to the seasons, his knowledge of seasons. The stars may tell humans when to harvest; but this mortal knowledge of temporality is only a shadow of the sun on water in comparison to the seasons of God's comprehension.

At this point – an argument of broadest generality and scope – Gregory brilliantly brings it all back to an intensely personal expression, his own life and his own authority, with two lines of simple power, a rhetorical switch of startling impact (43–4):

> ὡς καὶ Χριστὸς ἄριστος ἐμοῦ βιότοιο δικαστής.
> οὗτος ἐμὸς λόγος ἐστὶν ἀνάστερος, αὐτοκέλευθος.

> So too Christ is the best judge of my life.
> This is my story: no stars, self-driven.

Picking up on the simplicity of the confession 'But I know this' (34), Gregory makes his own life his example. His *logos* – account, story, rationale, his Christ-directed life: the ambiguity is pregnant[21] – has 'no stars'. The adjective *anasteros* is very rare but appears twice in Aratus' *Phainomena* (228, 349)[22] in the expected sense of 'starless', a sky without visible stars. Gregory, in his very denial of astrology, picks up and redrafts the vocabulary of the iconic literary astrologer. *Autokeleuthos*, 'self-driven', 'going one's own way', is also a very rare word. (It is also the marked oddity of the vocabulary here that makes his personal turn so striking, so *self-driven*.) *Autokeleuthos* indicates the free will of the human agent, the commitment to moral choice that astrology appears to deny in its fatalism. *Keleuthos* is a common term since Homer for the 'path' of song as well as the path of life, and here, applied to *logos*, overlaps the journey of life and its telling, both self-driven. This term, *autokeleuthos*, motivates the paragraph that follows, which sarcastically rehearses the argument against determinism (49–52). 'If the circle (*kuklos*) runs everything . . . then the circle runs

---

[21] See Schwab (2009) *ad loc.*
[22] Also in Ps-Manetho 4.528, as is noted by both Schwab (2009) *ad loc.* and Moreschini and Sykes (1997) *ad loc.* (who both oddly note only Aratus 349, and not 228).

my will itself, and there will be no drive in me that leads me towards the better, in decision or understanding'. The astrologer's 'circles' bind humans into an imposed inability not just to escape their fate but also to change for the better.[23] With Nonnus, we saw repeatedly how the circles of heavenly transits expressed time as a circling pattern: here we see a theological caution around such vocabulary. The order of the celestial bodies must not be allowed to tie a human life into an amoral determinism. Indeed, *autokeleuthos* is echoed by Nonnus when Lazarus stumbles out of the grave, on his own path, again (*Par.* 11.161). This is the only occasion in the *Paraphrase* that *autokeleuthos* occurs – Lazarus' rebirth as a model, a type, for the rebirth of every Christian into the moral choices of their lives. But the term also takes on significance from the regular repetition of a near homonym *autokeleustos*, 'self-ordered', which Nonnus uses nine times in the *Paraphrase* in a strikingly systematic manner that emphasizes precisely the dynamics of choice and agency in Christ's story. Jesus explains that what he says is not 'self-ordered' but comes from God; Jesus goes to his death, however, *autokeleustos*, unforced and self-willed. The believer in Jesus will act *autokeleustos*, 'of his own free will'.[24] The same term appears in Cyril, also to specify the necessary agency of human choice.[25] Gregory too in the third of the *Aporrhēta*, 'On the Spirit', celebrates the Spirit as 'divine might, coming from the Father, a self-determined being (*autokeleuston* (7))'.[26] There is a strong case that Nonnus' discourse is informed by a reading of Gregory, a theological backstory to his fascination with 'self-movement'.[27]

Gregory's poem on providence ends with an astral singularity. 'Let's have silence about the great glory of Christ, that messenger star' (53). This star, he claims, and as we saw in the last chapter, is not the sort of star that astrologers deal with, but 'strange, and not appeared before' (57). It had been foretold in the Hebrew Bible (a commonplace of Christian polemics, as we also discussed), but this star marked a unique moment in time. Whereas all the other stars follow the road that Christ ordered for them – call them 'fixed, wandering, retrograde, as they term it' – his lack of interest

---

[23] On the importance of this sense of growth in Gregory, see Beeley (2008) 111.

[24] 3.108; 5.54; 7.107; 10.64; 12.94; 16.40; 16.102; 18.26; 19.80 – ironically in the mouth of the priests, denigrating Jesus as king. Note too that *artiphaēs* 61 appears for the first time in extant Greek in Gregory, and for the second time in Nonnus *Par.* 9.88 (wrong reference in Moreschini and Sykes (1997) *ad loc.*) of the miracle of the blind man, who 'just now sees (and is seen)'.

[25] Cyril, *De ador.* 48 – virtue must be chosen; *Glaphyr. in Pent.* 16 – knowledge must be chosen; cf. *Expos. in Psal.* 25; *Contra Jul.* 43 – in combination with *rhopē* and *hormē* as in Gregory.

[26] Translation from Moreschini and Sykes (1997) *ad loc.*

[27] Simelidis (2016), with further bibliography, is the sharpest version currently available of this claim.

in the technical terminology of the science is performative, of course – 'this was the time when (*tēmos hote*) the cleverness of the science (*technē*) of the astrologers came crashing down' (83–4). Theodotus saw this star as the redesign of heaven; Gregory sees it as the destruction of the science of astrologers, who are now, like the Magi, to worship the Lord. Astrology cedes not so much to theology as to adoration, not to reason but to faith.

The poem ends with an envoi to star gazing that is typical of Gregory's rhetorical style. He waves off the long history of theory: it is of no concern if stars are made of some sort of fire – one time-honoured hypothesis – or 'some "fifth body" as they call it' – that is, specifically Aristotelian physics.[28] Such theories of the matter of stars do not matter to him. For 'we are taking our road upwards' – the 'our' is an emphatic gesture of complicity with his audience, together to rise above mere astrology; the road 'upwards' a neat transcendence of the claim of astrology to understand high matters. 'For we are hastening towards Rational Nature, which is both heavenly and tied to earth'. The next poem in the collection is 'On Rational Natures', so that this closing advertises what is immediately to come. But Gregory cannot resist the paradox: the rational (*logikē*), with its connection to the *Logos*, is privileged, but, as we have seen in the Theological Orations, inadequate without faith, and inevitably marred by human materiality. Hence it is both 'heavenly' – the *Logos* in man – but 'tied to the earth', *desmion aiēs*. The poem's final two words look towards the discussion of the Theological Orations and humanity's 'fleshly thickness', and opens a paradox of humanity's double and flawed knowing to be negotiated in the continuing exploration of the collection.

Indeed, this closing expression, *desmion aiēs*, is given a revelatory exposition particularly in the seventh poem in the collection 'On the Soul' – which looks back to Gregory's language of *mixis* that we encountered earlier. A human, argues Gregory, is made up of a heavenly soul and an earthly body: 'wherefore I love one part of my life because of earth (*dia gaian*) but in my heart I have a desire for another way because of my divine part (*theian dia moirēn*)' (7.76–7). This standard expression is described, however, as the *desis*, 'the binding' of the original human being. The one who mixed (*ho mixas*) the elements 'bound (*dēsato*) an image in earth' (80–1). This physical binding – which Gregory goes on to expound at length – is also a moral compulsion. God sends man on his road, and 'neither does he send him forth free, nor did he entirely bind him (*dēsato*)' (10). Choice

---

[28] Discussion in Moreschini and Sykes (1997) and Schwab (2009) *ad loc*. Proclus will pick this up in his commentary on Plato's *Timaeus* 3 152a–154e33 Diehl; 3.43–50 Baltzly.

remains endemic to humans. 'Tied to the earth', the final words of 'On Providence', are now explained as a multiple form of binding: humans are limited by their earthly condition; their very materiality is a binding, a binding *of* earth as well as *to* earth; and their lives are a tension between the freedom to choose and their moral binding to God's way. Gregory uses the poetic form to strain at the expressivity of language in his incremental attempt to capture the complexity – the ineffability – of how humans fit into God's creation.

It may at first sight seem surprising that in a collection of poems headed by a poem on First Principles and dealing in turn with the grandest of theological themes – the Son; the Spirit; the Universe; Rational Nature; the Soul; the Testaments and the Coming of Christ – there should be a poem on astrology. The great theologian Paul Tillich, however, paints this remarkable portrait of the era, which is offered in the standard edition of the poems by Donald Sykes as a framework specifically to understand the instrumentality of Gregory's poetry on the stars:

> In the late ancient world fate conquered providence and established a reign of terror among the masses; but Christianity emphasized the victory of Christ over the forces of fate and fear just when they seemed to have overwhelmed him at the cross. Here faith in providence was definitively established.[29]

Even from within Tillich's own intellectual formation as a committed Lutheran, this is a disturbingly, even crazily exaggerated picture – and it is a sign of theology's continuing hold on philology that this image of Tillich's should have been proffered without demurral by Donald Sykes as a guide to the significance of Gregory's poem. Even if it is taken that astrology to an extent replaces or supplements divination and oracles, all of which can be seen as attempts to produce a sense of control over the frightening vicissitudes of life, 'a reign of terror' scarcely reflects any accurate historical understanding of the functioning of astrology, which was also as much for the elite as for the masses. Nor did Christianity in any way succeed in 'conquering astrology' in the name of providence – astrology continued throughout late antiquity and was still going strong in Christian courts and intellectual circles throughout the early modern era too. It is certainly not adequate to see Gregory as a figure for this triumphalist fantasy – though fascinating to observe the continuing overwrought vehemence of a Christian

---

[29] Tillich (1951–63) I 294 (Tillich, of course, made *Kairos* [see chapter 2], central to his thought); Moreschini and Sykes (1997) 175.

resistance to astrology, even an astrology of the fourth century. Tillich's faith, followed by Donald Sykes, seems to have suppressed the historical evidence.

Yet Gregory's attack on astrologers is instrumental in developing Christian theology, and must be seen as indicative of a broad set of questions that go to the heart of Christian thinking about temporality, and human understanding of a life-time: how should the regularity of the time of the stars and the unfurling of a human life be understood together? It is telling, for example, that Gregory describes Jesus at the point of his birth with that opening word of Nonnus' *Paraphrase*, *achronos*, 'timeless', 'no time' (5.55). This is not so much putting Jesus beyond astrology – what is a birth horoscope for the timeless? – as a way of linking the discussion of the star of the Magi to the Christology of the earlier poems in the collection – and the Theological Orations on the Son, with their theoretical exposition of the theology of time we discussed in chapter 1.

In particular, the use of *achronos* looks back to the second of the *Aporrhēta*, 'On the Son', which is directly polemical in its antagonism to those suicidal theologians whose 'tongues are at war with divinity' (3–4). As so often in the arguments with Neo-Arians around the crisis of Nicaea, the opening salvo concerns time and the Trinity. The Son is immediately called *achronos*, 'timeless' (7) and 'equal in nature to his progenitor' (8), terms also familiar from the prologue to the *Paraphrase*, as well as anticipating its use in 'On Providence'. Gregory vividly distinguishes human birth, which involves 'flux and shamefully horrible cutting', the inevitable product of passion, because humans are 'bound' (*detos*, 15), which, as we have seen, is the watchword for the limitations of human materiality. God's birth by contrast is unconnected to passion because it involves the 'completely uncompounded, the unbodied' (*apēktos, asōmatos*, 16). *Apēktos* is an extremely rare adjective indeed and seems to be chosen to contrast with *detos*. But this contrast is only to make more comprehensible the temporal difference. 'As assuredly time (*chronos*) exists before me, time (*chronos*) is not before the Logos, whose progenitor is timeless (*achronos*)' (18–19). Gregory, as in 'On Providence' inserts himself at a crucial juncture as the example of the human: his time-bound existence contrasts with God's eternity. God the father, intensifies Gregory, is *anarchos*, which, in contrast to its use in 'On Providence', here indicates 'without beginning'. The syntax of this sentence, however, is significantly incomplete. It begins: 'When the Father is without beginning, with no lack or supplement to his divinity, then also the Son of the Father too, the son who has as a father a timeless beginning/principle' (18–20), but then the sentence drifts away into a series of subordinate clauses with no main verb for the Son. Gregory in the Third Theological Oration (29.3) is explicit that the language of 'when' and 'then' cannot possibly capture the timelessness of God, but we are forced to use such terms. Here, the syntax collapses around the 'when' and the 'then' as it travels towards an injunction not to split the Son and the Father from each other. The Son's Father is called an *achronos arkhē*, which I translated 'a timeless beginning/principle', in an attempt to keep the ambiguity (which, again, it is hard to imagine Nonnus reading without interest). Between *anarchos* and *achronos arkhē*, the question of an overlap between a 'timeless beginning' – another paradox for the faithful – or 'timeless principle', a principle of timelessness, seems to be left unresolved.

This irresolvability is articulated in the fourth of the *Aporrhēta*, 'On the Cosmos', where Gregory argues against the Manichaeans that the soul and the body are not to be seen as a primal compound of good and evil, but a compound – a *mixis* – to which evil is added only later. 'For me', he declares, with the personal turn again, 'God is one, without beginning, (*anarchos*), without conflict, one good light, the strength of simple and composite (*plektōn*) minds, the exalted in heaven and on earth' (4.39–41). The lack of a beginning is also a lack of conflict, a single goodness rather than a Manichaean opposition. This principle of the singleness of God places evil as man's invention: 'Self-destructive in corruption, I planted evil' (4.53) – Gregory as a human is fully implicated in original sin. The Cosmos itself was brought into being by God, and therefore is not eternal. God ruled over the 'empty aeons (*aiōsi*) before the universe was set in place and was adorned with forms' (4.62–3) – *kosmēthēnai*, 'adorned', is an etymological play on the other sense of *cosmos*, 'adornment' rather than 'universe'. God's mind was stirred to look at the forms (*tupoi*) of the universe about to come into being, but already present to God (66–8). 'All things are before God, what will be, what has come into being, and what is now present' – the timelessness of God – but 'for me time is divided like this, some before some after'. Again, Gregory presents himself, his time, as the counter-model to God's time. Human time is divided. But for God everything is *eis hen*, 'into one'.[30] The principle of the single nature of God – in contrast to the dualism of Manichaeism – and the timelessness of God are thus mutually implicative. For Gregory, although God is one, and his time one, humanity experiences itself as a *mixis* and time necessarily as divided.

[30] See Ayres (2004) 244–51 for this concept in the development of pro-Nicene theology.

Hence, to return to 'On the Son', the timelessness of the Father and the timelessness of the Son therefore require that we do not break apart (ἀπορρήξωμεν – a very aggressive term) the Father and the Son. As Gregory concludes, with his characteristically bold simplicity, conjoined with surprising vocabulary (26–7):

> ὃ γὰρ πάρος ἐστὶ Θεοῖο,
> ἢ χρόνος ἠὲ θέλησις, ἐμοὶ τμῆξις θεότητος.

> For what exists before God,
> Whether time or will, to me is a severance of divinity.

Gregory has explored how there can be no time before God, the Son, but here he adds 'will', *thelēsis*, a charged term he has not adumbrated in the poem – though it is extensively discussed in the Theological Orations on the Son, and a bone of contention in Christian polemic. If God 'willed' the birth of Jesus, it implies that Jesus is a result of God's objectification and thus subordinate. If God did not will Jesus' birth, how, then, did it come about? The co-temporaneous, equal nature of Father and Son is the answer to this question in Nicaean theology, but there is a certain silencing of such discussions about will in this poem until this bare reference. The final phrase is most striking, however. *Tmēxis*, which I translated 'severance', occurs only here in extant Greek literature. Its meaning is clear enough, but its strange form is arresting. As a human birth results in a 'shamefully horrible cutting', and as human time is 'divided', so divinity itself becomes like a human in the violence of the 'severance' that such theory forces between Father and Son. To cut apart the Father and the Son is to treat divinity as human.[31] Thus, after a long consideration of the qualities of the Son, Gregory concludes, 'You at any rate, do not dishonour divinity with these mortal attributes. Divinity made glorious the earthly form which, in gracious love for you, the immortal Son formed himself' (82–3). Gregory is determined that the divine attributes of Jesus transcend his human qualities (as has often been noticed in discussions of Gregory's Christology),[32] but here the culminating adjective which I translated as 'immortal' is

---

[31] So Prudentius accuses Marcion of daring *secare numen insecabile*, 'to cut an uncuttable divinity' (*Ham. Praef.* 61), and the anonymous *Carmen adversus Marcionitas* 2.14–15 (see Pollmann (1991)) accuses these heretics of *numen sine fine tremendum | dividere in partes*, 'dividing into parts the awesome divinity without end'.

[32] On Gregory's Christology the most helpful discussion is Beeley (2008) 115–52, with full bibliography, who writes (116) that Gregory 'is one of the chief architects of the language and concepts used in the Christological controversies that occupied the Church' over the next centuries. See also Winslow (1979); and with special emphasis on the problems of Gregory's Christology Norris (1991) 39–53.

aphthitos, the word associated most strongly with Achilles' all too human search for glory. In a similar way, Apollinaris' *metaphrasis* of the Psalms addresses God as *aphthite* – 'undecaying one'; the angel Gabriel in the *Vision of Dorotheos*, who Muse-like, along with Christ, inspires the poet's song, is called *aphthitos*.[33] *Aphthitos* is not a term that appears in the Septuagint or Patristic writing in general, though *aphtharsia* and *aphthartos* are key terms in the discourse of corruptibility and incorruptibility as Christian categories of the material body and of spirit. The poeticism is not just a synonym, however. As Gregory – we saw – appropriated Homer's boast of glory against Homer, here he turns the same vocabulary, which encapsulates a hero's struggle against mortality, to mark Jesus' transcendence of his mortal attributes. Writing not just against his Christian opponents but also against the heritage of pagan concepts of mortal and divine temporality, Gregory uses the strut of epic tradition to enforce his theological understanding of God's time.

*

Gregory's engagement with astrology, then, is deeply embedded in a series of theological debates about time and Christology, morality and the inadequacy of human knowledge, agency and transformation. This nexus of theological thinking takes a further turn, however, when Gregory, in his brief appointment as archbishop of Constantinople, faces his congregation in the church of the Holy Apostles on Christmas day 380,[34] and in a celebratory mood. *Oration* 38 is the first of his Epiphany Homilies, and unlike the works we have been discussing so far, it presents itself fully within a liturgical setting. It revels in the performativity of the occasion, and the prose is full of hymnic praise, celebratory incantations, and ecstatic pleasure in the celebration and in itself. It is the most Dionysiac of his orations. As the sermon runs its course, however, the polemical underpinnings of its theology shine forth, and its basis within a theology of time becomes clear.

The speech begins with a full-throated explosion of celebration: 'Christ is born, glorify him! Christ from heaven, meet him! Christ on earth, exalt him!'.[35] Even in such outbursts, Gregory constructs a texture of

---

[33] *P. Bodmer* 29.169, discussed very briefly by Agosti (2011) 289–91.

[34] It has been debated if the day is 25 December or 4 January and the year 379 or 380: see Moreschini and Gallay (1990) 13–22. For my purposes, what matters is that the day is the day called Christ's epiphany or birthday, as Gregory (3.1) explicitly declares, and I am happy to go with the traditional dating of 25 December 380. There is no evidence before the fourth century for celebrating Christmas: Beckwith (1996) 71.

[35] See Spira (2007) 197–200 for the rhetoric of this opening; also Spira (2007) 215–18.

argumentation: 'Christ in the flesh! Exult in trembling and joy! In trembling because of sin, in joy because of hope'. The quotation from the Psalms (2.11) (where the sense of the Hebrew is uncertain) – 'rejoice with trembling!' – is extended and glossed with the standard matrix of Christian emotional expectation: fear at sin and joy in hope. So too the liturgical and theological come together in his incantatory repetitions: 'Again (*palin*) darkness is undone, again (*palin*) light established, again (*palin*) Egypt is punished with darkness, again (*palin*) Israel is illumined by the pillar'. Each of the phrases marked by *palin* cites or alludes to a text from the Hebrew Bible: *palin* marks therefore the repetition that liturgy performs – it is again Christmas Day – but also the quotation (*palin*) which is a rewriting of the Old Testament as the New (*palin*) – as he cites from Paul 'The ancient has passed; see, everything has become new'. Yet this conversion (*antistrophē*, 4.8) too is a form of return: 'let us go forward to Christ, or *return* to him – this is the more proper expression' (4.3–4). There should not be talk of 'fashioning' (*plasis*, Ps. 118.73) but of 'refashioning' (*anaplasis*) (4.16–17). Thus, later in the speech, he concludes ringingly (16.18–19): 'In all this, there is one principle: my fulfilment (*teleiōsis*) and refashioning (*anaplasis*) and return (*epanodos*) to the first Adam'. With Nonnus' *Paraphrase* we emphasized the connection between the poetics of *palin* in paraphrase and the theology of *palin* in the Gospel. Gregory's opening repetitions of *palin* open a full theological and liturgical understanding that connects the performed repetition of the festival, which each year again repeats its celebration, and the re-quotation of scripture, with the newness of the New Testament promise to return to the lost Paradise. Time turning back on itself ...

These ecstatic opening passages also recall the Theological Orations and *Aporrhēta* poems, marking the continuity of religious thought across the genres. So, the Logos *pachunetai* (2.17) 'Thickens into flesh'; 'the invisible becomes visible', and, inevitably, 'the timeless begins', *ho achronos archetai*: paradox here seems to be part of the rhetoric of excited joy as much as it is a cue to faith over logic. Gregory also explains that the day can be called the Theophany because it was when God appeared (*phanēnai*) or the Genethlia because it is the day of God's birth (*gennasthai*), but this brief pause of glossing turns quickly into a long list of ways that their celebration is not to be like a pagan feast. This oscillation between more sober, scholarly comment and overwhelming displays of rousing enthusiasm repeatedly and performatively enacts the sense of joy bursting through in celebration. So, his introduction concludes with a chanted list of refused pleasures (5–21): 'we won't garland the halls; we won't set up choruses; we won't adorn the streets ... ', and so on for more than twenty lines. That's for the Greeks. The Christians will luxuriate rather in the divine law and Christian tales (6.7). This introduction – and it is clearly marked in the text as such by the transition into the argument of the sermon (6.16–21) – is programmatic in these glances towards theology which veer into joy and then return to theology. As we will see, this *performance* in itself will prove key to the agenda of the sermon. It is finally the performance of Christianity that will count.

The argument of the sermon – the promised *diēgēsis* – starts with the classical statement of God's time:

> Θεὸς ἦν μὲν ἀεὶ καὶ ἔστι καὶ ἔσται· μᾶλλον δὲ «ἔστιν» ἀεί. Τὸ γὰρ «ἦν» καὶ «ἔσται», τοῦ καθ' ἡμᾶς χρόνου τμήματα καὶ τῆς ῥευστῆς φύσεως.

> God was always and is and will be. Rather, 'he is' always. For 'was' and 'will be' are the severances of our time and of nature in flux.

Gregory repeatedly recognizes the insufficiency of such temporal language for God's time (as we have seen), but his terminology here specifically picks up on the poetry of the *Aporrhēta*. Human time is marked by the violence of its severances – *tmēmata* is from the same root as *tmēxis* – and by a nature in flux. *Rheustos*, 'in flux' is a philosophical term from the Greek tradition, an Empedoclean or Heraclitan sense of the instability of material form, here contrasted with the permanence and stability of God's existence – and an echo of the 'flux' (*rhusis*, *Apor.* 2.16) that is a physical sign of human birth into mortal time. In this infinity, God 'transcends our every conception of time and nature' (7.7–8). Humans can only grasp at God's essence 'all too dimly and indifferently', and the only thing that humans can conceive is his boundlessness, and only that because God has a simple nature (7.23–7). What is meant by a simple nature?[36] That which is not composed (*sunthetos*). As in the *Aporrhēta*, the threat of the composite is that it can be broken or severed. So the simple nature of God depends on the infinity of timelessness: he is *anarchon* (7.5), 'without beginning', and 'deathless and beyond destruction', *athanatos, anōlethros* – hence *aiōnos*, 'eternal'. 'Eternity', defines Gregory, is neither time (*chronos*) nor a part of time – it cannot be measured. Human time is measured by the celestial bodies like the sun, but eternity for divinity – and this is a very difficult idea, alluding to Gregory's technical, theoretical discussion of temporality – is 'stretched out alongside their existence, like a sort of "timely" movement and interval'. Gregory is

---

[36] See Ayres (2004) 278–301 for a discussion of simplicity and singleness in the thinking of the Cappadocians.

struggling to express what he has repeatedly said is a struggle for humans to comprehend, namely, the idea of eternity, and at this point of awkwardness, he just announces 'That's enough philosophy about God for now'. As we will see, this *aposiōpēsis*, this break into silence about philosophy, is a significant part of the performativity of the sermon. But what he has offered is the foundation of a story, God's time as the ground on which the narrative to come is to be structured.

For what follows is no less than the history of the universe up to the birth of Jesus, narrated in an overwrought mixture of summary and expansive emotional glosses.[37] So chapter 9 delivers an account of the creation of the angels. Chapter 10 moves on to the material and visible world, always no more than a microcosm of the invisible world of the spirit. At this point, he worries his audience is becoming impatient to get back to the festivities, but, he explains, that is exactly his route. So chapter 11 gives us the creation of man. Up to this point in the history of the universe there had been no *mixis* of opposites (11.5), but now God creates humanity as a mixture of the material and the spirit, which prompts another excited list, this time of mediated polarizations, 'earthly (*epigeion*), yet heavenly; of the moment (*proskairon*) yet immortal' (11.17–18). Chapter 12 sets man in the Paradise of Eden, equipped with free will, which leads to the original sin. After the Fall, Adam and Eve put on animal skins, described with full theological purchase as 'perhaps the thicker (*pachuteran*) flesh, mortal, antitype (*antitupos*)': the skins they wear are perhaps a type, a negative, reversed image of their previous life in Paradise – now their flesh is even more *pachus*, 'thick', bound to the earth, a block to knowledge; their bodies are now tied to mortality in the full sense, forced to labour and give birth in blood and pain; a constant contrast to the different life of the body in Paradise, towards which Gregory constantly longs to return. Chapter 13 begins with the educational punishments that follow the Fall, for different causes over different times, none of which succeed in changing the course of humanity: 'reason, law, prophecy, benefits, threats, blows, floods, fires, wars, victories, defeats, signs from heaven, signs from the air, from earth, from sea, unexpected revolutions of men, of cities, of nations'.[38] This rolling, rhetorical list is a priamel to God's final cure for man's ills, the birth of Jesus, the Logos – which prompts another roll-call of performed brilliance – and Nicaean theology (13.14–19):

[37] Fulford (2012) is an attempt to set this in the context of exegetical preaching.
[38] There are similar summary lists of disasters in Orosius (*aeterna perdito, nunc quoque*) without the sense of education: see Walter (2020) 210–20.

the one before the centuries (*proaiōnios*), the invisible, the ungraspable (*aperilēptos*), the incorporeal, the beginning/principle (*archē*) from the beginning/principle, the light from the light, the fountain of life and immortality, the impress (*ekmageion*) of the archetypal beauty, the unmoving seal, the unshifting image, the definition and the reason (*logos*) of the Father.

The temporality of Nicaean theology, which was Gregory's mission to promote in Constantinople, is emphatic: the Logos is 'before the centuries', timeless; it is a beginning out of a beginning, the spring of life which is also the immortality of Christian promise. Equally emphatic is Nicaean Christology: 'light from light' quotes from the creed,[39] and the repeated images of sameness, deny the subordination of the Son to the Father – the 'impress of the archetype', 'the unshifting image (*eikon*)',[40] the 'unmoving seal' – and culminate in the claim that the Logos is the 'definition and reason' of the Father. The circularity of defining the Logos as the logos demonstrates – enacts – the 'ungraspability' of the divine that is central to Gregory's epistemology. This extraordinary rhetoric where the chanted list of former admonitions is answered by the equally incantatory list of the attributes of Christ provides not just a theological framework for the birth of Jesus, but also its liturgical celebration.

Gregory continues with a narration of the mission of Jesus – and an angry rebuttal of any of the congregation who do not recognize divinity in the figure of Jesus: a continuation, that is, of his telling of the history of everything, framed by the threats of theological controversy (14–15). 'You will see', he announces with the plea for the immediacy of vision in experiencing this story (16.1), 'Jesus in the Jordan …' – and thus into a summary list of the miracles and sufferings, leading to the crucifixion, Resurrection and the Ascension, concluding in the lines we have already quoted, that 'In all of this there is one principle … my return to Adam'. The point of the story is the 'fulfilment' or 'perfection' (*teleiōsis*) of each member of the congregation, each 'I'. The story of Jesus must be a personal story of each Christian.

This conclusion is, however, also a crucial bridge to the close of the sermon. 'Dance!', enjoins Gregory, 'like David before the Tabernacle' (17.3–4). The birth of Jesus means that 'You have been released from the

---

[39] And illumination/light is a constant theme in Gregory: see Beeley (2008) 104–11.

[40] A common expression, found also in Athanasius *Contra Gentes* 41.1 (who is imitating Alexander of Alexandria, the opponent of Arius) – and who also contrasts the composite nature of man with the singleness of God: Ayres (2004) 41–50. *Aparallēktos eikon* is another phrase around which pro-Nicene theology formed.

bonds of your birth': *eluthēs* is a second-person singular, as Gregory addresses each congregant as an individual; the 'bonds' of birth recall the binding of earth, to earth, of the journey of life that marks out human existence. But to celebrate, each congregant is to find their place in the story. 'Be the ox in the manger', asserts Gregory, who 'knows his master' and is 'fit for sacrifice'; 'Run! Bring gifts with the Magi'. 'Adore with the shepherds'. Yet it is not just as a figure in the story of Jesus that each congregant should celebrate but also in an enactment of the life of Jesus himself. 'Travel blamelessly through all the ages and potentialities of Christ' (18.7–8). Here we see the power and paradox of *autokeleuthos*: the journey of the Christian must be a willed choice to follow the path of Jesus. So, concludes Gregory in yet another flowing list,

> Teach in the Temple and drive out the sacrilegious traders ... Submit to be stoned ... If you are scourged, ask what they have left out ... Taste gall because of the taste;[41] drink vinegar; hunt out spittings; accept blows, beatings; be crowned with thorns ... be crucified ... that you may rise with him ...

To celebrate Christ is to enact the *imitatio Christi*, which is 'to see God, as much as it is within reach (*ephikton*) for God to be seen' and 'to make God clear, as much as it is within reach (*ephikton*) for prisoners of flesh (*desmiois sarkos*)' (18.223–6). As ever, Gregory insists that even in this transcendent moment of fulfilment, the human being is still bound by the flesh of materiality.

The sermon that began with 'Christ is born, glorify him' ends with a powerful statement that each Christian has to imitate Jesus, to relive the life of Christ, to suffer with him: to *experience* Jesus' story. There is a thread of Nicaean theology running through the whole sermon, for sure, often forcefully indicated, but whenever the threshold of philosophical thought is approached, Gregory backs away – explicitly or performatively – into a more joyous expressivity. The demand to experience the story leads not to theological exegesis (though Gregory keeps letting the theology peep from behind the veil), but rather to performance – the rolling lists of attributes, the incantatory repetitions, the excited revelling in paradox, the expressions of celebration. The event of the birth of Jesus – Christmas Day – is therefore set in the narrative of time – the history of the universe from creation to its promise of eternity, told in sequence from the idea of God's timeless time through creation, Adam, the Fall, the history of human failings until the birth of Jesus, Jesus' story leading to the crucifixion and the promise of eternity. To

---

[41] This is typological: Gregory uses 'the taste' to refer to Adam's bite of the apple in Eden.

celebrate Christmas requires an understanding of time and the universe to comprehend the significance of the birth. To experience Christmas is to experience one's life within Christian time. That is the performance of Christianity demanded by the performance of the sermon.

*

Gregory of Nazianzus, like Gregory of Nyssa and Basil of Caesarea, discusses time from a technical philosophical perspective, as we indicated in chapter 1: this is how Gregory's take on temporality has been usually discussed by scholars, fitting him into the long tradition of Greek thinking about time. Indeed, Gregory writes himself into that tradition by his explicit citations of Aristotle and Plato and his engagement with Neo-Platonic thinking about such issues. It is clear that as for Augustine, so for this group of Cappadocian thinkers, linked by family, religion, space and controversy, time was a major theoretical issue of religion. What this chapter has set out to show, however, is how such theoretical argumentation, inflected as it constantly is by post-Nicaean theology, also informs less directly theoretical discussion and becomes part of the narrative and the performance of Christianity. It helps explain the fascination with and disparagement of astrology, an argument that opens out into a consideration of morality, agency and the role of providence in contradistinction to fate in Christian thinking. What is the regulation of time? For Gregory, the very interest of science in the accurate measurement of time reveals a flawed thinking about materiality, the human limitations of knowledge, what is within reach for prisoners of the flesh. But, as we finally saw, the understanding of temporality – both God's time and the history of humanity from creation forwards – is also a necessary foundation for the very acts of celebration, the festivities of the church. The performance of celebration in the experience of liturgy is a repetition that should entail an understanding of Christian life in time, where each life is also a repetition of the type of Jesus, an *imitatio Christi*. In the layers of typological thinking, each citation of scripture is a re-quotation that is also the supersessionist fulfilment of the newness of the New Testament; each liturgical celebration is a repetition that orders time into a Christian calendar, a repeat of earlier celebrations, a fulfilment of the order of worship; each life is lived as a fulfilment of the type, a repetition of original sin overcome for the faithful by the promise of grace: the fulfilment that is a return, time turning back on itself, palintropic. Gregory sets out to redefine for Christians how a life-time is to be experienced *as* time. To celebrate Christmas Day with Gregory is to conceptualize one's life as a paradigm of how to live in Christian time.

CHAPTER 14

Day to Day

For Gregory of Nazianzus, then, Christmas Day is to be experienced as a celebration of the history of the universe and as a living recognition of the transformative epiphany of Jesus Christ, an epiphany that changes how time is lived and perceived by the Christian faithful. Gregory wants to redefine how time is counted, recounted, experienced. Ambrose of Milan at around the same time, over in the West, is rather more modest in his vision. At least at first sight. His hymns are designedly simple and easily memorable in form, though, like William Blake's lyrics, they are far from simple in their linguistic depth and significance – and they proved extraordinarily influential in the invention of Christian time as well as of the Christian hymnic tradition. These hymns are to be sung by a congregation, and there are reports not just of people singing lustily and of the hymns spreading across Italy, but also of annoyance by Ambrose's opponents at their success in inculcating particular doctrinal views in the singers.[1] Hymns, that is, are to work not by a preacher telling his audience what to think but by a congregation's absorption of ideas through repeated performance, by the pleasure of singing.[2] The hymns of Ambrose construct a different relation with worshippers from Gregory's preaching. As we saw with Augustine's description of reading a Psalm, the repetition of performance becomes a form of embodiment through physical and mental memory. Each of Ambrose's fourteen hymns[3] has thirty-two lines – eight stanzas. There is no other lyric collection in antiquity where every poem has the same length, the same form.[4] This repetition provides a physical rhythm

---

[1] The best introductions remain Fontaine et al. (1992) 11–102; Dunkle (2016); on the history of liturgical reception see Marie-Hélène Julien in the same volume (109–14). Fontaine et al. (1992) is written by a team of ten scholars: the notes below give simply Fontaine et al. (1992) rather than specifying each author. Augustine, *Conf.* 9.6.14 records his tears at listening to Ambrose's hymns in Milan. See Zerfass (2008) 29–36.

[2] On the acceptable pleasure of psalmody see den Boeft (2008).

[3] Taking as authentic the poems included in the standard edition of Fontaine et al. (1992).

[4] Prudentius' *Dittochaeon* has 49 quatrains, each designed to be the title to a picture in church: again, in this later Christian collection form and ideological purchase are intimately connected.

and horizon of expectation – a bodily and mental pattern – that every performance enforces. You know, when you sing Ambrose, how long it lasts, when the end is coming. Regularity of time is a formal quality of Ambrose's writing – embodied in its performance by each member of his community.

The first of Ambrose's hymns is the most famous; it is not only still part of liturgy but is also often put to music in other settings. It begins:

> Aeterne rerum conditor,
> noctem diemque qui regis,
> et temporum das tempora
> ut alleves fastidium

> Eternal maker of the world,
> You who rule night and day,
> And give the times of times,
> To alleviate weariness

The metre of these hymns is an eight-syllable iambic form (technically, iambic dimeter acatalectic), and the flexibility and expressivity of Ambrose's use has been much imitated and discussed.[5] It is very easy to set to music and sing, and thus to repeat and absorb. The syntax and vocabulary are aggressively simple in comparison with Gregory's theologically inflected paradoxes, technical terminology and long periodic prose (or verse). God is addressed first as the eternal 'founder' of the universe. *Conditor* is a very familiar term from Roman political literature. Rome is dated from its foundation. The *Aeneid* announces that its subject is how great an effort it was *Romanam condere gentem*, 'to found the Roman nation'. Augustus is celebrated by Livy – in a remarkable passage of source criticism, unparalleled in Roman historiography – as the *templorum omnium conditor ac restitutor*, 'the founder and restorer of all temples'.[6] Unlike the roll-call of Roman founders of cities and cults, God is eternal and the founder of *rerum* – matter, or the universe (the general term which is the subject of Lucretius and other natural scientists in their search to understand the world). Orosius sums up this shift of sense when he writes that his Christian universal history starts *ab orbe condito*, 'from the foundation of the world': the punning rewriting of the standard historiographical dating, *ab urbe condita* is eloquent (as is its rewriting of Roman puns on *urbs* and *orbis*).[7] *Conditor* is rare in the Latin Bible, however; it occurs only

---

[5] See Dunkle (2016) 86–90 with extensive bibliography; Fontaine et al. (1992) 77–92.
[6] Livy 4.20.7: see Sailor (2006).   [7] Orosius, *Hist.* 1.1.14: see Walter (2020) 211.

once in reference to God in Hebrews 11.10, where Abraham, dweller in tents, is said to look forward to the city whose architect and maker (*conditor*) is God. *Conditor* in this case translates the Greek *dēmiourgos*, the Platonic term for the maker of the universe, and the contrast between even Abraham's hopeful work and God's establishment is stark. In Jerome's smart translation, *conditor* makes Paul's contrast speak to Latin political horizons. The phrase *rerum conditor* itself is found in Tertullian and Lactantius,[8] however, and *conditor* is common enough in Augustine and in the *City of God*, perhaps in echo of that passage of Paul. It is also repeated in Prudentius, who read Ambrose closely.[9] The word has a particular charge in Christian Latin in that it inherently contrasts God's creation with human order. So, here, God rules 'night and day' – a first gloss on *aeterne*. For Manilius, *fata regunt orbem, certa stant omnia lege | longaque per certos signantur tempora casus*, 'Fate rules the world, and everything is set by its fixed law | and the extent of time is marked by fixed events/descents'– that is, as time is defined by the descent (*casus*) and rise of the celestial bodies, so human life is determined from birth to death by the stars.[10] Unlike Manilius, here it is God who 'rules day and night' – a phrase which implies both always and the alternations of time – and, in an expression that is hard to translate neatly, God 'gives the times of times'. The polyptoton, as critics have noted, indicates that God both creates the hours and the seasons (*tempora*) and delineates eternity (*temporum*) by such divisions of ordinary time.[11] If *saecula saeculorum* becomes the liturgical equivalent of the Hebrew *le'olum va'ed*, as an expression of eternity, *tempora temporum* is the mundane equivalent for the regular unfurling of daily time – the will to see 'in the eternal the whole temporality of each day'.[12] The polyptoton itself, as Bert Pranger has argued, requires an audience to disambiguate the different senses of *tempus* – a weak version of the paradoxes of faith paraded by Gregory – and to engage with the construction of meaning, thus performing what Dunkle calls the mystagogic actualization of the hymn's dramatization of time.[13] As we will see, time is actualized at a threefold level:

---

[8] Tert. *Adv. Marc.* 1.10.1 and *universitatis conditor*, *De spect.* 2.4. Lact. *Inst.* 5.1.1; 6.9.14 see Fontaine et al. (1992) *ad loc*. Ambrose himself in *De virg.* 3.6.34 uses *rerum conditor* in a prayer.

[9] E.g. *Cath.* 4.9 (*rerum conditor*); *Ham.* 2; *Ham.* 346.

[10] Man. 4.14–6; on the Stoic background here see Volk (2009) 62–6, 220–1; Habinek (2011) and Green (2011) with further bibliography.

[11] Pranger (2007); den Boeft (2003). Proclus in his commentary on Plato's *Timaeus* twice quotes a Chaldaean Oracle fr 185, *chronou chronos*, referring to the sun, explaining that it means 'it makes manifest that time which is most primary and the cycle of the seasons is brought about in accordance with it' (4.256a Diehl; 55.32–56.1 Baltzly); see also 4.249f Diehl; 36.22 Baltzly.

[12] Fontaine et al. (1992) 40. 'Time and eternity are in the grip of each other' (den Boeft (2003) 57).

[13] Pranger (2007) in response to den Boeft (2003); Dunkle (2016), especially chapter 3 – this is the most sophisticated overall account of Ambrose's Hymns.

the representation of Christian time through the daily rituals of Christianity and their significance, the dramatization of the moment of the song, and the 'time' of metrical singing, the materiality of the verse. All three levels are expressed in the gift of *temporum tempora*.

The final clause expresses the purpose of God's division and regulation of time: 'to alleviate *fastidium*'. *Fastidium* is a surprising word. In my translation above, I have given what has become a standard Christian understanding. In one of the most popular nineteenth-century translations as a modern hymn, William Copeland writes 'weary mortals to relieve'; Gerard O'Daly similarly suggests 'to alleviate weariness'.[14] (John Henry Newman, who wrote the other common hymnic version used in liturgy, simply leaves this phrase out altogether.) It has even been translated as 'tedium' or 'boredom': Dunkle offers 'monotony'.[15] But *fastidium* is a much stronger and more emotive word. It means 'disgust' that can be prompted by a physical revulsion, or 'disdain' prompted by self-assertion or haughtiness.[16] In Jerome's Latin, it is used in Ezekiel (16.31) for the whore who raises her price out of *fastidium*, 'a harlot . . . who scornest hire', as the King James Bible puts it ('monotony' is not the issue). *Fastidium* is a surprise because it raises a question: how does the division of time, the separation of night and day, stop feelings of disgust or haughtiness? The translation 'weariness' – which turns the surprise into a cliché – ignores how these opening lines provide a programmatic agenda not just to the poem's discussion of time, but also to its moral normativity. The division of night and day will turn out to be a model for the repentance of sin, a rejection of worldly values in favour of Christian piety and purity. The disgust or haughtiness of humanity is to be alleviated in Christian worship. *Fastidium* epitomizes the sinners' emotions that will be alleviated by the morning's fulfilment of vows to God (32). Self-disgust is acknowledged and exorcised by singing with heart and soul.

The following three stanzas establish the significance of the division of day from night in a gradual transformation from the sequence of temporality to the morality of repentance, which retrospectively turns

---

[14] O'Daly (2012) 56. William Copeland (1804–85), an Oxford movement cleric who edited Newman's sermons, produced the aptly titled *Hymns for the Week and Hymns for the Seasons* in 1848; John Henry Newman (1801–90) published the hymn several times from 1850 to 1880.

[15] Dunkle (2016) 221.

[16] Kaster (2001) is a very full and persuasive treatment of the word. Septimius Severus describes himself in his cell experiencing a paradigmatic *praesentium fastidium*, 'disgust with present', in a list of a monk's emotions *Ep.* 2.1. Franz (1994) 193 rightly translates *Ekel*, but backs away from it as he proceeds!

each expression of night and day into versions of evil and good, ignorance and illumination (4–8):

> Praeco diei iam sonat,
> noctis profundae pervigil,
> nocturna lux viantibus,
> a nocte noctem segregans.

> The herald of the day now sounds,
> the watchman of deep night,
> nocturnal light to travellers,
> separating night from night.

The 'herald' of the day, an expression for what will be specified as the cock crowing, also opens the possibility of what O'Daly unnecessarily insists must be understood as a 'symbol', but not an allegory or type: Fontaine is more productive with his carefully open-ended adjective, 'allegorisable'.[17] The form of expression is suggestive rather than determinative. The *iam* gives a *hic et nunc* of performance: the cockcrow is the time of the poem, whenever it is sung, whenever it sounds out. This 'herald' is also a watchman of the deep night – a line that recalls both Virgil's underworld, 'the deep night' of death,[18] and the language we discussed from the Gospels of waiting, watching and the injunction not to sleep as 'You do not know the hour' when epiphany will come. This figure does not sleep at night but watches (*pervigil*). As such – and here the 'symbol' slips into a wider sense of allegorical reading – the watchman can be called a 'nocturnal light for travellers'. This verse recalls John 8.12, 'I am the light of the world and who follows me will not walk in darkness': the light is 'the light of life'. The night that is separated from night, therefore – the polyptoton echoing the polyptoton of the first stanza – is to be understood in a broadly metaphoric (allegorical, typological) and moral sense; indeed by splitting the signifier, segregating one sense of *nox* from another, it demands we dawn on a reading that is open to reading otherwise. The regulation of day and night is also the separation of sin and goodness, religious truth and the darkness of ignorance: hence the time of 'night' is also and always the 'night' of spiritual darkness. So the third stanza juxtaposes the rising of the sun and the dispelling of the demons of evil (9–12):

> Hoc excitatus Lucifer
> solvit polum caligine,

---

[17] O'Daly (2012) 24; better, Dunkle (2016) 78–83. On allegory, type and symbol see above, pp. 107–12.

[18] In Dido's anticipation (*Aen.* 4.26) and in Aeneas' regret for her suicide (*Aen.* 6.462) the same line ending, *noctemque profundam* links the events in a pattern of death without salvation.

hoc omnis errorum chorus
vias nocendi deserit.

By him aroused, the Morning Star
Releases the sky from darkness,
By him, the whole chorus of strayings
Deserts the ways of destructiveness.

If the first two lines of this stanza can be understood as no more than
a poetic expression for sunrise (even with the possible moral sense of *solvit*),
the second two lines (linked as they are by the repetition of *hoc* 'by him')
offer a boldly moral paraphrase.[19] The dawn chorus – a different congregation of singers from the performers of this hymn – is made up of *errorum*,
'strayings', 'losing the way', which may refer to the mistakes of demons or
to heretics – they are, of course, models of each other – who are forced by
the light to desert 'the way of destructiveness', which is both an image of
brigands and thieves forced away from the highways by the exposure
of day, and an image of the sinners of the world repenting. *Nocendi*,
which I translated as 'destructiveness', may remind us that the ancient
etymology of night (*nox*) found its root in *nocere*, 'to harm'.[20] The
cockcrow announces the dispersal of the harm of night's darkness.

The fourth stanza finally ties the image of the cock crowing precisely to
the Gospels 13–16):

Hoc nauta vires colligit
pontique mitescunt freta;
hoc ipse petra ecclesiae
canente culpam diluit.

By him, the sailor gathers strength,
And the straits of the sea calm;
By him, the very Rock of the Church,
At the singing washes away his sin.

The calming of the sea and the regathering of the sailor's strength
inevitably looks towards the miracles of Jesus, as if the 'symbol' of the
cockcrow could be instrumental, or stand for the power of Jesus. But the
final image evokes the one cockcrow of the Gospels. Jesus foretold that
Peter, *Petros*, the Rock, would deny him three times before the cock

---

[19] Ambrose's poem follows his more directly homiletic material closely, especially here *Hex.* 5.88–9: see
den Boeft (2003).
[20] Varro, *Ling.* 6.6, based on Catullus. On Isidore of Seville who also records the etymology (5.31), see
Henderson (2007).

crowed. Peter, threatened by the mob, denies, three times, that he is a disciple of Jesus. A cock crows. At the sound, Peter recognized what he had done, and in tears repented and never denied Jesus again.[21] Peter's tears are made the equivalent of baptism or a ritual of purification as they 'wash away' his sin. So later in the poem (28) we are encouraged to 'free' (*solvitur*, as the morning star 'frees', *solvit*, darkness) our sin, *culpam*, by 'weeping', *fletu* – making Peter's repentance the paradigm for each human.

The first four stanzas, then, trace the transformation of the 'now' of the early cock's crow into a repeated performance of the transformation from darkness to light, as a religious conversion of repentance and the rejection of the darkness of sin. The second four stanzas play out the implications of this transformational encouragement for the congregation of singers, and the pivot from the first four stanzas to the last four is the poem's central line, *surgamus ergo strenue*, 'let us rise therefore in strength' (as the poem turns from the second-person address to the embrace of the third-person collective). As so often in the language of the Bible, and especially in Christianity with the miracle of raising Lazarus or healing the paralysed man, the injunction 'rise up' is both literalized in images of standing and walking and climbing mountains, and always imbued with the moral fervour of religious transformation and exaltation.[22] So here the cockerel not only excites those who lie down (*excitat*, as Lucifer was *excitatus*), and upbraids the sleepy, but also he 'refutes the deniers', *negantes arguit* (20). The enemies of the Lord are those who cannot hear the sound of the wake-up call (scholars have pointlessly argued who the deniers must be, anti-Nicenes or others). So, as the cockerel sings (*canente*, 21), an echo of his singing to mark Peter's repentance (*canente* 16), as now the congregation sings, 'hope returns, health is restored to the sick, the robber's knife is sheathed, and faith returns to the fallen (*lapsis*)' (21–4). The palintropic poetics of salvation are heard in the repetition of *re-*: *redit*; *refunditur*; *revertitur*. As with Gregory, to rise up is to go back.

So, bringing the second-person and the third-person plural together, Jesus is asked to 'regard those of us who are stumbling (*labantes*), correct us with your gaze; if you do look, our falls (*lapsus*) fall away, and our sin is freed by weeping' (25–8). The 'fallen' *lapsis* (23) were once the 'tottering', *labantes* (24). But, with Jesus' help, our 'falls' (*lapsus*) fall away (*cadunt*). What stops the tottering (*labo*) becoming the falling (*labens*) is the correct-ive gaze of Jesus, which makes such slips 'fall away', a nice paradox of the

---

[21] Mat. 26.73–5; Luke 22.59–62; John 18 13–27. Discussed in Ambrose's prose at *Luc.* 10.74; *Hex.* 5.
[22] Franz (1994) 235–40.

fall. Dunkle, however, like several of the manuscripts, prints *labentes* 'slipping', 'falling', for *labantes* 'tottering' (though he translates it 'who stumble').[23] As Fontaine notes, *labentes*, which would pick up *lapsis* and *lapsus* in a single image, is metrically false, and, he adds, *labantes* is also richer theologically in that it allows for 'wavering', and does not just focus on those who have sinned. The Latin seems to anticipate the textual problem in its slippery language. Jesus 'corrects' by looking at *labantes* – by not allowing it to become *labentes*, and by 'solving' (*solvitur*) the error in the washing of tears.[24] Jesus is asked to look and correct the text of our lives, to keep *labo*, 'wavering', from (in a self-reflexive play) becoming *labentes*, 'falling' – or *lapsis* 'the fallen' or *lapsus* 'our falls'. Textual criticism of the soul . . . The *lapsus calami* that leads Dunkle to print *labentes* is also a theological *lapsus*, an oversight.

The final stanza brings the imagery full circle with a return to the language of the opening stanza (29–32):

> Tu lux refulge sensibus
> mentisque somnum discute,
> te nostra vox primum sonet
> et vota solvamus tibi.

> You, light, shine again on our senses,
> And cast off the sleep of our minds.
> May our voice sound you first
> And may we fulfil our vows to you.

As the light comes every day, so Jesus – called 'light', as so often in the post-Nicene period – is asked to shine again on their senses. The *re-* of *refulge* should not be ignored.[25] It links the repeated daily appearance of the sun, to the daily work of faith, performed in this hymn, and heralded in the sixth stanza's language of salvation. Nor should *sensibus* be translated 'soul'.[26] The contrast is between the physical observation of the sun, its impact on the senses, and the sleep of the mind, which does not allow a person to recognize the *temporum tempora*, the significance of God's eternal time in the daily routines of restoration. Now 'our voice' – the voice

---

[23] And 'tottering' in the body of the book, Dunkle (2016) 94. The reading *labentes* is defended at length by Franz (1994) 259–63, including the difficult claim that the incorrect metrics is a *lectio difficilior*. *Hex.* 5.88 has *Iesu titubantes respicit*, 'Jesus looks caringly at the waverers'.

[24] For the moral vocabulary of textual criticism, see Tarrant (2016) 30–48.

[25] As by O'Daly (2012) 56 and Dunkle (2016). Fontaine et al. (1992), as so often, is spot on in its note *ad loc*.

[26] O'Daly (2012) 57; *sensibus* is central to Dunkle's account of the hymns as a set: 'to form the congregation's senses in ecclesial perception' (Dunkle (2016) 115). By contrast see the deep distrust of the senses in Prudentius, *Ham.* 298–307 with Malamud (2011) 119–21.

that is singing the hymn, the *hic et nunc* of worship – is to sound out God first and as originator – founder – of all things. The cockcrow, which sounded (*sonat*) at the opening of the poem, has become the hopeful choir 'sounding out' (*sonet*) this hymn: they have become the heralds. Thus they can hope to fulfil their vows to God. *Solvamur*, 'let us fulfil', 'free ourselves', picks up the morning star's 'freeing' (*solvit*, 10) of darkness from the sky, the 'freeing' of sin by our tears (*solvitur*, 28): the vow is repaying what has been promised, the language fulfilling the imagery of the poem, its promise. The performance of the poem itself becomes the fulfilled vow of recognizing God's time in daily time, captured in the time of the poem. The ring composition of the poem enacts its palintropic poetics: we begin again. And do so every time the hymn is sung, adding *tempora* to *tempora*, time and time again.

This first hymn of Ambrose already indicates through its repetitions across stanzas, its polyptota and other word plays, and the careful slipperiness of its 'allegorisable' expressivity, just how dense the poetic texture of the apparent simplicity of Ambrose's poetry is. It asks us to see the immensity of God's time in the simplicity of cockcrow announcing the sun about to rise: 'to see the morning with new eyes'.[27] The hymn is about the necessary, daily reperformance of our recognition of God's time, and as a hymn it enacts this reperformance as it is sung again, daily, any day. Even if it is only read once, its imaginary is regular repetition, day by day, syllable by syllable. The chorus of the 'we' of the poem is a self-fulfilling prophecy of community, whose boundaries exclude the deniers, the fallen, the errant. The return of the poem at its end to its beginning is its own demonstration of temporality, a demonstration that sets itself against what Daniel Mendelsohn called the elusive melancholia of ring composition which recognizes that things can cohere, temporarily. For Ambrose and his choir, it is rather that things cohere temporally, *temporum tempora*, everlastingly. (Late Romantic melancholia was always an abreaction against religion.) For Ambrose, what is at stake in his hymns is a 'sanctification of time' which is 'a Christianization of the everyday', every day.[28]

*

Before we can turn later in this chapter to further hymns of Ambrose and then to their influence on Prudentius, in order to explore this Christian sanctification of the calendar as a regulation of time, we need some further

---

[27] Dunkle (2016) 98.   [28] Phrases taken from Dunkle (2016) 114; O'Daly (2012) 18.

brief background to appreciate the power of Ambrose's liturgical intervention, as 'the true inventor of Christian lyric poetry'.[29] The history of hymns and the history of the Christianization of time turn out to march hand in hand (if not always in time).

In the long and intricate history of the gradual development of Christianity as the central institutional force of the empire, there are three particular vectors that are integral to the impact of Ambrose's hymns. The first concerns the liturgical use of composed hymns as opposed to psalms. There is a long suspicion of the inherently pagan and inherently seductive nature of singing. Clement of Alexandria contrasted *skolia* (drinking songs) and erotic poetry with decent Christian hymns – but the early church 'consciously avoided original hymnody at least partly because of a caution about the music's potential to entice and deceive the crowds'.[30] Ambrose follows this easy opposition too when he attacks *mortiferi cantus acroamatum scaenicorum, quae mentem emolliant ad amores,* 'the death-bringing songs of theatrical performances which soften the mind for sexual affairs'. The Greek words *acroamatum scaenicorum* imitate Roman distaste for Greek effeminizing art, despite the fact that Ambrose was instrumental in bringing Eastern Greek theology to a broad Western audience. He sets such lethal singing against the pious pleasures of the Psalms: a gift of David 'like a heavenly conversation'.[31] In the same way, Athanasius in his celebrated *Letter to Marcellinus* sees in the Psalms a therapy for the emotions. This polarization of pious pleasure and pagan corruption masks a more complex internal division in Christian thinking, however. For the extreme asceticism of the growing monastic movement insisted on the Psalms as the royal road to prayer. 'The defining characteristic of monasticism at this time was psalmody'.[32] As John Chrysostom writes, 'In the monastery there is a holy choir of angelic hosts, and David is there first, middle and last'.[33] Epiphanius of Salamis expresses the extreme bluntly: 'the true monk must have prayer and psalmody in his heart without ceasing'.[34] We will return shortly to the demand that prayer should be 'unceasing', but the continuous recitation of psalms became a sign of a certain Egyptian monastic ideal, projected as a paradigm for a person's relationship to God. Vigils over the whole night in which psalms were recited are lovingly described as liturgical high points.[35] When the fourth-century pilgrim Egeria describes her trip to Jerusalem, she focuses

---

[29] den Boeft (2008) 427.   [30] Clem. *Paed.* 2.4; Dunkle (2016) 21.
[31] Ambr. *Hex,* 3.1.5; *Expl. Psalm.* 12.   [32] McKinnon (1991) 50.
[33] John Chrysostom, *De poenitentia.*   [34] *Apophthegmata Patrum Epiphanius* 3.
[35] Basil, *Ep.* 207 is a good case. See McKinnon (1991), (1994); Taft (1986).

her long description on the singing of psalms that dominated the liturgy in the Anastasis. This history has led James McKinnon to identify a 'later fourth-century psalmodic movement',[36] an ideological investment in the repeated performances of the Psalms as an expression of Christian faith. To institutionalize one's own hymns in the church is thus a polemical, ecclesial act.

This leads directly to the second vector, the politicization of liturgy. Athanasius attacked Arius – liturgy becomes a weapon in pro- and anti-Nicene rows – by accusing him of writing 'drinking songs'.[37] He describes Arius' *Thalia* in the clichéd critical language of professional rhetoric: effeminate, dissolute, corrupting. Gregory of Nyssa called Eunomius' poetry 'Sotadean'.[38] This accusation has led to some inconsequential attempts to judge the scansion of Arian verse.[39] Gregory's point rather is to associate his opponents with pagan tradition by any insult that will stick: a heretic's poetry must be bad poetry, corruptingly pagan. Socrates' *Church History* (6.8) gives a vivid account of Arians entering the city of Antioch and singing antiphonal songs all night with 'insulting verses' that had been adapted to an anti-Nicene agenda. John Chrysostom organized counter-choruses of singers, 'a good plan', comments Socrates, 'but it resulted in riots and danger'. The singers ended up fighting, and people were killed.[40] Competing performances of hymn singing become the mode of political conflict between competing religious groups. Ambrose as bishop of Milan became embroiled in such conflicts.[41] Although he 'handled the "Arian" legacy with tact',[42] the clash of his power as bishop with the emperor and the court was crucial to the spectacular 'politics of heresy' in this era. Nonetheless, Ambrose's hymns, notwithstanding a certain grumbling from his enemies, did not aggressively proclaim their evident and informing pro-Nicene ideology – especially in contrast with the Cappadocian writing we have been discussing – and did not result in the sort of tumult that Socrates describes in Antioch. It was possible for the hymns not to enflame what was 'the diffuse set of allegiances' in Milan.[43] The success of the hymns is testimony also to their political savvy.

The third vector is the most complex and the most salient, and concerns the times of prayer in Christian liturgy. There is a string of passages in the New Testament which suggest that continual prayer is an ideal. It is the message of the parable of the widow and the judge according to Luke (18.1); Paul puts

---

[36] McKinnon (1994).   [37] *Apol. contra Arian.* 1.5; *De synod.* 1.15.2–3.   [38] *Contra Eun.* 1.1.17.
[39] Stead (1978) is a classic example (and includes further bibliography).   [40] See Stanfill (2019).
[41] See in particular McLynn (1994) and the forceful Williams (2017), both with extensive bibliography.
[42] Brown (2012) 123.   [43] Williams (2017) 8.

the demand most strongly: 'Pray at all times in the Spirit' *en panti kairōi en Pneumati* (Eph. 6.18), and – directly – 'Pray continually (*adialeiptōs*)' (1 Thess. 5.17), an extension of the commands to be ever watchful. As we have seen, Epiphanius recognized continual prayer therefore as an ideal of monastic living. So, St Martin, a founding figure of Western monasticism, in the hagiographical prose of Sulpicius Severus, which we will discuss in the next chapter, fulfils the obligation literally: 'Never did a single hour or moment pass in which he was not engaged in prayer or focused on reading; even while reading or if it happened he was engaged in something else, he never relaxed his mind from prayer'.[44] Origen explores what this might mean, however. 'He prays without ceasing who combines his prayer with necessary work, and suitable activities with his prayer': indeed, he continues, 'the only way we can take Paul's saying 'Pray continually' to be possible is if we can say the whole life of a saint is one mighty integrated prayer'.[45] So Origen both requires that prayer is integrated into a holy life, and understands 'continually' to mean at points throughout the day. His first example is Daniel who, 'when great danger threatened him, prayed three times a day'. By the fourth century, it was usual in the monasteries to say psalms and pray at morning, night and at the third, sixth and ninth hours (terce, sext and none), as Egeria carefully records. John Cassian, who toured Egypt and helped transfer monastic ideals to the West, considered this to be a 'tempering of the perfection of the Egyptians'.[46] Fixed times of prayer was both a regulation and a compromise.

Indeed, fixed times seems to have been a requirement of Christian order from early days.[47] Clement of Rome's *Letter to the Corinthians* specifies 'set times (*kairoi*) and hours (*hōrai*)' for services, though it is far from clear what this would have meant in the first century. Tertullian in the second century specifies third, sixth and ninth hours for prayer,[48] and Hippolytus of Rome in the third century in his *Apostolic Tradition* lays out (41) a full day of prayer and ritual, starting with prayer and a teaching at dawn, allowing prayer at home at the third hour (and in your heart if you are out and about), and moving through to nighttime rituals: if your wife is not baptized, you should at midnight wash and retreat to another room to pray. Each of these fixed hours is also invested with a fixed symbolism. So Hippolytus explains the sixth hour in this way (41.7):

> Pray also at the sixth hour. Because when Christ was attached to the wood of the cross, the daylight ceased and became darkness. Thus you should pray

---

[44] Sulp. Sev. *VM* 26.3.   [45] Origen, *On Prayer* 12.2.   [46] *Inst.* 3.2.
[47] For this history see Bradshaw (1981); Taft (1986).   [48] *On Prayer* 25.

a powerful prayer at this hour, imitating the cry of him who prayed and all creation was made dark for the unbelieving Jews

The daily prayer at its specified hour is to recall the specified hour of the crucifixion, and its temporal miracle, including the cry of Jesus on the cross, a key moment in Christology's understanding of the humanity of Jesus, and a contrast with the Jews – a typical polarizing self-definition. Clement of Alexandria takes a slightly different perspective. The prayer at night is necessary to fulfil the biblical injunction to be vigilant and wait. 'A sleeping man is of no more use than a dead one. Therefore at night we ought to rise often and bless God.'[49] And, quoting Paul, Clement concludes that Christians are 'sons of light and sons of the day. We are not of the night or of darkness', darkness that is ignorance. It is against such a tradition that Ambrose's first hymn resounds, both as a hymn for the moment before dawn and as an explication of the significance of light and darkness in the performance of Christian worship. Regular and specified times of prayer by the fourth century had become fully institutionalized in the Christian churches, and observed with especial fervour in monastic establishments.

It is intriguing that Robert Taft SJ, in what is still the standard overview of the history of the liturgy of the hours, begins with the blithe assertion that 'these are the natural prayer hours in any tradition'.[50] From within his own commitment to his own tradition, and with little anthropology or history of systems of time, Taft normalizes his own practice as what is natural, or should be natural, 'in any tradition'. Yet in the Greek and Roman societies in which Christianity was shaped, there was simply no equivalent of such regulation. There were *no* daily prayers, either personal or civic, and certainly no suggestion that there should be a fixed time for sacrifice or libation.[51] Christianity's sanctification of the everyday and its hours as a moralized memorial of the New Testament narrative is markedly alien to the long history of Greek and Roman civic and personal practice. What is more – and this is equally traditional, alas, and equally ideological – Taft selectively misrepresents Jewish cultic practice, out of which Christianity also developed, and systematically silences its theoretical discussions of temporality.

---

[49] *Paed.* 2.9.   [50] Taft (1986) 11.

[51] This generalization is not undermined by the claim of Marinus – a fifth-century CE Palestinian student of Proclus – that Proclus prayed to the sun morning, noon and night (*Vit. Procl.* 22), nor by Plato's claim that Greeks and barbarians worship the sun at morning and evening *Laws* 10, 887e. For the Roman sense of the diurnal pattern, see Ker (2019); Martial (discussed above pp. 80–1) has none of the ritual rigour at stake here.

The history of Jewish practice in terms of its temporal commitments is especially complex because of the rupture in time forced by the destruction of the Temple of Jerusalem in 70. Judaism is unique in the ancient world in that its cultic practice was focused on a single site, a single Temple, in Jerusalem. This is where sacrifices and cultic celebrations took place, both as daily events and as the pilgrim festivals. Jews across the Mediterranean paid an annual tax to the Temple; but their local practice, even in Alexandria, is very hard even to evidence.[52] After the destruction of the Temple, cultic practice had to change and placed greater emphasis on its local centres (synagogues) and family-based events (the home). Events that could only take place in the Temple had to be re-conceptualized and relocated or ended. This long process – the development of rabbinic Judaism – also involves retrospective rewriting of history to make the contingencies of the contemporary more invested with the authority of tradition.[53] What is more, the development of rabbinic Judaism also increasingly privileged the study-hall as a place of religious activity, to the extent that study of texts could be said – by committed teachers and students of these texts – to be more important than prayer.[54] Study has no time regulation, although it can appropriate the same self-serving idealism that dismisses all other activity as a distraction from what should be a continual process of learning. In short, there is a changing social, political and religious map of practice, both over time and between different communities – which develops not only within the contours of its internal dynamics but also in interaction with the surrounding communities and their practices. Across this changing map, there is also a changing and intense discussion about temporality.

Now, it is indeed the case that the Pentateuch specifies that there should be a morning and evening sacrifice every day at the Temple, with additional sacrifices for sabbaths, new moons and festivals, as well as recognizing the contingent sacrifices of sin offerings, or personal vows, and the like. So, too, it is reported that a particular Psalm was specified for each day of the week to be sung in the Temple by the Levites, and that specified sets of Psalms were sung on festivals, the Hallel at Passover for example, the fifteen Psalms of Ascent, which may have been sung on the fifteen steps of the Temple during the pilgrim festivals.[55] The calendar, as we have already indicated, was a major source of theoretical and political debate across

---

[52] For the variety and difficulty of the evidence for prayer in the Second Temple era, see Flusser (1984).
[53] See e.g. Schwartz (2001), (2009).   [54] Reif (1993) 95–103.
[55] For the range of evidence of types of prayer material, see Bradshaw (1981) 1–23; Flusser (1984); Reif (1993).

Jewish history in antiquity, to which texts from Hellenistic Judaism, such as Jubilees, or, from later, the Seder Olam, indicate the range of detailed engagement of scholars and religious thinkers with the scope of time.[56] Taft takes such material – which he passes over in barely a handful of pages[57] – as a sign that Judaism is no more than a poorly articulated and thus barely influential model of liturgical time for the Christian divine office. Yet it is fundamental that the temporal requirements for sacrifice or prayer in the Pentateuch never indicate a specific hour beyond 'morning', 'evening', or 'when you rise', 'when you lie down'. Indeed, by the time that the Talmud is edited and written down in late antiquity, this openness is the prompt for an extended discussion of how time is to be experienced.

This discussion makes the boundaries and limitations of duration, rather than the fixing of the hour, its necessary subject. We discussed in chapter 3 how the timing of the evening prayer of the *shema* is discussed in the Talmud by constructing analogies with other ritual and social processes, and how its initial questions open out into an extended debate about the marking out of temporal zones and their boundaries and repetitions. We can now add a further example from that Talmudic discussion which will serve to mark the difference between the Christian and Jewish regulation of time. The first *mishnah* of the tractate *bBerakhot* offers a *ma'aseh* case, that is, a narrative example of what actually happened which is thus determinative: 'It happened (*ma'aseh*) that Rabbi Gamliel's sons came back from a taverna. They said "we have not yet said *shema*". He said to them, "if the light of dawn has not risen, you are obligated"'.[58] The story is told as evidence of Gamliel's assertion that the evening *shema* can be said until dawn, against Rabbi Eliezer's claim that it can be said only until the end of the first watch. Gamliel practised his theory. The text will go on to worry further what the rising of the light of dawn actually is, as the detail of what 'evening' as a time-frame could be taken to mean is explored.

There are two striking ways in which this text is different from the Christian regulation of time we have been examining. First, it embeds such discussion not in a monastery or a church but in the scene of a father waiting for his children to come home late from the pub, a father who happens to be a major rabbi. When the children see him, they admit they have not yet said *shema* and he indicates they still have time. It is a sharp and amused dramatic scene that demonstrates how even in a rabbinical

[56] See above pp. 173–7 and on Seder Olam, Milikowsky (2014) and on Jubilees VanderKamm (2018).
[57] Taft (1986) 5–11; a much more judicious and fuller discussion in Bradshaw (1981) 1–23.
[58] *bBerakhot* 2a.

household there has to be a certain flexibility about time within the family dynamics of an urban social life. The tale is told without express moralizing, and, unlike the domestic framing of many martyr lives, there are no consequences of the children's behaviour or their father's admonition. Second, the issue of time is precisely not about defining a fixed time for prayer and sticking to it, but working out what the duration of 'evening', 'when you lie down', is, and thinking about the extension of such a *time-frame* in the lived experience of domestic life. It is a matter of boundaries rather than precision of the fixed hour. So, paradigmatically, the halachic conclusion of the Talmud is that the morning *shema* should ideally be said before the end of the third hour, and even then it grants a longer allowance for the *Amidah*, the central prayer of the morning service – and, as 'ideally' indicates, there are provisions made for the grey area of the 'what if it is said later'. The insistence of Christian theological writers in their search for a liturgy of hours that prayers should be said at dawn, third, sixth, ninth hours and at the evening and in the night, far from embodying what is 'natural to any tradition', is distinctively different both from Greek and Roman cultural norms, and from the practice and theory of Jewish temporality.

Ambrose's hymn for the dawning hour is therefore also a telling moment in the history of liturgical worship as an expression of time. Now, for the Christian, time is to be organized and regulated to the hour in the liturgy of the hours. The hour requires prayer embodied in a communal song, which is not a Psalm, but which plays a public part in the contentiousness of theologically informed religious and political power. The song is not just praise of God, but a reminder of the narratives of the Gospels, tinged by theological import, and designed by its repetitive performance to construct a community of worshippers – and above all to mark out time as Christian rather than Greek, Roman or Jewish: a performance of Christian temporality.

*

The fourteen hymns of Ambrose combine three interlinked orderings of time. The first four hymns are for times of the day, the hours of prayer: the second hymn is for the rising of the sun, the third for the third hour, and the fourth for night time. The fifth, seventh and ninth hymns are for Christmas, Epiphany and Easter, that is, major festivals of the church which celebrate the events of Jesus' narrative. The remaining hymns each celebrate a saint's day, starting with the evangelist John (Hymn 6) and including Agnes (8), the passions of the apostles (12), Laurence (13), and the new saints of Milan, Victor, Nabor and Felix (10), and Protasius and Gervasius (11), and ending with All Saints (14). Overlapped, therefore, are the history of the Church from the birth of Jesus through the Gospels, the Apostles, the martyrs up to the most recent additions of the canon of Saints in Milan itself; the calendar of ritual over the year, Christmas, Epiphany, Easter; and the daily expression of worship, from before dawn to night. The three orderings are mutually implicative and normative. Not only are the calendars of daily worship, major festivals and saints' days expressions of the continuity of Christian time, where each moment and level of chronology implies each other – daily time and eternity, *temporum tempora*, interlinked – but also the exemplars of the saints which imitate the model of Jesus become a paradigm for the worshipper's faith. The collected hymns articulate a Christian ideology of time.

This ideology is expressed both in the continuing use of vocabulary and expression from the first hymn that creates an incremental semantic network between the different hymns, and in the explicit language of temporality. In the opening stanza of Hymn 2, Jesus is celebrated with the language of the Nicene creed as *de luce lucem* (2), 'light from light', and *lux lucis* (3), 'light of light', which is glossed as *dies dierum illuminans* (4), which can mean both 'shining as day of days', or the 'day illuminating of days'. The first hymn made the cockcrow 'the herald of the day', which had its allegorical sense; here Jesus is the day – as opposed to the usual image of the sun or the light. Thus Jesus is celebrated both as a transcendent light ('day of days') and as the day which gives light to the pattern of the days, the unfurling of time. The ambiguity is telling, and turns a standard Hebrew linguistic pattern ('king of kings', 'lord of lords') into a more semantically challenging Latin expressivity. Similarly, the Sun/Jesus is asked to descend – *inlabere* (5) – as Apollo in the *Aeneid* is asked to 'descend into our spirits' (*Aen.* 3.89), but the term takes on a heightened sense after the play between slipping and falling in the sixth and seventh stanzas of the first hymn: the descent of the sun is also to help humans against the *casus asperos*, the 'harsh falls', 'the tough events', of life (15). In the same way, as the first hymn encouraged us to rise up *strenue*, 'with strength', here God is asked to fill 'our acts with strength' (*strenuos*, 13). So, the final two stanzas play out the connection of each and every day with the idea of Jesus as the 'day of days' with the same turn to a moralized time that we saw in the first hymn. *Laetus dies transeat!*, 'Let the day pass in joy!', seems a standard morning prayer for the coming day to go well, but the joyfulness is a moral life: 'May modesty be like dawn! May faith be like noon! May our minds not know twilight!' The day is not just metaphorized – pure like the first rays; faithful with the ardour of the sun at noon; resisting the doubtfulness of half-light – but constructed as a tripartite eschatological sequence that is fully pro-Nicene in its theology (29–32):[59]

> Aurora cursus provehit;
> aurora totus prodeat
> in Patre totus Filius,
> et totus in Verbo Pater.
>
> Dawn advances its course.
> May the complete dawn come forth,
> The complete Son in the Father,
> And the complete Father in the Word.

The dawn, like the 'day in joy', is a marker of the time of the poem: it advances its *cursus*, a term which recalls the use of *dromos* and *kuklos* that we have seen in the Greek discussions of the stars and astrology – the regular circle of time passing. But *aurora* is also a symbol of the resurrection and a metonymy of Christ[60] – hence 'complete', 'perfected', 'whole'. *Aurora*, a feminine noun, has a masculine adjective *totus*, as for Peter *ipse*, 'himself' (masculine) qualified the Rock, *Petra* (feminine) (1.15): some copyists tried to correct the grammar, but the assimilation here to the Son and the Father, as with the association of Peter with his etymologized and symbolic name, twists grammar to an ideological – allegorizable – agenda. So this grammatically imperfect but 'perfect dawn' finds a fully Nicene Trinitarian expression: Father, Son and Word are as one. The return of each dawn is thus to be viewed as the visible sign of the resurrection of Jesus and of the Nicene theology that explains it. Again, the ring composition that insists on its own circular language (*cursus*), takes us back to the opening line of the poem, *splendor paternae gloriae*, 'Splendour of the Father's glory', which can now be fully appreciated not just as indicating the light of the sun as the visible emanation of God's glory, but also more specifically as the overlap between Jesus as 'the glory of God' – the passion of Jesus is called *doxa theou* – and Jesus as the sun. Ambrose wishes to make each moment of the day embedded in a religious understanding, a Christian vision of temporality.

The third hymn announces *iam surgit hora tertia*, 'Now the third hour rises' (1) – the *hic et nunc* of liturgical worship. But this hour matters because 'This hour (*haec hora*) is the one that gave the end to the ancient wicked crime' (9–10), that is, it is the time of day of the crucifixion that

---

[59] As noted by Fontaine et al. (1992) *ad loc.*   [60] See Fontaine et al. (1992) *ad loc.*

transformed the original sin into the Christian promise to transcend death. *Haec hora*, the very mark of 'now', is valued – each and every day – as the mark of the transcendence of time. Hence, the final hymn concludes by 'asking now (*nunc*)' that Jesus joins his praying servants – those singing the hymn – in the fellowship of martyrs 'for the centuries of eternity', *in sempiterna saecula*. This last poem began by praising the *aeterna Christi munera*, 'the eternal gifts of Christ'. The final ring composition is the circle of eternal time, which is begged for now in the moment of hymning. The singers are striving to fulfil the imitation of the saints they celebrate, to turn their now into a fulfilment of the promise of eternity.

The meaning of a saint's day is also explored in a fascinating passage from Ambrose's prose treatise on virginity. Once again he overlaps light with faith, and night or mist with uncertainty or the heresy of dissent. But he also takes the role of a day and its relation to celebration in a different direction – a remarkable gloss on what *dies dierum illuminans* (2.4) or *diei iam sonat* (1.5) might signify:

> Bona lux quae perfidiae discussit caliginem, fidei diem fecit. Dies factus est Petrus, dies Paulus, ideoque hodie natali eorum spiritus sanctus increpuit dicens: Dies diei eructat verbum, hoc est ex intimo thesauro cordis fidem praedicant Christi.

> This is the good light, which shattered the mist of faithlessness and made the day for faith. Peter has been made a day, Paul has been made a day, and thus on today their feast-day the Holy Spirit cried out saying: 'Day speaks forth the word to day', that is, they preach the faith of Christ from the deepest storehouse of their heart.[61]

The thought that light makes the day, and changes *perfidia* to *fides*, 'faithlessness' to 'faith', leads to the instrumental power of the examples of Peter and Paul, whose martyrdom – and faith – is central to Hymn 12. Peter 'became a day, Paul a day': the saints are transformed into their days, the day that the light of faith makes. The day, its time, is the embodiment of the saint (the day made saint). On each saint's day, the Holy Spirit declares that this day – the *hodie* of the liturgical time of celebration – sends forth *verbum* – the word of faith – to the other days, day, as it were, speaking to day, announcing (*praedicant*) faith. The sentence *Dies diei eructat verbum* is a quotation from Psalm 18 (19).3, a Psalm which we will shortly see is paraphrased in Ambrose's Hymn 5, and thus the Holy Spirit in quoting the Psalm performs the fulfilment of scripture. The days are linked in a conversation where the matter is the word

---

[61] *De Virgin.* 19.125.

of God, *verbum*.[62] In Ambrose's vision, the festal calendar becomes a *catena* of embodied sound proclaiming the *Verbum* of God – the chain of God's word in time, day to day.

These hymnic strategies that repeat words in changed senses to build up an incremental network of theologized meaning, that use ring composition to reinforce a formal palintropic poetics of redemption, and that transform narrative in and out of the here and now of performance and the permanent moral struggle of faith or the cosmological temporality of Christian history, recur throughout the hymns, and although I have analysed my examples here in the order of the hymns' modern edition, the strategies work across the whole collection in whatever order the hymns are sung, and sung again, and again. Yet there is one further aspect of Ambrose's writing that links these hymns into my earlier discussion of Nonnus. For what I have called the hymns' palintropic poetics is linked, as in Nonnus, to the practice of paraphrase.

The most patent appropriation of scripture is in Hymn 6, a celebration of the evangelist John, where the fisherman is described as hooking the word of God. This leads to a quotation of the first lines of the Gospel of John, fitted, word for word, into a stanza. It requires a certain *laissez faire* treatment of metrics, but has the effect of tying the hymns into the authoritative tradition of the Gospels, so that singing the hymn here is also reciting the Gospel, indeed the Gospel's foundational expression of time, *in principio erat* – with the result that reading the Gospel will echo with Ambrose's music and recontextualization of the verses. Hymn 5, however, is a remarkable demonstration of Ambrose's paraphrastic writing at work.[63] The first five lines reads (5.1–5):

> Intende, qui regis Israel
> super Cherubim qui sedes,
> appare Ephraem coram, excita
> potentiam tuam et veni.
> Veni redemptor gentium.
>
> Give ear, you who rule Israel,
> Who are enthroned upon the Cherubim,
> Appear before Ephraim,
> Rouse your power and come.
> Come, redeemer of the nations.

---

[62] Singing Psalms, for Ambrose, is a 'a model of heavenly conversation for us', *caelestis nobis instar conversationis*: *Expl. Psal.* 12.

[63] What follows builds on the excellent analyses of Fontaine et al. (1992) *ad loc.* and Dunkle (2016) 122–4.

These lines closely follow Psalm 79 (80) in the Latin translation:

> Qui regis Israel, intende qui deducis velut ovem Ioseph, qui sedes super Cherubim, appare coram Ephraim et Benjamin et Manasse, excita potentiam tuam et veni, ut saluos facias nos.

> You who rule Israel, give ear, you who leads Joseph like a flock, who are enthroned upon the Cherubim, appear before Ephraim, Benjamin and Manasseh, rouse your power and come, so that you may make us safe.

Ambrose removes the image of the shepherd marshalling Joseph like a flock, although it motivates the mention of Ephraim and Manasseh, Joseph's sons and the founders of the tribe of Joseph, as well as Benjamin, his favourite brother and another founder of a tribe of Israel. As Fontaine notes, this redrafting is not just for brevity or to fit the metrics. The Psalm is a plea to save us – this and the next verse, 'show us your favour that we may be delivered', are emphatically repeated three times in the Psalm – and 'us' is the tribes of Israel set against the enemy who are the nations. Ambrose has turned the Psalm's national plea into a universal assertion of the redemptive power of God. The paraphrase, here in a hymn on the nativity of Jesus, is also expressing the supersessionist insistence that Christianity is a promise to all mankind. Ambrose 'embeds a universalizing theology' in his hymn, which 'expresses this rupture' with Judaism that, for him, the birth of Jesus encapsulates.[64]

This adaptive paraphrasing of the Psalms continues in the hymn. Here are verses 17–24:

> Procedat[65] e thalamo suo,
> pudoris aula regia,
> geminae gigas substantiae
> alacris ut currat viam.

> Egressus eius a Patre,
> regressus eius ad Patrem;
> excursus usque ad inferos,
> recursus ad sedem Dei.

> Let him come forth from his chamber,
> the royal palace of modesty,
> a giant of twin substance,
> swift to run the race.

---

[64] Dunkle (2016) 123; Fontaine et al. (1992) 279 *ad* 5.5.

[65] Dunkle (2016) reads *procedit*; some manuscripts, in parallel to the Psalm, *procedens*. I follow Fontaine et al. (1992) here.

He went out from his Father,
He came back to his Father.
He journeyed to the dead,
He journeyed back to the throne of God.

These lines paraphrase Psalm 18 (19), though the gestures of appropriation here are far more aggressive.

> Et ipse tamquam sponsus procedens de thalamo suo, exsultavit tamquam gigas, ad currendam viam; a summo caelo egressio eius. Et occursus eius usque ad summum eius.

> And he comes forth like a bridegroom from his chamber, and rejoices in running the race, like a giant; his coming forth is from the height of the sky. His circuit is to the height of it.

The subject of the Psalm (*ipse*) is the sun, which appears like a groom from his marriage chamber; and like a 'giant' (a strong and motivated translation, as we will see, of the Hebrew *gibor*, 'hero', 'mighty') who runs a race. The sun's journey is a circuit across the heavens. This depiction stems from the Psalm's opening, 'The heavens declare the glory of the God.' But Ambrose has made the subject of his verses Jesus (an easy transition after the association of Jesus and the sun that we have already seen in the hymns). Jesus comes forth like a bridegroom, but his chamber now, with the expected ascetic sexuality, must be qualified immediately as a palace of chastity (*pudor*). Jesus is a 'giant', but in an extraordinary exegetical gloss, a giant 'of twin substance'.[66] The salient reference here is to Genesis 6.4, where the mysterious *Nephilim* appear. This is the time on earth when there are divine beings who sleep with humans: it is the moment which leads God to express his wrath that results in the flood. The *Nephilim* are described in Hebrew as *giborim*, which is translated as 'giants', *gigantes*, in the Septuagint,[67] rather than its usual rendering as 'mighty one', 'champion', 'hero', which may explain why *gibor* in Psalm 18 has been translated in Latin and Greek as *gigas*, 'giant', with the correspondingly strange image of a giant 'running a race'. The *Nephilim* are divinities who couple with mortals and produce offspring. Jesus is both divine and human.[68] Hence with

---

[66] A striking phrase quoted by Augustine *Contra sermonem Arianorum* 8.6. Augustine also notes the singing of this hymn in his *Serm.* 372.3: see Daley (1993) 492 n. 40.

[67] Rashi notes that *Nephilim* in Hebrew means 'giants'. He also notes its sense 'fallen', which led to *Nephilim* being seen as angels. Rabbinic interpretation of this passage stretches from cross-class marriage to angels sleeping with humans.

[68] Ambrose explains this sense also as *De incar. domin. sacr.* 5.35: *consors divinitatis et corporis*. Augustine explains the word *gigas* in the same way, *Enarr. in Ps. XVIII* and elsewhere see Daley (1993) 492 fn. 41.

an arresting combination of intertextual play between Genesis and the Psalm, on the one hand, and Christological theology on the other, Jesus here is a 'giant of twin substance', a phrase designed to proclaim the essential human and essential divine combined in Jesus, a principle of Nicene theology.[69] In the Psalm, the sun goes across the vault of heaven in its circuit. But in Ambrose's hymn, Jesus goes from the Father and back to the Father, and circuits down to the underworld and back up to the throne of God. The repetitions of *e-* and *re-* along with the repetition of *-cursus*, 'circuit', enact the palintropic poetics we have been tracing: the *occursus* of the Psalm is tellingly transformed into *excursus* and *recursus*. The return of Jesus is integral to the narrative of Jesus, and must be emphasized in this way. The paraphrase of the Psalm rewrites the sun's journey as Jesus' resurrection and the triumph over time.

This intermingling of narrative and theology is hard to read without complicity or rejection. Brian Dunkle SJ, who, as I have said, has produced the most sophisticated literary reading of Ambrose's hymns, is certainly on the side of complicity. 'The Christological reading of the Psalms ... does not compromise their integrity as texts about the Lord of the Old Testament', he writes, 'Ambrose retains a plain interpretation of the texts even as he offers a Christological or allegorical interpretation for his particular catechetical ends'.[70] When an image of the sun from several hundred years before the birth of Jesus is read as the figure of Jesus, how can 'the integrity' of a text be said to remain uncompromised? How can the 'plain interpretation' of a text about the sun be reconciled with a Christological – a supersessionist – interpretation? As if Dunkle using the term 'Old Testament' itself is not already fully complicit with this supersessionist ideology. Who is outside the choir when the hymn is sung? According to the self-serving rhetoric of the hymns themselves, only the deniers, the ignorant, those who live in darkness ... Such is the power and success of the hymns: to create the world of worship where singing from this hymn sheet is naturalized as the horizon of expectation for the community. For Dunkle, subscribing as he does to Ambrose's agenda, the compelling supersessionism of Ambrose is silently passed over in the easy assumption of the Christian vocabulary of 'Old Testament' and the comfortable assertion of its integrity, its plainness. As with Paul Tillich or Robert Taft, faith determines reading, and the violent logic of Christian supersessionism continues to distort scholarship.

*

[69] Daley (1993) 481–2. [70] Dunkle (2016) 125.

Prudentius read Ambrose intently. Prudentius, who combined his poetry with a career in imperial service, was roughly contemporary with Ambrose, though he lived longer than the bishop; he was engaged in the political and theological encounters that also dominated Ambrose's career. Ambrose wrote two letters against Symmachus, the leading non-Christian senator who had proposed re-establishing the altar of Peace in the Senate – a sacrifice at this altar opened the senatorial sessions, and it had therefore great symbolic valence – and Prudentius also wrote two lengthy hexameter poems *Against Symmachus* (with prefaces in asclepiads and glyconics, good Horatian metres): Prudentius' case depends on setting Roman imperial history into a new Christian eschatological narrative. The *Peristephanon* is a series of poems which dramatize martyr's bloody deaths, and the poet's response to their suffering.[71] One of the longest of these poems (2) relates the passion of St Laurence, as does Ambrose's Hymn 13;[72] the twelfth treats the martyrdom of Peter and Paul as does Ambrose Hymn 12. The growing cult of martyrs links the two poets' liturgical and spiritual interests.[73] Prudentius' *Psychomachia* dramatizes the internal struggles of the soul in the form of a classical and violent battle of the gods;[74] and in barely less bloodthirsty and certainly no less lurid a form, his *Hamartigenia*, 'Origin of Sin', depicts the sins of humans stemming from Satan's rebellion.[75] Ambrose's insistence on 'internal vision', the education of the senses towards a spiritual comprehension of the physical world, and the need to avoid the 'falls' of sin, offer a similar theological framework. But it is in Prudentius' *Cathemerinon* that his specific engagement with Ambrose is paraded.

The *Cathemerinon* is a collection of twelve hymns.[76] Several have the same subjects as Ambrose's hymns; four are in the same metre; Ambrose's language is echoed both boldly and subtly. *Cathemerinon* is a Greek word meaning 'daily' poems, or poems 'for the day by day', and each poem, and the collection as a whole, is engaged with setting a sense of daily time into a theological framework. Although sections of these works have made their way into modern liturgical hymns,[77] they are long poems, in multiple metres, and do not seem to have been designed, as Ambrose's hymns are,

---

[71] See Palmer (1989); Roberts (1993).   [72] See Conybeare (2002); Walter (2020) 194–208.
[73] Roberts (1993); Palmer (1989). On martyrs see Shaw (1996); Brown (2014); Grig (2004).
[74] Nugent (1985); Mastrangelo (2008); and now Pelttari (2019).
[75] Malamud (2011) is the best introduction to this text, along with Dykes (2011).
[76] O'Daly (2012) is a necessary starting point.
[77] O'Daly (2012) 383 notes examples. The most important, if distant descendant is perhaps Keble's *The Christian Year*, which was the highest-selling poetry book in the nineteenth century, with nearly 400,000 copies sold in Britain alone, and which has a similar 'day by day' lyric format. See Blair ed. (2004).

for repeated liturgical use. Where Ambrose's hymns are for a congregation led by a bishop, Prudentius' *Cathemerinon* is probably written for a more restricted readership.[78] 'Christian identity meant re-probing and re-surveying the boundary between public and private', writes Bowes, and she notes that after Nicaea there developed 'an increasingly shrill debate about how the private was to be valued in the face of a new kind of public'.[79] Prudentius' poetry also marks a porous boundary between acts of worship and acts of reading, exegesis and literary performance. At multiple levels, Prudentius' collection is an extended reflection on Ambrose's project. Its language is far more violent, even brutal in its martyr narratives; his politics are more strident; his focus on the suffering of this world is insistent and more aggressive than any of Ambrose's depictions of sin and punishment; his parade of classical learning is more explicit. (All of which no doubt helps explain why in the modern institutions of classics Prudentius is part of the canon, while Ambrose rarely features.) In the case of both Nonnus and Gregory we saw a struggle over what the register of Christian language should be, and over what audiences were projected by such Christian writing. Between Ambrose, whose hymns become part of the everyday worship they depict, and Prudentius, whose long, classically formed poetry engages in literary exposition of *exempla* from the Bible, we can see a similar strain between more popular and more learned expressivities, particularly signalled by Prudentius' express engagement with Ambrose's project.[80] Prudentius shares Ambrose's concern to construct Christian daily time, but shifts not just the focus of comprehension but also the mode of intellectual engagement. Prudentius is more present as a figure in his poems, and far more interested in the suffering self, the internal anguish of self-recognition. These are poems for reading, for personal absorption, absorption in the celebrating, suffering self. As Mastrangelo has argued at length, Prudentius' Christian hero is an individual struggling in his soul and for his soul, for whom the martyr is the prime figure of imitation and admiration.[81]

---

[78] Hershkowitz (2017) following Bowes (2008) calls him a 'villa-poet'.

[79] Bowes (2008) 216. Ambrose (Bowes (2008) 203) was opposed to *occulta consilia in domibus*, and in this rhetoric 'private worship was the mark of a heretic' (198).

[80] On Prudentius' classicism, a greatly discussed topic, for examples see Mastrangelo (2008) 14–40 (Virgil); Malamud (2011) 120–3 (Lucretius); Lühnken (2002) (Horace and Virgil); Malamud (1989); van Assendelft (1976); and also on Horace see e.g. Pucci (1991) with bibliography back to Breidt in 1887. Pucci's conclusion that Prudentius was a great poet who happened to be a Christian, however, severely distorts Prudentius' agenda and poetics, but is typical of one strand of criticism within Classics.

[81] Mastrangelo (2008).

This shift is evident from the Preface, which was probably attached to a large-scale edition of his poetry, but also can act as a significant prelude to the hymns. It is a 45-line personal life history and *confessio*. 'I have lived for five decades, if I am not mistaken', it begins, 'and, on top, the seventh pole (*cardo*) turns the year, while we enjoy the circling (*volubilis*) sun'. Fifty-seven years old, then, but the familiar language of astronomical time will turn out to be programmatic for the collection. Prudentius runs through his decades with summary tales of sinful development – at school he learned to lie; in youth, lustful brazenness; then a career in law, leading to government and imperial preferment. This life is but a priamel, however, for his impending death. Now at the very end (*fine sub ultimo*, 33), his sinful soul can cast off its stupidity and he can celebrate God at least with his voice. Consequently, he prays to let his soul 'link the days with hymns', *hymnis continuet dies* (37). As Ambrose desired to have 'day speak to day', Prudentius too will construct a chain of days through their hymns, organize time through celebration of God and the martyrs.[82] And Prudentius is clear about his project's polemical stance: 'Let my soul fight against heresies, debate the catholic faith, trample the rites of the nations, bring destruction, Rome, on your idols' (38–41). Ambrose rejected the 'deniers', 'those in ignorance'; Prudentius, with a characteristic intensification of his polemical language, is up for a fight (*pugnat, conculcat, labem . . . inferat*). Yet his literary self-reflexivity is ever present. He ends (43–5) by hoping to be 'free of the chains' of his physicality and let his voice go wherever it will until its last sound. *Liber*, 'free', puns on the word 'book', *liber*, often in Latin poetry encouraged to go where it must, and links the finality of composition with the end of the poet's life.[83] The last word of the poem is *ultimo*, 'last', recalling the *fine ultimo* he anticipates. Unlike Ambrose, whose self-representation is absent from his hymns, the voice of the poet Prudentius is placed first and last. As a preface, this poem has been read as indicating Prudentius' connections with the classical tradition, especially with Horace.[84] It also acts as a programmatic opening to the hymns, demanding that the collection should be read as the evidence of a life, a perspective on a life, and that the 'day by day' is to be comprehended as the threat and promise of the experience of human time, for each and every reader as an individual. It is life as a protreptic: an *exemplum* of what is to matter in the day to day.

[82] Pelttari (2019) translates *continuet* as 'fill', which misses the sense of the *catena* of days. Every medievalist will know why I emphasize *catena*.
[83] Malamud (1989) 77. On the tongue see Ballenger (2009) 91–125.
[84] See e.g. Pucci (1991) 679–85.

One organizing structure of the *Cathemerinon* seems clear enough. The twelve poems are structured in six pairs of poems, where the first three pairs focus on specific types and times of daily activity, and the second celebrate forms of religious practice and their temporality, moving from weekly activity, to the broadest cosmological time, to specific festivities of the calendar.[85] So the first pair of hymns echoes Ambrose, with a hymn for cockcrow and a hymn for the sunrise/morning. These are also in Ambrosian metre and full of direct echoes of Ambrose's language.[86] The second pair is on 'before eating' and 'after eating', linking worship to 'our daily bread' and its significance.[87] The third pair is for 'lamp-lighting' and 'just before sleep', with the expected anxiety about the darkening of the mind. The fourth pair, starting the second sequence, is on fasting and after fasting. After 'daily bread' comes the necessity of fasting. As Peter Brown demonstrated, fasting is a way of stopping the time of the body (removing menstruation, sexual desire) in order to reverse time back towards the garden of Eden, the time before the Fall.[88] Prudentius' hymns praise fasting but are also an argument against the extremes of ascetic fasting, in the name of a regulated temperance: here the fasting seems to be probably for one day a week. The fifth pair is a hymn for 'every hour', *omnis horae*, and a hymn for the burial of the dead. The hymn for every hour is a joyous song of praise that catalogues the works of Christ from the foundation of the world, through his miracles, crucifixion, descent to hell and return from it, and adds the fathers and saints who followed Jesus. Its conclusion is that Jesus has conquered death and returned humanity to life, and hence joyful celebration should continue. It is a hymn for every hour because every hour instantiates the full history of time. As Gregory Nazianzus placed Christmas in the history of time to explain its significance as a day of celebration, so here Prudentius insists that every human – he catalogues choirs of praise (109–11) – and nature itself (112–13), must celebrate together 'for ever and ever', *saeculorum saeculis*. The traditional closural formula takes on further significance after this narrative precisely of the *saecula saeculorum*.[89] The hymn is significantly in juxtaposition therefore with the hymn for burial of the dead. The 'era will quickly come', *venient cito saecula* (36), when the resurrection of the body will justify the care taken over burial.[90] As with Lazarus, this 'triumph over black death'

---

[85] Ludwig (1977) 318–21 is more sensible than the thesis of the whole article.
[86] See e.g. Herzog (1966) 65–7; O'Daly (2012) 57–62.
[87] Herzog (1966) 43–9 emphasizes the allegorical elements of food.   [88] Brown (1988).
[89] Mastrangelo (2008) 67–74 demonstrates the importance of *saeculum* as a term in Prudentius' typological discourse.
[90] See Rebillard (2012).

(128) is the Christian promise. After the vision of the centuries in the hymn for every hour, the eternity of the afterlife follows. The final two hymns proclaim the celebration of Christmas and Epiphany, linked in liturgy and theology in the fourth century. Again, both poems have the subjects and metre of hymns of Ambrose, and, as we saw, were the subjects of paired sermons by Gregory Nazianzus. The *Cathemerinon* understands its agenda of 'day by day' in four interlocking ways, then: first, the hours of the day as scenes of prayer and self-reflection; second, the day as structured around 'daily bread' or the self-controlled abstinence from food: what the physical needs of the body mean for the spiritual life of the faithful; third, the cosmological ordering of time, which is instantiated in every hour, and grounds the theological understanding of death, as an expression of the redrafting of eternity; fourth, the festal calendar of celebration. These four frames are mutually implicative, interlocking and integrate the collection as a collection. Prudentius' poetic project, it has been recognized, is the formation of the internally struggling Christian individual within a new Christian history: 'A true Christian conversion through the reading of [his poetry] forms a true Roman citizen'.[91] The poems of the *Cathemerinon* insist, however, that this conversion is a process that is shaped and continues in and through the experience of time, day by day.

Although I discussed how the repeated performance of Ambrose's hymns, all the same length and metre, encouraged a certain embodiment of faith in the choir of celebrants, Prudentius is concerned with time-bound physicality in a quite different manner. First of all, his vision of human life is distressingly physical in its suffering (4.81–4):

> Vexamur, premimur, malis rotamur;
> Oderunt, lacerant, trahunt, lacessunt;
> Iuncta est suppliciis fides iniquis.
>
> We are harassed, crushed, turned on the wheel of evil;
> They hate us, torment us, drag us, attack us;
> Faith is tied to unjust punishment.

Prudentius' 'we' is deeply and violently oppressed by an unnamed 'they' who assault them, with the result that faith itself is integrally linked to suffering that is not fair. Life for humans is a physical anguish. The body itself, however, is also the source of the wrong it suffers. As he states in the *Hamartigenia* (562): *gignimus omne malum proprio de corpore nostrum*, 'We beget all our evil from our own body'. So even cures must be taken 'with

---

[91] Mastrangelo (2008) 57, summarizing much contemporary criticism of Prudentius.

a full mouth and woven into your internal veins' (*Cath.* 4.89–90). The fleshly thickness that in Gregory Nazianzus made a bar to human understanding of God, becomes in Prudentius a defining physicality that is punished, damaged, and transcended by the acceptance of pain.

So his description of martyrs is stridently more vivid and brutal than Ambrose's, in the *Cathemerinon* and in the *Peristephanon*. When Ambrose tells of the martyrdom of young Agnes, he exclaims, 'Struck! With what dignity she bore it!' (8.25) – and that is the only description of her physical suffering. The next eight lines, the close of the poem, dwell on how she maintained her modesty as she fell, making sure her body was covered, so that she 'fell with a decorous fall', *lapsu verecundo cadens* (8.32) – a death which also shows how the 'fall' (*lapsus/cadunt*) feared in the first hymn can be transcended. Prudentius by contrast has Agnes demand penetration of her 'nipples by the whole blade', and offers to take 'the force of the sword deep into her chest' (*Peri.* 14.86–7), before she is decapitated. In the *Cathemerinon*'s description of the massacre of the innocents Prudentius dwells even more attentively on the physicality of violence (12.117–20):

> O barbarum spectaculum!
> inlisa cervix cautibus
> spargit cerebrum lacteum
> oculosque per vulnus vomit; . . .
>
> O savage display!
> A head dashed on the stones
> Scattered the milky brains
> And vomits the eyes through the wound.

The baby's brains are called 'milky' both because of their colour or consistency, but also because the child has been grabbed from its mother's breast. The precision of the eye being 'vomited' through the wound is designed to raise disgust. This is only one of several such stanzas.

Human life is thus a trial, and the enemies of Christianity, especially Satan, are amazed when 'crumbling clay can bear such toil', *posse limum tabidum tantum laboris sustinere* (7.191–2). Flesh is as fragile as mud, and only its capability of surviving violence is remarkable. A life-time therefore is a passage of distress turned to celebration only by the promise of life after death and the spectacle of suffering borne. It is not by chance that fasting and the burial of the dead play so big a role in Prudentius, and none in Ambrose's hymns. To inhabit the body in human time is for Prudentius a necessarily terrible ordeal. As he declared in the Preface to the collection, 'what of use have I achieved in the space of so much time?' (*tanti spatio temporis*) (5), when 'in time (*iam*), whatever it is that I was, death will have wiped it out', *cum iam quidquid id est quod fueram, mors aboleverit.* Whereas Horace could wryly recall his own past failures as the subject of the city's gossip – *heu ... fabula quanta fui*, 'Oh, ... what a story I was', Prudentius' past is a demonstration of existential irrelevance. For Prudentius, it would seem, the experience of human time is a torture alleviated only by faith, which can allow the transcendence of the suffering of human physicality.

The celebration of the hope that such faith brings, is itself also powerfully expressed as the transformation of time, in a way which recalls Gregory's sermon for Christmas Day, though for Prudentius the hope turns into an aggressive outpouring of religious hatred. *Cathemerinon* II is his hymn of Christmas Day. It begins with the sun:

> Quid est quod artum circulum
> sol iam recurrens deserit?
> Christusne terries nascitur,
> qui lucis auget tramitem?
>
> Heu quam fugacem gratiam
> festina volvebat dies!
> quam paene subductam facem
> sensim recisa extinxerat!
>
> Why is it that the sun is now running back
> And leaving behind its narrow circle?
> Is Christ born on earth
> Who increases the path of light?
>
> Alas! How transient the thanks
> The racing day was rolling round.
> How nearly it had extinguished its withdrawn torch,
> As it was cut back gradually!

The winter solstice as a celestial event – with grandiloquent rhetoric – is imaged as the sun expanding its journey across the sky, again, after its near disappearance – a bold overstatement – in the shortest day of the year. The language is familiar in its vocabulary if not its excess: the circle of the sun's course; its running back; the rolling of time. But it is all predicated on the birth of Christ who 'increases the path of light'. The slippage between the language of divine illumination and the language of the day is evident, and recalibrates the celestial signs.

Jesus' birth opens into a narrative of history, which takes us back to the foundation (*condidit*, 21) of the heavens when Jesus was already the causal force. Nonetheless – and this continues a well-known theological debate about why Jesus was not born earlier – 'as the ages (*saeculis*) were ordered, and the status of the universe (*rerum*) set in place, their founder and architect stayed in the Father's bosom' (25–8). Christmas Day requires a perspective back to the very foundation of time, with even more theological pointedness than in Gregory's sermon. Jesus stayed with his Father (29–32):

> Donec rotata annalium
> transvolverentur milia,
> atque ipse peccantem diu
> dignatus orbem viseret.

> Until thousands of years in cycles
> Were rolled across,
> And he thought fit to visit
> A world long sinful.

The long sweep of time, which turns in cycles (*rotata*), is captured in the impressive two-word line *transvolverentur milia*, 'thousands were rolled across', large words for the expanse of the centuries.[92] It is only after such a long time (*diu*) marked by human sin, that Jesus makes the decision to come to the world. The result of this decision, the birth that 'breaks the chain (*catena*) of death' (46), inaugurates a new golden age: *novellum saeculum . . . et lux aurea*, 'a new age and a golden light'. For Prudentius, as he seeks to map the contours of the daily, the single turning point of the winter solstice, which is the turning point of the year, marks the trans-formational singular event of the birth of Jesus, and must therefore be comprehended as the turning point in the history of the centuries.

The perspective is also forwards. The poem ends (89–116) with a vitriolic attack on the Jews for not recognizing Jesus. (It is often now claimed that such attacks are figurations of internal divisions between Christian groups, but in a discourse which moves so easily between figuration and direct description it is hard to delimit the hate speech – or its consequences – with such certainty. Indeed, it represses the long history of Christian anti-Semitism.) In his attack, the Jews will recognize the error of their ways at the end of time (104–7):

> Cum vasta signum bucina
> terris cremandis miserit
> et scissus axis cardinem
> mundi ruentis solverit.

---

[92] Ambrose uses the same trick: *multiplicabatur magis*, 'it was multiplied the more' (7.25).

> When the huge trumpet sends forth
> Its signal for the burning of the earth
> And the axis breaks and dissolves
> the pole of the collapsing world.

Ambrose's Fifth Hymn, as we saw, performed in celebration of
Christmas Day, adapted Psalms to his universalist model and imaged
Jesus as the sun, and equal to his Father. Prudentius makes his Nicene
rhetoric more explicit and determinative, and moves the adaptation of
Hebrew scripture into an assault on Jews who will 'experience the lightning
bolt of the cross' (112–13), a turn from the illumination of the sun to the
violence of celestial punishment. Time is now both the foundation of the
Christmas story and a weapon with which to crush your enemies. Like
Tertullian, Prudentius enjoys imagining the end of days as the bitter end of
his opponents.

The final poem of Prudentius' collection is for the celebration of
Epiphany, and it includes another arresting example of the blend of
astrology and eschatology that is the distinctive rhetoric of Christian
reflection on the story of the Magi. The opening, as we have come to
expect with the hymns of both Ambrose and Prudentius, is a theologically
laden construction of a dramatic present that overlays the moment of
celebration with the deep history and symbolism of the Christian narrative
(12.1–4):

> Quicumque Christum quaeretis,
> oculos in altum tollite:
> illic licebit visere
> Signum perennis gloriae.

> All you who are seeking Christ,
> Lift your eyes to the sky:
> There it will be possible to see
> The sign of eternal glory.

The 'you' embraces the celebrants, the readers recalling or anticipating
the celebration, and the Magi, who travel to find Jesus. What will be
visible – and the tense is important here – is a sign of 'eternal' glory. The
present search for the coming sign of the everlasting allows time to slide
between now and the widest perspective. So, the narrator prays that the
comet and all the stars that burn with the heat of the Dog Star 'may now
fall, destroyed by the light of God', *iam dei sub luce destructum cadat* (23–4).
The 'light of God' is both the star of Bethlehem and the empowering truth
of faith (as in Ambrose's metaphorics of light); the maleficent stars are to

fall (*cadat*) before it – not the 'descent' of astrological movement but the 'fall' of defeat. But what is the time of *iam*? Does it imply that the stars have 'already' been destroyed? Or is it a prayer that 'now', since the new star has appeared, they are to fall? Or is it the 'now' of the prayer itself, the liturgical present? A constant present of triumph? The following stanza begins, *En Persici* . . ., 'Look! Persians . . .': a dramatic present recalling the scriptural narrative of the Magi. The past is constantly present in the here and now, as the sign announces a future to come – and both past and future are but shadows of the eternity of God's time. The slippage in tenses – in time-frames – is part of Prudentius' theological vision of time, prompted here by the stars, the markers of time.

'The sign' to be seen is the star that 'conquers the circle of the sun in beauty and light' (4–5) – the star that led the Magi to Bethlehem. This star is not beholden to the moon's monthly course (*lunam menstruam*, 10), but on its own 'possesses the sky and controls the course of the days', *caelum possidens cursum dierum temperat* (11–12). The star is now in imperial ownership of the sky, and, like the emperor's management of the *cursus honorum*, it controls the very subject of the poems, the *cursus dierum*, the cycle of day to day. So, what the Magi see is the *regale vexillum*, the 'royal standard' (27); the constellations that flee ask (32–5):

> 'quis iste tantus', inquiunt
> 'regnator astris imperans,
> quem sic tremunt caelestia
> cui lux et aethra inserviunt?'

> 'Who', they ask, 'is this great commander
> Who is emperor to the stars,
> Who makes the celestial bodies tremble,
> To whom light and air are servants?'

Prudentius, unlike the other descriptions of the astrology of the star we have discussed, makes it not just a sign that outshines the other stars, or leads the Magi (as does Ambrose (7.10–12)), but an imperial and instrumental force controlling the stars, to whom the very matter of the heavens is a servant. So, the constellations that 'revolve in unchanging motions',[93] *in se retortis motibus* (14), refuse to follow their usual routes and hide, and the 'other globes of astral significance', *ceteri . . . signorum globi* (29–30) retreat.[94] Gregory, it will be recalled, saw the appearance of this star as the

---

[93] O'Daly's translation.
[94] For *globi* of stars see Ambrose, Hymn 7.2 *micantium astrorum globi*, 'the globes of shining stars'.

end of the science of astrology, Theodotus a redesign of heaven's map. Prudentius, the official of the state, represents the star as a military commander storming the heavens and compelling the stars into abject retreat.

This astrological conceit is linked to the star's status in time. *Hoc sidus aeternum manet*, 'This star stays eternally', announces Prudentius (17), it 'never sinks'. So the Magi declare (36–9):

> 'inlustre quiddam cernimus
> quod nesciat finem pati,
> sublime, celsum, interminum,
> antiquius caelo et chao.'

> 'We see something glorious,
> That does not know how to experience an end,
> Sublime, on high, without boundary,
> Older than the sky or chaos.'

The Magi, the viewers in the poem, know how to see. They do not merely see something bright but something whose brightness is glory. They see something that is without end, not Virgil's Roman empire, but God's; its position in the sky is also a transcendence; its course is 'without boundary' (*interminus*: a rare word that looks forward to Nonnus' fascination with what is *atermōn*). Most strikingly, this star – appearing only now – is nonetheless 'older than the sky and chaos'. This phrase implies not just what is above the earth and 'the world below' the earth, as O'Daly translates,[95] but the foundational moments of creation in the Christian and Greek tradition: in Genesis, in the beginning God creates the heaven (*caelum*); in Hesiod, the beginning of everything is *Chaos*. This star, beyond time, thus, in the eyes of the Magi, *is* the king of the nations, Jesus – *hic ille rex est gentium* (41) – who indeed was born 'before Chaos', *ante Chaos genitus*.[96] As Prudentius makes the star not just the sign of the birth of Jesus but (a figure of) Jesus, so the star's timeline follows Nicene theology and exists before the creation of the sky in which it is. This king is the king promised by scripture, the Magi conclude, the star that will arise from the stock of Jesse.

The ability to see beyond the physical, as the Magi do, is a fulfilment of the injunction of Paul, 'So we fix our eyes not on what is seen, but on what is unseen, since what is seen is temporal (*proskaira*) but what is unseen is

---

[95] O'Daly (2012) 355.

[96] *Ham.* 44 – a powerful pro-Nicaean and anti-Marcian passage. See Malamud (2011) 156–9. Paulinus of Nola encourages Jovius to write biblical epic starting from *chaos ante diem* 22.151.

eternal (*aiōnia*)' (2 Cor. 4.18). So the most visible object of all, the shining star, is itself a sign of the eternal, a fulfilment of prophecy, a figure of eschatological promise: in it they *see* the king. Ambrose quotes these lines of Paul twenty-one times.[97] Prudentius dramatizes Paul's injunction, Ambrose's inspiration, in this extended paraphrase of the brief scriptural narrative of the Magi. Crucial to this vision of the invisible is typology. As Catherine Conybeare comments on Prudentius' interpretative demands in the *Hamartigenia*, reading becomes 'an interpretative act and hence an ethical one': 'Not only what you read but *how* you read affects your *fides*.'[98] So, the hymn continues with the long description of the massacre of the innocents as the frame for the journey of the Magi, but after its all too vivid account of violence, from which I quoted earlier, it determines that this massacre is like the edict of Pharaoh to kill the male children of the Israelites. Moses escaped that slaughter, thus 'prefiguring Christ', *Christi figuram praeferens* (143). To make his point, Prudentius offers a lengthy typological reading of Moses encouraging us to 'recognize Jesus in the example of so great a man' (157–8), a typology already rehearsed at length in the fifth hymn.[99] Because of this typology, concludes Prudentius (181–4):

> iure ergo se Iudae ducem
> vidisse testantur Magi,
> cum facta priscorum ducum
> Christi figuram pinxerint.

> Therefore it is right that the Magi testify
> That they have seen the king of Judah,
> When the deeds of earlier leaders
> Portray the figure of Christ.

What enables the Magi to bear witness that they have seen the king of Judah, when they look at a star or a baby in a manger, is not so much prophecy or the guidance of a star but the history of typological understandings of Jesus that underpin the Christian interpretation of the Hebrew Bible as the Old Testament. The narrative that began with the looking to the star as a sign, ends with the fulfilment of the sign by seeing in the baby the scriptural history that has always already been the sign of his birth. Typology turns the linearity of aetiology into a continuity of repeated figuration. As Jonas Grethlein writes (of Augustine), typology redrafts 'the sequential narrative of the past through a view that is ... aligned with God's timeless

---

[97] Counted by Dunkle (2016).   [98] Conybeare (2007) 229; 239.
[99] *Cath.* 5.31–137, on which see Mastrangelo (2008) 107–12.

perspective'.[100] Prudentius' hymn sets out to make 'its audience, as it follows the text in its linear development, part of this non-linear mode of time',[101] to form what Catherine Conybeare nicely calls 'the typological imaginary'.[102]

These longer scriptural and theological narratives, with their startling *enargeia* and forceful rhetoric – that certainly distinguish Prudentius' hymns from Ambrose's – evoke a model of reading, which transforms Ambrose's collective hymns of worship into a reader's reflective engagement with the temporality of Christianity. The sleeper in the first hymn is implored to give up his 'sickly, sleepy, lazy' bed and be 'chaste, upright and sober' (1.5–7). Prudentius' Christian harbours 'dark, hidden thoughts' (2.13–16), ever 'conscious of sin' (2.10). 'We' merit nothing because of our sins (6.117–18). Prudentius as narrator leads the prayer, *precor*, 'I pray' (3.6). Prudentius' rhetoric repeatedly draws on the satires of Horace and Juvenal, as well as Christian homiletics, in turning a critical moral gaze at the reader. The repeated criticisms of human behaviour, along with first-person confessions, second-person injunctions and first-person plural complicities, construct a strident moral framework focused on sin and death, and the long narratives – the very time of reading – create a dynamic of narrative exemplarity which place his moral imperatives within a broader theological framework. 'The reader is called to account' … the poetry 'calls into being a responsible reader', makes her 'a participant in the poem's negotiation'.[103] As in the satires of Juvenal or Horace, whoever else is pilloried in the poetry, the faithful reader as much as the *hypocrite lecteur* is provoked towards intense self-scrutiny.

The connection between typology and the human experience of the present – how a past example *tempora nostra figurat*, 'prefigures our times' (*Psych.* 67)[104] – is captured in summary and with a full theological perspective in Prudentius' *Apotheosis*, a poem on the nature of the divine, where he writes (*Apoth.* 309–11):

> Christus forma patris, nos Christi forma et imago.
> condimur in faciem domini bonitate paterna
> venturo in nostrum faciem post saecula Christo, …

> Christ is the figure of the Father, we are the figure and image of Christ.
> We are founded in the image of the Lord by the goodness of the father,
> As Christ was to come into our image after centuries, …

---

[100] Grethlein (2013) 346.
[101] Walter (2020) 207 – she is talking here of *Peristephanon* 2 on Laurence.
[102] Conybeare (2007) 238. This is also the central thrust of Mastrangelo (2008).
[103] Dykes (2011) 18, 19, talking of the *Hamartigenia*.
[104] See Mastrangelo (2010), an expansion of his discussion in Mastrangelo (2008).

These are hard lines to translate as the grammar as much as the vocabulary contributes to the intricate vision of overlapped past, present and future. Christ is the *forma* of the Father: I have translated *forma* as 'figure', but after Nicaea it also includes an implication of substance, 'form'. We – human beings – are the 'figure and image' of Christ, that is, humans are to imitate the model of Jesus, as well as recognize our human body and divine soul. This divine form is immediately glossed from Genesis. We are made – *condimur*, as we have discussed, is also a politicized term – in God's image. The Latin Bible in Genesis uses the word *imago*, but Prudentius uses *facies*. The one thing that cannot be seen of God, however, is his face (*facies*). Our creation 'in the image' of God also recognizes the unknowability of God, the repeated principle of Prudentius' apophatic theology. This creation is in the present tense: creation is a continuing drama. The ablative absolute that follows does not indicate a precise relation between the previous sentence and its own assertion (I translate it with 'as' as the least marked option). The future participle could be causal, however: 'Because Christ is to come . . .'. Our very humanity is predicated at creation on the coming of Jesus. This future arrives *post saecula* 'after the aeons'. On the one hand, there is a linear narrative: there is creation, and after many years, there is the Incarnation. On the other hand, as Jesus is in the form or image of the Father, so we are to Jesus, but Jesus is also in the image of us. What is to come is also the always already. This is the productive tension of typology, which recognizes the centuries of time only to fold them back into an eternal figuration, a 'God's eye view of the past, present and future' as *determining* models of each other.[105]

It has often been noted that 'at the grave of the martyr the temporal distinction of then, the time of the passion, and now is abolished':[106] the saint, as Peter Brown argued, becomes the cultic *locus* which mediates between heaven and earth, the past, the present and the future.[107] The insistence on the constant present in the tenses of the hymns, however, has its more specific ethical impulse too. Everyone, writes Prudentius, is embarrassed or ashamed by the thoughts of night by the morning. So, the morning becomes the model of now (2.33–6):

> Nunc, nunc severum vivitur,
> nunc nemo temptat ludicrum,
> inepta nunc omnes sua
> vultu colorant serio.

---

[105] Mastrangelo (2008) 49 where he also discusses these lines.   [106] Roberts (1993) 13.
[107] Brown (2014). See also the essays in Howard-Johnston and Hayward (1999) and Grig (2004).

Now, now life is lived severely;
Now nobody tries a joke,
Everybody now colours their idiocies
In a serious face.

Humans put a brave face on their own moral failures. The fourfold repetition of *nunc*, 'now', emphasizes what Paul called the *proskairon*, the temporary, the moment, the mundane – the failure to look towards the eternal. So, continues Prudentius, *haec hora cunctis utilis*, 'This hour is useful for everyone' (2.37). The usefulness is sarcastic. The morning hour is useful for 'the soldier, lawyer, sailor, workman, ploughman, shop-owner' – for the trivialities of daily work, the sort of life of business he recorded for himself in his preface. Prudentius prays for something more: *intende nostris sensibus, vitamque totam dispice*, 'attend to our inner feelings; examine our whole life' (57–8). His hope is that 'a whole day', *tota dies* (101), can pass without sin. Whole days may stretch in a chain to make a whole life: this is the ethical logic of the *Cathemerinon*. To transform the *nunc, nunc* of mundane, fragmented living into the recognition that each moment is replete with the promise of eternity. For Prudentius, each bite of food is a memory of the taste of the apple that Adam took, which has been undone by Jesus as the second Adam; each act of fasting is a control of the flesh, which is the 'triumph of the emperor of the spirit', *triumphet imperator spiritus* (7.200). The annual festivals of the day of Christmas and Epiphany invoke the full scope of cosmological eternity and the transformation of the unfurling of time that is signalled by the birth of Jesus. So, too the hymn of *omnis horae*, 'every hour' tells of *omnia saecula saeculorum*, as the ceremony of the burial of the dead prompts a reflection on the mortality of the flesh and its transcendence in the Christian promise. Thus the collection that began with the poet anticipating his own death, concludes in its last line with *iam nemo posthac mortuus*, 'now nobody hereafter is dead' (12.207). The 'now' is the recognition of the end of the journey of the hymns: it is to this point that his poetry has been travelling. It is also a recognition that it is only 'now' after the birth of Jesus that such a claim can be made. The liturgical, theological present and the performative present overlap. Anything the poet might have been, claimed the preface, will have been wiped out by death: but the poetry itself challenges this. The immortality of poetic production, so often rehearsed at the end of Latin works – *non omnis moriar* – is echoed and redrafted here, now, now within a Christian eschatology.

*

Whereas Ambrose provides short hymns of identical length and metre for the regulated and public performance of liturgy within the context of the cathedral Christianity under the control of a bishop, hymns designed to produce a community of worshippers, tied together in faith and in the temporal experience of daily observance, Prudentius provides long poems of different lengths and metres, designed, it seems, not for liturgical, communal performance, but, like Callimachus' Hymns, as a reflection of such performance for a community of individual readers. His fascination is with the sinful, suffering self, how it is modelled by the martyrs, and how it experiences the daily time of the body and its pains along with such internal anguish. Whereas Ambrose uses the paradoxes of Christian rhetoric to open a broader and deeper perspective of how Christian time frames the regular observance of the hours of prayer and the festal calendar, Prudentius' remarkable poetic texture turns such paradoxes into a fierce and memorable language of self-scrutiny and moral fervour, where the violence of imagery and violence of pain are intertwined into a fierce and shocking beauty. Ambrose looks towards – helps form – the church as institution, and holds an instrumental place in the long construction of the liturgy of the hours, and the modelling – and the enacted experience – of the day as a regulated order of prayer that structures both cathedral Christianity and, above all, the monastery, which, as we saw, are both integral to the historiography of Western time.[108] Prudentius marks out a different space where the committed Christian can reflect on the extremes of asceticism and the commitment of the martyr as part of self-understanding, mediated through the experience of the day by day. Violence towards heretics, towards Jews, is matched by violence towards sin, and violence towards the sinning self, but tempered by the celebration of the Christian promise of transcendence of both pain and death, a promise that is articulated and experienced through the daily round of prayer and festal celebration. Prudentius, like most Christians, is neither an ascetic monk nor a martyr (nor a priest). His poetry is instrumental in imaging a Christian day that recognizes the alluring force of asceticism and martyrdom, and yet remains distanced from both. His poetry of self-scrutiny enacts thus a profoundly sophisticated poetics of imitation, embodiment *and* difference. If Ambrose is leading the choir's singing, Prudentius' positionality is the Christian lying in tears in front of the picture of a martyr, while on a mission to Rome, reflecting on his own position on the ground, crying.

[108] See above, pp. 81–4.

In the representations of day-by-day Christianity in the hymns of Ambrose and Prudentius, then, we can see the dynamic tensions within the development of Christianity around the turn of the fifth century – different parameters of the organization of time, as I discussed in the first section of this book. Ambrose and Prudentius articulate how the 'now' is recalibrated against timelessness, how a life is to be narrated as an ethical struggle, how time is to be regulated, how the past of the saints is to be memorialized, all predicated on an understanding of God's time and the Christian promise of life after death. To remake the day by day. I have so far emphasized Prudentius' engagement with Ambrose because it is this contrast which reveals most pointedly the agenda of Prudentius' writing and the direction of his project. It is also necessary, however, because so much recent writing on Prudentius by classicists has focused on how his poetry echoes and redrafts the language of the classical tradition, especially Virgil and Horace. It has become a mantra to repeat Bentley's judgement that Prudentius was 'the Virgil and Horace of the Christians', *Maro et Flaccus Christianorum*. There is no doubt that Prudentius was well read in the Latin poetry of the Republic and early Empire: he had a serious education. Nor is there any doubt that he uses this furniture of the mind in a significant manner. As has been demonstrated with some rigour, Prudentius appropriates Virgil's defining image of Rome and its future and uses it against itself to project a Christian imperial gaze; he takes up Horace's genial and ironic self-representation of failure and hesitation – and sense of daily life – and transforms it into his more self-lacerating ideal of self-awareness. Lucretius' Epicurean rejection of religion is inverted. The sharpness of Prudentius' rhetoric is informed by the language of Roman satire – Juvenal as much as Horace – as it is by Tertullian.[109] 'An allusive poetics of transformation, of the making new of traditional models, corresponds to the substantive message of personal renewal and transformation'.[110] Yet it is telling that the most recent discussion of the *Cathemerinon*, from which I have just quoted, a discussion which is the epitome of this classicist understanding of intertextuality, also proceeds without any reference to Ambrose (or any other Christian writers, bar St Paul).[111] Reading classically keeps the poetry away from any practice or politics of religion. We have seen in this book moments when theological commitments bury philological rigour: in this reduction of Prudentius to a late but honoured place within the classical tradition – business as usual for classicists,

---

[109] See above, n. 8.　　[110] Hardie (2019) 23.
[111] Hardie (2019) 216–22. So too van Assendelft (1976) barely mentions Ambrose's hymns.

ironically enough for Prudentius who is so scathing of such a lack of self-scrutiny – classical philology aggressively performs its institutional severance from theology. Prudentius suffers for it.

This contrast between Ambrose and Prudentius is instructively on display in Ambrose's twelfth hymn and the twelfth poem of Prudentius' *Peristephanon*, which are both dedicated to the celebration of the day of the martyrdoms of Peter and Paul in Rome. Ambrose's hymn begins *Apostolorum passio | diem sacravit saeculi*, 'The Passion of the Apostles | has made holy this day of our era'. *Saeculum* means the time of the mundane, 'this age' in contrast to God's time. In his commentary on the Psalms, Ambrose writes: *dies saeculi mali sunt quia saeculum in maligno positum est*, 'The days of this era are bad, because the era is set to be disposed towards evil'.[112] What had been no more than one more day of the ordinary calendar has become other, thanks to the passion of the martyrs: *sacravit*, 'has made holy', marks 'this transformation or this valorization of time', and the position of the verb between the two words of time, *diem* and *saeculi*, 'emphasizes this sacralization of time: the fact that from now on time can be a time of salvation and liturgy'.[113] The hymn moves from a portrayal of the two saints' deaths, to a description of the city of Rome in celebration, to close with a rousing acclamation of the city (31–2): *electa, gentium caput! | sedes magistri gentium!*, 'Chosen one, leader of the nations! | Throne of the master of the nations!'. The hymn travels from a recognition of the day's sacrality to participation in the day's celebration, as the hymn ends with its performers' acclamation of Rome as a Christian city.

Prudentius' twelfth *Peristephanon*, by contrast, opens with a conversation. An unnamed interlocutor asks what the unusually large and joyful celebration is, and receives the answer that it is the feast of Paul and Peter (3–6):

> Festus apostolici nobis redit hic dies triumphi,
>   Pauli atque Petri nobilis cruore.
> unus utrumque dies, pleno tamen innouatus anno,
>   uidit superba morte laureatum.

> This day we have the festival of apostles' triumph coming back,
>   A day made celebrated by the blood of Paul and Peter.
> The one same day – but returned anew after a full year –
>   Saw each win the crown of martyrdom in a splendid death.

---

[112] Ambr. *In Psal.* 36.32, cited by Fontaine et al. (1992) *ad loc.*, who also notes that Cyprian (*Laps.* 2) opposes *dies saeculi* and *terrena tempora* to *aeternitas*. Augustine's most extensive discussion of the sense of *saeculum* is *Enarr. in Ps. IX*, 6.

[113] Fontaine et al. (1992) *ad loc.*

The speaker explains that the two martyrs were killed on the same day in the calendar a year apart. The day returns – the cycle of time – the year is 'made new', and each return of the day is made an occasion of joy and celebration by virtue of its memory of the martyrdom.[114] So, in the narrative that follows which describes the martyrdoms, the second death takes place (21–3):

> ut teres orbis iter flexi rota percucurrit anni
>   diemque eundem sol reduxit ortus,
> evomit in iugulum Pauli Nero fervidum furorem, . . .

> when the round wheel of the turning year had run the course of its circle
>   and the rising sun brought back the same day,
> Nero vomited forth his hot rage against the throat of Paul . . .???

The overload of words for the turning of time – *teres, orbis, flexi, rota* – is emphatic. Nor is it merely a frame for the heated anger of the emperor. The sacredness of the day is because of this repetition, and is enacted in the repetition of the annual festivities. The narrator goes on to describe how the topography of Rome too has been made sacred (*sacrata*, 30; *sacravit*, 47; *sacra*, 63) by the two churches that are dedicated to the two martyrs, either side of the Tiber, and encourages his listener to join him in going to the celebrations. 'It is enough', he concludes (65–6), 'to have learned this at Rome: go home and remember to celebrate this double-festival day'. Prudentius dramatizes two men observing others at a festival – the standard pose of the *theōros*, the visitor to a cult site familiar from Pausanias or Plutarch or Ovid and already parodied in Theocritus 14 where two less educated women are shown watching a festival performance at the palace of Ptolemy. There is an intellectual distance in the exegesis of the landscape and the reasons behind the ceremonials. The final encouragement is to 'remember and celebrate too' – not so much joining the performative acclamation that closed Ambrose's hymn, as an enjoining to absorb what has been taught, and to repeat 'the day' for oneself.[115] The poem becomes both the description of a day of worship and an injunction to fulfil the day of celebration as a day of celebration. The persuasive strategies and projected audiences of Prudentius and Ambrose are again different, for all that both insist on turning the *dies saeculi* into a *dies sacratus*. They dramatize different styles of experiencing the day.

---

[114] On *renovatio* see Hardie (2019) 135–62.
[115] The same injunction, *memento*, occurs at *Cath.* 6.125; *Perist.* 10.835, but without this aetiological implication.

In chapter 3 we discussed the importance of the zealous regulation of time that the Rule of Benedict and other monastic systems demanded, and the significance of such regulation for the measuring and comprehension of time's order in Western history. With Ambrose and Prudentius, we can see how a poet who is a bishop and a poet who is a state official together help formulate an understanding of Christian time as a day-by-day experience: the articulation of the times of the day, and the relation of such articulation to the festal calendar, to the Christian ethical memory of the martyrs, and to the broadest cosmological time of theology, as expressions of being Christian. Here, too, is the story of how time becomes structured, how time is (to be) experienced. Such poetry, circulated, performed, read, reveals the formation of the temporal imaginary.

CHAPTER 15

'We Are the Times'
Making History Christian

The first chapter of this book opened my investigation into the discourse of temporality in late antiquity with the discussion of how a specific mistranslation of the Hebrew Bible by the Septuagint was a designed intervention in the cultural reception of the text. I start this last chapter with another question about a surprising intervention by the translators of the Septuagint, which will lead into an extraordinary rabbit-hole of exegesis and commentary – to emerge, once again, in Augustine's specific and remarkable obsession with time. This journey starts with a figure who will be a mainstay of this chapter, Sulpicius Severus, a Christian from Aquitaine who was born probably around 355. Sulpicius Severus is a celebrated author – though scarcely a familiar figure in the classical canon – famous because he wrote the Life of St Martin, which became a hugely influential text in the Middle Ages, and which has a claim to be the first Latin biography of a Christian saint who was not martyred – a hagiography, that is, rather than a martyr narrative.[1] We will discuss this text and its generic affiliations later. He also wrote dialogues about St Martin and letters (Paulinus of Nola was an especially significant correspondent), and a *Chronicon*.[2] This *Chronicon* is a history of the world from the creation to the Priscillian heresy, which contrasts strikingly with Eusebius' *Chronicon* (which we discussed in chapter 7) in its aim and form. Fully three quarters of Sulpicius Severus' *Chronicon* is a curt paraphrase of the Hebrew Bible (although he probably read it in Latin). In the *Chronicon*, it is always a surprise when Sulpicius Severus himself allows his authorial voice to emerge in what is otherwise a studiously objective stance, for all that his editing and redrafting is aggressive. Consequently, it is arresting that, when he comes to the story of Noah, he offers a personal

---

[1] The best general introduction remains Stancliffe (1983).
[2] There are hundreds of manuscripts of the *Vita Martini* but only a single manuscript of the *Chronicon*. For the *Vita Martini* the best commentary is Burton (2017); for the *Chronicon*, see Senneville-Grave (1999). Further bibliography below.

interpretation. When Noah correctly judges that the waters are receding, he takes action (1.2.1):

> corvum primum explorandae rei gratia, eoque non revertente – ut ego conicio, cadaveribus detentum – emisit columbam.

> He sent out first a raven to explore the situation, and when it did not come back – detained by the corpses, as I imagine – he sent out a dove.

To a reader familiar with the Hebrew text, both the action of the raven and the interpretation are challenging. The Hebrew text reads *va'yetze yatzov vashov*, which is standardly translated 'he went to and fro' until the waters receded.[3] The first bird, that is, can find no place to land but travels around and around. This leads to some delightful and bizarre midrashim. In *bSandhedrin* 108b, a conversation is imagined between the raven and Noah in which the raven rejects the task offered with a knock-down argument. He must be hated by God and by Noah, the raven argues, to have been selected for the mission, because, since they are unclean animals, there are only two ravens in the ark. If he died from heat or cold, therefore, the species of ravens would be wiped out. He adds that he suspects Noah of wanting to get rid of him so that he can have sex with the raven's wife! (Noah retorts angrily that since he has observed the prohibition of sex on the ark with his own wife he is scarcely likely to have sex with a raven.) Hence, however, the fearful raven will only fly round and round the ark. *Midrash Rabbah Bereshit* imagines a different conversation – the idea of a conversation comes from an etymological play on the Hebrew verbs. Noah asserts blithely that the raven can be sent because as an unclean animal he is no good for food or for a sacrifice, only to be reminded by God that ravens would feed Elijah in the desert (Kings 1.17.6) – a paradigmatic demonstration of how the narrative of the Talmud is informed by God's omniscient (always already) time. Therefore this midrash too explains why the raven stays close to the ark's sanctuary, to be safe for his future role. There is no reason given in the Hebrew Bible for why Noah sends out this first bird, and he draws no explicit conclusion from the bird's reaction; these gaps prompt imaginative stories about why the raven is chosen and what is at stake in his circling round and round. In both the Hebrew text of Genesis and in its long commentary tradition in Hebrew and Aramaic the raven leaves the ark and 'goes to and fro'. It does *not* 'go out and not come back', as Sulpicius Severus has it.

[3] Gen. 8.7.

The translation of the Septuagint, however, takes a strikingly different tack, which is the likeliest source for the tradition that Sulpicius Severus represents. The Septuagint reads 'He sent out the raven to see if the waters had receded. It went out and did not return until the waters had dried from the land'. The Septuagint adds, first of all, a reason for Noah's actions: the bird is let out to see 'if the waters had receded'. In the Hebrew Bible, this reason is given for the release of the dove, and the lack of such a reason in its logical place for the first bird has prompted critics who are so minded to suggest that the raven story is an interpolation either by the priestly source (or the earlier compositional level known as J) or from Mesopotamian literature, though the reasons given for such an interpolation are speculative at best.[4] Origen already marked this phrase with obeli as not being in the Hebrew text.[5] It demonstrates that the Septuagint is trying to edit the story, as it translates, to introduce a more explicit and clearly understandable motivation. But the Septuagint has also decided that the raven 'did not return'. One might generously assume that this is no more than a literalist understanding of 'going to and fro', that is, the raven circled and did not thus actually go back to the ark. But the addition of this negative opens a long history of textual doubt.

John Calvin, at the heart of the Reformation, sums it up succinctly: 'I wonder whence a negation, which Moses has not in the Hebrew text, has crept into the Greek and Latin version, since it entirely changes the sense'.[6] Usually assertive in his knowledge, Calvin is baffled about the source of the negative – a bafflement easy to share. Calvin continues: 'Hence the fable has originated, that the raven, having found carcasses, was kept away from the ark and forsook its protector. Afterwards futile allegories followed ...'.[7] Calvin is right that the Vetus Latina and the Vulgate both standardly read 'egrediebatur et non revertebatur', though Jerome, scholarly as ever, in his *Hebrew Questions on Genesis* adds *et de corvo aliter dicitur: 'emissit corvum et egressus est, exiens et revertens, donec siccarentur aquae de terra'*, 'About the raven there is another reading, "He sent out the raven and it went out, going away and returning until the waters dried from the earth"', a good translation of the Masoretic Hebrew, despite what his authorized

[4] See Moberly (2000) and Marcus (2002) for discussion and bibliography.
[5] Marcus (2002) 71 discusses this.
[6] Calvin (1948) *ad loc.* Calvin despised the Council of Trent's insistence on the authority of the Vulgate, its 'barbarous' declaration 'that Scripture should only signify to us whatever dreaming monks might choose' (Calvin (1958) 76), a quotation contextualized briefly in Hendel (2016) 277 (see 271–329).
[7] Calvin (1958) 76.

translation transmits. Philo, who read the Septuagint and may not have known Hebrew, is the earliest to offer one of the 'futile allegories', whereby the black raven stands for sin and the white dove for virtue, a reading happily extended by Ambrose and Prudentius (as well as by Greek Christians).[8] Josephus, by contrast, who did read Hebrew, simplifies the story, and has the raven go and come back as a sign that the world is still too wet, but then tells of only one dove, who returns with mud on its toes and an olive branch in its beak.[9] The mud anticipates the rabbinical under-standing of Genesis 8.9, where, when the first dove returns, 'Noah, putting out his hand, took it into the ark'. 'Putting out his hand' is understood in later commentaries precisely as holding the bird and inspecting it for signs of a recovered world, especially mud on its toes. The Greek and Latin texts and commentaries, with the exception of Josephus' unique version, seem consistently to read a different text from the Hebrew that they translate, and that the Hebrew commentators use. The Peshitta, the Syriac Bible which usually follows the Hebrew, in this case follows the Septuagint, and – since there is always one exception – *Pirkei de Rabbi Eliezer*, written as late as the turn of the eighth to ninth centuries, is the only text from within the Jewish tradition that assumes that the raven does not come back.[10] The great polyglot bibles of the Renaissance print Hebrew and Septuagint versions next to each other without demurral.[11] The King James Bible, however, translates the Hebrew ('which went forth, to and fro'), and most modern Christian commentaries discuss (translations of) the Hebrew text without recognition of the long history of Christian understanding of another version.

The doubt over whether the raven goes and comes back, or goes and does not come back, is most vividly evidenced in the oldest extant manu-script of the whole Vulgate, Codex Amiatinus, copied in the Benedictine monastery at Wearmouth-Jarrow in the north of England around the year 700. This beautiful manuscript has only one addition, written in a non-scribal hand. Above Genesis 8.7, which in this manuscript reads

---

[8] Philo, *Questions and Answers on Genesis* II.35; Ambrose, *On Mysteries* II; Prudentius, *Ditto*. 3. See also John Chrysostom, *Homilies on Genesis*.

[9] *AJ* 1.3.5. Josephus, unlike others who wish to separate the mythic Deucalion from the scriptural Noah, with the assimilationist zeal we saw in chapter 1, concedes that the story is well known to pagans, and singles out Berossus (*AJ* 1.3.6) – whose story, unlike Josephus, has three birds, like the Hebrew Bible's three missions of the dove.

[10] *Pirkei de Rabbi Eliezer* 23.

[11] Notwithstanding Fabricius' comment that 'variants are arts of the Devil' (letter to Buxtorf, 24 August 1625). Van Boxel (2006) gives an interesting account of Bellarmine's Christian Hebraist engagement with the problem; see also Hendel (2016) 279–82.

egrediebatur et revertebatur, the word *NON* has been added, and scholars have reason to believe it is in the hand of the Venerable Bede himself. It is as if the text and its correction materially dramatize Jerome's own awareness of the two readings. In an uncanny repetition of such ambivalence or correction, *Vulgate.org*, an online interlinear text and translation of the Vulgate, gives an incorrect Latin text *qui egrediebatur et revertebatur*, but translates it 'which went forth and did *not* return' – a difference – no doubt a providential misprint – which captures the history of the problem nicely enough. The silence in the Hebrew Bible about why Noah sends the raven out, or what his 'going to and fro' means, allows not just for the imaginative constructions of midrash but also for the interventions of editing as well as the work of exegesis.

Augustine, however, reveals how potent the Septuagint translation can become as an exegetical tool – in a route of thinking that is far from the midrashim. In his *Enarrationes in Psalmos* 102, he writes this stunning paragraph (*Ennar. in Ps.* 102.16):

> Sunt enim qui praeparant conversionem, et differunt, et fit in illis vox corvina, 'Cras, cras'. Corvus de arca missus, non est reversus. Non quaerit Deus dilationem in voce corvina, sed confessionem in gemitu columbino. Missa columba reversa est. Quamdiu: Cras, cras? Observa ultimum cras: quia ignoras quod sit ultimum cras, sufficiat quod vixisti usque ad hodiernum peccator. Audisti, saepe soles audire, audisti et hodie: quam quotidie audis, tam quotidie non corrigeris. *Tu* enim *secundum duritiam cordis tui et cor impoenitens, thesaurizas tibi iram in die irae et revelationis iusti iudicii Dei, qui reddet unicuique secundum opera sua* [Rom. 2.5–6].

> There are those who make preparations for their conversion, and delay; in them comes into being the voice of a raven, 'tomorrow, tomorrow' [*cras/cras* = caw/caw]. The raven was sent from the ark, and did not return. God does not seek delay in the voice of a raven, but confession in the moaning of a dove. The dove, sent out, returned. How long: tomorrow, tomorrow. Look to the last tomorrow. Because you do not know what the last tomorrow is, let it be enough that you have lived as a sinner until today. You have heard, you are used to hearing often, you have heard today too: as many times as you hear, you will not change. For 'according to the hardness of your heart and your unrepentant heart, you are storing up wrath against you in the day of wrath and the revelation of the just punishment of God, who will repay each man according to his deeds'.

The raven is a model of the continuing deferral of the hesitant convert. He 'goes out but does not return' – where 'return' is the *termus technicus* for the repentance required by *conversio* – *teshuvah* in Hebrew. *Conversio* requires *revertere*. The dove, by contrast, goes out and comes back. But this symbolic contrast is expressed in terms of voice. The dove moans in penitence; but the raven caws, which in Latin sounds *cras, cras,* 'tomorrow, tomorrow'.[12] As we discussed in chapter 9, *cras, cras* is precisely what Augustine himself declared in his own hesitant journey towards his own conversion in the *Confessions*. *Quamdiu* 'How long?' was his redrafting of the biblical language of pleading into his desperate awareness of the waiting for grace. Augustine models his own experience through the raven, hears his own dilatory travel towards God in the cry of the bird. 'You have heard . . .' he repeats, and his repetition overlaps three sorts of hearing: his injunction against the failed hearing of the sinner; his insistence on the revelatory pun of *cras, cras* ('hear!'); and the remembrance of his own text, the *Confessions* ('you have heard . . .'). He goes on to dismiss the sinner's deferral of repentance with a quotation from St Paul (Rom. 2.5.6), a reminder that the Day of Judgement will end such delays with terrible punishment. But this text also affords Augustine another telling pun. The problem of the hesitant repentant is precisely in his *duritiam* **cordis** and **cor** *impoenitens*. The repetition of *cor* is obscured in the King James translation ('after thy hardness and impenitent heart', which follows the Septuagint), but here is surely made to sound out significantly. The **corvus** lurks in the *cor* of the hesitant, as the *vox* of the raven is heard in the *cras, cras* of delay, Augustine's own voice of despair. Noah's raven who does not return becomes a potent image of the time-bound incapable convert, waiting for, but resisting, a tomorrow of grace.

The Septuagint's redrafting of the Hebrew Bible's language becomes in the hands of Augustine a route to express the failure of the sinner's everyday waiting and deferral of change. The bird which does not return (repent), the *corvus*, lives in the *cor* of the stumbling Christian who finds his very language sounding out the temporality of his sin of hesitancy, *cras, cras*, in the *vox corvina*. What appears to be a strategic editorial decision of the translators of the Septuagint, for this Christian of late antiquity, has become another story of how humans inhabit time.

*

Sulpicius Severus with his representation of Martin provides an intense, polemical and personally committed perspective on the contrast we saw in the last chapter between Ambrose and Prudentius. Like Prudentius, Sulpicius Severus had a career before committing to a thoroughgoing

---

[12] Augustine uses the same trope repeatedly: e.g. *Sermons on the Liturgical Season* 224.4; *Contra Faustum* 12.20.

Christian way of life. He practised successfully as a lawyer and had the literary and rhetorical training that one would expect for such a position.[13] His writing reflects such an education, and for a classically trained reader the influences of Sallust, Tacitus and the Roman historiographical and biographical tradition are evident, pervasive and integral to Sulpicius Severus' language. Sulpicius Severus, like Paulinus of Nola with whom he corresponds, became a committed Christian – it is not certain if he came from a Christian family, though it is likely – by 'selling all and following' the example of Paulinus of Nola and Martin as well as Jesus. Paulinus writes about their similar but different circumstances: 'You were nearer the prime of life, more richly praised, less burdened with inherited wealth, but no poorer in intellectual resources; and you were still immersed in the bustle of the law courts, that theatre of the world, where you enjoyed an outstanding name for eloquence'.[14] Sulpicius Severus, like Paulinus, became an icon of the success of the church's policy of conversion among the Roman landed classes.

With a brief worry about the principle, Sulpicius Severus did keep the income from one estate which allowed him to establish a community of like-minded men at a place called Primuliacum in Aquitaine. This establishment is right at the start of Western monastic traditions, where prayer, readings and study, rather than agricultural or other labour, formed the structure of the everyday – a similar comfortable set-up to Augustine's at Cassiacum. As with Augustine, dialogue is one of the privileged genres of publication for such educated, withdrawn communities, not only creating a sophisticated Christian rejoinder to the philosophical tradition inaugurated by Plato and Xenophon, but also providing a model of exchange and community. The dialogues of Sulpicius Severus depict the return of a friend, Postumianus, who has been to the East. He promises stories; Gallus, another disciple of Martin, is allowed to make up a third, and eventually local monks, hearing of what is going on, make up an audience. The dialogues discuss the ascetic practices of eastern Christians, including Jerome, in comparison with the church in Gaul and Martin's exemplary spiritual and practical inheritance.[15] As with John Cassian, whom we discussed earlier,[16] these dialogues are an instrumental sign and symptom of the spread of monasticism in the West. The form of these exchanges in their elegance, commitment, shared criticism and good-feeling are *part* of

---

[13] Stancliffe (1983) 15–110 records what we know about Sulpicius' *vita*.
[14] Paulinus, *Ep.* 5.5. See Conybeare (2000); Trout (1999).
[15] On the role of Origen, the most discussed part of this dialogue, see van Andel (1980).
[16] See pp. 82–4.

the argument about how normative Christianity should be lived, as much as such questions constitute the substance of their debates.

Martin, by contrast, was 'uneducated', *illiteratus*, according to Sulpicius Severus, his disciple and biographer. Martin came from a pagan, military family, and at the age of fifteen he was enrolled in the Roman army by his father, who had risen to the rank of military tribune. Martin was part of the Imperial Guard under Constantius and Julian. It was as a soldier – and not yet 'reborn in Christ'[17] – that he performed the act through which he is most often depicted in later iconography. He was dressed in uniform when he passed a destitute beggar whom nobody else was pitying. Martin took what he had – his own cloak – and split it in two and gave half to the beggar. It made some of the bystanders laugh, because it made his uniform look absurd, but that night Martin had a dream vision in which he saw Jesus wearing the half cloak. (Later depictions, in which Martin is fully enclosed in grand medieval body armour and riding a horse, make the gesture seem no more than symbolic and scarcely the basis for a future ascetic life.) Eventually, the story goes, on the eve of a battle, Martin declared to the emperor 'I am Christ's soldier; I am not allowed to fight'. The emperor's rage is deflected by the enemy's surrender, and Martin succeeded in leaving the army, travelling to Gaul to visit Hilary of Poitiers, the mainstay of Nicene orthodoxy in the country, and to commit himself to a Christian life.

Martin returned from his travels to Tours where he was elected bishop by popular demand. He founded a monastery nearby and, until his death, continued both as bishop and as the head of the monastery – living in austere asceticism with his disciples, and, according to Sulpicius, performing a stunning series of miracles. Martin, then, like Ambrose, is a bishop. But Martin's ill-educated asceticism – he wrote nothing – contrasts with Ambrose's worldly and very literate career. Martin generally dismissed the opportunity his celebrated holiness gave him to meet the great and the good (except for one dinner with Maximius (*VM* 20)), whereas Ambrose was a figure at court. Ambrose indeed became the very figure of the worldly authority of the church when he required the emperor Theodosius to perform repentance in front of him, a spectacle of power that epitomized the changing position of the church in the empire.[18] When Martin did enter politics, specifically around the Priscillianist controversy, it was a failed attempt to persuade the emperor not to put Priscillian to death

---

[17] *Necdum . . . regeneratus in Christo*, VM 2.8.
[18] The contrast between Ambrose and Martin in these terms is made explicitly in Sulp. Sev. *Ep.* 1.25.

for his heresy – the first time a Christian emperor had punished a Christian with death for his views.[19] Ambrose for his part was deeply and successfully involved in political intrigues between the Arian court and the Nicaean cathedral. Whereas Sulpicius Severus' life of Martin is full of his miracles, Paulinus of Milan's biography of Ambrose has a series of public and private political successes, but includes only two miracles, both explicitly typological: a woman touches his robe and is cured;[20] he lies over a dead child who recovers.[21] Neither event is mentioned in Ambrose's own writings, notes Paulinus of Milan, and they are very much a footnote to what makes Ambrose worthy of praise in the eyes of his hagiographer.[22] Whereas Ambrose has left us hymns, sermons, treatises – a normative, systematic engagement with the Christian imagination – Martin, like Theophilus who destroyed the Serapeion in Alexandria, violently assaulted monuments and buildings of the religious traditions he despised, forcing the local inhabitants to accept his destructiveness. As the role of bishop in the church is very much under formation in this period, Martin and Ambrose indicate two different trajectories, for all that they were both elected by popular acclaim and shared anti-Arian stances.

Indeed, the intense political activity around Priscillian (to which we will inevitably return when we look at Sulpicius Severus' *Chronicon*, our main source for the controversy) allows us to nuance what appeared in my last chapter as a contrast between Ambrose's institutional embedding and Prudentius' 'villa Christianity'. The Council of Saragossa, held in 380, in response to the Prisicillian heresy, passed as its fourth decree that Christians were 'not to be concealed in houses, nor to stay on estates, nor to head for the mountains, nor to walk in bare feet, but to flock to the churches'.[23] This decree appears to target private groups in small city houses (seen as an insidious political threat since the fifth century BCE), the 'villa-Christians' such as Sulpicius Severus on his estate at Primuliacum (or Prudentius), anchorites or hermits who went into the wilderness alone, and mendicant ascetics. They are all enjoined by the bishops, who drafted the decree, to come into the churches, and thus to submit to the authority of the bishops. Indeed, the council even criticized ascetic practice itself and private reading, which might be thought to be mainstays in the self-representation of Christian virtue. The council proposed thus an aggressive policy of institutional centralization. Priscillian was attacked in part

---

[19] Burrus (1995).   [20] Paulinus, *Vit. Ambr.* 10: see John 14.12.
[21] Paulinus, *Vit. Ambr.* 28: see 2 Kings 4.   [22] Noted both at *Vit. Ambr.* 10 and *Vit. Ambr.* 28.
[23] Cited and discussed by Burrus (1995) 37–8.

because he seemed to work, like Prudentius, 'outside a church context',[24] included women in his groups (hence the accusations against him of seductiveness and moral corruption), and bypassed the formal institutions of the church.[25] Yet Martin too was caught up in such accusations, though he had no truck with women: although he was a bishop, he also followed an ascetic life-style, and asserted his own authority separate from the wrangles of church doctrine. His charismatic persona – and performance – was distrusted by these established authorities.

This controversy reveals that in the late fourth century the political arguments about the status of the bishops and the reach of the institutional church focused not just on the dogmatic disputes over Arianism and the Nicene creed, but also on regulating the boundaries between public and private authority. The individuality of the saint's ascetic practices and miraculous power could be seen as a threat to the established church as well as an example; Martin's withdrawn monasticism provided a contrast to the worldly bishops. How a Christian life was to be lived – in what forms of community, with what commitments to civic institutions, with what political engagement – remained a sharply contested normative argument. There are many stories of the requirements and rejections of institutional expectations in the name of something purer or more committed – counter-stories, expressed repeatedly as the proclamation of what is more properly Christian. Gregory of Nazianzus, in the East, wrote of his bitter disappointment with the politicking in Constantinople, returned to his backwater, and took a vow of silence for a year, during which time he wrote poetry – different styles of public withdrawal (and yet publicity). Synesius, when he took up the role of bishop, refused to separate from his wife, insisting that his married life would continue, in its propriety – a view expressed in a letter to his bishop, Theophilus, which circulated as a document beyond its recipient, as a semi-public statement of principle and a self-representation. Jerome, in his very public rejection of his former public life in Rome, withdrew to the desert in the Holy Land, but took his library and staff and remained fully active in the doctrinal exchanges of the era – one of the sights to visit on pilgrimage, a figure whose influence thereby stretched beyond his disseminated publications. Such negotiations were articulated as the question of how to engage with the *saeculum*, the 'age', the 'time' in which they lived: as Sulpicius says of Martin, all his talk

---

[24] Palmer (1989) 3. Victorinus (Aug. *Conf.* 8.2.4) was upbraided by Simplicianus: 'I shall not believe or count you a Christian till I see you in a Church of Christ'. But he replied with a laugh, 'So do walls make Christians?'.

[25] The issue of gender is particularly well discussed by Burrus (1995).

was how the *saeculi onera*, 'the burdens of this age', 'time's weight', must be relinquished.[26]

Sulpicius Severus' *Vita Martini* is our main source for Martin's career as a bishop. Its exemplary status is broadcast by Paulinus of Milan's biography of Ambrose where he reports that Augustine instructed him to write the work as 'Athanasius had penned the life of Anthony, Jerome the life of Paul, and Sulpicius Severus, the servant of God, also composed in a highly polished style, a *Life* of Martin, the venerable bishop of Tours'. The *Life of Anthony*, as we saw, played a key role in Augustine's own narrative of conversion and was a fundamental text for the ascetic movement of Christianity; Jerome's life of Paul is also focused on asceticism and its sexual anxieties. The *Vita Martini* in turn constructs the life of a bishop as an ascetic, miracle-working model of faith and virtue.[27] Unlike the lives of Anthony and Paul, it was completed before its subject's death. Unlike most hagiographies and – of course – unlike all martyr narratives, it has no representation of the saint's death, and thus no culmination in the perfected end, the triumph over this life in the transfiguration of death. Partly for this reason, its generic affiliation has been much discussed by scholars, who have seen the influence of Suetonius and Sallust much in evidence.[28] The *Vita Martini* is best seen, however, on the one hand, as a hinge between the multiform classical tradition of biography and what will become the genre of Christian hagiography: a foundational text which does not so much epitomize as seed the generic tradition to come. On the other hand, it is also to be regarded as experimental, one of several roughly contemporary attempts to write a Christian life – such as the autobiographical examples of Augustine's *Confessions*, Gregory of Nazianzus' *Poem on Himself*, Synesius *Epistle* 105, or the more general models of a Christian day, discussed in the previous chapter. It is as the expression of how a life can be narrated that the *Vita Martini* is salient for my argument.

The opening chapter of the *Vita* is a sophisticated programmatic statement. The work is prefaced by a letter to the dedicatee, Desiderius, which expresses Sulpicius Severus' humble agreement to send his humble work into the world anonymously, with all due apologies for its lack of style (a generic modesty).[29] Sulpicius had hoped to keep the work private, but allows it to go to Desiderius with the promise it would go no further, or if it

---

[26] *VM* 25.4. A distant echo of Catullus 14.22–3 *saecli incommoda, pessimi poetae?*

[27] Harper (1965). See, by way of contrast, Williams (2008).

[28] Discussion and bibliography in Burton (2017), and for the influence of these Latin writers on the *Chronicon* van Andel (1976).

[29] Conybeare (2000) 41–59 discusses how Christians strove to enact a 'spiritualization of the aristocratic habit of forming and maintaining connections by letter'.

did, to travel anonymously: the negotiation of what is private and what is public is carefully and openly articulated as an elegant but transcended hesitation. This performance of modesty is immediately belied by the bold and stylish opening of the *Vita* itself (1.1):

> Plerique mortales, studio et gloriae saeculari inaniter dediti, exinde perennem, ut putabant, memoriam nominis sui quaesierunt, si vitas clarorum virorum stilo illustrassent.

> Many human beings, vainly committed to the pursuit of worldly glory, have sought consequently to make what they think is an everlasting memorial for their own name, if they might render famous the lives of glorious men by their pen.

This opening generalization about humanity – a grand frame for his project – utilizes the language of the classical tradition against itself. Sallust's *Catiline* and his *Jugurtha*, biographies both, start with similar broad references to mankind (*omnes homines/genus humanum*), and the phrase *multi mortales* is a repeated phrase in *Jugurtha*.[30] Sallust also declares the glory of writing of famous deeds (*Cat.* 3.1–2). Tacitus' *Agricola* opens with a sarcastic comment about the *clarorum virorum facta* in his own times, the 'deeds of glorious men'. The search for *perennem memoriam*, 'everlasting memorial' echoes Horace's famous claim to have erected a *monumentum aere perennius*, a 'monument more everlasting than bronze' – a self-fulfilling (in all senses) claim on posterity. Yet such references to the long tradition of the pursuit of glory in the expectation of immortal fame – an ideology we discussed in chapter 1 – are undermined by Sulpicius' normative dismissiveness. Such dedication is vain (*inaniter*), its immortality is a false judgement (*ut putabant*). Even the claim to be *clari*, 'glorious', may contain an 'implicit contrast between the glory sought by distinguished (but mortal) Roman heroes of old, and the glory of God'.[31] Both the pursuit of worldly glory and the hope to immortalize it in writing – the hero and the bard – are set up to be dismissed.

With a display of nuance, Sulpicius Severus allows that the classical tradition has allowed 'some return on their expectation: not an everlasting return, but a little nonetheless' (1.2). This is because such writers – whose language he is imitating – have indeed 'spread abroad the memory of themselves', which is, Sulpicius Severus immediately qualifies, 'albeit in vain'. The examples of great men have also stimulated in readers

---

[30] See for references and discussion Burton (2017) *ad loc.*   [31] Burton (2017) 140 *ad loc.*

a considerable desire for emulation (*aemulatio*). Sulpicius Severus recognizes both the strength of the classical tradition of historiography or biography – the fame of Suetonius, Sallust, Tacitus, whose names, of course he will not cite to add to their *memoria* – as he recognizes the express purpose of such writing: to provide the *exempla* that motivate virtue. Yet this recognition is no more than a foil for his rejection of it: *Sed tamen nihil ad beatam illam aeternamque vitam haec eorum cura pertinuit,* 'Yet all this effort of theirs has no bearing at all on that life which is blessed and eternal' (1.2). The writings of the past are doomed to triviality because their aim is not defined by Christian time, the life which is 'blessed and eternal'. In *beatam illam* we can hear the familiar trope of Roman writing, *beatus ille, qui ...,* 'happy is he who ...', the repeated expression of what a good life is, often indeed associated with a noble death. In the Psalms, from the very first words, *beatus vir,* this same motif is common; but now to be *beatus* must be understood through *aeterna vita,* the Christian promise of everlasting life. Biography is defined by its sense of time.

Indeed, for Sulpicius Severus the Christian vision of time vitiates the heroes of the past: 'What good did glory (*gloria*) do even to the heroes themselves, when it is destined to perish with the age (*saeculo*) of their biographers?' (1.3). The eternal life that Christianity offers dwarfs the mere human era (*saeculum*), which is the condition of pagan biography. Nor, indeed, is exemplarity, the second purpose of such pagan life stories, of any value to posterity. Sulpicius picks out for criticism Hector and Socrates – that is, Homer and Plato, on the one hand, the two privileged authorities against whom Christianity sets itself, the wisdom of epic and philosophy, poetry and prose; or, on the other hand, military success or intellectual success, two routes to glory: the one *pugnantem,* 'fighting', the other, *philosophantem* 'philosophizing'. Both Hector and Socrates are useless exempla, argues Sulpicius, because they rested their hopes in mere stories, myths (*fabulis*) – epic, dialogues – and gave 'their souls to tombs', misunderstandings explicable because 'they reckoned the value of human life solely in terms of actions in the present'. Even Socrates' commitment to the immortality of the soul and Hector's search for immortal fame are no more than 'actions in the present', because they do not look towards true immortality, and must thus be restricted to the here and now of this *saeculum.* They trusted they would live on – but only in human memory. But, concludes Sulpicius, 'It is human duty (*officium*) to seek not an everlasting memory (*perennem memoriam*) but rather everlasting life (*perennem ... vitam*)' (1.4). How is this to be done? 'Not by writing, or fighting, or philosophizing, but by pious, holy, religious living'. Everlasting life contrasts with mere memory that continues, and it is won – a parallel contrast – not by the activities of a biographer or by a military hero or by a philosopher but by a form of living – a Christian life. This ideal is not merely a general template but also is programmatic for the life of Martin to come. He wrote nothing; he was a soldier but rejected such a career for a life as a soldier of Christ; he was saintly but did miracles not philosophy. He will exemplify what it means to have a life that is lived 'piously, venerably, religiously'.

This opening paragraph of the *Vita Martini* establishes a new framework for biography. A life is still to be exemplary. This life (as Sulpicius goes on to assert) will lead, however, towards *vera sapientia*, 'true wisdom', as opposed to mere philosophy, and *divina virtus*, 'divine virtue', as opposed to bravery in battle, which is indeed *militia caelestis*, 'military service for heaven'.[32] Exemplarity depends on values, and the opposition between Christian and pagan ideals is articulated starkly. Yet such moral values are justified and determined by the temporal framework of eternal life: how life is to be led depends on this understanding of eternal life. Such (eternal) life is opposed to (eternal) memory which is all that pagan celebration and aspiration can offer. Christian time and Christian biography are mutually implicative – and transcend pagan achievements.

Sulpicius takes a full second paragraph to spell out his reasons for writing the biography of Martin. Martin is an exemplary Christian, fit for imitation, and thus deserving not to be forgotten. The author hopes humbly for eternal reward, though neither Martin nor the author seek fame or praise. Yet even in this strategic modesty, there is a polemic taking shape. In Sulpicius' second dialogue, a discussion of asceticism, miracles, and the less than satisfactory behaviour of bishops, Postumianus recalls that Martin often used to say to Sulpicius Severus that once he was a bishop, 'he did not possess the gift of working miracles in anything like the same degree as he could remember possessing it previously' as a monk.[33] Philip Rousseau adds that the bishops of Apphy of Oxyrhynchus and Netras of Pharan are quoted as saying 'God's gift has not deserted me because I am a bishop'. He concludes that Sulpicius is 'convinced that exact correspondence could be achieved between the demands of spirituality and an ecclesiastical career', and that Martin fulfilled that balance perfectly.[34] 'Our people here', declares Postumianus, 'should not press the example beyond the limits observed by Martin' (2.8). In the context of the arguments about

---

[32] *VM* 1.6.   [33] *Dial.* 2.4.   [34] Rousseau (1978) 150–1.

Priscillian, and the debates over the boundaries between private and institutional religiosity, the representation of Martin as a model 'contains always an implicit reproach to the majority of bishops (at least in Gaul) who fall short of this ideal'.[35] A *beata vita* depends on *aeterna vita*, but how the *beata vita* is to be lived remains integrally contested between the Christian writers of the fourth century. Martin's exemplarity is also polemical.

The representation of the life of Martin is fascinating, and sets the agenda for future hagiographies of Christian saints. His early life – by which is meant his life before his conversion – is barely adumbrated. We are told in a sentence where he was born and bred and who his family was. A second sentence places him in a military career. But almost immediately – and at greater length – we are reassured that he wanted to enter a monastery at age 10, and was always thinking of a religious life. (There is nothing of the complex psycho-drama of Augustine's *Confessions*, nor any reflection on the relation between youthful experience and later life, as, say, in Plutarch's life of Alexander.) His military career is described through his preparation for baptism – by helping the poor, culminating in his gift of half of his cloak to the poor beggar, and his consequent vision of Jesus.[36] And the final story of his pre-conversion life is his resistance to fighting and his eventual release from the army. Only three of the twenty-seven chapters of the biography, then, are dedicated to his life before his career in the church, and even these are almost entirely taken up with his incipient Christian faith. Scholars have been much vexed by the chronological contradictions between the *Vita* and stories in the dialogue and letters, and have argued without adequate conclusion about how long Martin served in the army, and what the gap was between his baptism and discharge.[37] It might be better to recognize that Sulpicius Severus turns his gaze away from this period of Martin's life except inasmuch as it provides exemplary tales of the future saint's Christianity. The seminal story of the beggar is introduced with *quodam itaque tempore*, 'so at some time . . .'. In a story focused on *aeterna vita*, the mundane business of dates and human causality is of only trivial notice.

Indeed, the remaining 24 chapters of the *Vita* are hard to locate in any significant chronological order. After his first significant miracle, he was regarded as a *beatus vir* and *sanctus*, 'a blessed man', 'holy'. This is marked

---

[35] Burton (2017) 140 *ad* 1.7.
[36] See Praet (2016) with further bibliography; Roberts (1994) for the afterlife.
[37] A discussion started by the seminal work of Babut (1912); see e.g. Stancliffe (1983) 111–48; Barnes (1996), (2010) 205–8; Burton (2017) 9–25.

by *ab hoc primum tempore*, 'from this time, first . . .'. The next chapter opens *nec multo post*, 'Not long after'. The next '*sub idem fere tempus*', 'At about the same time . . .'. Other chapters begin '*insequenti tempore*', 'in the course of time'; and repeatedly with 'about the same time', 'at the same time', 'meanwhile'. It is even vaguer than the vague chronology in the more teleological model of the Gospels.[38] Each chapter contains a singular, often miraculous event, until the final summary chapters.[39] It is striking that the letters and dialogues add extra miracle stories with a similar lack of concern for where they fit into a life-time. The *Life* has no death scene, and thus ends with a general eulogy of Martin's character (and a declaration of Sulpicius Severus' own sincerity). In the second *Letter*, Sulpicius Severus confesses he had a vision of Martin rising to heaven holding the *Vita* in his hands – and awoke to the message of Martin's death. He was – declares Sulpicius – a martyr, although he was not killed for his faith, because 'he suffered for the hope of eternity'.[40] Martin's *Vita* is a set of juxtaposed stories, to which other stories can – and for Sulpicius, should – be added. We are told that Martin served under Constantius and Julian, but there are barely any other standard markers of the *saeculum*. The death of Maximius is registered only as the fulfilment of a prophecy by Martin. Unlike the historiographers, Sulpicius Severus offers no dates or dating devices. What counts in the life is not its location in what we could call secular history, but its repeated performance of sanctity. The life is constituted by a series of discrete scenes of displayed holiness.

The Talmud's representation of a Jewish life, as we discussed in chapter 9, fractured the continuity of time into a series of halachic choices, singular scenes where the exercise of debate over the right thing to do could be performed. This narrative style changed the possibility of causality, agency, regulation and the experience of time. In the *Vita Martini*, narrative is fractured into a series of singular stories, for all that the *Life* begins with a birth, and, through later texts at least, adds the telling of a death. But the individual scenes do not focus on moments of debate or choice. The moment of conversion converts the narrative: before baptism, the focus is on the barriers to conversion overcome, and the early signs of the religious life to be chosen. After conversion, what counts is not the order of a life – not the development of a person, career advancement, an education into new understanding, not even growing old. In this life of waiting, all that is deemed significant is the *exercise of sanctity*. Hence the

---

[38] Luke 3.1 dates the preaching of John the Baptist to the fifteenth year of Tiberius' reign.
[39] Harper (1965).   [40] See the discussion of 'living martyrs' above, p. 202.

vita becomes a series of exemplary stories, each self-contained, each the performance of holiness, with little explicit link between them. The narrative form is an expression of the rubric of evaluation: to live *pie, sancte, religioseque*, 'piously, holily, and religiously'. Each story reveals those qualities, another *exemplum* narrated for the faith of the reader, led by the author and his sincerity. The programmatic force of *aeterna vita* requires a life that is 'constant prayer and reading', interspersed with miracles, and, as a sign of his sanctity, fights with the devil, visits from angels, and conversations with Agnes, Thecla and Mary (something, we are told by Sulpicius, pushing against the boundaries of faith, that even the monks in the monastery doubted to be true). It is these singular signs of his status as a holy man that gives the narrative its episodic structuring and its moral force. This form of a *Life* is an expression of how living is to be shaped by the promise of eternity. The *Life of Martin* indeed becomes the foundation for later hagiographies, instrumental in modelling the Christian experience of a life-time. The *Vita* tells its reader how to live a life, to tell a life, to inhabit time.

*

The dialogue with Postumianus (sometimes divided by editors into two dialogues) ends suitably enough with Martin's views on the end of the world. Martin states that Nero will come to power in the West and make the worship of pagan idols compulsory; meanwhile the Antichrist will have power in the East in Jerusalem, and he will compel men to deny Christ and he will set himself up as Christ, and order all men to be circumcised. The Antichrist will go on to defeat Nero, and the whole world will be under his sway until the Second Coming when he will be defeated by Christ. '"There was no doubt too that Antichrist was already born; his conception was the work of an evil spirit. He is now a child and will take over supreme power when he comes of age"'.[41] This particular version of the end of days, condemned by Jerome as a heresy, and, consequently, omitted from some manuscripts of the dialogues,[42] is given in Martin's name by Gallus, who comments 'you can judge how soon this fearful future may be upon us'. Although Nero had died centuries earlier, the terror of the imminent persecution heralding the end of days is strikingly vivid. If *aeterna vita* is the principle through which the *Vita* is conceived, here in this dialogue, Sulpicius Severus' supplement to the *Life* after Martin's death, the prospect of the end of time is the necessary perspective provided

---

[41] *Dial.* 2.14.   [42] See Babut (1912) cclxix–cclxx. In general, Vaesen (1988).

to the dialogue's tale of faith and miracles. But Gallus' story is interrupted in full flow by a slave who announces that another priest is at the door – which ends the dialogue. The following dialogue, led also by Gallus on the following day, continues the discussion – but never returns to Martin's apocalyptic vision. That the story of the imminent end of days is interrupted and receives no commentary, and then is passed over in continuing silence, may be significant. In the *Chronicon*, Sulpicius carefully frames such stories with the rhetoric of doubt. When he comes to Nero, he calls him the 'most filthy not just of humans but also of animals' because he was the first to persecute Christians. He continues: 'I do not know if he will be the last fulfilment too, if indeed, as the opinion of many holds, he will come before the Antichrist.'[43] He concludes the life of Nero with the fact that his body was not found, and that 'it is believed (*creditur*) that although he had stabbed himself with his sword, his wound was cured, as was written on this topic, "the blow of his death was cured" [Rev. 13.3], because he had to be sent back at the end of time (*sub saeculi fine*) to fulfil the mystery of his evil'. Nero is written into the book of Revelation – the reference is to one of the heads of the monster from the sea. But Sulpicius distances himself from the argument by ascribing it to the view of many and to belief (a historiographer's caution). The imminence of the end of days, so strong a conviction in Paul, had by the fourth century become a more carefully hedged topic, and Martin's prediction is left to hang unfulfilled.

The *Chronicon* indeed takes a strikingly different view of time from the *Vita Martini*. The prologue to this work also begins with a programmatic statement of principle (*Praef.* 1.):

> Res a mundi exordio sacris litteris editas breviter constringere et cum distinctione temporum usque ad nostram memoriam carptim dicere aggressus sum, multis id a me et studiose efflagitantibus, qui divina compendiosa lectione cognoscere properant.

> I have set out to give a condensed history of events from the beginning of the world, as laid out in scripture, and to tell them separately, with distinction of times up to the time of our own memory; many people have asked this from me earnestly; they were keen to know about divine things from a summary account.

It is worth lingering on this dense and precise statement. The subject of his writing is *res* – 'things', 'events' – which summons in its generality both

---

[43] *Chro.* 28.1.

the cosmographic epic of Lucretius (*De rerum natura*) and the *res gestae* of historiographical accounts.[44] But these *res* are specifically what has been written in scripture, a pointed indication of what should be memorialized in a history. We are not, therefore, to be treated to a version of the Persian Wars or the Peloponnesian War, nor the Social War or Hannibal, stuff of the *saeculum*. Or: the *thōmata* promised by Herodotus will now be superseded by the *thaumata* of divine miracle. This is a history that will take its start from the beginning of the world: not *ab urbe condita*, the foundation of the city from which Livy and others mark their narrative's purpose.[45] Yet, this account is to be a condensed version (*constringere*, since Cicero, is a technical term for such abridging).[46] The promise of the history from the beginning – all time – will not take too much time to tell. As I have discussed elsewhere, scale is an essential rhetorical element of the literature of late antiquity, where epyllion and epigram play so significant a role.[47] Sulpicius Severus' abridged scripture is another paraphrase, another rewriting of scripture, now in historiographical mode – a Christianizing of history and a historicization of Christianity. The blending of genres and the rhetoric of scale is a sophisticated self-placement between Christian and pagan literary tradition. Sulpicius continues that his narrative will be shaped *carptim*, which I take to mean 'separately', 'in discrete parts', that is, not through the parallel tabulation of multiple narratives that Eusebius pioneered, on the one hand – thus setting himself within the practices of Christian chronicle writing – and, on the other, as he specifies, *cum distinctione temporum*, 'with distinction of times'.[48] As we will see, this involves not just periodization, but also a concern for dating and ordering of events. Where the *Vita Martini* purposively resisted any such historiographical insistence on period and placement of action within a time-frame – its actions took place very much without *distinctione temporum* – the *Chronicon* announces its historiographical credentials from the beginning. If the *Dialogue* left open the question of the end of days, the *Chronicon* announces its end at the start: *ad nostram memoriam*. The *Vita Martini* dismissed *memoria* as the hope of pagan striving; the *Chronicon* makes Sulpicius Severus' *memoria* the end of the history. The *Chronicon* will aim to do two

---

[44] van Andel (1976) collects and discusses echoes of earlier historiography in the *Chronicon*.

[45] Socrates, *Hist. Eccl.* 5 pr. indicates that church history, by contrast, should start with the Incarnation. See van Nuffelen (2010) 167, and more generally van Nuffelen (2004).

[46] Festus, for example, provided an abridged history of Rome (*Breviarium*) as Solinus reworked and abridged Pliny and Pomponius Mela.

[47] Goldhill (2020) 38–70.

[48] *Carptim* echoes Sallust, again – *Cat.* 4.2 – *his* historiographical methodology.

things sometimes kept separate in the historiographical tradition; it will combine both the deep history *a mundi exordio* and a detailed contemporary history of conflict for which the author's authority as an eye-witness is crucial. Again, the careful recognition of – and placement within – generic expectations is articulated with self-aware reticence. The claim that many people have asked him to write is a cliché of modesty, but here it is used also to summarize his double purpose: first, to relate *divina*, 'divine things', a purposive re-expression of the opening *res* now as a charged insistence on Providence as the driving structure of history; second, to relate them in *compendiosa lectione*, an 'abridged text', a selection that is, as he specifies in the following sentence, two books containing the facts (*gestis*) culled from many volumes. The *Chronicon* thus, in its opening sentence, is programmatically announced not just as an abridged paraphrase of scripture leading to an eye-witness account of the church, but also as a self-aware generic intervention in how the Christian narrative of time is to be comprehended and enacted.

Sulpicius continues with the promise that despite being a paraphrastic summary, this abridgement has omitted 'almost nothing of the facts' (*gestis*). (The 'almost' opens a vista of selectivity that we will see to be performed repeatedly, as with the story of the raven and doves on the ark.) He also explains his combination of deep history and contemporary conflict. 'It seemed not absurd', he writes, 'after I had run through sacred history up to the crucifixion of Christ and the deeds of the Apostles, to add also the events (*gesta*) that happened afterwards'. If this bare self-justification looks back to *nostram memoriam*, the following sentence offers a fascinating gloss on *distinctione temporum* (*Praef.* 2):

> Ceterum illud non pigebit fateri, me, sicubi ratio exegit, ad distinguenda tempora continuandamque seriem usum esse historicis mundialibus atque ex his, quae ad supplementum cognitionis deerant, usurpasse . . .

> But I will have no hesitation in confessing that, whenever reason demanded it, I have made use of pagan historians for making my distinction of times and for maintaining an unbroken sequence; I have appropriated from them what was needed to supplement my knowledge . . .

Although he had announced that his history was extracted from scripture, he adds here that he has no embarrassment in admitting that he has used the pagan historians he has been alluding to already. This appropriation (*usurpare* may imply a philological *militia caelestis*) is necessary both to fulfil his aim of *distinctio temporum* and to produce a continuous narrative: there are swathes of history for which there are no scriptural sources, of course. But, with an appeal to *ratio*, 'reason', he also allows pagans to supply what he needs by way of knowledge. Much later, Sulpicius explains that the lack of reference in secular histories to the events of scripture is itself a sign of God's plan, to keep *sacrae voces* pristine of such influences.[49] Where there are any discrepancies in chronology, Sulpicius, like Jerome, blames copyists' errors.[50] As the *Chronicon* in its opening revealed a sophisticated intermingling of generic markers between Christian and pagan historiographic trajectories, here such rhetoric is given an epistemological grounding. The Christian cannot do without pagan knowledge. Christian history and historiography remain fully imbricated with pagan forms.

Sulpicius Severus' introduction closes with a defence of the voice of the paraphrast, which, as we saw with Erasmus' paraphrases, became a heated topic of debate in early modern Europe, and is expressed in late antiquity by the contrast between Ps-Apollinaris' explanatory preface to his metaphrasis of the Psalms, written in the voice of the author, and Nonnus' refusal of any self-representation in his paraphrase of the Gospel of John (though not in the *Dionysiaca*). Sulpicius insists that he does not present his abbreviation of scripture as an author (*auctor*). His aim is not that any reader should *omit* to read scripture; and he insists that any reader familiar with scripture will recognize what he reads. Thus his paraphrase cannot replace scripture (*Praef.* 3):

> Etenim universa divinarum rerum mysteria non nisi ex ipsis fontibus hauriri queunt.

> For they cannot draw up the universal mysteries of divine matters except from the sources themselves.

Sulpicius denies his own authority: readers should go back to scripture, where the highest level of truth lies. It is only in scripture that the mysteries that appertain to the universe, the perspective of God's time, can be discovered. What, then, is the purpose of this abbreviated history? Explicitly he claims it is to 'teach those who are unaware and bring conviction to the educated' (*Praef.* 2). As we will see, it offers more than this, not least when it moves from scripture to contemporary events. But the awkwardness is marked in the defensiveness. For whom, then, is such an abbreviation of scripture necessary? What is left out and what included?

---

[49] 2.14.3, a passage discussed by Williams (2011) 286–8.　　[50] 1.39.1.

What image of Christianity does Sulpicius Severus' historiographical Christianity provide?[51]

The preface to the *Chronicon*, then, asks who owns the past, to retell it. Whose knowledge is to count? Whose style of writing? On the one hand, in the voice of the paraphrast, Sulpicius Severus insists that he abbreviates, and does not write as an author, and almost every fact from scripture is there; on the other hand, Sulpicius admits he has intertwined pagan knowledge where reason and continuity demand it. There is a lack to be filled, his narrative needs its supplements from beyond scripture. This tension, we will see, is played out again and again in the history that follows. It is not resolved by insisting that the true mystery of divine matters is to be found only in the scriptures, which serves to locate authority elsewhere. What is particularly salient, however, is the different modelling of time in the *Vita Martini* and the *Chronicon*. Where the *Vita Martini* in the voice of the faithful disciple and hagiographer offers discrete scenes of sanctity with little interest in dates, sequence or world history, the *Chronicon* in the voice of the scholarly historiographer is committed to *distinctio temporum* and *series* – the framework, as it were, that is lacking in the *Vita*. Together, these two perspectives map complementary but different trajectories of Christian narratives of time, embodied in different narrative forms. As we will see, the contrast produces different ideas of how the *praesens,* the moment of the now in time, is to be comprehended.

That the *Chronicon* is much more than a mere abbreviation is evident from the narrative's opening sentences (1.1.1):

> Mundus a Deo constitutus est abhinc annos iam paene sex milia, sicut processu voluminis istius digeremus. Quamquam inter se parum consentiant qui rationem temporum investigatam ediderunt.

> The world was created by God, almost six thousand years ago, as we will set out in the course of this book. Although those who have published a researched calculation of the time all too little agree among themselves.

The expected opening – that God created the world – is immediately qualified not just by the addition of a time-scale – something no scriptural account hazards – but also by a placement of such a claim within the structure of his own work – a self-aware marking of composition – and within ongoing scholarly debate on chronology in general. The narrative includes internal commentary on its own proceeding. There is no before to

---

[51] See Williams (2011) for the problem of Sulpicius' authority in historiographical terms.

this history. From Herodotus and Thucydides onwards, the historiographer's commitment to an epistemology of evidence produces a rupture between myth and history, a speculated era of before and a knowable time of what can be researched and tested, a *distinctio temporum* that plays no role in this account. Even Eusebius started with Abraham. For Sulpicius, myth has no function. *Distinctio temporum* is not a distinction of *types* of time, as it can be in classical historiography, but the counting of years in one sequence (*series*) towards an end.

The six days of creation are not represented. Sulpicius Severus goes straight to creating humans and the Garden of Eden, from which humans pass to exile in 'our world', without – remarkably – the intervention of the snake, or the actions of Eve, and in nothing more than a subordinate clause ('when they tasted the tree forbidden to them'). The abbreviation is drastic. When we get to Abraham, it is telling what is included, what excluded. Abraham was born, we are told, 1,017 years after the flood – maintaining the chronicle's interest in dating and counting the *tempora*. We are told that his and Sarah's names were changed by God – but, he adds, 'the not insignificant mystery of this fact is not for this work to expound'. The explicit refusal to reveal the mystery is also an encouragement to seek beyond his text for its full meaning, as the preface had suggested. Similarly, he turns to the language of typology only once – with Deborah who is said to be a type of the church – though a certain historical typology is integral to the history, in that the turning point of the Incarnation provides 'the meaning that underlay all of the significant events recorded in the *Chronicle*'.[52] Sarah's story is almost entirely occluded: she is said to exile Hagar, but otherwise her laugh, her response to the angels, her beauty that leads her to be taken into the Pharaoh's harem, are all ignored. Lot's wife, however, unlike Eve, prompts a moralizing comment: when she turns back to look at Sodom and Gomorrah, it is a demonstration 'of human sinfulness, which is pained to restrain from what is forbidden'. When Lot is captured, and Abraham goes to rescue him, Sulpicius specifies that he took 318 men with him. It is marked that this detail of number is included (although the snake omitted from the Garden of Eden). The number of bishops at the Council of Nicaea was said to be 318 as a typological fulfilment of this story – hence the detail also has an ideological import, and needs to be included. So when Sulpicius indicates that he will offer no summary of Leviticus because it is full of laws (1.19.1), it is not only to focus on the narrative history of Israel,

---

[52] For Deborah (Judges 4) see 1.23.3. For historical typology, see Williams (2011).

but also in line with Christians' dismissiveness of what they call Jewish 'law-bound' religion. He also refuses to abbreviate the Gospels or the Acts of the Apostles lest his 'form of concision detracts from their dignity'. He does, however, specify the day and year of Jesus' birth, and who the consuls were, the standard Roman method of dating – something no Gospel contains. Sulpicius' *Chronicon*, like all histories, embeds its ideology in its narrative form.

The interest in dating and counting continues throughout, expressed, as with Jesus' birthday, with a certain scholarly precision. So, Joshua's death took place 3,884 years after the creation of the world, but – the occasional admitted doubt is a strategy of self-authorization – he adds with due caution, 'about the time (*tempore*) of his rule I can't be definitive. The usual opinion, however, is that he ruled the Hebrews twenty-seven years' (1.22.3). He tells us that thanks to his voluminous reading he found the sequence of Babylonian kings interpolated in an anonymous book, and it agrees with his own calculations (2.5.4). Getting the time right – the number of years of any period of rule and the relation between different stories – is a fundamental aim of the *Chronicon* – *ut temporum ordo consertus sit*, 'that the order of times should be fixed' (2.19.1) – and so the books of the later prophets are intercalated with the rise to power of Darius in Persia. This is important not only as the background to the book of Daniel but also in making the transition from scripture to the era of Greek rule, a transition made by noting that Alexander came to power by defeating a descendent of Darius on the Persian throne. Getting the time right – measuring, adding the years – is crucial. The 6,000 years of this human *saeculum* requires careful counting to anticipate the end of days. The repeated insistence on the time-scale of his history is not an antiquarian obsession but a necessarily Christian understanding of the unfurling of time and the place of the present in providential world history. Christianity's commitment to a moment of creation and an eschatology of the end of days makes historical time finite (here 6,000 years). The modern, shocking rediscovery of 'deep time', starting with geology and biology in the nineteenth century, is, as we saw, set precisely against this Christian delimitation of the beginning and end of time.

So, too, the moralizing increases as the narrative continues. When Sulpicius records that the Levites are not allowed to hold property, he expostulates that he cannot pass over this example in silence. Today, he complains, priests have completely forgotten such an injunction. 'These days (*hoc tempore*) so great a desire for possession has entered priests' minds like a cancer' (1.22.2). He goes on to list the signs of their rapaciousness, and concludes by expressing his 'shame and distaste for our era (*temporum nostrorum*)'. The *exemplum* of the past holds a critical mirror to the present. This despair at the present is the heading for the final section of the *Chronicon*: *sequuntur tempora aetatis nostrae gravia et periculosa*, 'There follow the times of our era, fraught and dangerous' (2.46.1). This anatomy of the present follows from the Arian controversy, where Severus certainly does not conceal the dogmatic position from which he writes, and culminates in his account of the Priscillian controversy.[53] He summarizes his era as 'a perpetual war of discord', and attacks the confusion and corruption brought about now (*nunc*) by the bishops and their 'hatred or partiality, fear, unreliability, envy, factiousness, lust, greed, arrogance, somnolence, laziness' (2.51.5). The triumph of the church is not celebrated, as it is in Ambrose or Prudentius. Rather, for Sulpicius Severus, there have been nine persecutions (he numbers them for us).[54] The tenth, he notes, will be when the Antichrist comes to power, as prophesized by the book of Revelation. It is a commonplace of classical rhetoric and especially of moralizing historiography that the present of Rome is slipping into degeneracy brought about by empire's riches, a grim contrast with the hard and honest virtues of the early Fathers. Sulpicius' narrative echoes such rhetoric, not least in that fine list of modern sins in the church. But this rhetoric is not the same as the classical models it echoes: it has a different epistemological basis. This sinfulness is a sign of the coming of the end. His despair is not merely a lament that casts his enemies as the epitome of all corruption and depravity. It is also a necessary stage in world history. The history from the beginning of the world, *abhinc annos iam paene sex milia*, 'now almost six thousand years ago', has an end point. Both the *iam* and the *paene* in this first act of dating are significant. In the 'almost' is the teleology of eschatology that makes the *iam*, the present, an insistent concern. The final four verbs of this last paragraph of the *Chronicon* are in the imperfect, when Sulpicius has usually used the perfect or historic present in his narrative. Senneville-Grave comments astutely that Sulpicius here follows a habit of Sallust who uses the imperfect in this way 'to indicate to his readers the end of a world'.[55] For Senneville-Grave the end is the collapse of the church into corruption. Sulpicius' imperfects, however, are also an invitation to consider the further and final end, the end of days.

---

[53] Burrus (1995) is the most interesting discussion of this.

[54] For a detailed comparison of Sulpicius' list of persecutions with other such lists, see van Andel (1976) 117–42.

[55] Senneville-Grave (1999) 491 *ad* 51.5.

To see the present as an ominous moment in Christian time is one agenda of Sulpicius Severus' *Chronicon*. To achieve this, he needs to tell the history of the world from its creation because from its creation it is destined for destruction. He needs to count each year because each year matters in creating the *ordo temporum*. It places the time of our age, *tempora aetatis nostrae*, and us, necessarily and integrally within this time-frame. In this vision of history, the historiographer can use pagan sources because there is no outside of Christian time. He can summarize and paraphrase because it is the *ordo temporum* that is being constructed. Specific moral *exempla* are highlighted in preparation for the modern history of persecution, triumph and heresy – the now which demands our comprehension. Its imperfect ending opens a vista on the end to come. His present is a grim moment before the foretold end. Sulpicius Severus' *Chronicon* does not have the intellectual reach of Augustine's *City of God*, certainly, nor does it have the impact or scholarly working of Eusebius' and Jerome's *Chronica*. But, in the same era as Augustine and Jerome, Sulpicius Severus' *Chronicon* also aims to provide a comprehensive history of Christian time, where the complicity of *nostrae aetatis*, 'our age', does not allow for any other perspective. His history is a Christian vision of the whole of time.

*

Nonnus' *Paraphrase*, as we discussed at some length, constructs through its paraphrastic poetics a theologically framed discourse of beginnings, based on the paradoxes of a Nicene, Christological understanding of temporality; in the *Dionysiaca*, by contrast, with its Dionysiac poetics, Nonnus transforms the mystery of origin into a polyphony of competing stories and theoretical perspectives from Greek philosophy, cosmology, myth, literature, as beginning becomes a site not of fixed and explanatory origin but of explosively creative difference and transformation. Gregory of Nazianzus celebrates Christmas by taking his congregation back to the origin of time, in order to understand the Incarnation. Augustine's *Confessions* is driven by his bafflement at creation – not just the origin of the world but original sin, where he himself comes from, the whence of grace: timelessness is the frame and incomprehensible paradox of God in man's coming into being. Augustine's *City of God*, as we also outlined, is the most sophisticated and influential treatment of how history is, from the beginning and at source, formed by Providence, and thus destined to end. For the Christianities of late antiquity, a discourse of origin is fundamental and integral, and it is within this discourse that Sulpicius Severus' *Chronicon* is shaped, with its history *a mundi exordio*, and its final tense imperfects waiting for the foretold end. As the language and institutions of power are changing across empire, so too the narrative of the authority of origins is being redrafted, as Christian authors seek to place Christianity always already at the fountain-head of cultural value. There is a shared agenda to rewrite history's primal scene as Christian.

Paulus Orosius was commissioned to participate in this historiographical agenda. His *Historiae*, the final text to be briefly analysed in this book, was requested by Augustine, while he was writing *City of God*, to be a demonstration of the grim truth of past, pagan history and of the glorious prospects of the Christian era, a case aimed both at pagan opponents and at Christians in need of rhetorical ammunition – or self-justificatory historical understanding, necessary after the shock of the sack of Rome by the Visigoths in 410. Orosius was Spanish, probably from the town that is now Braga in Portugal; he visited Augustine in Hippo, and, with Augustine's introduction, Jerome in Palestine, but little else is known of him apart from the few years of this intellectual pilgrimage in the second decade of the fifth century.[56] He was asked by Jerome to speak on behalf of orthodoxy against Pelagian dogmatists at the Synod of Jerusalem in 415, where he was accused of heresy by archbishop John and consequently wrote his *Liber Apologeticus* in his own defence. Orosius also wrote for Augustine the *Commonitorium*, an account of what Priscillianists and Origenists believed, asking for Augustine's theological opinion. He was involved, that is, with the same issues as Sulpicius in Spain, and engaged with the leading writers of Christian theory around the Mediterranean. Orosius (like Augustine in *City of God*) writes his *Historiae* specifically after and in response to the sack of Rome by the Visigoths under Alaric in 410 – an event which shocked and horrified the empire and its Christian authorities.[57] While Sulpicius' *Life of Martin* was hugely influential on later hagiography, Orosius' histories proved a model of equal importance for medieval historiography, despite much modern disdain for his intellect and historiographical credentials. Because Augustine commissioned the *Historiae*, it has become a commonplace of modern criticism to compare the *City of God* with the *Historiae*, usually with the result that the sophistication and scale of vision in Augustine convict Orosius not just of superficiality and naivety, but also

---

[56] Bare facts judiciously recorded in Fear (2010a) 1–6; Zecchini (2003). Seminal, general overview in Momigliano (1963).

[57] 'Stunned and stupefied . . . hanging there, between hope and despair. The brightest light of all the lands was extinguished . . . the whole world had perished in a single city' is Jerome's dramatic account of his reaction *On Ezechiel* 1 *pr.* (trans. O'Donnell (2004) 208).

of being a henchman – Orosius calls himself Augustine's dog – who 'never understood a tithe of what Augustine said to him'.[58] The critical arguments over Orosius – it is a familiar story with much of the literature of late antiquity – have consequently turned on claim and counterclaim of whether Orosius is 'good' or not as a historian.[59] It is time, as Greensmith argues for the similarly reviled and counter-defended Quintus of Smyrna, to move on from rehearsing such tropes – and clichés – of evaluation.[60] Orosius, for my argument, provides a telling contrast to the eschatology and abbreviated narration of Sulpicius – a different modelling of what Christian history, Christian understanding of a placement in time, can be. It is with Orosius that we will end.

Orosius' sense of a beginning is quite remarkable. 'Pretty well all published scholars', he writes, 'in Greek or Latin, who for the sake of lasting memorial (*memoriam*) have passed down in their words the deeds of kings and peoples, have constructed the beginning of their account from Ninus, the son of Belus, king of the Assyrians' (1.1.1). This is arresting not least because it seems to be self-evidently false of pretty well all extant ancient historians, though it is true at least of Justin, one of Orosius' main sources. How can Orosius come to make such a statement? His dismissal of such a starting point is that it denies that there is a beginning to the world, a story of creation – a 'blind ignorance' which thus reduces all of earlier humanity (*humanum genus*) to the status of anonymous animals (*pecudes*), as if only with Ninus was humanity awoken into a new state of awareness. Before, Ninus, then, was there nothing of note? Again, *spatium mythicum* or Censorinus' 'the unclear, through ignorance' is to have no place in Christian history, for which there is to be no outside, no beyond knowing. Ninus, or, significantly, any other such start – Polycrates, say – can only be arbitrary. 'I' – he begins his counter-case – 'have decided to trace the beginning of human misery from the beginning of human sin'. Since the theme of Orosius' history is to be the miseries of the human race, to start with the original sin is a necessary beginning. The arbitrariness of starting history with Ninus is a foil to the integral logic of an origin in the original sin of Eden.

Orosius adds quickly that he has put together just a few extracts – like Sulpicius, he is committed to *brevitas*, although his seven books of the

---

[58] O'Donnell (2004) 208 – a judgement delivered with the full authority of the President of the Society of Classical Studies; see, most recently, Hartog (2020) 105–16.

[59] See, for example, Corsini (1968), especially ch. 9; Goetz (1980) 136–46; Cobet (2009) 86–7; van Nuffelen (2012) repeatedly discusses such issues of value with acumen (and on Augustine specifically, 198–205).

[60] Greensmith (2020).

Historiae are far longer than the *Chronicon*. This editing is demonstrated all too vividly in what immediately follows, which is as startling as his opening claim of what nearly everyone writes (1.1.5):

> Sunt autem ab Adam primo homine usque ad Ninum, magnum ut dicunt, regem, quando natus est Abraham, anni III.CLXXXIIII, qui ab omnibus historiographis vel omissi vel ignorati sunt.

> From Adam the first man, however, until King Ninus, whom they call 'the great', when Abraham was born, there were 3,184 years, which are either passed over by historians, or unknown to them.

The story of the Garden of Eden is even more briefly alluded to here than in Sulpicius. It is no more than a starting point in order to count the years until Ninus, when, we are told in a subordinate clause, Abraham was born – a recognition of synchronicity, which, we will see, is fundamental to Orosius' sense of history. What could be the starting point of a history of the church – Abraham's turn to God – is passed over as subordinate to the ignorance or ignoring of what preceded it, the main clause. So much for the book of Genesis. For the Christian apologist Orosius, in sharp contradistinction to the continuities projected by Sulpicius' *Chronicon*, Genesis, it seems, is not pressingly relevant as historiography.

This repression of the scriptural material, which dominated the narrative of the *Chronicon*, is continued when Orosius returns to start the narrative of history after the celebrated chapter 2 has provided a full geography of the world.[61] (Grethlein and Schlögel would both be quick to note the long history of interaction between the languages of time and space.)[62] When Orosius returns to Genesis, he refers to the creation – *post fabricam ornatumque mundi huius*, 'after the fabrication and decoration of this world' – without any specification of the biblical six days; and goes on to describe the original sin in highly moralized vocabulary but without any narration of what the sin is. There is no mention of Adam, Eve, the tree, the snake, let alone Cain, Abel and so forth: this is a programmatic, generalized statement of a pattern of man doing wrong and being punished by God. It sets the paradigm for the history to come. The flood is the next biblical event Orosius alludes to, but the authority for the tale is given as *veracissmi scriptores*, 'the most reliable authors'. Is this the Bible? It is telling that Orosius goes on to give evidence for the flood from the presence of shells on mountain tops, something even those 'ignorant of the times past

---

[61] See the excellent discussion in Merrills (2005) 64–99.   [62] Grethlein (2013); Schlögel (2016).

and the author of those times' can witness. When, a few paragraphs later, he comes to Sodom and Gomorrah, the next biblical story included in the history, his source is expressly given as Tacitus – not scripture – who is quoted, and then supplemented with a more moralized explanation of the destruction. Similarly for Joseph, the source offered is expressly Justin's epitome of Pompeius Trogus (who makes Moses the son of Joseph), as – later on – the plagues are given their first summation from quotations from both Justin and Tacitus. Unlike Sulpicius, who abbreviated and recapitulated biblical writings to produce three-quarters of his history, Orosius includes only the barest of references to the Hebrew Bible; he never gives scripture expressly as his authority; he insists wherever possible on citing and even quoting well-known pagan historians as authorities, whom he expands and corrects in line with his programmatic agenda of demonstrating the moral punishment of human sin. Pagans are appropriated as authorities to be displayed but criticized – or allowed, from their own mouths, to list the horrors of the pre-Incarnation past. Orosius thus – unlike Sulpicius – identifies dates consistently through non-Christian time-scales – each date is given as 'before' or 'since the foundation of the city', and within this Roman framework, he braids the classical tradition and scriptural authority together, intercalating stories, identifying parallel events between the scriptural history and other narratives. So, thirty years after the plagues in Egypt, the fifty daughters of Danaus killed their husbands, and Procne and Philomela took their revenge on Tereus, at the same time as Perseus went from Greece to Asia. For Orosius, a universal history, like his geography, covers the map of myth and history of different cultures. Pagan authors are made to speak a charged Christian history of the miseries of a pre-Christian era. Orosius 'baptize[s] secular history'.[63]

Tempora Christiana, 'the Christian era', is Orosius' description of the times in which he lives. At the turn of the fifth century, the present – 'an elastic concept' which could stretch back to Constantine[64] – is increasingly referred to as *tempora Christiana*.[65] Although the term is used with multiple colourings, it is rarely simply positive. Peter Brown states that by this phrase, Christian writers 'meant, not the stability of the Constantian order, but a new age, overshadowed by a crisis of authority, that led to renewed barbarian raids throughout the Roman provinces of the West'.[66] *Tempora Christiana*, that is, does not refer to a fixed and agreed time, but marks the

---

[63] Fear (2010b) 185. [64] Markus (2000) 200, speaking of Augustine's use of *tempora Christiana*.
[65] Markus (1988) with the necessary qualifications of Madec (1975) and the rejoinder of Markus (2000); Brown (2013).
[66] Brown (2013) 86.

space of a question, an anxiety: how or in what ways has the adoption of Christianity by the empire changed the understanding of time? What does it mean to live in a Christian empire? What self-understanding is required by the recognition that to live now is to inhabit Christian history? Simply: how different is the Christian now? It should not be underestimated how extraordinary a claim it is that all of history is to be comprehended according to the assertion of a religious allegiance. From Hesiod onwards, the 'ages of man', with a lost and longed-for golden age, is a familiar trope of mythic cosmology; historians conceptualize Athens of the fifth century BCE as a 'classical' city since even the fourth century BCE;[67] recognizing the difference between the Republic and the imperial age is a standard ordering of Roman historical time; the very rise of the Roman empire requires explanation, especially to Greek historians, educated to another past. Notwithstanding all these strategies of periodization, it is only with *tempora Christiana* that the structuring of time is articulated through what can be called a religion. The dynamics of belonging now depend on faith or belief, orthodoxy and heresy. The logic of *tempora Christiana* makes outsiders of pagans or Jews, all non-Christians, who evidence the untimeliness of not accepting the new of the good news. (Modern institutions in the West have not yet fully disabused themselves of these gestures of exclusion.) The religious present of *tempora Christiana*, like the anthropological present, sets its others outside the time of the now, and without an inheritance in the future. The language of *tempora Christiana* not only orders history but also excludes its opponents from fully participating in time, except as excluded, punishable others.

'We are the times', declares Augustine, at the climax of his Sermon 80:

> mala tempora, laboriosa tempora, hoc dicunt homines. bene uiuamus, et bona sunt tempora. nos sumus tempora: quales sumus, talia sunt tempora.

> The times are bad, the times are burdensome, that's what men say. Let us live well, times are good. We are the times; what we are, the times are.

What humans do – the complicit 'we' of his projected Christian congregation – makes the world good or bad. And yet, Augustine frequently states, despite the fact that sin is punished, it is not simply that good deeds lead directly to salvation: the apophatic mystery of God's purpose remains opaque in the unfurling of events.[68] Augustine's

---

[67] Hanink (2014).
[68] See Murphy (2011) and, more generally, Markus (1988); Ayres (2009); Harrison (2000); Wetzel (1992).

insistence on the moral evaluation of human agency – which we saw in his strident rejection of astrology – also leads him to reject a naïve Christian triumphalism in *tempora Christiana*. His 'radical agnosticism about God's purposes in human history offered no comfort'.[69] For Augustine, 'We simply can't read God's purposes in history'.[70] For Sulpicius, the present, a brief step before the imminent end of days, was summarized in a list of sins and chaotic strife; his times are 'difficult and dangerous'. For Augustine, subject to God's inscrutable planning, and striving for good while marred by original sin, the now remains conditional.

Orosius too, following his master, determines the present of *tempora Christiana* as conditional. There are four strategic moves to Orosius' writing the present. They are all formed by his historiographical agenda, that is, by his express intention to counter pagan and Christian claims that the sack of Rome with its attendant physical and explanatory crises throws a challenge to Christianity and its powers of salvation. The sack of Rome made Christians, as well as those who did not see themselves as Christians, question not just the stability of the empire but also the implications of such instability for their theologically informed comprehension of history. Was Christianity and its destruction of the sites and practices of traditional worship to blame for the sack of Rome? Was the sack a punishment? Orosius is quite clear that 'the world and humanity (*homo*) is ruled by divine Providence, as good as it is just'.[71] He also confesses – a good rhetorical ploy – that he too, like others faced by Rome's sack, 'found myself in confusion. I had often thought that the disasters of our present times seemed to seethe beyond measure'.[72] The 'however' follows inevitably: 'However, I found that the days gone by were not only equally fraught (*graves*) as today, but even more awfully miserable the further they were removed from the remedy of the true religion.'[73] Orosius' argument requires that he sets out empirically that in the past things were worse – hence his regular return to pagan authorities so that he can fend off the charge of *parti-pris* sources – and that he shows how the true religion has made things better. Yet he also knows full well that, despite his celebration of religion's salvation, the end of days is integral to Christian history. 'An exception, of course – quite different – are the last (*novissimis*) days at the end of time (*saeculi*), and with the appearance of the Antichrist, and the final judgement.'[74] The torments of this final time are predicted in

---

[69] Markus (2000) 205.   [70] Murphy (2011) 601.
[71] *Hist.* 1.1.9, with Goetz (1980) and Fear (2005).   [72] *Hist. pr.* 13.
[73] *Hist. pr.* 13. Not really a conversion, as van Nuffelen (2012) 63 puts it.   [74] *Hist. pr.* 15.

Holy Scripture, and will be unlike either the present or the past. Sulpicius feels their imminence; Orosius, in the only other mention of the end of days in the *Historiae*, recognizes only that they 'are going to come some-time', *quandoque ventura.*[75] The delayed sufferings and punishment at the end of days might be declared different from any other human experience, and in the unknown future, but this alone is not enough to reconcile the two historical narratives, the one which makes Christianity the progressive betterment of human conditions, the other which leads inexorably towards the grim misery of the Antichrist's rise to power. What strategies, then, does Orosius use in order to depict the present with regard to the past and the future, to comprehend the now in time's passing?

The first strategy concerns synchronicity, which, as we saw with the Parian Marble or Polybius or Diodorus Siculus, had long been part of the arsenal of the historiographer's articulation of order. We have already noted how in Book 1 biblical and non-biblical events are intercalated to create parallels between the times of the different mythic traditions. The beginning of Book 7 of the *Historiae* is the fullest expression of this logic. Orosius combines numerology and synchronicity to create a fated pattern of history. Abraham was born forty-three years into the realm of Ninus, Jesus was born at the end of the forty-second year of Augustus' reign. So, too, the kingdom of Carthage lasted 700 years, as did the kingdom of Macedonia. With a considerable strain on the evidence, Babylon is said to have been the dominant empire for 1,164 years, as Rome from its founda-tion to the sack of Alaric also counts 1,164 years.[76] Most significantly, however – as has often been discussed – Orosius insists that the birth of Jesus comes in the reign of Augustus because Augustus alone created both the empire that was to be the proper home of Christianity, linking so many different peoples under one authority, and the time when 'the whole world' experienced 'the profoundest calm and a unified peace'.[77] Numerology and synchronicity come together to determine necessity, which brings the history of Rome and the history of Christianity onto the same tracks: 'The secular and the Christian are triumphantly made interdependent by a synchronism of the birth of Christ and the accession of power of the emperor Augustus.'[78] It is fundamental to Orosius, in a manner quite different from Sulpicius, that Rome is the condition of possibility for Christianity. Book 1 takes us from Adam to the foundation of Rome;

---

[75] *Hist.* 7.27.15.   [76] *Hist.* 2.3.2–3.

[77] *Hist.* 7.3.10. See Lacroix (1965) 171, who argues that Orosius' basic point is simple: God's commit-ment to Rome is a matter of Christ; Christ sets up his church at the heart of the empire.

[78] Fear (2010b) 182.

Book 7 covers Roman history from the birth of Christ to Orosius' own day. Books 2–6 – as dominant in his history as scripture is in Sulpicius – are made up by the history of Rome's growth, the seedbed of Christianity. As Andrew Fear memorably puts it: 'Orosius was the first author to give Roman history a purpose'.[79] Orosius' sense of time makes a continuity of secular and religious history, because the Roman empire and Christianity are integrally intertwined. The attempt of the opponents of Christianity to blame Christianity for the sack of Rome are undermined by Orosius' demonstration that Rome was always already on a path to Christian triumph.

The pattern of this historical continuity is structured in particular by a theory of the Four Empires, which Orosius takes up from the book of Daniel, essential reading for Christian apocalyptic reflections. This is his second strategy. Orosius determines that the four empires are Babylon in the East, Macedonia in the North, Carthage in the South, and Rome in the West – a universal geography of power. (Carthage is not usually one of the kingdoms, but, as van Nuffelen notes, this 'original concoction that is part of a long tradition and a contemporary vogue' thus keeps Rome firmly in the foreground.)[80] Babylon lasted as a city 1,400 years, Orosius states, and both Macedonia and Carthage were the dominant imperial powers for 700 years. Rome too has its sevens: in its 700th year a great fire destroyed fourteen districts. This self-marked repetition of seven embodies and blazons forth God's providence. It is not by chance that there are seven books of the *Historiae*. There is no overlap between these empires: as the last king of one dies, the first king of the next comes into the light. This bold superstructure of history again demands that Christianity is seen as an integral part of the divine plan for imperial authority. Christianity is Roman because its claim to be a worldwide religion requires a worldwide empire as its geography.

Yet one obvious conclusion from such a pattern might be that Rome too, as Sulpicius would have it, is approaching its end, and with it the end of the world. Is not the sack of Rome one sign of this inevitable and foretold fading of Roman power? This is where the crucial turning point – and third strategy – of Orosius takes shape. From early in the narrative, Orosius is intent on declaring that Rome is different from Babylon in this respect: it is not about to fall. He states this point as boldly and baldly as he can (2.3.6):

[79] Fear (2010b) 182. [80] van Nuffelen (2012) 49.

Ecce similis Babyloniae ortus et Romae, similis potentia, similis magnitudo, similia tempora, similia bona, similia mala; tamen non similis exitus similisve defectus. Illa enim regnum amisit, haec retinet, illa interfectione regis orbata, haec incolumi imperatore secura est.

Look! Babylon and Rome have a similar beginning, similar power, similar size, similar length of time, similar goods, similar evils; but their end or decline are not the same. Babylon lost its kingdom; Rome retains hers. Babylon was orphaned by the killing of her king; Rome is secure, its emperor unharmed.

The repeated injunctions to 'look!', 'see!' stab home the point. The similarity demanded by the theory of the four empires produces the crucial difference between Rome and Babylon. Rome is a Christian city. Babylon was undone because of the baseness of the lusts of their ruler; Rome was saved because of the 'most chaste self-control of the Christian religion in their ruler'. It was more than just the ruler's Christian goodness that saved the city, however. Unlike Babylon's lack of reverence for religion, in Rome 'There were Christians, who spared; Christians who were spared; and Christians because of whose memory and in whose memory, this sparing took place.' The sack of the city, which so traumatized Jerome in Palestine, has now become in Orosius' apologetics a scene of restraint, where the Christian Visigoths pitied the Christians of Rome – there is no mention of any tension between the Arian Visigoths and pro-Nicaean Romans – and this pity was enacted in the name of a glorified history of Christians, the *sancti* who make Christian history glorious. When Orosius comes to narrate the history he pre-announces here in Book 2, he highlights the story of a nun whose plea to a Visigoth warrior results in a procession of nuns carrying the ritual treasures of St Peter, protected from harm by the Visigoth army.[81] Far from being the blameworthy cause of Rome's disasters, Christianity is 'responsible for the salvation of Rome'.[82]

Orosius knows well the instability of empire, historiography's admonition to the powerful since Herodotus. He marks it as a cliché: 'not my project', he sniffs, 'to expatiate on the unstable nature of changeable things', and he seals the point with a quotation that is proven, he says, by the fall of Babylon: *quidquid enim est opera et manu factum, labi et consumi vetustate*, 'whatever is made by the work of man's hands, collapses and is consumed by age' (2.6.13–14). This *sententia* is a paraphrase of a passage from the very well-known speech of Cicero, *Pro Marcello*. This

[81] 7.49.3–14.   [82] Merrills (2005) 42.

speech is about pardon – Cicero is celebrating Caesar's pardon of Marcellus – delivered at a crisis point that led to the collapse of the Republic, and it marks a moment of maximum disjunction in the shifting relationship of Caesar and Cicero. *Pro Marcello* can stand as an icon of the *res mutabiles* Orosius does not want to discuss. Orosius could have found many passages from scripture to make such a familiar point – 'Unless the Lord builds the house, those who build it, build in vain', 'he who builds on sand . . .' and so forth – but by appropriating Cicero, he not only invokes a non-Christian proof, once again, for his Christian argument, but also allows classical Rome, the mutability of its power and its rulers' relationships, to stand as its own example of the moral he is quoting Cicero to prove. Orosius quotes from authoritative tradition to undermine the tradition, from the inside, as it were. Such quotations are 'not just literary flourishes, but ideological statements about what drives history: Jesus, not Caesar; the Church, not Augustus'.[83] For Orosius, literary quotations from classical authorities form part of his drive to narrate Christian time.

'Old age', *vetustas*, we may remember from Septimius Severus' inscription on the Pantheon. Empires, it seems, like buildings, have a life expectancy, as does a human. An intense sermon of Augustine makes this association of the naturalness of a human's ageing even with the age of the world itself: 'The world is perishing; the world is growing old, the world is fainting, the world is labouring with the panting breath of old age (*vetustas*). Do not fear. Your youth will renew itself like the eagle.'[84] Augustine can imagine the gradual decline of the world towards the end of time, which he contrasts with the Christian promise of rebirth. But Orosius repeatedly pushes away the possibility of an imminent end. Orosius embodies not so much 'a confidence in the divine permanence of the Christian empire'[85] as a recognized potential to keep staving off the end of empire by an active Christian goodness. A 'flawed temporal state',[86] or fragile continuity is held forth as the answer. Everything human decays. Yet Orosius' eschatology is 'situated in an unknown future', and the Church remains 'the true centre of historical action'.[87] Modern critics of Orosius have been swift to dismiss what they see as a naïve theodicy driving his historiography. The pattern of human corruption leading to divine punishment is, however, fully in service of a more cautious persuasive normativity in Orosius' narrative. Humanity, constantly threatened by its own sinfulness and punishment, which history repeatedly reveals, must

---

[83] van Nuffelen (2012) 190.   [84] *Serm.* 81.8.   [85] Merrills (2005) 57.
[86] Merrills (2005) 57. See Herzog (2002b).   [87] van Nuffelen (2012) 197.

turn again and again to a Christian morality with its hope of God's mercy, if the fragile potential of a Christian empire is to be maintained. This is how Orosius understands the present in relation to the past and future.

The past, for Orosius, is exemplary. It is with *exempla* that Orosius' fourth strategy takes shape. *Exempla* – or *paradeigmata* in Greek – as we previously outlined, are integral to classical rhetoric in theory and practice; they forge a relation to the past that is marked by genealogy and disruption: the genealogy that asserts a privileged connection between the present case and what has gone, and a disruption that recognizes that the past has passed and is different from the present in its very instructiveness.[88] *Exempla* provided the glue of ancient rhetoric, and, as such, are integral not just to the performance of rhetoric but also to the rhetoric of self-performance. For Christian writing, classical exemplarity therefore needed negotiation: what is to be made of the great men and deeds of the past by the new normativity and new parameters of achievement? For Basil – to extract one strategy from his educational principles – the examples of the past can be appropriated for present purposes: Odysseus, naked on the beach before Nausicaa, can be a model of virtue. Augustine, as has been extensively discussed, used 'the cultural expectations and literary training he shares with classically educated members of his audience to destabilize their perceptions of their own history, and to substitute a Christian interpretation of events'[89] – he redesigned familiar exemplarities to a new template. In the *City of God* in particular, the examples of the past – both the exemplary force of history and the historical usage of *exempla* – became a route to reformulating an understanding of how living in today's time could be comprehended.

Orosius sees danger for the Christian in the *exempla* of classical tradition. He is quietly dismissive of the effect of such rhetorical exercise. He concludes his account of Athens and its war with Persia – the most classical of classical pasts and the source of the grandeur of the historiographical tradition – with a *praeteritio* and a trademark (mis)quotation of Virgil: *quis enim cladem illius temporis, quis fando funera explicet aut aequare lacrimis possit dolores?*, 'For who could describe the disaster of that time, who might express in speech its deaths or could equal with tears its grief?' (2.18.4).[90] Aeneas' description of the sacking of Troy (*Aen.* 2.361–2) is applied to the miseries now of Greece at war – a nicely understated rehearsal of Scipio's quotation of Homer's vision of the fall of Troy for the destruction of

---

[88] See above, pp. 100–12.
[89] Conybeare (1999) 63; see also Murphy (2011); Burns (2001); Ayres (2009).
[90] *quis cladem illius noctis, quis funera fando | explicet aut possit lacrimis aequare labores? Aen.* 2.361–2; note especially the change of *noctis* to *temporis* and *labores* to *dolores*, as well as the word order.

Carthage, the *locus classicus* for the epic memory of the rise and fall of cities and empires.[91] But Orosius manipulates this shift of perspective with a further twist (2.18.4):

> Verumtamen haec ipsa, quia multo interiectu saeculorum exoleverunt, facta sunt nobis exercitia ingeniorum et oblectamenta fabularum.

> In reality, these very events because they have faded through the interval of centuries, have become for us exercises of intellect and the pleasure of storytelling.

The triumphs and horrors of classical Greece have become no more than what he has just demonstrated: an exercise of rhetorical skill in the pleasure of storytelling. We know that Virgil's Aeneas, when he tells his story of disaster, is also beguiling Dido: there are further dangerous pleasures hinted at. Rhetoric and literature have produced an emotional deadening towards the past. Time itself has dulled our feelings. If only his readers would apply their minds more intently, continues Orosius, and observe as if from a height above the spectacle, they would be able to measure each era (*tempus*) according to its own qualities, and then they would realize that neither the past would have been so terrible without the punishing rage of God, nor the present so full of good without the kindness and mercy of God. Providence revealed ... Orosius, moving between the example of classical Athens and the exemplary epic of Virgil, demands that his readers evaluate a time for what it is, and to see in the contrast of present and past the working of Providence. His readers need to be cajoled into discovering again the horrors of the past and the beneficence of the present.

The preface to Book 4 of the *Historiae* generalizes this case with a certain flair. Orosius begins with Virgil again, this time with an acknowledged quotation of a line from the first book of the *Aeneid*: 'Perhaps these sufferings too will at some time hence be a pleasure to recall' (*Aen.* 1.203).[92] The story of the foundation of Rome is the foundation of this historiography. Orosius explains that the worse events are, the more pleasurable it is to recall them later; that everyone thinks the future will be better; and that in the moment, present sufferings always loom largest in the imagination. These three rather trite observations open into a discussion of how people always think their own troubles trump the past. He takes the case of a man who gets up from his warm bed to find it is

---

[91] See above, p. 174.

[92] McCormack (1998) is good on Virgil's role in Augustine; on such allusiveness in historiography see also Kelly (2008).

icy outside. If he says, 'It's cold today', no problem: that's normal behaviour. If, however, he shouts that there has never been such cold even when Hannibal crossed the Apennines and lost his elephants, Orosius would personally drag him from his incriminatingly warm sheets and show him children playing in the frost to reveal the moaner's 'childish dissoluteness', *puerilibus licentiis* (4. *pr.* 10). His pain comes not from the 'violence of the time' (*in tempore violentiam*) but from his own laziness. Orosius sardonically mocks the man's use of the *exemplum* of Hannibal crossing the Alps as no more than a corrupt and lazy gesture of self-satisfaction. He makes the man an example for his trivial use of the past. This story is a parable for the historiographer. And for anyone who thinks the Christian present is not better than the past. By contrast, Orosius sets out to give us the full terribleness of what has happened in the past, so that we will neither underestimate the evil of the past nor use the past trivially to emphasize the evil of the present in a false comparison. Exemplarity is integral to Orosius' historiography: his argument 'questions the creation of an exemplary past in all three aspects, that is, its actual reality, its creation by ancient historians, and its impact on contemporary perception'.[93] For Orosius, to understand the narrative of Christian time requires this moralizing redesign of his inherited tradition of the temporality of classical exemplarity. What the past can say has been transformed.

*

Sulpicius Severus, writing before the sack of Rome, nonetheless hears time's chariot rushing near. His vision of Christian time demands that scripture fills our perspective: three-quarters of history is subsumed by what is written in the Hebrew Bible; and the New Testament provides the model for understanding both the pattern of the Hebrew Bible and the sorry state of current affairs which justify his anticipation of the end of days. When Sulpicius cites ancient historians or uses their language, it is to supplement the chronology of his biblical narrative, or to transcend the lures of the *saeculum*, the burden of time's corrupting normality. He performatively turns his eyes from everlasting glory towards everlasting life. Orosius, by contrast, allows the end of days to be 'going to come sometime'. His narrative also goes back to original sin for its starting point, to see a repeated pattern of human wrongdoing and divine punishment, but this is a historical narrative that embraces North, South, East and West, the four empires that structure secular power: a history far beyond

---

[93] van Nuffelen (2012) 70. The previous paragraph owes a good deal to this excellent discussion.

the geography of scriptural narrative. All four empires are destined to rise and fall according to Providence, but, for Orosius, the past was a far worse place. To evidence his claim, he tells a (baptized) secular history, which is made to convict itself, by quotation and interpretation. The *exempla* of the classical past are refuted as false models, misrepresented by traditional historiography, and insidious in their distortion of the self-understanding of the present. Yet the indefinite future of *quandoque ventura*, the 'going to come sometime', also allows Orosius to defer the end of Rome, and to locate the continuity of deferral in the shared frame of God's beneficence and human – Christian – moral choices. Christians are not responsible for the sack of Rome, therefore, but the very agents of its salvation. The present for Sulpicius Severus is summed up in a fiercely incremental list of moral corruptions; for Orosius, similarly at the very conclusion of his narrative, the present – *tempora Christiana* – can be positively distinguished from the confusion of disbelief that is embodied in the conflicts of human history (*conflictationes saeculi*), because of the increasing presence of Christ's Grace: *Christianis temporibus propter praesentem magis Christi gratiam ab illa incredulitatis confusione discretis* (7.43.19). This is a different *distinctio temporum*, a different sense of what the present is, of what is at stake in distinguishing *tempora Christiana*. Orosius comprehends Christian time differently from Sulpicius, and the narrative that embeds this definition takes a quite different form. Orosius' continuing potentialities of Christian agency in time contrast tellingly with Sulpicius' sense of an ending. Shifting conceptualizations of the end allow strikingly contrastive narratives of time and differing understandings of how a Christian may inhabit it. This juxtaposition of Sulpicius and Orosius offers two distinctive, productive routes into the map that medieval historiography will draw. Historiography played a major role in the first section's discussion of how a self-placement in time was constructed through writing the narrative of the past, and how such historiographical structures are predicated on a roster of concepts about temporality. Christianity changes history. Through Sulpicius and Orosius together, we can see the shape of Christian time as a historiographical project being formed. It will last for centuries.

Coda
Writing in the Time of Sickness

It has been a strange time in which to write about time. This book was written during a year marked and scarred across the world by the Coronavirus, Covid-19. It would be nice to imagine that the comments I am about to make will quickly be out of date, but at the time of writing not only is there no immediate sign of any lessening of the pandemic, but also the glib slogans of public discourse 'build back better', 'the new normal', seem peculiarly hollow. It seems unlikely that this time of pandemic will be forgotten in the near future, or its impact unfelt.

For those who have lost friends and family, and for those whose lives have been made miserable by long-term sickness, physical or mental, by the loss of income, by the loss of employment, or by the collapse of social exchange, this time will be remembered through mourning, regret, despair and relief at its passing inasmuch as such losses pass – never without a residue. One of the strangest feelings of being in this time, however, is the transformation of time itself. For everyone, routines, long lived, have changed. What it means to go through a day has altered under lockdowns and the restrictions of travel and contact. Even the pattern of the week has slipped, as the difference between the weekend and the weekday has slid. The pattern of work and holiday, the expectations and practices of travel – where time and speed are so pressing – have had to mutate. Festival celebration – religious, familial, communal – along with the release and joy of the leisure activities of sport and theatre or music – the distinctions of time – are silenced or muted. The sense of the future has become differently insistent. As with Augustine, 'how long?' has become the cry of a desperate sense of waiting. The promise or threat of last week's figures, next week's rise, has become a new shared public and private anxiety. Yet the hours and days march on. As Augustine also wrote (*Conf.* 4.8.13), *non vacant tempora*, 'time does not take holidays'. To live in a time of sickness is to become acutely and differently aware of time.

Yet it is also telling that at the same time in different parts of the world there has been developing a violent political attempt to grasp hold of how an understanding of time is shaped. In Poland, following the model of Hungary, laws have been passed, designed to control what accounts of history will be acceptable – legal – from now on, and only accounts that conform to the government's self-serving projection of patriotism will be allowed. In India, the government of Modi has projected a fantasy of the past based on a nationalist purity that has veered into the absurd, but with murderous consequences. Far right groups in America, not dis-encouraged by President Trump's understanding of how stories about the past can be true or false, have increasingly disseminated poisonous models of a once celebrated and now threatened racial purity, with a parallel dismissiveness towards the history of slavery and violent imperi-alism. In both Russia and China – with quite different inflections – the government uses the state media to insist on an authorized version of the past that suits its own agenda – and, we should also note, the deep complexity of the Russian public discourse of the past, so often simplified in the Western media, has been brilliantly explored – for example – by the oeuvre of Svetlana Alexievich, Nobel Prize winner for literature, espe-cially in *Second-Hand Time*. We could add examples, no less in need of sophisticated analysis, from arenas of conflict from the Middle East to Britain's Brexit debates. History has always been manipulated for ideo-logical purposes, certainly, not only by the victorious, and the totalitarian states of the Third Reich and the Stalinist Soviet Union remain the paradigms of such authoritarian regulation of the narrative of the past. Nonetheless, for anyone even cautiously committed to the values of a plural, liberal democracy, this trend towards increasingly aggressive, arrant and aggrandizing acts of control can only feel deeply worrying, not least with an eye to the consequences of such actions in the past.

It is, then, an insistently grim time to think about how time is repre-sented and experienced. This history of the transformation of temporality in late antiquity has uncovered its fair share of extreme ideologues, not to mention their strident and unpleasantly consequential projections of how the past must be understood. The Christian invention of time cannot be decently separated from the violent politics with which it marches, hand in hand. Yet the hope remains that perhaps a more nuanced and developed understanding of what is at stake in 'inhabiting time', how it can be debated and transformed over time, what factors go into its comprehension and experience, might contribute not just to understanding our current temporalities, but also to resisting the oversimplification and distorted genealogies of the past that make violent politics possible.

Cambridge, Lent Term, 2021.

Bibliography

Aarslef, H. (1982) *From Locke to Saussure: Essays on the Study of Language and Intellectual History*, London.

Abraham, N. and Torok, M. (1994) *The Shell and the Kernel: Renewals of Psychoanalysis I*, trans. N. Rand, Chicago.

Accorinti, D. (1995/6) 'L'etimologia di Βηρυτός: Nonn. Dion. 41.364–7', *Glotta* 73: 127–33.

(1996) *Parafrasi del Vangelo di S. Giovanni. Canto XX*, Pisa.

(2004) *Nonno di Panopoli. Le Dionisiache. Volume Quarto (Canti XL–XLVIII)*, Milan.

Accorinti, D. ed. (2016) *Brill's Companion to Nonnus of Panopolis*, Leiden.

Adam, B. (1990) *Time and Social Theory*, Cambridge.

(1995) *Timewatch: The Social Analysis of Time*, Cambridge.

Adams, S. (2020) *Greek Genres and Jewish Authors: Negotiating Literary Culture in the Greco-Roman Era*, Waco, TX.

Agamben, G. (2005) *The Time that Remains: A Commentary on the Letter to the Romans*, Stanford.

Agnosini, M. (2020) *Nonno di Panopoli: Parafrasi del Vangelo di San Giovanni*, Rome.

Agocs, P. (2019) 'Speaking in the Wax Tablets of Memory', in Castagnoli and Ceccarelli eds. (2019): 68–92.

Agosti, G. (2001) 'L'epica biblica nella tarda antichità greca: autori e lettori nel IV e V secolo', in Stella ed. (2001): 67–104.

(2003) *Nonno di Panopoli: Parafrasi del Vangelo di San Giovanni: Canto Quinto*, Florence.

(2004) *Nonno di Panopoli. Le Dionisiache. Volume Terzo (Canti XXV– XXXIX)*, Milan.

(2009) 'Cristianizzazione della poesia greca e dialogo interculturale', *Cristianesimo nella Storia* 31: 313–35.

(2011) 'Usurper, imiter, communiquer: le dialogue intercultural dans la poésie grecque chrétienne de l'antiquité tardive', in Beylache and Dubois eds. (2011): 275–99.

(2015) 'Chanter les dieux dans la société chrétienne: les *Hymns* de Proclus dans le contexte culturel et religieux de leur temps', in Belayche and Pirenne-Delforge eds. (2015): 183–212.

(2018) '*Versus de limine* and *in limine*: Displaying Greek *Paideia* at the Entrance of Early Christian Churches', in van Opstall ed. (2018): 254–81.

(forthcoming) 'Metrical Inscriptions in Late Antiquity: What Difference did Christianity Make', in Hadjittofi and Lefteratou eds. (forthcoming).

Alberdina Houtman, A. and Schwartz, J. eds. (1998) *Sanctity of Time and Space in Tradition and Modernity*, Leiden.

Alcock, S. (1996) 'Landscapes of Memory and the Authority of Pausanias', in Bingen ed. (1996): 241–67.

(2002) *Archaeologies of the Greek Past: Landscape, Monuments and Memories*, Cambridge.

Alcock, S., Cherry, J. and Elsner, J. eds. (2001) *Pausanias: Travel and Memory in Roman Greece*, Oxford.

Alexander, L. (1993) *The Preface to Luke's Gospel: Literary Convention and Social Context in Luke 1 1–4 and Acts 1.1*, Cambridge.

Alexander, P. (1992a) 'Pre-Emptive Exegesis: Genesis Rabbah's Reading of the Story of Creation', *Journal of Jewish Studies* 43: 230–45.

(1992b) 'Targum, Targumim', *Anchor Bible Dictionary* 6: 320–31.

Alexander, P., Lange, A. and Pillinger, R. eds. (2010) *In the Second Degree: Paratextual Literature in Ancient Near Eastern and Ancient Mediterranean Culture and Its Reflections in Medieval Literature*, Leiden.

Allen, D. (1996) 'A Schedule of Boundaries: An Exploration Launched from the Water-clock of Athenian Time', *Greece and Rome* 43: 157–68.

Allen, T. ed. (2018) *Time and Literature*, Cambridge.

Allen-Hornblower, E. (n.d.) 'Revisiting the Apostrophes to Patroclus in *Iliad* 16', *Center for Hellenic Studies*, www.chs.harvard.edu/CHS/article/display/4702.

Anderson, O. (1981) 'A Note on the 'Mortality' of Gods in Homer', *Greek, Roman, and Byzantine Studies* 22: 323–7.

Ando, C. (1990) 'Augustine on Language', *Révue des études augustiniennes* 40: 45–78.

Andrews, J. (2020) '*Kairos*: The Appropriate Time, Place and Degree in Protagoras' Myth of Origins', in Bierl, Christopoulos and Papachrysostomou eds. (2020): 267–84.

Annas, J. (1975) 'Aristotle, Number and Time', *The Philosophical Quarterly* 25: 97–113.

(1992) 'Aristotle on Memory and the Self', in Nussbaum and Rorty eds. (1992): 297–311.

Appadurai, A. (1981) 'The Past as a Scarce Resource', *Man* 16: 201–19.

Arendt, H. and Heidegger, M. (1998) *Briefe 1925–1975 und andere Zeugnisse*, ed. U. Lutz, Frankfurt am Main.

Arrighetti, G. (2006) *Poesia, poetiche, e storia nella reflessione dei Greci*, Pisa.

Asad, T. (2003) *Formations of the Secular: Christianity, Islam, Modernity*, Stanford.

Assmann, A. (2013) *Ist die Zeit aus den Fugen? Aufstieg und Fall des Zeitregimes der Moderne*, Munich.

Assmann, J. (2000a) *Religion und kulturelles Gedächtnis*, Munich.

(2000b) *Das kulturelle Gedächtnis: Schrift, Erinnerungund politische Identität in frühen Hochkulturen*, Munich.

Atack, C. (2020) 'Plato's Queer Time: Dialogic Moments in the Life and Death of Socrates', *Classical Receptions Journal* 12: 10–31.

Ayres, L. (2004) *Nicaea and its Legacy: An Approach to Fourth-Century Trinitarian Theology*, Oxford.

(2009) 'Into the Poem of the Universe: *Exempla*, Conversions, and Church in Augustine's *Confessiones*', *Zeitschrift für antikes Christentum* 13: 263–281.

Babut, E.-C. (2012) *Saint Martin de Tours*, Paris.

Backman, J. (2007) 'All of a Sudden: Heidegger and Plato's *Parmenides*', *Epoché* 11: 393–408.

Bader, G. (2018) 'Paula and Jerome: Towards a Theology of Late Antique Pilgrimage', *International Journal for the Study of the Christian Church* 18: 344–53.

Bajoni, M. (2003) 'A propos l'aition de Beyrouth dans les Dionysiaque de Nonnos de Panopolis', *Antiquité Classique* 72: 197–202.

Bakhtin, M. (1981) *The Dialogic Imagination: Four Essays*, trans. C. Emerson and M. Holquist, Austin and London.

Bakker, E. (2002) 'Khrónos, Kléos and Ideology from Homer to Herodotus', in Reichel and Rengakos eds. (2002): 11–30.

Balás, D. (1976) 'Eternity and Time in Gregory of Nyssa's Contra Eunomium', in Dörrie, Altenburger and Schramm eds. (1976): 128–55.

Ballenger, J. (2009) *The Wound and the Witness: The Rhetoric of Torture*, Albany.

Banerjee, P. (2006) *The Politics of Time: 'Primitives' and History-Writing in a Colonial Society*, Oxford.

Bannert, H. and Kroll, N. eds. (2017) *Nonnus of Panopolis in Context II: Poetry, Religion, Society*, Leiden.

Bär, S. (2007) 'Quintus Smyrnaeus und die Tradition des epischen Musenberufs', in Baumbach and Bär eds. (2007): 29–64.

Baragwanath, E. (2008) *Motivation and Narrative in Herodotus*, Oxford.

Baragwanath, E. and de Bakker, M. eds. (2012) *Myth, Truth and Narrative in Herodotus*, Oxford.

Barak, O. (2013) *On Time: Technology and Temporality in Modern Egypt*, Berkeley.

Barbanti, M.-G. and Manganaro, P. eds. (2002) *Henosis kai Philia: Unione e Amicizia. Ommaggio a Francisco Romano*, Catania.

Bardill, J. (2006) 'A New Temple for Byzantium: Anicia Juliana, King Solomon and the Gilded ceiling of the Church of St. Polyeuktos in Constantinople', *Late Antique Archaeology* 3: 339–70.

Barkan, L. (1986) *The Gods Made Flesh: Metamorphosis and the Pursuit of Paganism*, New Haven.

Barnes, J. ed. (1982) *Science and Speculation: Studies in Hellenistic Theory and Practice*, Paris.

Barnes, T. (1996) 'The Military Career of Martin of Tours', *Analecta Bollandiana* 114: 25–32.

(2010) *Early Christian Hagiography and Roman History*, Tübingen.

Baroway, I. (1935) 'The Hebrew Hexameter: A Study in Renaissance Sources and Interpretation', *English Literary History* 2: 66–91.

Barthes, R. (1975) *S/Z*, trans. R. Miller, London.

Bartky, I. (2000) *Selling the True Time: Nineteenth-Century Timekeeping in America*, Stanford.

Barton, C. (1993) *The Sorrows of the Ancient Romans*, Princeton.

Barton, T. (1994a) *Ancient Astrology*, London.

  (1994b) *Power and Knowledge: Astrology, Physiognomics and Medicine Under the Roman Empire*, Ann Arbor.

Bassi, K. (1993) 'Helen and the Discourse of Denial in Stesichorus' Palinode', *Arethusa* 26: 51–75.

Bauer, W. (1971 [1934]) *Orthodoxy and Heresy in Earliest Christianity*, trans. R. Kraft and G. Krodel, Philadelphia.

Baumbach, M. and Bär, S. eds. (2007) *Quintus Smyrnaeus: Transforming Homer in the Second Sophistic*, Berlin.

Bažil, M. (2009) *Centones Christiani: métamorphoses d'une forme intertextuelle dans la poésie latine chrétienne de l'Antiquité tardive*, Paris.

Beard, M. (2014) *Laughter in Ancient Rome: On Joking, Tickling and Cracking Up*, Berkeley.

Becker, E.-M. (2016) 'Shaping Identity by Writing History: Earliest Christianity in its Making', *Religion in the Roman Empire* 2.2: 152–69.

  (2017) *The Birth of Christian History: Memory and Time from Mark to Luke-Acts*, New Haven.

Beckwith, R. (1996) *Calendar and Chronology, Jewish and Christian: Biblical Intertestamental and Patristic Studies*, Leiden.

Bedouelle, G. (2002) 'The *Paraphrases* of Erasmus in French', in Pabel and Vessey eds. (2002): 279–90.

Beecher, D. (2004) 'Petrarch's "Conversion" on Mont Ventoux and the Patterns of Religious Experience', *Rennaissance and Reformation* 28.3: 55–75.

Beeley, C. (2008) *Gregory of Nazianzus on the Trinity and Knowledge of God: In Your Light We Shall See Light*, Leiden.

Beeley, C. ed. (2012) *Re-Reading Gregory of Nazianzus: Essays on History, Theology and Culture*, Washington.

Beer, G. (1983) *Darwin's Plots: Evolutionary Narrative in Darwin, George Eliot and Nineteenth-Century Fiction*, Cambridge.

Behr, J. (2004) *The Nicene Faith: True God of True God*, Crestwood, NY.

Bender, J. and Welberry, D. eds. (1991) *Chronotypes: The Construction of Time*, Stanford.

Ben-Dov, J. (2008) *Head of All Years: Astronomy and Calendars at Qumran in their Ancient Context*, Leiden.

Benedict XVI (2006) *Glaube und Vernunft: die Regensburger Vorlesung*, with commentary by G. Schwan, A. Khoury and K. Lehman, Freiburg.

Benoist, S. and Daguet-Gagey, A. eds. (2008) *Un Discours en images de la condamnation de mémoire*, Metz.

Bergren, A. (1980) 'Helen's Web: Time and Tableau in the *Iliad*', *Helios* 7: 19–34.

(1981) 'Helen's "Good Drug": *Odyssey* IV 1–305', in Kresic ed. (1981): 201–14.

(1982) 'Sacred Apostrophe: Representation and Imitation in the Homeric Hymns', *Arethusa* 15: 83–108.

(1983) 'Odyssean Temporality: Many (Re)turns', in Rubino and Shelmerdine eds. (1983): 38–73.

(1989) 'The Homeric Hymn to Aphrodite: Tradition and Rhetoric, Praise and Blame', *Classical Antiquity* 8: 1–41.

Bergson, H. (1911) *Laughter: An Essay on the Meaning of the Comic*, trans. C. Brereton and F. Rothwell, New York.

Bernabé, A. (2008) 'El mito órfico de Dioniso y los Titanos', in Bernabé and Casadesús eds. (2008): 591–607.

Bernabé, A. and Casadesús, F. eds. (2008) *Orfeo y la tradición órfica: un reencuentro*, 2 vols., Madrid.

Bernabé, A. and García-Gasco, R. (2016) 'Nonnus and Dionysiac-Orphic Religion', in Accorinti ed. (2016): 91–110.

Berns, G. (1976) 'Time and Nature in Lucretius' "De Rerum Natura"', *Hermes* 104: 477–92.

Bettini, M. (1991) *Anthropology and Roman Culture: Kinship, Time, Images of the Soul*, trans. J. van Sickle, Baltimore.

Bevegni, C. (2006) *Eudocia Augusta, Storia di san Cipriano*, Milan.

Beylache, N. and Dubois, J.-D. eds. (2011) *L'Oiseau et le poisson: cohabitations religieuses dans les mondes grec et romain*, Paris.

Beylache, N. and Pirenne-Delforge, V. eds. (2015) *Fabriquer du divin: constructions et ajustements de la representation des dieux dans l'Antiquité*, Liège.

Bickerman, E. (1968) *Chronology of the Ancient World*. 2nd ed., London.

(1976) "The Septuagint as Translation" in E. Bickerman, *Studies in Jewish and Christian History* 2 vols. I: 167–200, Leiden.

(1988) *The Jews in the Greek Age*, Cambridge, MA.

Bierl, A., Bouvier, D. and Cesca, O. eds. (2021) *Orality and Literacy in the Ancient World XII. Orality and Narration: Performance and Mythic-ritual Poetics*, Leiden.

Bierl, A. Christopoulos, M. and Papachrysostomou, A. eds. (2020) *Time and Space in Ancient Myth, Religion and Culture*, Berlin and New York.

Bingen, J. ed. (1996) *Pausanias Historien: Huit exposés suivis de discussions; Vandœuvres-Genève, 15–19 août 1994*, Geneva.

Birth, K. (2012) *Objects of Time: How Things Shape Temporality*, New York.

Bitton-Ashkelony, B. (2005) *Encountering the Sacred: The Debate on Christian Pilgrimage in Late Antiquity*, Berkeley.

Bjørnstad, H., Jordheim, H. and Régent-Susini, A. eds. (2019) *Universal History and the Making of the Global*, Abingdon.

Blair, A. and Goejing, S. eds. (2016) *For the Sake of Learning: Essays in Honor of Anthony Grafton*, Leiden.

Blair, K. (2012) *Form and Faith in Victorian Poetry and Religion*, Oxford.

Blair, K. ed. (2004) *John Keble in Context*, London.

Bloch, D. (2007) *Aristotle on Memory and Recollection*, Leiden.

Block, E. (1982) 'The Narrator Speaks: Apostrophe in Homer and Virgil', *Transactions of the American Philological Association* 112: 7–22.

Bloemendal, J. ed. (2016) *Erasmus Studies* 36.2 (2016) special edition on the *Paraphrases*.

Boatwright, M. (2013) 'Hadrian and the Agrippa Inscription of the Pantheon', in Opper ed. (2013): 19–30.

Boersma, H. (2012) 'Overcoming Time and Space: Gregory of Nyssa's Anagogical Theology', *Journal of Early Christian Studies* 20: 575–612.

Bogner, H. (1934) 'Die Religion des Nonnos von Panopolis', *Philologus* 89: 320–33.

Bohrer, K. H. (1981) *Plötzlichkeit: zum Augenblick des ästhetischen Scheins*, Berlin.

Børtnes, J. and Hägg, T. eds. (2006) *Gregory of Nazianzus: Images and Reflections*, Copenhagen.

Bostock, D. (1978) 'Plato on Change and Time in the *Parmenides*', *Phronesis* 23: 229–42.

Bouché-Leclercq, A. (1899) *L'Astrologie grecque*, Paris.

Bouchon, R., Brillet-Dubois, P. and Le Meur-Weismann, N. eds. (2008) *Hymnes de la Grèce antique: approches littéraires et historiques*, Lyon.

Bowen, A. and Rochberg, F. eds. (2020) *Hellenistic Astronomy: The Science in its Contexts*, Leiden.

Bowersock, G. (1990) *Hellenism in Late Antiquity*, Princeton.

Bowersock, G., Brown, P. and Grabar, O. (1999) *Late Antiquity: A Guide to the Post-Classical World*, Cambridge, MA.

Bowes, K. (2008) *Private Worship, Public Values and Religious Change in Late Antiquity*, New York.

Bowler, P. (1989) *The Invention of Progress: The Victorians and their Past*, Oxford.

Boyarin, D. (1994) *Intertextuality and Midrash*, Bloomington.

(1999) *Dying for God: Martyrdom and the Making of Christianity and Judaism*, Stanford.

(2009) *Socrates and the Fat Rabbis*, Chicago.

(2015) *A Travelling Homeland: The Babylonian Talmud as Diaspora*, Philadelphia.

Boyle, M. (1977) *Erasmus on Language and Method in Theology*, Toronto.

Bradshaw, P. (1981) *Daily Prayer in the Early Church: A Study of the Origin and Early Development of the Divine Office*, London.

Branham, B. (1989) *Unruly Eloquence: Lucian and the Comedy of Traditions*, Cambridge, MA.

Bregman, J. (1982) *Synesius of Cyrene: Philosopher-Bishop*, Berkeley.

Brine, K., Ciletti, E. and Lähnemann, H. eds. (2010) *The Sword of Judith: Judith Studies across the Disciplines*, Cambridge.

Brisson, L. (1998) *Plato the Myth Maker*, Chicago.

Brooks, P. (1992) *Reading for the Plot: Design and Intention in Narrative*, Cambridge, MA.

Brown, P. (1988) *The Body and Society: Men, Women and Sexual Renunciation in Early Christianity*, New York.

(1992) *Power and Persuasion in Late Antiquity: Towards a Christian Empire*, Madison, WI.

(1998) *Late Antiquity*, Cambridge, MA.

(2012) *Through the Eye of a Needle: Wealth, the Fall of Rome and the Making of Christianity in the West*, Princeton.

(2013) *The Rise of Western Christendom: Triumph and Diversity, A.D. 200–1000*, Malden.

(2014) *The Cult of the Saints: Its Rise and Function in Latin Christianity*, 2nd ed., Chicago.

Brown, S. (2005) *Ovid: Myth and Metamorphosis*, Bristol.

Brubaker, L. (2011) 'Talking about the Great Church: Ekphrasis and the *Narration on Hagia Sophia*', *Byzantinoslavica* 69: 80–7.

Buch-Hansen, G. (2018) 'The Johannine Literature in a Greek Context', in Lieu and de Boer eds. (2018): 138–56.

Buckland, A. (2013) *Novel Science: Fiction and the Invention of Nineteenth-Century Geology*, Chicago.

Buckley, J. (1966) *The Triumph of Time: A Study of Victorian Concepts of Time, History, Progress, and Decadence*, Cambridge, MA.

Burns, P. (2001) 'Roles of Roman Rhetorical *Exempla* in Augustine's *City of God*', *Studia Patristica* 38: 31–40.

Burnyeat, M. (1977) 'Socratic Midwifery, Platonic Inspiration', *Bulletin of the Institute of Classical Studies* 24: 7–16.

Burridge, J. (1992) *What Are the Gospels: A Comparison with Greco-Roman Biography*, Cambridge.

Burrow, J. (1981) *A Liberal Descent: Victorian Historians and the English Past*, Cambridge.

Burrus, V. (1995) *The Making of a Heretic: Gender, Authority and the Priscillianist Controversy*, Berkeley.

(2004) *The Sex Lives of Saints: An Erotics of Ancient Hagiography*, Philadelphia.

(2006) 'Life After Death: the Martyrdom of Gorgonia and the Birth of Female Hagiography', in Børtnes and Hägg eds. (2006): 153–70.

(2019) *Ancient Christian Ecopoetics: Cosmologies, Saints, Things*, Philadelphia.

Burton, P. (2017) *Sulpicius Severus' Vita Martini*, Oxford.

Butler, J. (2009) *The Frames of War*, London.

Buxton, R. (2012) 'Instructive Irony in Herodotus', *Greek, Roman, and Byzantine Studies* 52: 559–86.

Buxton, R. ed. (1999) *From Myth to Reason: Studies in the Development of Greek Thought*, Oxford.

Cain, A. (2009) *The Letters of Jerome: Asceticism, Biblical Exegesis and the Construction of Christian Authority in Late Antiquity*, Oxford.

(2013) *Jerome's Epitaph on Paula*, Oxford.

Calhoun, C., Juergensmeyer, M. and van Antwerpen, J. eds. (2011) *Rethinking Secularism*, Oxford.

Calvin, J. (1948) *Commentaries on the First Book of Moses, Called Genesis*, trans. J. King, Grand Rapids.

(1958) 'Acts of the Council of Trent: with the Antidote' [1547] in *Tracts and Treatises in Defense of the Reformed Faith*, trans. H. Beveridge, Grand Rapids.

Cambiano, G. (2007) 'Problemi della memoria in Platone', in Sassi ed. (2007): 1–23.

Cameron, A. (1965) 'Wandering Poets: a Literary Movement in Byzantine Egypt', *Historia* 14: 470–509.

(1993) *The Greek Anthology: From Meleager to Planudes*, Oxford.

(2016) *Wandering Poets and Other Essays on Late Greek Literature and Philosophy*, Oxford and New York.

Cameron, Averil (1991) *Christianity and the Rhetoric of Empire: The Development of Christian Discourse*, Berkeley.

Cameron, E. ed. (2018) *The Cambridge History of the Bible III*, Cambridge.

Cancik, H. (1998) 'Lucian on Conversion: Remarks on Lucian's Dialogue *Nigrinus*', in Collins ed. (1998): 26–48.

Canevaro, M. (2019) 'Memory, the Orators, and the Public in Fourth-Century BC Athens', in Castagnoli and Ceccarelli eds. (2019): 136–57.

Carleton-Paget, J. and Gathercole, S. eds. (2021) *Celsus and His World*, Cambridge.

Carleton-Paget, J. and Schaper, J. eds. (2013) *The New Cambridge History of the Bible*, vol. 1, Cambridge.

Carruthers, M. (1990) *The Book of Memory: A Study of Memory in Medieval Culture*, Cambridge.

(1998) *The Craft of Thought: Meditation, Rhetoric, and the Making of Images, 400–1200*, Cambridge.

Casey, R. (1934) *The Excerpta ex Theodoto of Clement of Alexandria*, London.

Cassin, B. (2014) '*Topos/Kairos*: Two Modes of Invention', in *Sophistical Practice: Towards a Consistent Relativism*, trans. O. Feltham, Fordham: 87–101.

Castagnoli, L. (2010) *Ancient Self-Refutation: The Logic and History of the Self-Refutation Argument from Democritus to Augustine*, Cambridge.

(2019a) 'In Memory of the Past? Aristotle on the Objects of Memory', in Castagnoli and Ceccarelli eds. (2019): 236–57.

(2019b) 'The *Phaedo* on Philosophy and the Soul', in Fine ed. (2019): 183–206.

Castagnoli, L. and Ceccarelli, P. (2019) 'Introduction', in Castagnoli and Ceccarelli eds. (2019): 1–52.

Castagnoli, L. and Ceccarelli, P. eds. (2019) *Greek Memories: Theories and Practices*, Cambridge.

Castelli, E. (2004) *Martyrdom and Memory: Early Christian Culture Making*, New York.

Castriota, D. (1993) *Myth, Ethos and Actuality: Official Art in Fifth-Century Athens*, Madison.

Ceccarelli, P. (2019) 'Economies of Memory in Greek Tragedy', in Castagnoli and Ceccarelli eds. (2019): 93–114.

Chadwick, H. (1991) *Augustine: The Confessions*, Oxford.

(2001) *The Church in Ancient Society: From Galilee to Gregory the Great*, Oxford.

Chadwick, O. (1950) *John Cassian: A Study in Primitive Monasticism*, New York and Cambridge.

Chakrabarty, D. (1997) 'The Time of History and the Times of Gods', in Lowe and Lloyd eds. (1997): 35–60.

(2000) *Provincializing Europe: Postcolonial Thought and Historical Difference*, Chicago.

Chaplin, J. (2000) *Livy's Exemplary History*, Oxford.

Charlesworth, J. (1977) 'Jewish Astrology in the Talmud, Pseudepigrapha, the Dead Sea Scrolls and Early Palestinian Synagogues', *Harvard Theological Review* 70: 183–200.

Chomarat, J. (1981) *Grammaires et rhétoriques chez Érasme*, 2 vols., Paris.

Christ, G. and Morche, F.-J. eds. (2018) *Cultures of Empire: Rethinking Venetian Rule, 1400–1700*, Leiden.

Chrubasik, B. and King, D. eds. (2017) *Hellenism and Local Communities of the Eastern Mediterranean: 400* BCE*–250* CE, Oxford.

Chuvin, P. (1991) *Mythologie et géographie dionysiaques: recherches sur l'oeuvre de Nonnos de Panopolis*, Clermont-Ferrand.

(2014) 'Revisiting Old Problems: Literature and Religion in the *Dionysiaca*', in Spanoudakis ed. (2014): 3–18.

Cimakasky, J. (2017) *The Role of Exaíphnes in Early Greek Literature: Philosophical Transformation in Plato's Dialogues and Beyond*, Lanham.

Clark, C. (2019) *Time and Power: Visions of History in German Politics from the Thirty Years' War to the Third Reich*, Princeton.

Clark, E. (1977) *Clement's Use of Aristotle: The Aristotelian Contribution to Clement of Alexandria's Refutation of Gnosticism*, New York.

(1999) *Reading Renunciation: Asceticism and Scripture in Early Christianity*, Princeton.

Clark, G. (1993) *Augustine's Confessions*, Cambridge.

Clarke, J. (2020) 'Pain, Speech and Silence in Prudentius *Peristephanon* 5 and 9', *Vigiliae Christianae* 74: 4–28.

Clarke, K. (2008) *Making Time for the Past: Local History and the Polis*, Oxford.

Clarke, M. and ní Mhanoaigh, M. (2020) 'The Ages of the World and the Ages of Man: Irish and European Learning in the Twelfth Century', *Speculum* 95: 467–500.

Clay, D. (2004) *Archilochos Heros: The Cult of Poets in the Greek Polis*, Cambridge, MA.

Clay, J. (1981–2) 'Immortal and Ageless Forever', *Classical Journal* 77: 112–17.

(1989) *The Politics of Olympus: Form and Meaning in the Major Homeric Hymns*, Princeton.

(1997) *The Wrath of Athena: Gods and Men in the Odyssey*, Lanham.

(2003) *Hesiod's Cosmos*, Cambridge.

Coakley, S. ed. (2003) *Rethinking Gregory of Nyssa*, Oxford.

Cobet, J. (2009) 'Orosius' Weltgeschichte: Tradition und Konstruktion', *Hermes* 137: 60–92.

Coetzee, J. M. (1980) *Waiting for the Barbarians*, London.

Cohen, B. ed. (1995) *The Distaff Side: Representing the Female in Homer's Odyssey*, Oxford.

Cohen, S. (1979) *Josephus in Galilee and Rome*, Leiden.

Cohen, S. and Schwartz, J. eds. (2007) *Studies in Josephus and the Varieties of Ancient Judaism*, Leiden.

Coleman, S. and Elsner, J. (1995) *Pilgrimage: Past and Present in the World Religions*, Cambridge, MA.

Collins, A. Y. ed. (1998) *Ancient and Modern Perspectives on the Bible and Culture: Essays in Honour of Hans Dieter Betz*, Atlanta.

Collins, J. (1998) *The Apocalyptic Imagination: An Introduction to Jewish Apocalyptic Literature*, 2nd ed., Grand Rapids.

Collobert, C., Destrée, P. and Gonzalez, F. eds. (2012) *Plato and Myth: Studies on the Use and Status of Platonic Myths*, Brill.

Connerton, P. (1989) *How Societies Remember*, Cambridge.

Conybeare, C. (1999) '*Terrarum Orbi Documentum*: Augustine, Camillus and Learning from History', in Vessey, Pollman and Fitzgerald eds. (1999): 59–74.

(2000) *Paulinus Noster: Self and Symbols in the Letters of Paulinus of Nola*, Oxford.

(2002) 'The Ambiguous Laughter of St Lawrence', *Journal of Early Christian Studies* 10: 175–202.

(2007) '*Sanctum, lector, percense volumen*: Snakes, Readers and the Whole Text in Prudentius' *Hamartigenia*', in Klinghirn and Safran eds. (2007): 225–40.

(2016) *The Routledge Guidebook to Augustine's Confessions*, New York.

Conybeare, C. and Goldhill, S. eds. (2020) *Classical Philology and Theology: Disavowal, Entanglement and the God-Like Scholar*, Cambridge.

Coope, U. (2005) *Time for Aristotle: Physics 4 10–14*, Oxford.

Cooper, J. (1995) *T. S. Eliot and the Ideology of the 'Four Quartets'*, Cambridge.

Cooper, K. (1999) *The Virgin and the Bride: Idealized Womanhood in Late Antiquity*, Cambridge, MA.

Corke-Webster, J. (2019) *Eusebius and Empire: Constructing Church and Rome in the Ecclesiastical History*, Cambridge.

Corsini, E. (1968) *Introduzione alle 'Storie' di Orosio*, Turin.

Cox Miller, P. (1983) *Biography in Late Antiquity: The Quest for the Holy Man*, Berkeley.

Craig, J. (2002) 'Forming a Protestant Consciousness? Erasmus' *Paraphrases* in English Parishes, 1547–1666', in Pabel and Vessey eds. (2002): 313–60.

Cramer, F. (1954) *Astrology in Roman Law and Politics*, Philadelphia.

Crawford, M. and Ligota, C. eds. (1995) *Ancient History and the Antiquarian: Essays in Memory of Arnaldo Momigliano*, London.

Csapo, E. (2008) 'Star Choruses: Eleusis, Orphism and New Musical Imagery and Dance', in Revermann and Wilson eds. (2008): 262–90.

Cullmann, O. (1946) *Christus und die Zeit: die urchristliche Zeit- und Geschichts-Auffassung*, Zurich.

Cumont, F. (1912) *Astrology and Religion among the Greeks and Romans*, New York.

Currie, B. (2005) *Pindar and the Cult of Heroes*, Oxford.

Currie, M. (2007) *About Time: Narrative, Fiction and the Philosophy of Time*, Edinburgh.

Dagron, G. (1984) *Constantinople imaginaire: études sur le receuil des 'Patria'*, Paris.

Daley, B. (1993) 'The Giant's Twin Substances: Ambrose and the Christology of Augustine's "Contra Sermonem Arianorum"', in Lienhard, Muller and Teske eds. (1993): 477–95.

(2006) *Gregory of Nazianzus*, London.

(2012) 'Systematic Theology in Homeric Dress', in Beeley ed. (2012): 3–12.

Dames, N. (2007) *The Physiology of the Novel: Reading, Neural Science, and the Form of Victorian Fiction*, Oxford and New York.

Daniélou, J. (1948) *Origène*, Paris.

(1950) *Sacramentum Futuri: études sur les origines de la typologie biblique*, Paris.

Darbo-Peschanski, C. ed. (2000) *Construction du temps dans le monde grec ancien*, Paris.

Daston, L. and Galison, P. (2007) *Objectivity*, New York.

Davidson, J. (2006) 'Revolutions in Human Time: Age Class in Athens and the Greekness of Greek Revolutions', in Goldhill and Osborne eds. (2006): 29–67.

Davis, K. (2008) *Periodization and Sovereignty: How Ideas of Feudalism and Secularization Govern the Politics of Time*, Philadelphia.

Dawson, J. (1992) *Allegorical Readers and Cultural Revision in Ancient Alexandria*, Berkeley.

(2002) *Christian Figural Reading and the Fashioning of Identity*, Berkeley.

Dawson, Z. (2019) 'Does Luke's Preface Resemble a Greek Decree? Comparing the Epigraphical and Papyrological Evidence of Greek Decrees with Ancient Preface Formulae', *New Testament Studies* 65: 552–71.

De Blasi, A. (2017/18) *Gregorio di Nazianzo εἰς τὰ ἔμμετρα (carme II 1, 39) per un'edizione e comment*, MA thesis, University of Padua.

Degani, E. (1961) *AION da Omero ad Aristotele*, Padua.

De Jong, I. and Nünlist, R. eds. (2007) *Time in Greek Literature*, Leiden.

De Jong, I. and Sullivan, J. eds. (1994) *Modern Critical Theory and Classical Literature*, Leiden.

De Maniquis, R. (1985) 'The Dark Interpreter and the Palimpsest of Violence: de Quincey and the Unconscious', in Snyder ed. (1985): 109–39.

Den Boeft, J. (2003) '*Aeterne rerum conditor:* Ambrose's Poem about "Time"', in García Martinez and Luttikhuizen eds. (2003): 27–40.

(2008) 'Delight and Imagination: Ambrose's Hymns', *Vigiliae Christianae* 62: 425–40.

Den Boeft, J. and Hilhorst, A. eds. (1993) *Early Christian Poetry: A Collection of Essays*, Supplements to *Vigiliae Christianae* 22, Leiden.

Denzey Lewis, N. (2020) 'Hellenistic Astronomy in Early Christianities', in Bowen and Rochberg eds. (2020): 551–71.

De Quincey, T. (1998) *Suspiria de Profundis* [1848], in *Thomas de Quincey: Confessions of an Opium Eater and Other Writings*, ed. G. Lindop, Oxford: 183–233.

Derow, P. and Parker, R. eds. (2003) *Herodotus and his World*, Oxford.

Derrida, J. (1967) *L'Écriture et le différence*, Paris.

De Stefani, C. (2002) *Nonno di Panopoli: Parafrasi del Vangelo di S. Giovanni. Canto 1*, Bologna.

Destrée, P. and Hermann, F.-G. eds. (2011) *Plato and the Poets*, Brill.

de Temmermann, K. and Demoen, K. eds. (2016) *Writing Biography in Greece and Rome: Narrative Technique and Fictionalization*, Cambridge.

Detienne, M. (1981) *L'Invention de la mythologie*, Paris.

Detienne, M. and Vernant, J.-P. (1974) *Les Ruses de l'intelligence: la mètis des Grecs*, Paris.

Detienne, M. and Vernant, J.-P. eds. (1989) *The Cuisine of Sacrifice among the Greeks*, trans. P. Wissing, Chicago.

de Vore, D. (2020) '"The Only Event Mightier Than Everyone's Hope": Classical Historiography and Eusebius' Plague Narrative', *Histos* 14: 1–34.

Dewald, C. and Marincola, J. eds. (2006) *The Cambridge Companion to Herodotus*, Cambridge.

de Wet, C. and Mayer, W. eds. (2019) *Revisioning John Chrysostom: New Approaches, New Perspectives*, Leiden.

Dickson, K. (2019) '*Kairios* and *Kairos*: Walls and ways in Homer', *Helios* 46: 97–114.

Dietz, M. (2005) *Wandering Monks, Virgins, and Pilgrims: Ascetic Travel in the Mediterranean World, A.D. 300–800*, University Park, PA.

Dijkstra, J. (2016) 'The Religious Background of Nonnus', in Accorinti ed. (2016): 75–90.

Dillery, J. (2018) 'Making Logoi: Herodotus' Book 2 and Hecataeus of Miletus', in Harrison and Irwin eds. (2018): 18–51.

Dillon, J. (2002) 'An Unknown Platonist on God', in Barbanti and Manganaro eds. (2002): 237–45.

Dillon, S. (2007) *The Palimpsest: Literature, Criticism, Theory*, London.

Dimant, D. (2014) *History, Ideology and Bible Interpretation in the Dead Sea Scrolls*, Tübingen.

Dingley, R. (2000) 'The Runs of the Future: Macaulay's New Zealander and the Spirit of the Age', in Sandison and Dingley eds. (2000): 15–33.

Dinshaw, C. (2012) *How Soon Is Now? Medieval Texts, Amateur Readers and the Queerness of Time*, Durham, NC.

Dodaro, R. and Lawless, G. eds. (2000) *Augustine and his Critics*, London.

Doherty, L. (1995) *Siren Songs: Gender, Audiences, and Narrators in the Odyssey*, Ann Arbor.

Dohrn-van Rossum, G. (1996) *History of the Hour: Clocks and the Modern Temporal Order*, trans. T. Dunlap, Chicago.

Doroszewski, J. (2014) 'Judaic Orgies and Christ's Bacchic Deeds: Dionysiac Terminology in Nonnus' Paraphrase of St. John's Gospel', in Spanoudakis ed. (2014): 287–302.

(2016) 'The Mystery Terminology in Nonnus' *Paraphrase*', in Accorinti ed. (2016): 327–51.

Dörrie, H. (1976) 'Une exégèse néoplatonicienne du prologue de l'Evangile selon St Jean', in H. Dörrie, *Platonica Minora*, Munich: 491–507.

Dörrie, H., Altenburger, M. and Schramm, U. eds. (1976) *Gregor von Nyssa und die Philosophie*, Leiden.

Douglass, S. (2005) *Theology of the Gap: Cappadocian Language Theory and the Trinitarian Controversy*, New York.

Driver, S. (2013) *John Cassian and the Reading of Egyptian Monastic Culture*, New York and London.

Dueck, D. (2020) 'A Lunar People: The Meaning of an Arcadian Epithet, or, Who Is the Most Ancient of Them All', *Philologus* 164: 133–47.

Duff, T. (1999) *Plutarch's Lives: Exploring Virtue and Vice*, Oxford.

Dummett, M. A. E. (1954) 'Can an Effect Precede its Cause?', *Proceedings of the Aristotelian Society* 28: 27–44.

(1964) 'Bringing About the Past', *Philosophical Review* 73: 338–59.

Dunkle, B. (2016) *Enchantment and Creed in the Hymns of Ambrose of Milan*, Oxford.

Dunn, F. (2007) *Present Shock in Late Fifth-Century Greece*, Ann Arbor.

Dunn, J. (2011) 'Seeing in and through Time', in Lianeri ed. (2011): 307–14.

Dupont, A., Boodts, S., Partoens, G. and Leemans, J. eds. (2018) *Preaching in the Patristic Era: A New History of the Sermon*, vol. 6, Leiden.

Durkheim, E. (1915) *The Elementary Forms of Religious Life*, trans. J. Swain, London.

Dykes, A. (2011) *Reading Sin in the World: The Hamartigenia of Prudentius and the Vocation of the Responsible Reader*, Cambridge.

Eadie, J., Ober, J. and Bender, E. eds. (1985) *The Craft of the Ancient Historian: Essays in Honor of Chester Starr*, Lanham, MD.

Edelman, L. (2004) *No Future: Queer Theory and the Death Drive*, Durham, NC.

Edmondson, J., Mason, S. and and Rives, J. eds. (2005) *Flavius Josephus and Flavian Rome*, Oxford.

Edwards, A. (1984) *Achilles in the Odyssey: Ideologies of Heroism in the Homeric Epic*, Königstein.

(1988) 'ΚΛΕΟΣ ΑΦΘΙΤΟΝ and Oral Theory', *Classical Quarterly* 38: 25–30.

Edwards, C. (2007) *Death in Ancient Rome*, New Haven.

(2014) 'Death and Time', in Heil and Damschen eds. (2014): 323–41.

Edwards, M. (1996) 'Porphyry's "Cave of the Nymphs" and the Gnostic Controversy', *Hermes* 124: 88–100.

(2002) *Origen Against Plato*, Aldershot.

(2007) 'Socrates and the Early Church', in Trapp ed. (2007): 127–41.

Edwards, M. and Swain, S. eds. (1997) *Portraits: Biographical Representation in the Greek and Latin Literature of the Roman Empire*, Oxford.

Eidinow, E., Kindt, J. and Osborne, R. eds. (2016) *Theologies of Greek Religion*, Cambridge.

Eisen, U. and Moellendorf, P. von eds. (2013) *Über die Grenze: Metalepse in Text- und Bildmedien des Altertums*, Berlin.

Elias, N. (1992) *Time: An Essay*, trans. E. Jephcott, Chicago.

Eliot, T. S. (1920) *Poems 1920*, London.

Elm, S. (2006) 'Gregory's Women: Creating a Philosopher's Family', in Børtnes and Hägg eds. (2006): 171–91.

(2012) *Sons of Hellenism, Fathers of the Church: Emperor Julian, Gregory of Nazianzus and the Vision of Rome*, Berkeley.

Elsner, J. (1994) 'From the Pyramids to Pausanias and Piglet: Monuments, Travel and Writing', in Goldhill and Osborne eds. (1994): 224–54.

(1995) *Art and the Roman Viewer: The Transformation of Art from the Pagan World to Christianity*, Cambridge.

(2001) 'Describing Self in the Language of the Other: Pseudo (?) Lucian at the Temple of Hierapolis', in Goldhill ed. (2001): 123–53.

(2003) 'Iconoclasm and the Preservation of Memory', in Nelson and Olin eds. (2003): 209–31.

Elsner, J. and Hernández Lobato, J. (2017) *The Poetics of Late Latin Literature*, Oxford.

Elsner, J. and Rutherford, I. eds. (2005) *Pilgrimage in Graeco-Roman and Early Christian Antiquity: Seeing the Gods*, Oxford.

Erll, A. (2011) *Memory in Culture*, New York.

Eshleman, K. (2007/8) 'Affection and Affiliation: Social Networks and Conversion to Philosophy', *Classical Journal* 103: 129–40.

(2012) *The Social World of Intellectuals in the Roman Empire: Sophists, Philosophers and Christians*, Cambridge.

Fabian, J. (1983) *Time and the Other: How Anthropology Makes its Objects*, New York.

Falkner, T. and de Luce, J. eds. (1989) *Old Age in Greek and Latin Literature*, Buffalo.

Fantuzzi, M. and Hunter, R. (2004) *Tradition and Innovation in Hellenistic Poetry*, Cambridge.

Faulkner, A. (2008a) *The Homeric Hymn to Aphrodite: Introduction, Text and Commentary*, Oxford.

(2008b) 'The Legacy of Aphrodite: Anchises' Offspring in the *Homeric Hymn to Aphrodite*', *American Journal of Philology* 129: 1–18.

(2010) 'St Gregory of Nazianzus and the Classical Tradition: The *Poemata Arcana* qua Hymns', *Philologus* 154: 78–87.

(2014) 'Faith and Fidelity in Biblical Epic: the *Metaphrasis Psalmorum*, Nonnus and the Theory of Translation', in Spanoudakis ed. (2014): 195–210.

(2017) 'Nonnus' "Younger Legend": The Birth of Beroe and the Didactic Tradition', *Greece & Rome* 64: 103–14.

(2020) *Apollinaris of Laodicea: Metaphrasis Psalmorum*, Oxford.

Faulkner, A. ed. (2011) *The Homeric Hymns: Interpretative Essays*, Oxford.

Fauth, W. (1995) *Helios Megistos: zur synkretistischen Theologie der Spätantike*, Leiden.

Favro, D. (2014) 'Moving Events: Curating the Memory of the Roman Triumph', in Galinsky ed. (2014): 85–101.

Fear, A. (2005) 'The Christian Optimism of Paulus Orosius', in Hook ed. (2005): 1–16.

(2010a) *Orosius: Seven Books of the History against the Pagans*, Liverpool.

(2010b) 'Orosius and Escaping from the Dance of Doom', in Liddel and Fear eds. (2010): 176–88.

Feeney, D. (2007) *Caesar's Calendar: Ancient Time and the Beginnings of History*, Berkeley.

Feldherr, A. (2010) *Playing Gods: Ovid's Metamorphoses and the Politics of Fiction*. Princeton.

Felson Rubin, N. (1994) *Regarding Penelope: From Character to Poetics*, Princeton.

Ferrari, G. (1989) 'Plato and Poetry', in Kennedy ed. (1989): 92–148.

Ferrari, L. (1977) 'Augustine and Astrology', *Laval théologique et philosophique* 33: 241–51.

Ferrier, V. and Mantero, A. eds. (2006) *Les Paraphrases bibliques aux xvie et xviie siècles. Actes du Colloque de Bordeaux des 22, 23, 24 Septembre 2004*, Geneva.

Festugière, A. (1950) *La revelation d'Hermès Trismégiste I: L'astrologie et les sciences occultés*, Paris.

Fincher, J. (2017) 'The Tablets of Harmonia and the Role of Poet and Reader in the *Dionysiaca*', in Bannert and Kroll eds. (2017): 120–37.

Fine, G. ed. (2019) *The Oxford Handbook on Plato*, 2nd ed., Oxford.

Finkelberg, M. (1986) 'Is ΚΛΕΟΣ ΑΦΘΙΤΟΝ a Homeric Formula?', *Classical Quarterly* 36: 1–5.

(2007) 'More on ΚΛΕΟΣ ΑΦΘΙΤΟΝ', *Classical Quarterly* 57: 34–50.

Fitzgerald, W. (1987) *Agonistic Poetry: The Pindaric Mode in Pindar, Horace, and the English Ode*, Berkeley.

(2007) *Martial: The World of Epigram*, Chicago.

Fletcher, R. and Hanink, J. eds. (2016) *Creative Lives in Classical Antiquity: Poets, Artists and Biography*, Cambridge.

Florensky, P. (1997) *The Pillar and Ground of Truth: An Essay in Orthodox Theodicy in Twelve Letters*, trans. B. Jakim, Princeton.

Flower, H. (1996) *Ancestor Masks and Aristocratic Power in Roman Culture*, Oxford.

(2006) *The Art of Forgetting: Disgrace and Oblivion in Roman Political Culture*, Chapel Hill, NC.

Flower, R., Kelly, C. and Williams, M. eds. (2010) *Unclassical Traditions: Alternatives to the Classical Past in Late Antiquity*, 2 vols., *Cambridge Classical Journal* Supplements 34 and 35, Cambridge.

Flusser, D. (1984) 'Psalms, Hymns and Prayers', in Stone ed. (1984): 551–77.

Foley, H. (1995) 'Penelope as a Moral Agent', in Cohen ed. (1995): 93–115.

Foley, H. ed. (1993) *The Homeric Hymn to Demeter*, Princeton.

Fontaine, J. and Pietri, C. eds. (1985) *Le Monde latin antique et la Bible*, Paris.

Fontaine, J. et al. eds. (1992) *Ambroise de Milan: Hymnes*, Paris.

Foucault, M. (1969) *L'Archéologie de savoir*, Paris.

Foucher, L. (1996) 'Aion: le Temps absolu', *Latomus* 55: 5–30.

Fowler, R. (1998) 'Genealogical Thinking, Hesiod's *Catalogue*, and the Creation of the Hellenes', *Proceedings of the Cambridge Philological Society* 44: 1–19.

(2003) 'Herodotus and Athens' in Derow and Parker eds. (2003): 305–18.

(2006) 'Herodotus and his prose predecessors', in Dewald and Marincola eds. (2006): 29–45.

Franchi, R. (2013) *Nonno di Panopoli. Parafrasi del Vangelo di San Giovanni. Canto Sesto*, Bologna.

(2016) 'Approaching the 'Spiritual Gospel': Nonnus as Interpreter of John', in Accorinti ed. (2016): 240–66.

François, W. (2008) 'Erasmus' Plea for Bible Reading in the Vernacular: The Legacy of Modern Devotion?' *Erasmus of Rotterdam Society Yearbook* 28: 91–120.

(2009) 'The Condemnation of the Vernacular Bible Reading by the Parisian Theologians (1523–31)', in François and den Hollander eds. (2009): 111–39.

(2016) *Vernacular Bible and Religious Reform in the Middle Ages and Early Modern Era*, Leuven.

François, W. and den Hollander, A. eds. (2009) *Infant Milk or Hardy Nourishment? The Bible for Lay People and Theologians in the Early Modern Period*, Leuven.

Franek, J. (2016) '*Omnibus Omnia:* The Reception of Socrates in Ante-Nicene Christian Literature', *Graeco-Latina Brunensia* 21: 31–58.

Frank, G. (2000a) 'Macrina's Scar: Homeric Allusion and Heroic Identity in Gregory of Nyssa's Life of Macrina', *Journal of Early Christian Studies* 8: 511–30.

(2000b) *The Memory of the Eyes: Pilgrims to Living Saints in Christian Late Antiquity*, Berkeley.

Fränkel, H. (1960) *Wege und Formen frühgriechischen Denkens*, ed. F. Tietze, 2nd ed., Munich.

Franz, A. (1994) *Tageslauf und Heilsgeschichte: Untersuchungen zum literarischen Text und liturgischen Kontext der Tagzeitenhymnen des Ambrosius von Mailand*, St Otilien.

Freccero, C. (2006) *Queer/Early/Modern*, Durham, NC.

Frede, D. and Reis, B. eds. (2009) *Body and Soul in Ancient Philosophy*, Berlin.

Frede, M. (2006) 'The Early Christian Reception of Socrates', in Judson and Karasmanis eds. (2006): 188–202.

Fredriksen, P. (1986) 'Paul and Augustine: Conversion Narratives, Orthodox Traditions and the Retrospective Self', *Journal of Theological Studies* 37: 3–34.

(2018) *When Christians Were Jews: The First Generation*, New Haven.

Friedman, S. (2010) *Talmudic Studies: Investigating the Sugya, Variant Readings, and Aggada*, New York and Jerusalem.

Friese, H. ed. (2001) *The Moment: Time and Rupture in Modern Thought*, Liverpool.

Fritzsche, P. (2004) *Stranded in the Present: Modern Times and the Melancholy of History*, Cambridge.

Fulford, B. (2012) 'Gregory of Nazianzus and Biblical Interpretation', in Beeley ed. (2012): 31–48.

Funkenstein, A. (1989) 'Collective Memory and Historical Consciousness', *History and Memory* 1: 5–26.

Furley, D. and Nehemas, A. eds. (1994) *Aristotle's Rhetoric: Philosophical Essays*, Princeton.

Gaca, K. (2009) *The Making of Fornication: Eros, Ethics and Political Reform in Greek Philosophy and Early Christianity*, Berkeley.

Gafni, I. (1996) 'Concepts of Periodization and Causality in Talmudic Literature', *Jewish History* 10: 21–38.

Gager, J. (1975) *Kingdom and Community: The Social World of Early Christianity*, Englewood Cliffs.

Gagné, R. (2019) 'Cosmic Choruses', in Horky ed. (2019): 188–211.

  (2020) 'Whose handmaiden? "Hellenization" between Philology and Theology', in Conybeare and Goldhill eds. (2020): 110–25.

Galinsky, K. ed. (2014) *Memoria Romana: Memory in Rome and Rome in Memory*, Ann Arbor.

  (2016) *Memory in Ancient Rome and Early Christianity*, Oxford.

Galison, P. (2003) *Einstein's Clocks, Poincarés Maps: Empires of Time*, London.

Gallay, P. (1978) *Grégoire de Nazianz: Discours 27–31*, Paris.

Gallet, B. (1990) *Recherches sur kairos et l'ambiguité dans la poèsie de Pindare*, Bordeaux.

Gamble, C. (2021) *Making Deep History: Zeal, Perseverance and the Time Revolution of 1859*, Oxford.

Garcia, T. (2014) *Form and Object: A Treatise on Things*, trans. M. Ohm and J. Cogburn, Edinburgh.

García-Gasco Villarubia, R. (2008) 'Orfeo y el orfismo en las Dionisíacas de Nono', in Bernabé and Casadésus eds. (2008): 1575–1600.

García Martinez, F. (2013) *Between Philology and Theology: Contributions to the Study of Ancient Jewish Interpretation*, Leiden.

García Martinez, F. and Luttikhuizen, G. eds. (2003) *Jerusalem, Alexandria, Rome: Studies in Ancient Cultural Interaction in Honour of A. Hilhorst*, Leiden.

Gardner, G. (2008) 'Astrology in the Talmud: An Analysis of Bavli Shabbat 156', in Iricinschi and Zellentin eds. (2008): 314–38.

Gay, H. (2003) 'Clock Synchrony, Time Distribution and Electrical Time-Keeping in Britain 1880–1925', *Past and Present* 181: 107–40.

Gee, E. (2000) *Ovid, Aratus, and Augustus: Astronomy in Ovid's Fasti*, Cambridge.

  (2001) 'Cicero's Astronomy', *Classical Quarterly* 51: 520–36.

Geertz, C. (1973) *The Interpretation of Culture*, New York.

Gell, A. (1992) *The Anthropology of Time: Cultural Constructions of Temporal Maps and Images*, Providence and Oxford.

Georgiadou, A. and Oikonomopoulou, K. eds. (2020) *Space, Time and Language in Plutarch*, Berlin and New York.

Gerhold, V. (2018) 'Defeating Solomon: Intertextuality and Symbolism in the Legend of Hagia Sophia', *Scripta* 11: 11–38.

Gibson, R. and Whitton, C. eds. (forthcoming) *The Cambridge Critical Guide to Latin Literature*, Cambridge.

Gigli Piccardi, D. (2003) *Nonno di Panopoli. Le Dionisiache. Volume Primo (Canti I–XII)*, Milan.

(2009) 'Phanes ἀρχέγονος φρήν (Nonno D. 12.68 e orac. Ap. Didym. De Trin. II 27)', *Zeitschrift für Papyrologie und Epigraphik* 169: 71–8.

Gilby, E. (2006) *Sublime Worlds: Early Modern French Literature*, Abingdon and New York.

Glennie, P. and Thrift, N. (2009) *Shaping the Day: A History of Timekeeping in England and Wales 1300–1800*, Oxford.

Goetz, H. (1980) *Die Geschichtstheologie des Orosius*, Darmstadt.

Goldhill, S. (1984) *Language, Sexuality, Narrative: The Oresteia*, Cambridge.

(1986) *Reading Greek Tragedy*, Cambridge.

(1988) 'Reading Differences: Juxtaposition and the *Odyssey*', *Ramus* 17: 1–31.

(1991) *The Poet's Voice: Essays on Poetics and Greek Literature*, Cambridge.

(1994) 'The Failure of Exemplarity', in de Jong and Sullivan eds. (1994): 51–73.

(1999) 'Literary History Without Literature: Reading Practices in the Ancient World', *Sub-Stance* 28: 57–89.

(2002) *Who Needs Greek? Contests in the Cultural History of Hellenism*, Cambridge.

(2010) 'Idealism in the *Odyssey* and the Meaning of *Mounos* in *Odyssey* 16', in Mitzis and Tsagalis eds. (2010): 115–28.

(2016) *Very Queer Family Indeed: Sex, Religion, and the Bensons in Victorian Britain*, Chicago.

(2017) 'The Limits of the Case Study: Exemplarity and the Reception of Classical Literature', *New Literary History* 48: 415–35.

(2020) *Preposterous Poetics: The Politics and Aesthetics of Form in Late Antiquity*, Cambridge.

(2021) 'Shaping the Religious Debate from within Second Sophistic Culture', in Carleton Paget and Gathercole eds. (2021): 206–14.

(forthcoming a) 'Latin Literature and Greek', in Gibson and Whitton eds. (forthcoming).

(forthcoming b) 'The Personal Voice: Six Fragments of a Sentimental Education', in J. Billings and F. Budelmann eds. *Helios* special issue.

Goldhill, S. ed. (2001) *Being Greek Under Rome: Cultural Identity, the Second Sophistic and the Development of Empire*, Cambridge.

Goldhill, S. and Greensmith, E. (2020) 'Gregory of Nazianzus in the *Palatine Anthology*: The Poetics of Christian Death', *Cambridge Classical Journal* 66: 29–69.

Goldhill, S. and Morales, H. eds. (2007) *Dying for Josephus*, *Ramus* 31 Special Issue.

Goldhill, S. and Osborne, R. eds. (1994) *Art and Text in Ancient Greek Culture*, Cambridge.

(2006) *Rethinking Revolutions through Ancient Greece*, Cambridge.

Goldschmidt, N. and Graziosi, B. eds. (2018) *Tombs of the Ancient Poets: Between Literary Reception and Material Culture*, Oxford.

Golega, J. (1930) *Studien über die Evangeliendichtung des Nonnos von Panopolis*, Breslau.

(1960) *Der homerische Psalter: Studien über die dem Apollinarios von Laodikeia zugeschriebene Psalmenparaphrase*, Studia Patristica et Byzantina, Ettal.

Gonnelli, F. (2004) *Nonno di Panopoli. Le Dionisiache. Volume Secondo (Canti XIII–XXIV)*, Milan.

Gonzalez, F. (2019) 'Shattering Presence: Being as Change, Time as the Sudden Instant in Heidegger's 1930–31 Seminar on Plato's *Parmenides*', *Journal of the History of Philosophy* 57: 313–38.

Goodman, M. (2000) *State and Society in Roman Galilee, A.D. 132–212*, 2nd ed., London.

Gorham, G. (2014) 'Hobbes on the Reality of Time', *Hobbes Studies* 27: 87–103.

Gosse, E. (1907) *Father and Son*, London.

Gould, J. (2001) *Myth, Ritual, Memory and Exchange: Essays in Greek Literature and Culture*, Oxford.

Gould, T. (1990) *The Ancient Quarrel between Poetry and Philosophy*, Princeton.

Gowing, A. (2005) *Empire and Memory: The Representation of the Roman Republic in the Roman Empire*, Cambridge.

Graf, F. (2009) 'Serious Singing: The Orphic Hymns as Religious Texts', *Kernos* 22: 169–82.

Grafton, A. (1983–93) *Joseph Scaliger: A Study in the History of Classical Scholarship*, 2 vols., Oxford.

(1995) 'Tradition and Technique in Historical Chronology', in Crawford and Ligota eds. (1995): 15–31.

Grafton, A. and Williams, M. (2006) *Christianity and the Transformation of the Book: Origen, Eusebius and the Library of Caesarea*, Cambridge, MA.

Green, A. (2002) *Time in Psychoanalysis: Some Contradictory Aspects*, trans. A. Weller, London and New York.

Green, R. (2006) *Latin Epics of the New Testament: Juvencus, Sedulius, Arator*, Oxford.

Green, S. (2011) '*Arduum ad astra*: The Poetics and Politics of Horoscope Failure in Manilius' *Astronomica*', in Green and Volk eds. (2011): 120–38.

Green, S. and Volk, K. eds. (2011) *Forgotten Stars: Rediscovering Manilius' Astronomica*, Oxford.

Greensmith, E. (2020) *The Resurrection of Homer in Imperial Greek Epic: Quintus Smyrnaeus' Posthomerica and the Poetics of Impersonation*, Cambridge.

(forthcoming) 'The Miracle Baby: Zagreus and the Poetics of Mutation', in Verhelst ed. (forthcoming).

Greenwood, E. (2016) 'Futures Real and Unreal in Greek Historiography', in Lianeri ed. (2016): 79–100.

Grethlein, J. (2010) *The Greeks and their Past: Poetry, Oratory and History in the Fifth Century* BCE, Cambridge.

(2013) *Experience and Teleology in Ancient Historiography: 'Futures Past' from Herodotus to Augustine*, Cambridge.

(2016) 'Lucian's Response to Augustine. Conversion and Narrative in *Confessions* and *Nigrinus*', *Religion in the Roman Empire* 2: 256–78.

Grethlein, J. and Krebs, C. eds. (2012) *Time and Narrative in Ancient Historiography: The 'Plupast' from Herodotus to Appian*, Cambridge.

Gribetz, S. (2020) *Time and Difference in Rabbinic Judaism*, Princeton.

Griffin, J. (1980) *Homer on Life and Death*, Oxford.

Griffin, M. (1976) *Seneca: A Philosopher in Politics*, Oxford.

Griffith, M. (1977) *The Authenticity of the Prometheus Bound*, Cambridge.

Griffiths, A. (1999) 'Euenius the Negligent Nightwatchman (Herodotus 9.92–6)', in Buxton ed. (1999): 169–82.

Griffiths, F. (1979) *Theocritus at Court*, Leiden.

Grig, L. (2004) *Making Martyrs in Late Antiquity*, London.

Grosz, E. (2004) *The Nick of Time: Politics, Evolution and the Untimely*, Durham, NC.

Gruen, E. (2008) 'The *Letter of Aristeas* and the Cultural Context of the Septuagint', in Karrer and Kraus eds. (2008): 134–56.

Gundel, W. and Gundel, H. (1966) *Astrologoumena*, Wiesbaden.

Gunderson, E. (1996) 'The Ideology of the Arena', *Classical Antiquity* 15: 113–51.

Gurvitch, G. (1961) *The Spectrum of Social Time*, Dordrecht.

Güthenke, C. (2020) 'For Time Is/Nothing If Not Amenable: Exemplarity, Time, Reception', *Classical Receptions Journal* 12: 46–61.

Gutzwiller, K. (1983) 'Charites or Hiero: Theocritus *Idyll* 16', *Rheinisches Museum für Philologie* 126: 212–38.

Habinek, T. (2011) 'Manilius' Conflicted Stoicism', in Green and Volk eds. (2011): 32–44.

Hachlili, R. (2002) 'The Zodiac in Ancient Jewish Synagogal Art: A Review', *Jewish Studies Quarterly* 9: 219–58.

Hadas-Lebel, M. (2003) *Philon d'Alexandrie: un penseur en diaspore*, Paris.

Hadjittofi, F. (2011) '*Res Romanae*: Cultural Politics in Quintus Smyrnaeus' *Posthomerica* and Nonnus' *Dionysiaca*', in Baumbach and Bär eds. (2011): 357–78.

(forthcoming) *Nonnus of Panopolis: The Paraphrase of the Gospel according to John*, Berkeley.

Hadjittofi, F. and Lefteratou, A. (forthcoming) *The Genres of Late Antique Christian Poetry: Between Modulations and Transpositions*, Berlin and New York.

Hägg, T. (2012) *The Art of Biography in Antiquity*, Cambridge.

Hägg, T. and Rousseau, P. eds. (2000) *Greek Biography and Panegyric in Late Antiquity*, Berkeley.

Halberstam, J. (2005) *In a Queer Time and Place: Transgender Bodies, Subcultural Lives*, New York.

Halivni, D. W. (2005) 'Aspects of the Formation of the Talmud', in Rubenstein ed. (2005): 339–60.

Hall, C. (2020) 'Origen and Astrology', *Studia Patristica* 100: 113–22.

(forthcoming) *Origen and Prophecy*, Oxford.

Hall, E. and Stead, H. (2020) *A People's History of Classics: Class and Greco-Roman Antiquity in Britain and Ireland 1689–1939*, Abingdon and New York.

Halliwell, S. (2008) *Greek Laughter*, Oxford.

Hampton, T. (1990) *Writing from History: The Rhetoric of Exemplarity in Renaissance Literature*, Ithaca.

Hanink, J. (2014) *Lykurgan Athens and the Making of Classical Tragedy*, Cambridge.

(2018) 'Pausanias' Dead Poets Society', in Goldschmidt and Graziosi eds. (2018): 235–49.

Hannah, D. (2015) 'The Star of the Magi and the Prophecy of Balaam in Earliest Christianity, with special reference to the lost *Books of Balaam*', in Van Kooten and Barthel eds. (2015): 431–62.

Hannah, R. (2002) 'Imaging the Cosmos: Astronomical Ekphraseis in Euripides', *Ramus* 31: 19–32.

(2009) *Time in Antiquity*, London and New York.

(2020a) 'Methods of Reckoning Time', in Bowen and Hochberg eds. (2020): 24–40.

(2020b) 'The Sundial and the Calendar', in Bowen and Rochberg eds. (2020): 323–39.

Hanson, R. (1988) *The Search for the Christian Doctrine of God: The Arian Controversy 318–81*, Edinburgh.

(2002) *Allegory and Event: A Study of the Sources and Significance of Origen's Interpretation of Scripture*, Louisville.

Haraway, D. (1988) 'Situated Knowledges: The Science Question in Feminism and the Privilege of Partial Perspectives', *Feminist Studies* 14: 575–99.

Hardie, P. (1986) *Virgil's Aeneid: Cosmos and Imperium*, Oxford.

(1993) *The Epic Successors of Virgil: A Study in the Dynamics of a Tradition*, Cambridge.

(2019) *Classicism and Christianity in Late Antique Latin Poetry*, Berkeley.

Hardie, P. ed. (2002) *The Cambridge Companion to Ovid*, Cambridge.

Hardie, P., Barchiesi, A. and Hinds, S. eds. (1999) *Ovidian Transformations: Essays on Ovid's Metamorphosis and its Reception*, Cambridge.

Harnack, A. von (1892) 'Über das Verhältnis des Prologs des vierten Evangeliums zum gantzen Werk', *Zeitschrift für Theologie und Kirche* 2: 189–231.

Harper, J. (1965) 'John Cassian and Sulpicius Severus', *Church History* 34: 371–80.

Harrison, C. (2000) *Augustine: Christian Truth and Fractured Humanity*, Oxford.

Harrison, R. M. (1983) 'The Church of St. Polyeuktos in Istanbul and the Temple of Solomon', in C. Mango and O. Pritsak eds., *Okeanos: Essays Presented to Ihor Ševčenko on his Sixtieth Birthday by his Colleagues and Students*, Harvard Ukrainian Studies 7, Cambridge, MA: 276–9.

(1986) *Excavations at Saraçhane in Istanbul 1*, Princeton.

(1989) *A Temple for Byzantium: The Discovery and Excavation of Anicia Jujiana's Palace-Church in Istanbul*, London.

Harrison, S. (2000) *Apuleius: A Latin Sophist*, Oxford.

Harrison, T. (2000) *Divinity and History: The Religion of Herodotus*, Oxford.

Harrison, T. and Irwin, E, eds. (2018) *Interpreting Herodotus*, Oxford.

Harstrup, K. (1990) 'The Ethnographic Present: A Re-invention', *Cultural Anthropology* 5: 45–61.

Hartog, F. (1988) *The Mirror of Herodotus: The Representation of the Other in the Writing of History*, Berkeley, CA.

(2010) 'Polybius and the First Universal History', in Liddel and Fear eds. (2010): 30–40.

(2011) *Regimes of Historicity: Presentism and Experiences of Time*, New York.

(2020) *Chronos: L'Occident aux prises avec le Temps*, Paris.

Hatzimichali, M. (2017) 'Text and Wisdom in the Letter of Aristeas', in Chrubasik and King eds. (2017): 156–77.

Haubold, J., Steele, J. and Stevens, K. eds. (2019) *Keeping Watch in Babylon: The Astronomical Diaries in Context*, Leiden.

Haugen, K. (2012) 'Hebrew Poetry Transformed, or Scholarship Invincible Between Renaissance and Enlightenment', *Journal of the Warburg and Courtauld Institutes* 75: 1–29.

Hawkins, T. (2014) *Iambic Poetics in the Roman Empire*, Cambridge.

Hayward, C. (2013) 'The Aramaic Targums', in Carleton-Paget and Schaper eds. (2013): 218–41.

Heath, J. (2013) 'Greek and Jewish Visual Piety: Ptolemy's Gifts in the *Letter of Aristaeus*', in S. Pearce ed. (2013): 38–48.

(2020) *Clement of Alexandria and the Shaping of Christian Literary Practice: Miscellany and the Transformation of Greco-Roman Writing*, Cambridge.

Hedrick, C. (2000) *History and Silence: Purge and Rehabilitation of Memory in Late Antiquity*, Austin.

Heffernan, T. (2012) *The Passion of Perpetua and Felicity*, Oxford.

Hegedus, T. (2007) *Early Christianity and Ancient Astrology*, New York.

Heidegger, M. (1962) *Being and Time*, trans. J. Macquarrie and E. Robinson, Oxford.

Heil, A. and Damschen, G. eds. (2014) *Brill Companion to Seneca: Philosopher and Dramatist*, Leiden.

Heilen, S. (2005) 'Emperor Hadrian in the Horoscopes of Antigonus of Nicaea', in Oestemann, Rutkin and von Stuckrad eds. (2005): 49–67.

(2015) 'The Star of Bethlehem and Greco-Roman Astrology, especially Astrological Geography', in Van Kooten and Barthel eds. (2015): 297–357.

Heilen, S. and Greenbaum, D. (2016) 'Astrology in the Greco-Roman World', in Jones ed. (2016): 123–42.

Hell, J. (2019) *The Conquest of Ruins: The Third Reich and the Fall of Rome*, Chicago.

Hendel, R. (2016) *Steps to a New Edition of the Hebrew Bible*, Atlanta.

Henderson, J. ed. (2013) *The Unfolding of Words: Commentary in the Age of Erasmus*, Toronto.

Henderson, J. G. (1987) 'Lucan/Word at War', *Ramus* 16: 122–64.

(1997) 'The Name of the Tree: Recounting *Odyssey* XXIV 340–2', *Journal of Hellenic Studies* 107: 87–116.

(2007) *The Medieval World of Isadore of Seville: Truth From Words*, Cambridge.

(2011) 'Watch This Space (Getting around 1.215–46)', in Green and Volk eds. (2011): 59–84.

Henkin, D. (2018) 'Tick, Tock, Tuesday: Serial Timekeeping and the History of the Modern Week', *Nineteenth-Century Contexts* 40: 509–24.

Herbert-Brown, G. (2002) 'Ovid and the Stellar Calendar' in Herbert-Brown ed. (2002): 101–28.

Herbert-Brown, G. ed. (2002) *Ovid's Fasti: Historical Readings at its Bimillennium*, Oxford.

Herder, J. G. (1998) *Schriften zur Literatur und Philosophie, 1792–1800*, Berlin.

Hernández de la Fuente, D. (2011) 'The One and the Many and the Circular Movement: Neo-Platonicism and Poetics in Nonnus of Panopolis', in Hernández de la Fuente ed. (2011): 305–26.

(2014) 'Neoplatonic Form and Content in Nonnus: Towards a New Reading of Nonnian Poetics', in Spanoudakis ed. (2014): 229–50.

Hernández de la Fuente, D. ed. (2011) *New Perspectives on Late Antiquity*, Newcastle upon Tyne.

Hershkowitz, P. (2017) *Prudentius, Spain and Late Antique Christianity: Poetry, Visual Culture and the Cult of the Martyrs*, Cambridge.

Herzog, R. (1966) *Die allegorische Dichtkunst des Prudentius*, Munich.

(1976) *Die Bibelepik der lateinischen Spätantike: Formgeschichte einer erbaulichen Gattung*, Munich.

(2002a) 'Metapher-Exegese-Mythos: Interpretationen zur Entstehung eines biblischen Mythos in der Literatur der Spätantike', in R. Herzog, *Spätantike: Studien zur römischen und lateinisch-christlichen Literatur*, ed. P. Habermehl, Göttingen: 115–77.

(2002b) 'Orosius oder die Formulierung eines Fortschrittskonzepts aus der Erfahrung des Niedergangs', in R. Herzog, *Spätantike: Studien zur römischen und lateinisch-christlichen Literatur*, ed. P. Habermehl, Göttingen: 293–320.

(2002c) '"Wir leben in der Spätantike": eine Zeiterfahrung und ihre Impulse für die Forschung', in R. Herzog, *Spätantike: Studien zur römischen und latei-nisch-christlichen Literatur*, ed. P. Habermehl, Göttingen: 312–48.

Hesk, J. (2000) *Deception and Democracy in Classical Athens*, Cambridge.

Hesketh, I. (2011) *The Science of History in Victorian Britain*, London.

Heszer, C. (1997) *The Social Structure of the Rabbinic Movement in Roman Palestine*, Tübingen.

Hext, K. (2013) *Walter Pater: Individualism and Aesthetic Philosophy*, Edinburgh.

Higbie, C. (2013) *The Lindian Chronicle and the Greek Creation of their Past*, Oxford.

Hill, S. (1979) *Melito of Sardis: On Pascha and Fragments*, Oxford.

Himmelfarb, M. (2010) *The Apocalypse: A Brief History*, Chichester.

Hinds, S. (1987) *The Metamorphosis of Persephone: Ovid and the Self-Conscious Muse*, Cambridge.

Hobbs, A. (2000) *Plato and the Hero: Courage, Manliness and the Impersonal Good*, Cambridge.

Hochschild, A. (2012) *Memory in Augustine's Theological Anthropology*, Oxford.

Hofer, A. (2013) *Christ in the Life and Writing of Gregory of Nazianzus*, Oxford.

Hoffmann, H. (1968) 'Morum tempora diversa: Charakterwandel bei Tacitus', *Gymnasium* 75: 220–50.

Hölkeskamp, K.-J. (2014) 'In Defence of Concepts, Categories and Other Abstractions: Remarks on a Theory of Memory (in the Making)', in Galinsky ed. (2014): 63–70.

Holmes, B. (2020) 'At the End of the Line: On Kairological History', *Classical Receptions Journal* 12: 62–90.

Hölscher, L. (1999) *Die Entdeckung der Zukunft*, Frankfurt.

Honigman, S. (2003) *The Septuagint and Homeric Scholarship in Alexandria*, London.

(2007) 'The Narrative Function of the King and the Library in the Letter of Aristeas', in Rajak et al. eds. (2007): 128–46.

Hook, D. ed. (2005) *From Orosius to the Historia Silense*, Bristol.

Hopkins, K. (1983) *Death and Renewal*, Cambridge.

(1999) *A World Full of Gods: Pagans, Jews and Christians in the Roman Empire*, London.

Horkheimer, M. (1994) *Critique of Instrumental Reason*, trans. M. O'Connell, New York.

Horky, P. ed. (2019) *Cosmos in the Ancient World*, Cambridge.

Höschele, R. (2018) 'Poets' Corners in Greek Epigram Collections', in Goldschmidt and Graziosi eds. (2018): 197–214.

Houtman, A., de Jong, A. and Misset-van der Weg, M. eds. (2008) *Empsychoi Logoi: Religious Innovations in Antiquity*, Ancient Judaism and Christianity 73, Leiden.

Howard, D. (2018) 'The Old, the Antique, and the Venerable in Venetian Renaissance Architecture', in Christ and Morche eds. (2018): 63–89.

Howard-Johnston, J. and Hayward, P. eds. (1999) *The Cult of Saints in Late Antiquity and the Early Middle Ages: Essays on the Contribution of Peter Brown*, Oxford.

Hübner, W. (2020) 'The Professional Ἀστρόλογος', in Bowen and Rochberg eds. (2020): 297–322.

Hughes, J. (2014) 'Memory and the Roman Viewer: Looking at the Arch of Constantine', in Galinsky ed. (2014): 103–14.

Hunt, E. D. (1982) *Holy Land Pilgrimage in the Late Roman Empire*, Oxford.

Hunt, Emily (2003) *Christianity in the Second Century: The Case of Tatian*, London.

Hunter, R. (1995) 'Poetry and Philosophy in the *Phainomena* of Aratus', *Arachnion* 2: 1–34.

(1996) *Theocritus and the Archaeology of Greek Poetry*, Cambridge.

(2021) 'Homer in Origen *Against Celsus*', in Carleton Paget and Gathercole eds. (2021): 215–48.

Hutton, W. E. (2005) *Describing Greece: Landscape and Literature in the Periegesis of Pausanias*, Cambridge.

Iliffe, R. (2013) 'Newton's Religious Life and Work', www.newtonproject.ox.ac.uk /view/contexts/CNTX0001.

(2016) *Priest of Nature: The Religious Worlds of Isaac Newton*, Oxford.

Iricinschi, E. and Zellentin, H. eds. (2008) *Heresy and Identity in Late Antiquity*, Tübingen.

Ishibashi, Y. (2014) 'In Pursuit of Accurate Timekeeping: Liverpool and Victorian Electrical Horology', *Annals of Science* 71: 474–96.

Jacobus, H. (2020) 'Astral Divination in the Dead Sea Scrolls', in Bowen and Rochberg eds. (2020): 539–50.

Jaeger, W. ed. (1958–) *Gregorii Nysseni Opera*, Leiden.

James, W. (1890) *Principles of Psychology*, 2 vols., London.

Jameson, F. (2002) *A Singular Modernity: Essay on the Ontology of the Present*, New York.

(2003) 'The End of Temporality', *Critical Inquiry* 29: 695–718.

Jarick, J. (1990) *Gregory Thaumatourgos' Paraphrase of Ecclesiastes*, Atlanta.

Jensen, R. (2011) *Living Water: Images, Symbols and Settings of Early Christian Baptism*, Leiden.

(2012) *Baptismal Imagery in Early Christianity: Ritual, Visual and Theological Dimensions*, Grand Rapids.

Johnson, A. (2006) *Ethnicity and Argument in Eusebius' Praeparatio Evangelica*, Oxford.

(2014) *Eusebius*, London.

Johnson, D. (1991) 'Story and Design in Book Eight of Augustine's *Confessions*', *Biography* 14: 39–60.

Johnson, S. (2006) *The Life and Miracles of Thecla: A Literary Study*, Washington.

(2016) 'Nonnus' Paraphrastic Technique: A Case Study of Self-Recognition in John 9', in Accorinti ed. (2016): 267–88.

Johnson, S. ed. (2006) *Greek Literature in Late Antiquity: Dynamism, Didacticism, Classicism*, Aldershot and Burlington.

Johnson S. F. (2019) 'Lists, Originality and Christian Time: Eusebius's Historiography of Succession', in Pohl and Wiesser eds. (2019): 191–218.

Jones, A. (2017) *A Portable Cosmos: Revealing the Antikythera Mechanism, Scientific Wonder of the Ancient World*, Oxford.

(2019) 'Greco-Roman Sun-Dials: Precision and Displacement', in Miller and Symons eds. (2019): 125–57.

Jones, A. ed. (2016) *Time and Cosmos in Greco-Roman Antiquity*, Princeton.

Jones, C. P. (2014) *Between Pagan and Christian*, Cambridge, MA.

Jordheim, H. (2014) 'Introduction: Multiple Times and the Work of Synchronization', *History and Theory* 53: 498–518.

(2017) 'Synchronizing the World: Synchronism as Historiographical Practice, Then and Now', *History of the Present* 7: 59–95.

(2019) 'Making Universal Time: Tools of Synchronization', in Bjørnstad, Jordheim and Régent-Susini eds. (2019): 133–52.

Judson, L. and Karasmanis, V. eds. (2006) *Remembering Socrates: Philosophical Essays*, Oxford.

Kahlos, M. (2007) *Debate and Dialogue: Christian and Pagan Cultures c. 360–430*, Aldershot.

Kalmin, R. (2006) *Jewish Babylonia between Persia and Roman Palestine*, Oxford.

Kanaan, V. L. (2019) *The Ancient Unconscious: Psychoanalysis and the Ancient Text*, Oxford.

Karrer, M. and Kraus, W. eds. (2008) *Die Septuaginta: Texte, Kontexte, Lebenswelten*, Tübingen.

Kartschoke, D. (1975) *Bibeldichtung: Studien zur Geschichte der epischen Bibelparaphrase von Juvencus bis Otfried von Weißenburg*, Paderborn.

Kaster, R. (2001) 'The Dynamics of *Fastidium* and the Ideology of Disgust', *Transactions of the American Philological Association* 131: 143–89.

Katz, M. (1991) *Penelope's Renown: Meaning and Indeterminacy in the Odyssey*, Princeton.

Kaye, L. (2018) *Time in the Babylonian Talmud: Natural and Imagined Times in Jewish Law and Narrative*, Cambridge.

Keevak, M. (1995) 'Reading (and Conversion in) Augustine's *Confessions*', *Orbis Litterarum* 50: 257–71.

Keizer, H. (1999) 'Life Time Entirety: A Study of AION in Greek Literature and Philosophy, the Septuagint and Philo', PhD Amsterdam.

Kelley, N. (2006) 'Philosophy as Training for Death: Reading the Ancient Christian Martyr Acts as Spiritual Exercises', *Church History* 75: 723–47.

Kelly, C. (2010) 'The Shape of the Past: Eusebius of Caesarea and Old Testament History', in Flower, Kelly and Williams eds. (2010): II: 13–27.

Kelly, G. (2008) *Ammianus Marcellinus: The Allusive Historian*, Cambridge.

Kelly, M. (2004) 'On the Mind's Pronouncement of Time: Aristotle, Augustine and Husserl on Time-Consciousness', *Proceedings of the American Catholic Philosophical Association* 78: 247–62.

Kennedy, D. (2011) 'Sums in Verse?', in Green and Volk eds. (2011): 165–201.
  (2013) *Antiquity and the Meaning of Time: A Philosophy of Ancient and Modern Literature*, London.

Kennedy, G. ed. (1992) *The Cambridge History of Literary Criticism*, Cambridge.

Ker, J. (2009) 'Drinking from the Water-Clock: Time and Speech in Imperial Rome', *Arethusa* 42: 279–302.
  (2019) 'Diurnal Selves in Ancient Rome', in Miller and Symons eds. (2019): 184–213.

Kermode, F. (1966) *The Sense of an Ending*, Oxford.

Kern, S. (2003) *The Culture of Time and Space, 1880–1918*, Cambridge, MA.

Kessler, H. (1994) *Studies in Pictorial Narrative*, London.

Kierkegaard, S. (1980) *The Concept of Anxiety: A Simple and Psychologically Orienting Deliberation on the Dogmatic Issue of Hereditary Sin*, trans. and ed. R. Thomte, Princeton.

Kimpton, F. (2014) 'The *Fasti's* Celestial World and the Limitations of Astronomical Knowledge', *Classical Philology* 109: 26–47.

King, H. (1986) 'Tithonus and the Tettix', *Arethusa* 19: 15–35.

(1998) *Hippocrates' Women: Reading the Female Body in Ancient Greece*, London.

King, Martin Luther (1964) *Why We Can't Wait*, New York.

King, R. (2019) 'Memory and Recollection in Plato's *Philebus*: Use and Definition', in Castagnoli and Ceccarelli eds. (2019): 216–35.

Kirby, J. (1997) 'Aristotle on Metaphor', *American Journal of Philology* 118: 517–54.

Kirk, A. ed. (2005) *Memory, Tradition, and Text: Uses of the Past in Early Christianity*, Atlanta.

Kirsch, W. (1989) *Die lateinische Versepik des 4. Jahrhunderts*, Berlin.

Kister, M., Newman, H., Segal, M. and Clements, R. eds. (2015) *Tradition, Transmission and Transformation from Second Temple Literature through Judaism and Christianity in Late Antiquity*, Leiden.

Klein, K. L. (2000) 'On the Emergence of memory in Historical Discourse', *Representations* 70: 127–50.

Klinghirn, W. and Safran, L. eds. (2007) *The Early Christian Book*, Washington.

Klooster, J. (2013) 'Apostrophe in Homer, Apollonius and Callimachus', in Eisen and von Moellendorf eds. (2013): 151–73.

Koditscheck, T. (2011) *Liberalism, Imperialism, and the Historical Imagination*, Cambridge.

Koerner, J. (2003) *The Reformation of the Image*, Chicago.

Komorowska, J. (2004) 'A Vision of Chaos: *Dionysiaca* 1: 163–257', *Eos* 91: 294–312.

Koselleck, R. (2002) *The Practice of Conceptual History: Timing History, Spacing Concepts*, trans. T. Presner et al., Stanford.

(2004) *Futures Past: On the Semantics of Historical Time*, New York.

Kosmin, P. (2018) *Time and its Adversaries in the Seleucid Empire*, Cambridge, MA.

Kotzé, A. (2004) *Augustine's Confessions: Communicative Purpose and Audience*, Leiden.

Kreps, A. (2018) 'From Jewish Apocrypha to Christian Tradition: Citations of *Jubilees* in Epiphanius' *Panarion*', *Church History* 37: 345–70.

Kresic, S. ed. (1981) *Contemporary Literary Hermeneutics and the Interpretation of Classical Texts*, Ottawa.

Kristeva, J. (1981) 'Women's Time', *Signs* 7: 13–35.

Kroll, N. (2016) *Die Jugend des Dionysos: die Ampelos-Episode in der Dionysiaka des Nonnos von Panopolis*, Berlin.

Krueger, D. (2000) 'Writing and the Liturgy of Memory in Gregory of Nyssa's Life of Macrina', *Journal of Early Christian Studies* 8: 483–510.

(2004) *Writing and Holiness: The Practice of Authorship in the Early Christian East*, Philadelphia.

Kugel, J. (1997) *The Bible as It Was*, Cambridge, MA.

Kuhn-Treichel, T. (forthcoming) 'Poetological Name-Dropping: Explicit References to Poets and Genres in Gregory Nazianzen's Poems', in Hadjittofi and Lefteratou eds. (forthcoming).

Kurke, L. (1991) *The Traffic in Praise: Pindar and the Poetics of Social Economy*, Berkeley.

Kyle, D. (1998) *Spectacles of Death*, London.

Lacroix, B. (1965) *Orose et ses idées*, Montreal and Paris.

Ladner, G. (1959) *The Idea of Reform: Its Impact on Christian Thought and Action in the Age of the Fathers*, Cambridge, MA.

Laks, A. (1994) 'Substitution et connaissance: une interprétation unitaire (ou presque) de la théorie aristotélicienne de la métaphore', in Furley and Nehamas, eds. (1994): 283–305.

Lambert, D. (2015) *How Repentance Became Biblical: Judaism, Christianity and the Interpretation of Scripture*, Oxford.

Lamberton, R. (1986) *Homer the Theologian: Neoplatonist Allegorical Reading and the Growth of the Epic Tradition*, Berkeley.

(1992) 'The Neoplatonists and the Spiritualization of Homer', in Lamberton and Keaney eds. (1992): 115–33.

Lamberton, R. and Keaney, J. eds. (1992) *Homer's Ancient Readers*, Princeton.

Landes, D. (1983) *Revolution in Time: Clocks and the Making of the Modern World*, Cambridge, MA.

Landow, G. (1985) 'Lawrence and Ruskin: The Sage as Word-Painter', in Meyers ed. (1985): 35–50.

Lane, M. (2001) *Plato's Progeny: How Socrates and Plato Still Captivate the Modern Mind*, London.

Lane Fox, R. (2015) *Augustine: Conversions to Confessions*, London.

Lang, H. (1980) 'On Memory: Aristotle's Corrections of Plato', *Journal of the History of Philosophy* 18: 379–93.

Langlands, R. (2011) 'Roman *Exempla* and Situation Ethics: Valerius Maximus and Cicero *De Officiis*', *Journal of Roman Studies* 101: 100–22.

(2015) 'Roman Exemplarity: Mediating Between General and Particular', in Lowrie and Lüdemann eds. (2015): 68–80.

(2018) *Exemplary Ethics in Ancient Rome*, Cambridge.

Langworthy, O. (2019) 'Theodoret's Theologian: Assessing the Origin and Significance of Gregory of Nazianzus' Title', *Journal of Ecclesiastical History* 70: 455–71.

Lapin, H. (2012) *Rabbis as Romans: The Rabbinic Movement in Palestine, 100–400 CE*, Oxford.

Laplanche, J. (1999a) *Essays on Otherness*, trans. J. Fletcher and L. Thurston, London and New York.

(1999b) 'Notes on Afterwardsness', in Laplanche (1999a): 264–9.

(1999c) 'Time and the Other', in Laplanche (1999a): 238–63.

(2017) *Après coup*, trans. J. House and L. Thurston, New York.

Lardinois, A. and McClure, L. eds. (2001) *Making Silence Speak: Women's Voices in Greek Literature and Society*, Princeton.

Larsen, T. (2014) *The Slain God: Anthropologists and the Christian Faith*, Oxford.

Latour, B. (1993) *We Have Never Been Modern*, trans. C. Porter, Cambridge, MA.

Lauritzen, D. and Tardieu, M. eds. (2013) *Le voyage des legendes: homage à Pierre Chuvin*, Paris.

Lawrence, D. H. (1994) *The Complete Poems of D. H. Lawrence*, Ware.

Leach, E. (1961) *Rethinking Anthropology*, London.

Lefteratou, A. (forthcoming) 'Jesus' Baptism in the Scamander: Eudocia's Homeric Centos and Christian Ritual', in Bierl, Bouvier and Cesca eds. (2021).

Le Goff, J. (1980) *Time, Work and Culture in the Middle Ages*, trans. A. Goldhammer, Chicago.

Leich, R. (2006) *Astrologoumena Judaica: Untersuchungen zur Geschichte der astrologischen Literatur der Juden*, Tübingen.

Leigh, M. (1997) *Lucan: Spectacle and Engagement*, Oxford.

Lemon, L. and Reis, M. (1965) *Russian Formalist Criticism: Four Essays*, Lincoln, Nebraska.

Leushuis, R. (2016) 'Speaking the Gospel: The Voice of the Evangelist in Erasmus' *Paraphrases on the New Testament*', in Bloemendal ed. (2016): 163–85.

Levey, G. and Modood, T. eds. (2009) *Secularism, Religion and Multicultural Citizenship*, Cambridge.

Levi, D. (1944) 'Aion', *Hesperia* 13: 269–314.

Levine, L. (1975) *Caesarea under Roman Rule*, Leiden.

(2012) *Visual Judaism in Late Antiquity: Historical Contexts of Jewish Art*, New Haven.

Lévi-Strauss, C. (1963) *Structural Anthropology*, New York.

Levitin, D. (forthcoming) 'From Palestine to Göttingen (via India): Hebrew Matthew and the Origins of the Synoptic Problem'.

Lévy, C. (1998) *Philon d'Alexandrie et le langage de la philosophie*, Turnhout.

Levy, H. (1979) 'Homer's Gods: A Comment on their Immortality', *Greek, Roman, and Byzantine Studies* 20: 215–18.

Lewis, W. (1927) *Time and Western Man*, London.

Leyerle, B. (1996) 'Landscape as Cartography in Early Christian Pilgrimage', *Journal of the American Academy of Religion* 64: 119–43.

Li, Y. (2018) 'The Silence of the Muse', *Arethusa* 51: 91–115.

(unpublished) 'Being Late and Being Mistaken in the Homeric Tradition', PhD Yale.

Lianeri, A. ed. (2011) *The Western Time of Ancient History: Historiographical Encounters with the Greek and Roman Past*, Cambridge.

(2016) *Knowing Future Time in and through Greek Historiography*, Berlin and New York.

Liddel, P. and Fear, A. eds. (2010) *Historiae Mundi: Studies in Universal History*, London.

Lieberman, S. (1942) *Greek in Jewish Palestine: Studies in the Life and Manners of Jewish Palestine in II–IV Century* CE, New York.

(1950) *Hellenism in Jewish Palestine: Studies in the Literary Transmission, Belief and Manners in the I Century B.C.E.–IV Century C.E.*, New York.

Lienhard, J., Muller, E. and Teske, R. eds. (1993) *Augustine: Presbyter Factus Sum*, New York.

Lieu, J. (1996) *Image and Reality: The Jews in the World of the Christians in the Second Century*, London and New York.

(2004) *Christian Identity in the Jewish and Graeco-Roman World*, Oxford.

Lieu, J. and de Boer, M. eds. (2018) *The Oxford Handbook of Johannine Studies*, Oxford.

Lifschitz, A. and Squire, M. eds. (2017) *Rethinking Lessing's Laocoon: Antiquity, Enlightenment, and the 'Limits' of Painting and Poetry*, Oxford.

Lightfoot, J. (2003) *Lucian: On the Syrian Goddess*, Oxford.

  (2014) 'Oracles in the *Dionysiaca*', in Spanoudakis ed. (2014): 39–54.

  (2016) 'Nonnus and Prophecy: Between 'Pagan' and 'Christian' Voices', in Accorinti ed. (2016) 625–43.

  (2017) 'In the Beginning Was the Voice', in Bannert and Kroll eds. (2017): 141–55.

  (2020) *Ps-Manetho, Apotelesmatica. Books 2, 3, and 6*, Oxford.

Ligota, C. and Quantin, J.-L. eds. (2006) *History of Scholarship; A Selection of Papers from the Seminar on the History of Scholarship Held Annually at the Warburg Institute*, Oxford and New York.

Lim, R. (1995) *Public Disputation: Power and Social Order in Late Antiquity*, Berkeley.

Limberis, V. (2011) *Architects of Piety: The Cappadocian Fathers and the Cult of the Martyrs*, Oxford.

Lindheim, H. and Morales, H. eds. (2015) *New Essays in Homer: Language, Violence and Agency. Ramus* 44 Special Issue.

Livrea, E. (1989) *Nonno di Panopoli. Parafrasi del Vangelo di S. Giovanni, Canto XVIII*, Naples.

  (2000) *Nonno di Panopoli. Parafrasi del Vangelo di San Giovanni, Canto B*, Bologna.

Lloyd, G. (1968) 'Plato as a Natural Scientist', *Journal of Hellenic Studies* 88: 78–92.

  (1987) *The Revolutions of Wisdom: Studies in the Aims and Practices of Greek Science*, Berkeley.

Long, A. (1982) 'Astrology: Arguments Pro and Contra', in Barnes ed. (1982): 165–92.

  (2019) *Death and Immortality in Ancient Philosophy*, Cambridge.

Loraux, N. (1981a) *Les Enfants d'Athéna: idées sur la citoyenneté et la division des sexes*, Paris.

  (1981b) *L'Invention d'Athènes: histoire de l'oraison funèbre dans la 'cité classique'*, Paris.

  (1986a) 'Le Corps vulnérable d'Arès', in Malamoud and Vernant eds. (1986): 335–54.

  (1986b) *The Invention of Athens: The Funeral Oration in the Classical City*, trans A. Sheridan, Cambridge, MA.

Low, P. (2020) 'Remembering, Forgetting, and Rewriting the Past: Athenian Inscriptions and Collective Memory', *Histos* Supplement 11: 235–68.

Lowe, L. and Lloyd, D. eds. (1997) *The Politics of Culture in the Shadow of Capital*, Durham, NC.

Lowrie M. and Lüdemann S. eds. (2015) *Exemplarity and Singularity: Thinking Through Particulars in Philosophy, Literature, and Law*, London and New York.

Lowth, R. (1787) *Lectures on the Sacred Poetry of the Hebrews*, 2 vols., trans. G. Gregory, London [*De sacra poesi Hebraeorum Praelectiones Academicae*, (1753) Oxford.]

Ludlow, M. (2007) *Gregory of Nyssa, Ancient and (Post)modern*, Oxford.

(2015) 'Texts, Teachers and Pupils in the Writings of Gregory of Nyssa', in van Hoof and van Nuffelen eds. (2015): 83–102.

Ludwig, W. (1977) 'Die christliche Dichtung des Prudentius und die Transformation der klassischen Gattungen', in *Christianisme et formes littéraires de l'antiquité tardive en occident*, Entretiens sur l'antiquité classique 23, Fondation Hardt, Geneva, 1976: 303–72.

Luhmann, N. (1975) 'Weltzeit und Systemgeschichte: über Beziehungenzwischen Zeithorizonten und sozialen Strukturen gesellschaftlicher Systeme', in Luhmann, N. ed. *Soziologische Aufklärung Bd. 2: Aufsätze zur Theorie der Gesellschaft*, Opladen: 103–33.

(1982) 'The Future Cannot Begin: Temporal Structures in Modern Society', in N. Luhmann ed. *The Differentiation of Society*, New York: 271–88.

Lühnken, M. (2002) *Christianorum Maro et Flaccus: zur Vergil- und Horazrezeption des Prudentius*, Göttingen.

Luraghi, N. ed. (2001) *The Historian's Craft in the Age of Herodotus*, Oxford.

Lyell, C. (1837) *Principles of Geology*, 5th ed., London.

Lynn-George, M. (1988) *Epos: Word and Narrative in the Iliad*, Atlantic Highlands.

Lysack, K. (2019) *Chronometres: Devotional Literature, Duration and Victorian Reading*, Oxford.

Ma, J. (2002) *Antiochus III and the Cities of Western Asia Minor*, 2nd ed., Oxford.

(2015) *Statues and Cities: Honorific Portraits and Civic Identity in the Hellenistic World*, Oxford.

Mackay, E. ed. (2008) *Orality, Literacy, Memory in the Ancient Greek and Roman World*, Leiden.

McCabe, M. M. (2013/14) 'Seven Characters in Search of a Teacher: Process and Progress in the *Euthydemus*', *Révue de métaphysique et de morale* 80: 491–505.

McCormack, S. (1998) *The Shadows of Poetry: Vergil in the Mind of Augustine*, Cambridge.

McEachnie, R. (2018) 'Zeno, Chromatius and Gaudentius. Italian Preachers amid Transition', in Dupont et al. eds. (2018): 454–76.

McGill, S. (2005) *Virgil Recomposed: The Mythological and Secular Centos in Antiquity*, Oxford.

McGing, B. and Mossman, J. eds. (2006) *The Limits of Ancient Biography*, Swansea.

Maciver, C. (2012) *Quintus Smyrnaeus' Posthomerica: Engaging Homer in Late Antiquity*, Brill.

(2020) 'Triphiodorus and the Poetics of Imperial Greek Epic', *Classical Philology* 115: 164–85.

McKinnon, J. (1991) 'The Book of Psalms, Monasticism and the Western Liturgy', in van Deusen ed. (1991): 43–58.

(1994) 'Desert Monasticism, and the Later Fourth-Century Psalmodic Movement', *Music and Letters* 75: 505–21.

McLynn, N. (1994) *Ambrose of Milan: Church and Court in a Christian Capital*, Berkeley.

(1999) 'Augustine's Roman Empire', *Augustinian Studies* 30.2: 29–44.

(2009) *Christian Politics and Religious Culture in Late Antiquity*, Farnham.

(2015) 'Gregory's Governors: *Paideia* and Patronage in Cappadocia', in van Hoof and van Nuffelen eds. (2015): 48–67.

MacMullen, R. (1997) *Christianity and Paganism in the Fourth to Eighth Centuries*, New Haven.

McNamara, M. (2010) *Targum and Testament Revisited: Aramaic Paraphrases of the Hebrew Bible. A Light on the New Testament*, Grand Rapids and Cambridge.

Madec, G. (1975) '"*Tempora christiana*": expression du triomphalisme chrétien ou recrimination païenne?', in Mayer and Eckermann eds. (1975): 112–36.

Mahmood, S. (2015) *Religious Difference in a Secular Age: A Minority Report*, Princeton.

Maier, F. (2018) 'Past and Present as *paradoxon theōrēma* in Polybius', in Miltzios and Tamiolaki eds. (2018): 55–74.

Maier-Katkin, D. and Maier-Katkin, B. (2007) 'Love and Reconciliation: The Case of Hannah Arendt and Martin Heidegger', *Harvard Review* 32: 34–48.

Malamoud, C. and Vernant, J.-P. eds. (1986) *Le Temps de la réflexion: Corps des dieux*, Paris.

Malamud, Margaret (2016) *African Americans and the Classics: Antiquity, Abolition and Activism*, London.

Malamud, Martha (1989) *A Poetics of Transformation: Prudentius and Classical Mythology*, Ithaca.

(2011) *The Origin of Sin: An English Translation of the 'Hamartigenia'*, Ithaca, NY.

Mandelbrote, S. (2016) 'When Manuscripts Meet: Editing the Bible in Greek during and after the Council of Trent', in Blair and Goejing eds. (2016): 251–67.

(2018) 'The Old Testament and its Ancient Versions in Manuscript and Print in the West, from c. 1480 to c. 1780', in Cameron ed. (2018): 82–109.

Mango, C. (1992) 'Byzantine Writers on the Fabric of Hagia Sophia', in Mark and Cakmak eds. (1992): 41–56.

Mansfeld, J. (1992) *Heresiology in Context: Hippolytus' Elenchos as a Source for Greek Philosophy*, Leiden.

Marasco, G. ed. (2003) *Greek and Roman Historiography in Late Antiquity: Fourth to Sixth Century* AD, Leiden.

Maraval, P. (2002) 'The Earliest Phase of Christian Pilgrimage in the Near East (before the 7th Century)', *Dumbarton Oaks Papers* 56: 63–74.

Marcus, D. (2002) 'The Mission of The Raven (Genesis 8.7)', *Journal of Ancient Near Eastern Studies* 29: 71–80.

Marincola, J. (1991) *Authority and Tradition in Ancient Historiography*, Cambridge.

Mark, R. and Cakmak, A. eds. (1992) *Hagia Sophia from the Age of Justinian to the Present*, Cambridge.

Markus, R. (1988) *Saeculum, History and Society in the Theology of St Augustine*, Cambridge.

(1990) *The End of Ancient Christianity*, Cambridge.

(2000) 'Tempora *Christiana* Revisited', in Dodaro and Lawless eds. (2000): 199–212.

Marmodoro, A. and McLynn, N. eds. (2018) *Exploring Gregory of Nyssa: Philosophical, Theological and Historical Studies*, Oxford.

Martens, P. (2008) 'Re-visiting the Allegory/Typology Distinction: The Case of Origen', *Journal of Early Christian Studies* 16: 282–317.

Martindale, C. (1993) *Redeeming the Text: Latin Poetry and the Hermeneutics of Reception*, Cambridge.

Martindale, C., Evangelista, S. and Prettejohn, E. eds. (2017) *Pater the Classicist: Classical Scholarship, Reception and Aestheticism*, Oxford.

Martindale, C. and Thomas, R. eds. (2006) *Classics and the Uses of Reception*, Malden, MA.

Mason, S. (1998) *Understanding Josephus: Seven Perspectives*, Sheffield.

Masters, J. (1992) *Poetry and Civil War in Lucan's Bellum Civile*, Cambridge.

Mastrangelo, M. (2008) *The Roman Self in Late Antiquity: Prudentius and the Poetics of the Soul*, Baltimore.

(2010) 'Typology and Agency in Prudentius' Treatment of the Judith Story', in Brine, Ciletti and Lähnemann eds. (2010): 153–68.

Mayer, C. and Eckermann, W. eds. (1975) *Scientia Augustiniana: Studien über Augustinus, den Augustinismus, und den Augustinerorden. Festschrift Adolar Zumkeller*, Würzburg.

Mbembe, A. (2017) *Critique of Black Reason*, trans. L. du Bois, Durham, NC.

Meijering, E. (1979) *Augustin über Schöpfung, Ewigkeit und Zeit: das elfte Buch der Bekenntnisse*, Leiden.

Meinel, F. (2009) 'Gregory of Nazianzus' *Poemata Arcana*: ἄρρητα and Christian Persuasion', *Cambridge Classical Journal* 55: 71–96.

Mellor, D. H. (1981) *Real Time*, Cambridge.

Mendels, D. (2004) *Memory in Jewish, Pagan and Christian Societies of the Greco-Roman World*, Library of Second Temple Studies 48, London and New York.

Mendelsohn, D. (2020) *Three Rings: A Tale of Exile, Narrative and Fate*, Charlottesville and London.

Merrills, A. (2005) *History and Geography in Late Antiquity*, Cambridge.

Meyers, J. ed. (1985) *D. H. Lawrence and Tradition*, Amherst, MA.

Miguélez Cavero, L. (2008) *Poems in Context: Greek Poetry in the Egyptian Thebaid, 200–600 A.D.*, Berlin and New York.

Mikalson, J. (2003) *Herodotus and Religion in the Persian Wars*, Chapel Hill.

Milikowsky, C. (2014) *Seder Olam: Commentary and Introduction*, Jerusalem [in Hebrew].

Miller, J. (1986) *Measures of Wisdom: The Cosmic Dance in Classical and Christian Antiquity*, Toronto.

Miller, J. J. (2003) 'Time in Literature', *Daedalus* 132: 86–97.

Miller, K. and Symons, S. eds. (2019) *Down to the Hour: Short Time in the Ancient Mediterranean and Near East*, Leiden.

Miltzios, N. (2016) 'Knowledge and Foresight in Polybius', in Lianeri ed. (2016): 141–54.

Miltzios, N. and Tamiolaki, M. eds. (2018) *Polybius and his Legacy*, Berlin and New York.

Mitchell, W. (1984) 'The Politics of Genre: Space and Time in Lessing's *Laocoon*', *Representations* 6: 98–115.

Mitralexis, S. (2016) 'Maximus the Confessor's 'Aeon' as a Distinct Mode of Temporality', *Heythrop Journal* 57. DOI: 10.1111/heyj.12319.

Mitzis, P. and Tsagalis, C. eds. (2010) *Allusion, Authority and Truth: Critical Perspectives on Greek Poetic and Rhetorical Praxis*, Berlin and New York.

Moberly, R. W. (2000) 'Why Did Noah Send out a Raven?', *Vetus Testamentum* 50.3: 345–56.

Modood, T. (2019) *Essays on Secularism and Multiculturalism*, London.

Moles, J. (1978) 'The Career and Conversion of Dio Chrysostom', *Journal of Hellenic Studies* 98: 79–100.

(1996) 'Herodotus Warns the Athenians', *Papers of the Liverpool Latin Seminar* 9: 259–84.

(2011) 'Luke's Preface: The Greek Decree, Classical Historiography and Christian Redefinitions', *New Testament Studies* 57: 461–82.

Möller, A. (2001) 'The Beginning of Chronography: Hellanicus' *Hiereiai*', in Luraghi ed. (2001): 241–62.

Momigliano, A. (1963) 'Pagan and Christian Historiography in the Fourth Century A.D.', in Momigliano ed. (1963): 79–99.

(1977) *Essays in Ancient and Modern Historiography*, Chicago.

(1983) *The Development of Greek Biography*, Cambridge, MA.

(1985) 'The Life of St. Macrina by Gregory of Nyssa', in Eadie, Ober and Bender eds. (1985): 443–58.

(1994) *Studies on Modern Scholarship*, ed. G. Bowersock and T. Cornell, Berkeley.

Momigliano, A. ed. (1963) *The Conflict Between Paganism and Christianity in the Fourth Century*, Oxford.

Moore, C. ed. (2019) *Brill Companion to the Reception of Socrates*, Leiden.

Moreschini, C. and Gallay, P. (1990) *Grégoire de Nazianze. Discours 38–41*, Paris.

Moreschini, C. and Sykes, D. (1997) *St Gregory of Nazianzus: Poemata Arcana*, Oxford.

Morgan, K. (2000) *Myth and Philosophy from the Pre-Socratics to Plato*, Cambridge.

Morris, J. (2005) *F. D. Maurice and the Crisis of Christian Authority*, Oxford.

Mosshammer, A. (2008) *The Easter Computus and the Origins of the Christian Era*, Oxford.

Most, G. (2007) *Doubting Thomas*, Cambridge, MA.

(2011) 'What Ancient Quarrel Between Philosophy and Poetry', in Destrée and Hermann eds. (2011): 1–20.

Muehlberger, E. (2012) 'Salvage: Macrina and the Christian Project of Cultural Reclamation', *Church History* 81: 273–97.

Mueller, M. (2007) 'Penelope and the Poetics of Remembering', *Arethusa* 30: 337–62.

Munson, R. V. (2012) 'Herodotus and the Heroic Age: The Case of Minos', in Baragwanath and de Bakker eds. (2012): 195–212.

Murnaghan, S. (1992) 'Maternity and Mortality in Homeric Poetry', *Classical Antiquity* 11: 242–62.

Murphy, A. (2011) 'Augustine and the Rhetoric of Roman Decline', *History of Political Thought* 26: 586–606.

Nagel, A. and Wood, C. (2010) *Anachronic Renaissance*, New York.

Nagy, G. (1979) *The Best of the Achaeans: Concepts of the Hero in Archaic Greek Poetry*, Baltimore.

Najman, H. (2010) *Past Renewals: Interpretive Authority, Renewed Revelation and the Quest for Perfection*, Leiden.

Nasrallah, L. (2005) 'Mapping the World: Justin, Tatian, Lucian and the Second Sophistic', *Harvard Theological Review* 98: 283–314.

(2010) *Christian Responses to Roman Art and Architecture: the Second-Century Church Amid the Spaces of Empire*, Cambridge.

(2019) *Archaeology and the Letters of Paul*, Oxford.

Need, S. (2008) *Truly Divine and Truly Human: The Story of Christ and the Seven Ecumenical Councils*, London and Peabody, MA.

Nehemas, A. (1998) *The Art of Living: Socratic Reflections from Plato to Foucault*, Princeton.

Neils, J. (2006) *The Parthenon Frieze*, Cambridge.

Neis, R. (2013) *The Sense of Sight in Rabbinic Culture: Jewish Ways of Seeing in Late Antiquity*, Cambridge.

Nelson, R. and Olin, M. eds. (2003) *Monuments and Memory: Made and Unmade*, Chicago.

Neusner, J. (2004) *The Idea of History in Rabbinic Judaism*, Leiden.

Newman, H. (1998) 'Between Jerusalem and Bethlehem: Jerome and the Holy Places of Palestine', M. Alberdina Houtman and J. Schwartz eds. (1998): 216–27.

Newman, J. H. (1833) *The Arians of the Fourth Century*, London.

Newton, I. (1846) *Mathematical Principles of Natural Philosophy*, London.

Niehoff, M. (2001) *Philo on Jewish Identity and Culture*, Tübingen.

(2005) '*Creatio ex Nihilo* Theology in *Genesis Rabbah* in the Light of Christian Exegesis', *Harvard Theological Review* 99: 37–64.

(2011) *Jewish Exegesis and Homeric Scholarship in Alexandria*, Cambridge.

(2012) *Homer and the Bible in the Eyes of Ancient Interpreters*, Leiden.

(2018) *Philo of Alexandria: An Intellectual Biography*, New Haven.

Nietzsche, F. (1997) *Untimely Meditations*, ed. D. Breazeale, trans. R. J. Hollingdale, Cambridge.

Nightingale, A. (1995) *Genres in Dialogue. Plato and the Construction of Philosophy*, Cambridge.

(2011) *Once Out of Nature: Augustine on Time and the Body*, Chicago.

Nock, A. (1933) *Conversion: The Old and the New in Religion from Alexander the Great to Augustine of Hippo*, Oxford.

(1935) 'A Vision of Mandulis Aion', *Harvard Theological Review* 27: 53–104.

Norris, F. (1991) *Faith Gives Fullness to Reason: The Five Theological Orations of Gregory Nazianzen*, Leiden.

Nouhaud, M. (1982) *L'Utilisation de l'histoire dans les orateurs attiques*, Paris.

Nowotny, H. (1989) *Eigenzeit: Entstehung und Strukturierung eines Zeitgefühls*, Frankfurt am Main. [see Nowotny 1994]

(1992) 'Time and Social Theory: Towards a Social Theory of Time', *Time and Society* 1: 421–54.

(1994) *Time: The Modern and Postmodern Experience*, trans. N. Plaice, Cambridge. [see Nowotny 1989]

Nugent, G. (1985) *Allegory and Poetics: The Structure and Imagery of Prudentius' 'Psychomachia'*, Frankfurt am Main.

Nünlist, R. (2009) *The Ancient Critic at Work: Terms and Concepts of Literary Criticism in Greek Scholia*, Cambridge.

Nussbaum, M. and Rorty, A. eds. (1992) *Essays on Aristotle's De Anima*, Oxford.

O'Daly, G. (1977) 'Time as *distentio* and St. Augustine's Exegesis of Philippians 3:12–14', *Revue d'études augustiniennes* 23: 265–71.

(2012) *Days Linked by Song: Prudentius' Cathemerinon*, Oxford.

O'Donnell, J. (2004) 'Late Antiquity: Before and After', *Transactions of the American Philological Association* 134: 203–13.

Oestermann, G., Rutkin, H. and Stuckrad, K. von eds. (2005) *Horoscopes in the Public Sphere: Essays on the History of Astrology*, Berlin and New York.

Ogle, V. (2015) *The Global Transformation of Time, 1870–1950*, Cambridge, MA.

Ogren, B. ed. (2015) *Time and Eternity in Jewish Mysticism*, Leiden.

Onians, R. (1988) *The Origins of European Thought about the Body, the Mind, the Soul, the World, Time and Fate*, 2nd ed., Cambridge.

Opper, T. ed. (2013) *Hadrian: Art, Politics and Economy*, London.

Orlinsky, H. (1975) 'The Septuagint as Holy Writ and the Philosophy of the Translators', *Hebrew Union College Annual* 46: 89–114.

Osborn, E. (1997) *Tertullian: First Theologian of the West*, Cambridge.

(2005) *Clement of Alexandria*, Cambridge.

Osborne, R. (1987) 'The Viewing and Obscuring of the Parthenon Frieze', *Journal of Hellenic Studies* 107: 98–105.

Otlewska-Jung, M. (2014) 'Orpheus and the Orphic Hymns in the *Dionysiaca*', in Spanoudakis ed. (2014): 76–96.

Otten, W. and Pollmann, K. eds. (2007) *Poetry and Exegesis in Premodern Latin Christianity: The Encounter Between Classical and Christian Strategies of Interpretation*, Leiden.

Ousterhout, R. ed. (1990) *The Blessings of Pilgrimage*, Urbana.

Pabel, H. and Vessey, M. eds. (2002) *Holy Scripture Speaks: The Production and Reception of Erasmus' Paraphrases on the New Testament*, Toronto.

Page, D. (1981) *Further Greek Epigrams*, Cambridge.

Palmer, A.-M. (1989) *Prudentius on the Martyrs*, Oxford.

Parker, D. (2010) *Codex Sinaiticus: The Story of the World's Oldest Bible*, London.

Pascal, B. (1963) *Œuvres complètes*, Paris.

Pater, W. (1900) *Marius the Epicurean*, London.

(1910) *The Renaissance: Studies in Art and Poetry*, London.

Patrich, J. (2011) *Studies in the History and Archaeology of Caesarea Maritima*, Leiden.

Pattison, G. (2002) *Kierkegaard, Religion, and the Nineteenth-Century Crisis of Culture*, Cambridge.

Pearce, S. (2007) 'Translating for Ptolemy: Patriotism and Politics in the Greek Pentateuch', in Rajak, Pearce, Aitken and Dines eds. (2007): 165–81.

Pearce, S. ed (2013) *The Image and its Prohibition in Jewish Antiquity*, Oxford.

Pelling, C. (2019) *Herodotus and the Question Why*, Austin.

Pelttari, A. (2019) *The Psychomachia of Prudentius: Text, Commentary, and Glossary*, Norman.

Pender, E. (1992) 'Spiritual Pregnancy in Plato's *Symposium*', *Classical Quarterly* 42: 72–86.

Peradotto, J. (1990) *Man in the Middle Voice: Name and Narration in the Odyssey*, Princeton.

Perkins, J. (1995) *The Suffering Self: Pain and Narrative Representation in the Early Christian Era*, London and New York.

Peroli, E. (1997) 'Gregory of Nyssa and the Neo-Platonic Doctrine of the Soul', *Vigiliae Christianae* 51: 117–39.

Phillips, J. (2002) '*Sub evangelistae persona:* The Speaking Voice in Erasmus' *Paraphrase on Luke*', in Pabel and Vessey eds. (2002): 127–50.

Phillips, T. (2020) *Untimely Epic: Apollonius Rhodius' Argonautica*, Oxford.

Plass, P. (1980) 'Transcendent Time and Eternity in Gregory of Nyssa', *Vigiliae Christianae* 34: 180–92.

Platt, V. (2011) *Facing the Gods: Epiphany and Representation in Greco-Roman Art, Literature and Religion*, Cambridge.

(2018) 'Silent Bones and Singing Stones: Materializing the Poetic Corpus in Hellenistic Greece', in Goldschmidt and Graziosi eds. (2018): 21–50.

Pohl, W. and Wiesser V. eds. (2019) *Historiography and Identity I: Ancient and Early Christian Narratives of Community*, Turnhout.

Polk, D. (1991) 'Temporal Impermanence and the Disparity of Time and Eternity', *Augustinian Studies* 22: 63–82.

Pollmann, K. (1991) *Die 'Carmen adversus Marcionitas': Einleitung, Übersetzung, und Kommentar*, Göttingen.

(2017) *The Baptized Muse: Early Christian Poetry as Cultural Authority*, Oxford.

Porter, J. (2016) *The Sublime in Antiquity*, Cambridge.

The Postclassicisms Collective (2019) *Postclassicisms*, Chicago.

Praet, D. (2016) '"The Divided Cloak as *redemptio militiae*": Biblical Stylization and Hagiographical Intertextuality in Sulpicius Severus' *Vita Martini*', in de Temmerman and Demoen eds. (2016): 133–5.

Pranger, B. (2007) 'Time and the Integrity of Poetry: Ambrose and Augustine', in Otten and Pollmann eds. (2007): 49–64.

(2010) *Eternity's Ennui: Temporality, Perseverance and Voice in Augustine and Western Literature*, Leiden.

Prendergast, C. (2019) *Counterfactuals: Paths of the Might Have Been*, London.

Pretzler, M. (2007) *Pausanias: Travel Writing in Ancient Greece*, London.

Price, R. and Gaddis, M. (2007) *The Acts of the Council of Chalcedon*, 3 vols., Liverpool.

Price, R. and Whitby, M. eds. (2009) *Chalcedon in Context: Church Councils 400–700*, Liverpool.

Pucci, J. (1991) 'Prudentius' Readings of Horace in the *Cathemerinon*', *Latomus* 50.3: 677–90.

Pucci, P. (1987) *Odysseus Polutropos: Intertextual Readings in the Iliad and the Odyssey*, Ithaca, NY.

(1992) *Oedipus and the Fabrication of the Father: Oedipus Tyrannus in Modern Criticism and Philosophy*, Ithaca, NY.

(2000) 'Le Cadre temporel de la volonté divine chez Homère', in Darbo-Peschanski ed. (2000): 33–48.

Purves, A. (2010) *Space and Time in Ancient Greek Narrative*, Cambridge.

(2019) *Homer and the Poetics of Gesture*, Oxford.

Raaflaub, K. (1987) 'Herodotus' Political Thought and the Meaning of History', *Arethusa* 20: 221–48.

Radstone, S. (2008) *The Sexual Politics of Time: Confession, Nostalgia, Memory*, London and New York.

Rajak, T. (2002) *Josephus: The Historian and his Society*, London.

(2009) *Translation and Survival: The Greek Bible of the Ancient Jewish Diaspora*, Oxford.

Rajak, T., Pearce, S., Aitken J. and Dines, J. eds. (2007) *Jewish Perspectives on Hellenistic Rulers*, Berkeley.

Rangos, S. (2014) 'Plato on the Nature of the Sudden Moment, and the Asymmetry of the Second Part of the *Parmenides*', *Dialogue* 53: 538–74.

Rapp, C. (2005) *Holy Bishops in Late Antiquity: The Nature of Christian Leadership in an Age of Transition*, Berkeley.

Rebenich, S. (2009) 'Late Antiquity in Modern Eyes', in Rousseau ed. (2009): 77–92.

Rebillard, E. (2012) *The Care of the Dead in Late Antiquity*, trans. E. Rawlings and J. Routier-Pucci, Ithaca.

Redfield, J. (1975) *Nature and Culture in the Iliad: The Tragedy of Hector*, Chicago.

(1985) 'Herodotus the Tourist', *Classical Philology* 80: 97–118.

Reed, A. Y. (2004) 'Abraham as Chaldaean Scientist and Father of the Jews: Josephus *Ant.* I. 154–168, and the Greco-Roman Discourse about Astronomy/Astrology', *Journal for the Study of Judaism* 35: 119–58.

(2015) 'Retelling Biblical Retellings: Epiphanius, the Pseudo-Clementines and the Reception-History of Jubilees', in Kister et al. eds. (2015): 304–21.

Rees, M. (2003) *Our Final Hour: A Scientist's Warning: How Terror, Error and Environmental Disaster Threaten Humankind's Future in This Century – on Earth and Beyond*, London.

Reichel, M. and Rengakos, A. eds. (2002) *Epea Ptereonta*, Stuttgart.

Reif, S. (1993) *Judaism and Hebrew Prayer: New Perspectives on Jewish Liturgical History*, Cambridge.

Revermann, M. and Wilson, P. eds. (2008) *Performance, Iconography, Reception*, Oxford.

Reydams-Schils, G. ed. (2003) *Plato's Timaeus as Cultural Icon*, Notre Dame.

Richardson, E. (2016) *Classical Victorians: Scholars, Scoundrels and Generals in Pursuit of Antiquity*, Cambridge.

Richardson, N. (1978) *The Homeric Hymn to Demeter*, Oxford.

(2011) 'The Homeric *Hymn to Demeter*: Some Central Questions Revisited', in Faulkner ed. (2011): 44–58.

Ricoeur, P. (1984) *Time and Narrative*, 3 vols., trans. K. McLaughlin and D. Pellauer, Chicago.

Riggsby, A. (2003) 'Pliny in Space (and Time)', *Arethusa* 36: 167–86.

(2009) 'For Whom the Clock Drips', *Arethusa* 42: 271–8.

Rimell, V. (2008) *Martial's Rome: Empire and the Ideology of Epigram*, Cambridge.

Ripat, P. (2011) 'Expelling Misconceptions: Astrologers at Rome', *Classical Philology* 106: 115–54.

Robbins, J. (1985) 'Petrarch Reading Augustine: "The Ascent of Mont Ventoux"', *Philological Quarterly* 64: 533–53.

Roberts, M. (1985) *Biblical Epic and Rhetorical Paraphrase in Late Antiquity*, Liverpool.

(1993) *Poetry and the Cult of the Martyrs: The Liber Peristephanon of Prudentius*, Ann Arbor.

(1994) 'St Martin and the Leper: Narrative Variation in the Poems of Venantius Fortunatus', *The Journal of Medieval Latin* 4: 82–100.

Robinson, J. (2001) *Redating the New Testament*, Eugene.

Robson, E. (2004) 'Scholarly Conceptions and Quantifications of Time in Assyria and Babylonia c750–250 BCE', in Rosen ed. (2004): 45–90.

Rochberg, F. (2020) 'The Babylonian Contribution to Greco-Roman Astronomy', in Bowen and Rochberg eds. (2020): 147–159.

Rood, T. (2007) 'Polybius', in De Jong and Nünlist eds. (2007): 165–81.

(2016) 'Horoscopes of Empires: Future Ruins from Thucydides to Macaulay', in Lianeri ed. (2016): 339–60.

Rood, T., Atack, C. and Phillips, T. (2020) *Anachronism and Antiquity*, London.

Rooney, D. and Nye, J. (2009) '"Greenwich Observatory Time for the Public Benefit": Standard Time and Victorian Networks of Regulation', *British Journal for the History of Science* 42: 5–30.

Rosen, R. ed. (2004) *Time and Temporality in the Ancient World*, Philadelphia.

Rosen, S. (1988) *The Quarrel Between Poetry and Philosophy: Studies in Ancient Thought*, New York and London.

Rosenberg, D. and Grafton A. (2010) *Cartographies of Time: A History of the Time Line*, Princeton.

Ross, D. (1991) 'Time, the Heaven of Heavens and Memory in Augustine's Confessions', *Augustinian Studies* 22: 191–205.

Rotstein, A. (2016) *Literary History in the Parian Marble*, Hellenic Studies Series 68, Washington.

Rous, S. (2019) *Reset in Stone: Memory and Reuse in Ancient Athens*, Madison.

Rousseau, P. (1978) *Ascetics, Authority and the Church*, Oxford.

(1985) *Pachomius: The Making of a Community in Fourth-Century Egypt*, Berkeley.

(1994) *Basil of Caesarea*, Berkeley.

Rousseau, P. ed. (2009) *A Companion to Late Antiquity*, Oxford.

Rubenstein, J. (2003) *The Culture of the Babylonian Talmud*, Baltimore.

(2007) 'Talmudic Astrology: *Bavli Shabbat* 156a–b', *Hebrew Union College Annual* 78: 109–48.

Rubenstein, J. ed. (2005) *Creation and Composition*, Tübingen.

Rubino, C. and Shelmerdine, C. eds. (1983) *Approaches to Homer*, Austin.

Rudolf, K. (2014) 'Propaganda for Astrology in Aramaic Literature', *Aramaic Studies* 12: 121–34.

Rudwick, M. (1992) *Scenes from Deep Time: Early Pictorial Representations of the Prehistoric World*, Chicago.

(2005) *Bursting the Limits of Time: The Reconstruction of Geohistory in the Age of Revolution*, Chicago.

(2008) *Worlds Before Adam: The Reconstruction of Geohistory in the Age of Reform*, Chicago.

Rummel, E. (2002) 'Why Noel Béda Did Not Like Erasmus' *Paraphrases*', in Pabel and Vessey eds. (2002): 265–79.

Runia, D. (1986) *Philo of Alexandria and the 'Timaeus' of Plato*, Leiden.

(2012) *Philo of Alexandria: An Annotated Bibliography, 1997–2006*, Leiden.

Rüpke, J. (1995) *Kalender und Öffentlichkeit: die Geschichte der Repräsentation und religiösen Qualifikation von Zeit in Rom*, Berlin and New York.

(2006) *Zeit und Fest: eine Kulturgeschichte des Kalendars*, Munich.

Sacks, K. (1981) *Polybius on the Writing of History*, Berkeley.

Saffrey, H. (1984) 'La Dévotion de Proclus au soleil', *Philosophies non chrétiennes et christianisme: Annales de l'Institut de Philosophie et de Sciences Morales*: 73–86.

Sailor, D. (2006) 'Dirty Linen, Fabrication, and the Authorities of Livy and Augustus', *Transactions of the American Philological Association* 136: 329–88.

Salzman, M. (1990) *On Roman Time: The Codex-Calendar of 354 and the Rhythms of Urban Life in Late Antiquity*, Berkeley.

(2004) 'Pagan and Christian Notions of the Week in the Fourth-Century C.E. Western Roman Empire', in Rosen ed. (2004): 185–211.

Samuel, A. (1972) *Greek and Roman Chronology: Calendars and Years in Classical Antiquity*, Munich.

Sandison, A. and Dingley, R. eds. (2000) *Histories of the Future: Studies in Fact, Fantasy and Science Fiction*, Basingstoke.

Sandwell, I. (2011) *Religious Identity in Late Antiquity: Greeks, Jews, and Christians in Antioch*, Cambridge.

Sassi, M. ed. (2007) *Tracce dell mente: teorie della memoria da Platone ai moderne*, Pisa.

Schäfer, P. (1996) 'Jews and Gentiles in Yerushalmi Avodah Zarah', in Schäfer ed. (1996): 336–52.

(2008) 'Bereshit Bara Elohim: Bereshit Rabba, Parashah 1, Reconsidered', in Houtman, de Jong and Misset-van der Weg eds. (2008): 267–89.

Schäfer, P. ed. (1996) *The Talmud Yerushalmi and Graeco-Roman Culture III*, Tübingen.

Schäublin, C. (1985) 'Konversionen in antiken Dialogen?', in Schäublin ed. (1985): 117–31.

Schäublin, C. ed. (1985) *Katalepton: Festschrift für Bernhard Wyss zum 80. Geburtstag*, Basel.

Schein, S. (2008) 'Divine and Human in the *Homeric Hymn to Aphrodite*', in Bouchon, Brillet-Dubois and Le Meur-Weissmann eds. (2008): 295–312.

Scheindler, A. (1881) *Nonni Panopoli Paraphrasi S. Evangelii Joannei*, Leipzig.

Schiesaro, A. (2002) 'Ovid and Professional Discourse of Scholarship, Religion, Rhetoric', in Hardie ed. (2002): 62–75.

Schivelbusch, W. (1986) *The Railway Journey: The Industrialization of Time and Space in the Nineteenth Century*, Berkeley.

Schleiermacher, F. (1998) *Hermeneutics and Criticism*, trans. A. Bowie, Cambridge and New York.

Schlögel, K. (2016) *In Space We Read Time: On the History of Civilization and Geopolitics*, New York.

Schottenius Cullhed, S. (2015) *Proba the Prophet: The Christian Vergilian Cento of Faltonia Betitia Proba*, Leiden.

Schutz, A. (1962) *The Phenomenology of the Social World*, Chicago.

Schwab, A. (2009) *Gregor von Nazianz: Über Vorsehung, Περὶ Προνοίας*, Tübingen.

(2012) 'From a Way of Reading to a Way of Life: Basil of Caesarea and Gregory of Nazianzus about Poetry in Christian Education', in Tanaseanu-Döbler and Döbler eds. (2012): 147–62.

Schwartz, S. (2001) *Imperialism and Jewish Society 200 BCE to 640 CE*, Princeton.

(2009) *Were the Jews a Mediterranean Society? Reciprocity and Solidarity in Ancient Judaism*, Princeton.

Scodel, R. (2008) 'Social Memory in Aeschylus' *Eumenides*', in Mackay ed. (2008): 115–41.

Scourfield, J. ed. (2007) *Texts and Culture in Late Antiquity: Inheritance, Authority, Change*, Swansea.

Secord, J. (1986) *Controversy in Victorian Geology: The Cambrian–Silurian Dispute*, Princeton.

(2000) *Victorian Sensation: The Extraordinary Publication, Reception, Secret Authorship of Vestiges of the Natural History of Creation*, Chicago.

Secunda, S. (2013) *The Iranian Talmud: Reading the Bavli in its Sasanian Context*, Philadelphia.

Sedley, D. (1989) 'Teleology and Myth in the *Phaedo*', *Proceedings of the Boston Area Colloquium in Ancient Philosophy* 5: 359–83.

(2007) *Creationism and its Critics in Antiquity*, Berkeley.

(2009) 'Three Kinds of Platonic Immortality', in Frede and Reis eds. (2009): 145–61.

Segal, C. (1971) *The Theme of the Mutilation of the Corpse in the Iliad*, Leiden.

(1994) 'Transition and Ritual in Odysseus' Return', in *Singers, Heroes and Gods in the Odyssey*, Ithaca, 65–84.

Sennett, R. (2005) *The Culture of the New Capitalism*, New Haven.

(2008) *The Craftsman*, New Haven.

Senneville-Grave, G. (1999) *Sulpice Sévère: Chroniques*, Paris.

Sens, V. (1992) 'Theocritus, Homer and the Dioscuri: Idyll 22 137–22', *Transactions of the American Philological Association* 122: 335–50.

Sevcenko, I. (1992) 'The Search for the Past in Byzantium around 800', *Dumbarton Oaks Papers* 46: 279–93.

Shaw, B. (1993) 'The Passion of Perpetua', *Past and Present* 139: 3–45.

(1996) 'Body/Power/Identity: Passions of the Martyrs', *Journal of Early Christian Studies* 4: 269–312.

(2011) *Sacred Violence: African Christians and Sectarian Hatred in the Age of Augustine*, Cambridge.

Shear, J. (2011) *Polis and Revolution: Responding to Oligarchy in Classical Athens*, Cambridge.

(2012a) 'The Politics of the Past: Remembering Revolution in Athens', in J. Marincola, L. Llewellyn-Jones and C. Maciver eds., *Greek Notions of the Past in the Archaic and Classical Eras: History without Historians*, Edinburgh: 276–300.

(2012b) 'Religion and the *Polis*: The Cult of the Tyrannicides in Athens', *Kernos* 12: 27–56.

(2013) '"Their Memories Will Never Grow Old": The Politics of Remembrance in the Athenian Funeral Oration', *Classical Quarterly* 63: 511–36.

(2020) 'An Inconvenient Past in Hellenistic Athens: The Case of Phaidros of Sphettos', *Histos* Supplement 11: 269–301.

Sheehan, J. (2005) *The Enlightenment Bible: Translation, Scholarship, Culture*, Princeton.

Sheffield, F. (2001) 'Psychic Pregnancy and Platonic Epistemology', *Oxford Studies in Ancient Philosophy*, ed. D. Sedley, Oxford: 1–33.

Sherman, S. (1996) *Telling Time: Clocks, Diaries, and English Diurnal Form, 1660–1785*, Chicago.

Shorrock, R. (2001) *The Challenge of Epic: Allusive Engagement in the Dionysiaca of Nonnus*, Leiden.

(2008) 'The Politics of Poetics: Nonnus' *Dionysiaca* and the World of Late Antiquity', *Ramus* 37: 99–113.

(2011) *The Myth of Paganism: Nonnus, Dionysus and the World of Late Antiquity*, London.

(2014) 'A Classical Myth in a Christian World: Nonnus' Ariadne Episode (*Dion.* 47.265–475)', in Spanoudakis ed. (2014): 313–32.

(2016) 'Christian Themes in the *Dionysiaca*', in Accorinti ed. (2016): 577–600.

Shuger, D. (1994) *The Renaissance Bible: Scholarship, Sacrifice and Subjectivity*, Berkeley.

Shuttleworth, S. (1984) *George Eliot and Nineteenth-Century Science: The Make-Believe of a Beginning*, Cambridge.

Sieber, F. (2016) 'Nonnus' Christology', in Accorinti ed. (2016): 308–26.

(2017) 'Words and their Meaning: On the Chronology of the Paraphrasis of St John's Gospel', in Bannert and Kroll eds. (2017): 156–65.

Simelidis, C. (2009) *Selected Poems of Gregory of Nazianzus*, Göttingen.

(2016) 'Nonnus and Christian Literature', in Accorinti ed. (2016): 289–307.

Simonetti, M. (1975) *La Crisi Ariana nel IV secolo*, Rome.

Sivan, H. (1988) 'Pilgrimage, Patronage and the Emergence of Christian Palestine', in Ousterhout ed. (1990): 54–65.

(1990) 'Holy Land Pilgrimage', *Classical Quarterly* 38: 528–35.

Skilton, D. (2007) 'Tourists at the Ruins of London: The Metropolis and the Struggle for Empire', *Cercles* 73: 93–119.

Sluga, H. (1993) *Heidegger's Crisis: Philosophy and Politics in Nazi Germany*, Cambridge, MA.

Smelik, W. (2013) *Rabbis, Language and Translation in Late Antiquity*, Cambridge.

Smith, M. (2018) *The Idea of Nicaea in the Early Church Councils*, AD *431–451*, Oxford.

Smith, P. (1981) *Nursling of Mortality: A Study of the Homeric Hymn to Aphrodite*, Frankfurt am Main.

Smith, R. (1995) *Julian's Gods: Religion and Philosophy in the Thought and Action of Julian the Apostate*, London.

Smith, Z. (2011) 'Killing Orson Welles at Midnight', *New York Review of Books*, 28 April.

Smolak, K. (1979) 'Beobachtungen zur Darstellungsweise in den Homerzentonen', *Jahrbuch der Österreichischen Byzantistik* 28: 29–49.

(2001) 'Die Bibeldichtung als "Verfehlte Gattung"', in Stella ed. (2001): 15–29.

Snyder, R. ed. (1985) *Thomas de Quincey: Bicentenary Studies*, Norman and London.

Sobel, D. (1996) *Longitude: The True Story of a Lone Genius Who Solved the Greatest Scientific Problem of his Time*, London.

Sorabji, R. (1983) *Time, Creation and the Continuum: Theories in Antiquity and the Early Middle Ages*, Chicago.

(2006) *Aristotle on Memory*, 2nd ed., Chicago.

Spanoudakis, K. (2013) 'The Resurrections of Tylus and Lazarus in Nonnus of Panopolis (*Dion.* XXV 451–52 and *Par.* Λ', in Lauritzen and Tardieu eds. (2013): 191–208.

(2014a) *Nonnus of Panopolis: Paraphrase of the Gospel of John 11*, Oxford.

(2014b) 'The Shield of Salvation: Dionysus' Shield in Nonnus *Dionysiaca* 25.380–572', in Spanoudakis ed. (2014): 333–74.

(2016) 'Pagan Themes in the *Paraphrase*', in Accorinti ed. (2016): 601–24.

Spanoudakis, K. ed. (2014) *Nonnus of Panopolis in Context: Poetry and Cultural Milieu in Late Antiquity with a Section on Nonnus and the Modern World*, Berlin and New York.

Spelman, H. (2018) *Pindar and the Poetics of Permanence*, Oxford.

Spira, A. (2007) *Kleine Schriften zu Antike und Christentum: Menschenbild-Rhetorik-Gregor von Nyssa*, ed. H. Drobner, Berlin.

Springer, C. (1988) *The Gospel as Epic in Late Antiquity: The Paschale Carmen of Sedulius*, Leiden.

Stancliffe, C. (1983) *St. Martin and his Hagiographer: History and Miracle in Sulpicius Severus*, Oxford.

Stanfill, J. (2019) 'The Body of Christ's Barbarian Limb: John Chrysostom's Processions and the Embodied Performance of Nicene Christianity', in de Wet and Mayer eds. (2019): 670–97.

Stead, G. C. (1978) 'The "Thalia" of Arius and the Testimony of Athanasius', *Journal of Theological Studies* 29: 20–52.

Stegemann, V. (1930) *Astrologie und Universalgeschichte: Studien und Interpretationen zu den Dionysiaka des Nonnos von Panopolis*, Leipzig and Berlin.

Steinbock, B. (2013) *Social Memory in Athenian Public Discourse: Uses and Meanings of the Past*, Ann Arbor.

Stella, F. ed. (2001) *La scrittura infinita. Bibbia e poesia in età medievale e umanistica*, Florence.

Stern, S. (1996) 'Fictitious Calendars: Early Rabbinic Notions of Time, Astronomy and Reality', *Jewish Quarterly Review* 87: 103–29.

 (2001) *Calendar and Community: A History of the Jewish Calendar 2nd Century BCE–10th Century CE*, Oxford.

 (2012) *Calendars in Antiquity: Empires, States and Societies*, Oxford.

Stewart, C. (1998) *Cassian the Monk*, Oxford.

Stock, B. (1998) *Augustine the Reader: Meditation, Self-Knowledge, and the Ethics of Interpretation*, Cambridge, MA.

Stocking, C. (2017) *The Politics of Sacrifice in Early Greek Myth and Poetry*, Cambridge.

Stone, M. ed. (1984) *Jewish Writings of the Second Temple Period*, Assen and Philadelphia.

Storin, B. (2019) *Self-Portrait in Three Colours: Gregory of Nazianzus' Epistolary Autobiography*, Berkeley.

Strang, C. and Mills, K. (1974) 'Plato and the Instant', *Proceedings of the Aristotelian Society*, Supplement 48: 63–96.

Struck, P. (2004) *The Birth of the Symbol: Ancient Readers at the Limits of Their Texts*, Princeton.

Stuckrad, K. von (2000) *Das Ringen von die Astrologie: jüdische und christliche Beiträge zum antiker Zeitverständnis*, Berlin.

 (2015) 'Stars and Powers: Astrological Thinking in Imperial Politics from the Hasmoneans to Bar Kokhba', in Van Kooten and Barthel eds. (2015): 387–98.

Swenson, A. (2013) *The Rise of Heritage: Preserving the Past in France, Germany and England, 1789–1914*, Cambridge.

Taft, R. (1986) *The Liturgy of the Hours in East and West: The Origins of the Divine Office and its Meaning for Today*, Collegeville, MN.

Talbert, R. (2017) *Ancient Portable Sundials: The Empire in Your Hand*, Oxford.

Tanaseanu-Döbler, I. and Döbler, M. eds. (2012) *Religious Education in Premodern Europe*, Leiden.

Tarrant, R. (2016) *Texts, Editors and Readers: Methods and Problems in Latin Textual Criticism*, Cambridge.

Taylor, C. (2007) *A Secular Age*, Cambridge, MA.

Taylor, J. and Hay, D. (2012) 'Astrology in Philo's *De Vita Contempletiva*', *Aram* 24: 293–309.

Tennyson, G. (1981) *Victorian Devotional Poetry: The Tractarian Mode*, Cambridge, MA.

Thalmann, G. (1991) Review of Clay (1989). *Classical Philology* 86: 144–7.

Thomas, E. (1997) 'The Architectural History of the Pantheon in Rome from Agrippa to Septimius Severus via Hadrian', *Hephaistos* 15: 163–86.

Thomas, E. and Witschel, C. (1992) 'Constructing Reconstruction: Claim and Reality of Roman Rebuilding Inscriptions from the Latin West', *Papers of the British School at Rome* 60: 135–77.

Thomas, R. (1989) *Oral Tradition and Written Record in Classical Athens*, Cambridge.

(2000) *Herodotus in Context: Ethnography, Science and the Art of Persuasion*, Cambridge.

(2019) *Polis Histories, Collective Memory and the Greek World*, Cambridge.

Thompson, E. (1967) 'Time, Work-Discipline, and Industrial Capitalism', *Past and Present* 38: 56–97.

Thomson, S. (2014) 'The Barbarian Sophist: Clement of Alexandria's *Stromateis* and the Second Sophistic', DPhil thesis, Oxford.

Tillich, P. (1951–63) *Systematic Theology*, 3 vols., Chicago.

Trachtenberg, M. (2010) *Building-in-Time: From Giotto to Alberti and Modern Oblivion*, New Haven.

Trapp, M. ed. (2007) *Socrates from Antiquity to the Enlightenment*, Aldershot.

Traub, V. (2016) *Thinking Sex with the Early Moderns*, Philadelphia.

Trédé-Boulmer, M. (2015) *Kairos: L'à propos et l'occasion. Le mot et la notion, d'Homère à la fin du IVe siècle avant J.-C.*, Paris.

Trout, D. (1999) *Paulinus of Nola: Life, Letters, and Poems*, Berkeley.

Turner, V. (1967) *The Forest of Symbols*, Ithaca.

(1969) *The Ritual Process*, Baltimore.

Usher, M. (1998) *Homeric Stitchings: The Homeric Centos of the Empress Eudocia*, Lanham.

Vaesen, J. (1988) 'Sulpice Sévère et la fin des temps', in Verbeke, Verhelst and Welkenhuysen eds. (1988): 49–71.

van Andel, G. (1976) *The Christian Concept of History in the Chronicle of Sulpicius Severus*, Amsterdam.

(1980) 'Sulpicius Severus and Origenism', *Vigiliae Christianae* 34.3: 278–87.

van Assendelft, M. (1976) *Sol Ecce Surgit Igneus: A Commentary on the Morning and Evening Hymns of Prudentius (Cathemerinon 1, 2, 5, and 6)*, Groningen.

van Boxel, P. (2006) 'Robert Bellarmine, Christian Hebraist and Censor', in Ligota and Quantin eds. (2006): 251–75.

van den Berg, R. (2001) *Proclus' Hymns: Essays, Translation, Commentary*, Leiden.

van der Ben, N. (1980) 'De Homerische Aphrodite-hymne 1 – de Aeneas Passages in de Ilias' *Lampas* 13: 40–77.

VanderKam, J. (2018) *A Commentary on the Book of Jubilees*, 2 vols., ed. S. White Crawford, Minneapolis.

van Deusen, N. ed. (1991) *The Place of the Psalms in the Intellectual Culture of the Middle Ages*, Albany.

van Gennep, A. (1960) *The Rites of Passage*, trans. M. Vizedom and G. Caffee, London.

van Hoof, L. and van Nuffelen, P. eds. (2015) *Literature and Society in the Fourth Century A.D.: Performing Paideia, Constructing the Present, Presenting the Self*, Leiden.

van Kooten, G. (2021) 'Response to Richard Hunter on Homer in the Polemics between Celsus and Origen', in Carleton Paget and Gathercole eds. (2021): 249–53.

van Kooten, G. and Barthel, P. eds. (2015) *The Star of Bethlehem and the Magi: Interdisciplinary Perspectives from Experts on the Ancient Near East, the Greco-Roman World and Modern Astronomy*, Leiden.

van Nuffelen, P. (2004) *Un heritage de paix et de piété: étude sur les Histoires de Socrate et de Sozomène*, Leuven.

(2010) 'Theology versus Genre? The Universalism of Christian History in Late Antiquity', in Liddel and Fear eds. (2010): 162–75.

(2012) *Orosius and the Rhetoric of History*, Oxford.

van Opstall, E. ed. (2018) *Sacred Thresholds: The Door to the Sanctuary in Late Antiquity*, Leiden.

van Winden, J. (1983) 'The World of Ideas in Philo of Alexandria: An Interpretation of *De Opificio Mundi* 24–5', *Vigiliae Christianae* 37: 209–17.

Varley-Winter, R. (2018) *Reading Fragments and Fragmentation in Modernist Literature*, Brighton.

Varner, E. (2004) *Mutilation and Transformation: Damnatio Memoriae and Roman Imperial Portraiture*, Leiden.

(2008) 'Memory Sanctions, Identity Politics, and Altered Imperial Portraits', in Benoist and Daguet-Gagey eds. (2008): 129–52.

Vassilopoulou, P. and Clark, S. (2009) *Late Antique Epistemology: Other Ways to Truth*, Basingstoke and New York.

Vasunia, P. (2001) *The Gift of the Nile: Hellenizing Egypt from Aeschylus to Alexander*, Berkeley.

Verbeke, W., Verhelst, D. and Welkenhuysen, A. eds. (1988) *The Use and Abuse of Eschatology in the Middle Ages*, Leuven.

Verghese, T. P. (1976) '*DIASTEMA* and *DIASTASIS* in Gregory of Nyssa: Introduction to a Concept and the Posing of a Problem', in Dörrie, Altenburger and Schramm eds. (1976): 243–58.

Verhelst, B. ed. (2020) *Nonnus of Panopolis in Context IV*, Leuven.

Vernant, J.-P. (1959) 'Aspects mythiques de la mémoire et du temps', *Journal de la Psychologie* 56: 1–29 [translated in Vernant 2006].

  (1965) *Mythe et pensée chez les grecs: étude de psychologie historique*, Paris.

  (1989) 'At Man's Table: Hesiod's Foundation Myth of Sacrifice', in Detienne and Vernant eds. (1989): 21–86.

  (1991) *Mortals and Immortals*, ed. F. Zeitlin, Princeton.

  (2006) *Myth and Thought among the Greeks*, New York.

Vessey, M. (1993) 'Jerome's Origen: The Making of a Christian Literary Persona', *Studia Patristica* 28: 135–45.

  (1998) 'The Demise of the Christian Writer and the Remaking of "Late Antiquity": From H.-I. Marrou's Saint Augustine (1938) to Peter Brown's Holy Man (1983)', *Journal of Early Christian Studies* 6: 377–411.

  (2004) '*Quid facit cum Horatio Hieronymus*? Christian Latin Poetry and Scriptural Poetics', in Otten and Pollmann eds. (2004): 29–48.

  (2015) 'Literary History: A Fourth-Century Roman Invention?', in van Hoof and van Nuffelen eds. (2015): 16–30.

Vessey, M., Pollmann, K. and Fitzgerald, A. eds. (1999) *History, Apocalypse and the Secular Imagination: New Essays on Augustine's City of God*, Augustinian Studies 30.2, Bowling Green, OH.

Veyne, P. (1976) *Le Pain et le cirque: sociologie historique d'un pluralisme politique*, Paris.

Vian, F. (1993) 'Préludes cosmiques dans les *Dionysiaques* de Nonnos de Panopolis', *Prometheus* 19: 39–52.

Vidal-Naquet, P. (1981) *Le Chasseur noir: formes de pensée et formes de société dans le monde grec*, Paris.

Vidas, M. (2014) *Tradition and the Formation of the Talmud*, Princeton.

Villa, D. (1996) *Arendt and Heidegger: The Fate of the Political*, Princeton.

Vinzent, M. (2014) *Marcion and the Dating of the Synoptic Gospels*, Leuven.

Vlastos, G. (1939) 'Disorderly Motion in the *Timaios*', *Classical Quarterly* 33: 71–83.

Volk, K. (2009) *Manilius and his Intellectual Background*, New York and Oxford.

Vout, C. (2008) 'The Art of Damnatio Memoriae', in Benoist and Daguet-Gagey eds. (2008): 153–72.

Walter, A. (2020) *Time in Ancient Stories of Origin*, Oxford.

Wardy, R. (1988) 'Lucretius on What Atoms Are Not', *Classical Philology* 83: 112–28.

Warren, J. (2006) 'Epicureans and the Present Past', *Phronesis* 51: 362–87.

Weber, M. (1981) [1923] *General Economic Theory*, New Brunswick.

Weil, S. (1939) *L'Iliad ou le poème de la force*, Paris.

(1962) [1940] 'The Great Beast: Some Reflections on the Origins of Hitlerism', in *Selected Essays*, trans. R. Rees, ed. N. Cameron and R. Stevens, London: 89–144.

Wessel, S. (2004) *Cyril of Alexandria and the Nestorian Controversy: The Making of a Saint and of a Heretic*, Oxford.

West, S. (1991) 'Herodotus' portrait of Hecataeus', *Journal of Hellenic Studies* 111: 144–69.

Westwood, G. (2020) *The Rhetoric of the Past in Demosthenes and Aeschines: Oratory, History and Politics in Classical Athens*, Oxford.

West-Pavlov, R. (2013) *Temporalities*, London.

Wetzel, J. (1992) *Augustine and the Limits of Virtue*, Cambridge.

Wetzel, J. ed. (2012) *Augustine, City of God: A Critical Introduction*, Cambridge.

Whitby, M. (2006) 'The St Polyeuktos Epigram (*AP* I.10): A Literary Perspective', in Johnson ed. (2006): 159–87.

(2007) 'The Bible Hellenized: Nonnus' Paraphrase of St. John's Gospel and "Eudocia's" Homeric Centos', in Scourfield ed. (2007): 195–231.

Whitmarsh, T. (2001) *Greek Literature and the Roman Empire: The Politics of Imitation*, Oxford.

Wiater, N. (2016) 'Shifting Endings, Ambiguity and Deferred Closure in Polybius' *Histories*', in Lianeri ed. (2016): 243–65.

Wiedemann, T. (1992) *Emperors and Gladiators*, London.

Wilcox, D. (1987) *The Measure of Times Past: Pre-Newtonian Chronologies and the Rhetoric of Relative Time*, Chicago.

Wilkinson, J. (1982) 'Paulinus' Temple at Tyre', *Jahrbuch der Österreichischen Byzantistik* 32: 553–61.

Williams, B. (2002) *Truth and Truthfulness: An Essay in Genealogy*, Princeton.

Williams, M. S. (2008) *Authorized Lives in Early Christian Biography: Between Eusebius and Augustine*, Cambridge.

(2011) 'Time and Authority in the *Chronicle* of Sulpicius Severus', in Lianeri ed. (2011): 280–97.

(2017) *The Politics of Heresy in Ambrose of Milan: Community and Consensus in Late Antique Christianity*, Cambridge.

Williams, R. (1987) *Arius: Heresy and Tradition*, London.

Wills, G. (1999) *St. Augustine*, New York.

Wilson, E. (2014) *The Greatest Empire: A Life of Seneca*, Oxford.

Wilson, J. (1980) '*Kairos* as Due Measure', *Glotta* 58: 177–204.

(1981) '*Kairos* as "Profit"', *Classical Quarterly* 31: 418–20.

Winkler, J. (1985) *Auctor et Actor: A Narratological Reading of Apuleius' Golden Ass*, Berkeley.

Winslow, D. (1979) *The Dynamics of Salvation: A Study of Gregory of Nazianzus*, Cambridge, MA.

Winter, E. (2013) *Zeitzeichen: zur Entwicklung und Verwendung antiker Zeitmesser*, 2 vols., Berlin.

Wiseman, P. (2014) 'Popular Memory', in Galinsky ed. (2014): 43–62.

Wishnitzer, A. (2015) *Reading Clocks Alla Turca: Time and Society in the Late Ottoman Empire*, Chicago.

Wittgenstein, L. (1922) *Tractatus Logico-Philosophicus*, London.

Wohl, V. (1993) 'Standing by the Stathmos: Sexual Ideology in the *Odyssey*', *Arethusa* 26: 19–50.

Wolfson, E. (2006) *Alef, Mem, Tau: Kabbalistic Musings on Time, Truth and Death*, Berkeley.

  (2011) *A Dream Interpreted within a Dream: Oneiropoiesis and the Prism of Imagination*, New York.

  (2015) 'Retroactive Not Yet: Linear Cicrcularity and Kabbalistic Temporality', in Ogren ed. (2015) 15–52.

Wolkenhauer, A. (2019) 'Time, Punctuality and Chronotopes: Concepts and Attitudes concerning Short Time in Ancient Rome', in Miller and Symons eds. (2019): 214–38.

Wood, D. (2001) *The Deconstruction of Time*, Evanston, IL.

Woodbury, L. (1967) 'Helen and the Palinode', *Phoenix* 21: 157–76.

Woolf, G. (1994) 'Becoming Roman, Staying Greek: Culture, Identity and the Civilizing Process in the Greek East', *Cambridge Classical Journal* 40: 116–43.

  (1996) 'Monumental Writing and the Expansion of Roman Society in the Early Roman Empire', *Journal of Roman Studies* 86: 22–39.

  (1998) *Becoming Roman: The Origins of Provincial Civilization in Gaul*, Cambridge.

Woolf, V. (1939/86) 'A Sketch of the Past' (1939), reprinted in *Moments of Being*, ed. J. Schulkind, New York.

  (1928) *Orlando: A Biography*, New York.

Worman, N. (2001) 'This Voice Which is Not One: Helen's Verbal Guises in Homeric Epic', in Lardinois and McClure eds. (2001): 19–37.

Yates, F. (1966) *The Art of Memory*, London.

Yerushalmi, Y. (1982) *Zakhor: Jewish History and Jewish Memory*, Seattle.

Young, D. (1993) ''Something Like the Gods': A Pindaric Theme and the Myth of *Nemean* 1', *Greek, Roman, and Byzantine Studies* 34: 123–32.

Young, F. (1997) *Biblical Exegesis and the Formation of Christian Culture*, Cambridge.

  (2010) *From Nicea to Chalcedon: A Guide to the Literature and its Background*, 2nd ed., London.

Ypsilanti, M. (2014) 'Image Making and the Art of Paraphrasing: Aspects of Darkness and Light in the *Metabole*', in Spanoudakis ed. (2014): 123–37.

Zecchini, G. (2003) 'Latin Historiography: Jerome, Orosius and the Western Chronicles,' in Marasco ed. (2003): 317–45.

Zeitlin, F. (1996) *Playing the Other: Gender and Society in Classical Greek Literature*, Chicago.

Zemka, S. (2011) *Time and the Moment in Victorian Literature and Society*, Cambridge.

Zerfass, A. (2008) *Mysterium Mirabile: Poesie, Theologie und Liturgie in den Hymnen des Ambrosius von Mailand zu den Christusfesten des Kirchenjahres*, Tübingen and Basel.

Zerubavel, E. (1981) *Hidden Rhythms*, Chicago.

(1985) *The Seven Day Cycle*, New York.

Zinn, P. (2016) 'Lucretius on Time and its Perception', *Kriterion: Journal of Philosophy* 30: 125–51.

Zornberg, A. (1995) *Genesis: The Beginning of Desire*, Philadelphia.

Zuckerberg, D. (2018) *Not All Dead White Men: Classics and Misogyny in the Digital Age*, Cambridge, MA.

Zuntz, G. (1992) *AION in der Literatur der Kaiserzeit*, Vienna.

Index Locorum

Subject Index

For EU product safety concerns, contact us at Calle de José Abascal, 56–1°,
28003 Madrid, Spain or eugpsr@cambridge.org.

www.ingramcontent.com/pod-product-compliance
Ingram Content Group UK Ltd.
Pitfield, Milton Keynes, MK11 3LW, UK
UKHW050728090126
466816UK00008B/45